SCHAUM'S OUTLINE OF

THEORY AND PROBLEMS

OF

GROUP THEORY

•

BY

BENJAMIN BAUMSLAG, Ph.D.
BRUCE CHANDLER, Ph.D.
Department of Mathematics
New York University

•

SCHAUM'S OUTLINE SERIES
McGRAW-HILL BOOK COMPANY
New York, St. Louis, San Francisco, Toronto, Sydney

04124

8 9 10 11 12 13 14 15 SH SH 8 7 6 5 4 3 2

Preface

The study of groups arose early in the nineteenth century in connection with the solution of equations. Originally a group was a set of permutations with the property that the combination of any two permutations again belongs to the set. Subsequently this definition was generalized to the concept of an abstract group, which was defined to be a set, not necessarily of permutations, together with a method of combining its elements that is subject to a few simple laws.

The theory of abstract groups plays an important part in present day mathematics and science. Groups arise in a bewildering number of apparently unconnected subjects. Thus they appear in crystallography and quantum mechanics, in geometry and topology, in analysis and algebra, in physics, chemistry and even in biology.

One of the most important intuitive ideas in mathematics and science is symmetry. Groups can describe symmetry; indeed many of the groups that arose in mathematics and science were encountered in the study of symmetry. This explains to some extent why groups arise so frequently.

Although groups arose in connection with other disciplines, the study of groups is in itself exciting. Currently there is vigorous research in the subject, and it attracts the energies and imagination of a great many mathematicians.

This book is designed for a first course in group theory. It is mainly intended for college and first year graduate students. It is complete in itself and can be used for self-study or as a text for a formal course. Moreover, it could with advantage be used as a supplement to courses in group theory and modern algebra. Little prior knowledge is assumed. The reader should know the beginnings of elementary number theory, a summary of which appears in Appendix A. An acquaintance with complex numbers is needed for some problems. In short, a knowledge of high school mathematics should be a sufficient prerequisite, and highly motivated and bright high school students will be able to understand much of this book.

The aim of this book is to make the study of group theory easier. Each chapter begins with a preview and ends with a summary, so that the reader may see the ideas as a whole. Each main idea appears in a section of its own, is motivated, is explained in great detail, and is made concrete by solved problems.

Chapter 1 presents the rudiments of set theory and the concept of binary operation, which are fundamental to the whole subject. Chapter 2, on groupoids, further explores the concept of binary operation. In most courses on group theory the concept of groupoid is usually treated briefly if at all. We have chosen to treat it more fully for the following reasons: (a) A thorough understanding of binary compositions is thereby obtained. (b) The important ideas of homomorphism, isomorphism and Cayley's theorem occur both in the chapter on groupoids and in the chapters on groups, and the repetition ensures familiarity.

Chapter 3 shows that the concept of group is natural by producing a large number of examples of groups that arise in different fields. Here are discussed groups of real and complex numbers, the symmetric groups, symmetry groups, dihedral groups, the group of Möbius transformations, automorphism groups of groupoids and fields, groups of matrices, and the full linear group.

Chapter 4 is concerned with the homomorphism theorems and cyclic groups. The concept of homomorphism is fundamental, and thus the theorems of this chapter are indispensable for further study.

Chapter 5 is on finite groups. The Sylow theorems are proved, the concept of external direct product is introduced, and groups up to order 15 are classified. The chapter concludes with the Jordan-Hölder theorem and a proof that most alternating groups are simple.

Chapter 6 is on abelian groups. Two important classes of abelian groups are treated: finitely generated and divisible groups. Undergraduates will probably find their needs are met by the material through Section 6.3. Graduate students will certainly want to continue.

Chapter 7 is on permutational representations and extensions. Chapter 8 is on free groups and presentations. Those who would like to study the theory of groups more deeply will find a guide to the literature at the end of the book.

Chapters 1-4 must be read in order, although, if desired, only the first three sections of Chapter 3 need be read at first (the other sections of Chapter 3 may be studied when they are needed). The order of reading Chapters 5-8 can be varied, although part of Chapter 7 is required for the last sections of Chapter 8.

The reader need not work all the solved problems; he should decide for himself how much practice he needs. Some of the problems are designed to clarify the immediately preceding text, and the reader will find that the solutions may overcome some of his obstacles. On the whole, however, it is advisable to attempt the problems before reading their solutions. The numerous supplementary problems, some of which are very difficult, serve as a review of the material of each chapter.

We thank Prof. Gilbert Baumslag for giving us access to several chapters of unpublished notes and for many useful suggestions. We thank Sister Weiss for reading two chapters of an early draft, Harold Brown for much helpful advice, Henry Hayden for typographical arrangement and art work, and Louise Baggot for the typing. Finally we express our appreciation to Daniel Schaum and Nicola Monti for their unfailing editorial cooperation.

B. BAUMSLAG
B. CHANDLER

June 1968

CONTENTS

CONTENTS

CONTENTS

Chapter 1

Sets, Mappings and Binary Operations

Preview of Chapter 1

This chapter begins with a few remarks about sets. A set is a collection of objects. For example, the real numbers form a set, the objects being the numbers.

The real numbers have an operation called addition. Addition essentially involves two numbers, for the addition of a single number is meaningless, while the addition of three or more numbers is repeated addition of two numbers. Because addition involves two numbers it is called a binary operation.

The main object of this chapter is to define precisely the notion of a binary operation. The concept of binary operation is required to define the concept of group.

We introduce the important ideas of cartesian product and mapping. Welding them together gives rise to an explicit definition of a binary operation. Another important idea is that of equivalence relation, which is a generalization of the idea of equality. The reader will also pick up much useful notation.

1.1 SETS

a. Basic notions

Set is synonymous with *collection.* The objects in a set are termed the elements of the set. Usually we denote sets by capital Latin letters, for example, B, G, T. We shall denote

(i) the set of positive integers $1, 2, 3, \ldots$ by P
(ii) the set of nonnegative integers $0, 1, 2, \ldots$ by N
(iii) the set of all integers by Z
(iv) the set of rational numbers by Q
(v) the set of real numbers by R
(vi) the set of complex numbers by C.

The elements of a set will usually be denoted by small Latin letters such as s, t, u, etc. By $s \in S$ we mean "s is an element of S" or "s belongs to S". In particular, $2 \in P$. If s is not an element of S, we write $s \notin S$ and read this as "s is not an element of S" or "s does not belong to S" or "s is not in S". For example, $-1 \notin P$.

In dealing with sets it is advantageous to abbreviate the phrase "the set whose elements are" by using braces. Thus, for example, we write $\{1, 2\}$ for the set whose elements are 1 and 2 and similarly we write $\{-1, 0, 1, 2, \ldots\}$ for the set whose elements are $-1, 0, 1, 2, \ldots$. A variation of this notation is useful to describe a set in terms of a property which singles out its elements. Thus we write $\{x \mid x \text{ has the property } \mathcal{P}\}$ for the *set of all those elements x which have the property $\mathcal{P}$.* Here $\mathcal{P}$ stands for some "understandable" property; to illustrate: $\{x \mid x \text{ is a real number}\}$ is R, the set of real numbers. ($\mathcal{P}$ here is the property of being a real number.) Notice that we read $\{x \mid x \text{ is a real number}\}$ as the set of all those elements x which have the property that x is a real number, or the set of all those elements x such that x is a real number.

We can now introduce some useful notation.

(i) If a and b are real numbers and $a < b$, then the *open interval* $\langle a, b \rangle$ is defined by $\langle a, b \rangle = \{x \mid x \in R \text{ and } a < x < b\}$.

(ii) Again if a and b are real numbers and $a < b$, then the *closed interval* $[a, b]$ is defined by $[a, b] = \{x \mid x \in R \text{ and } a \leqq x \leqq b\}$.

(iii) In coordinate geometry $(0, 0)$ denotes the origin, $(0, 1)$ the point A, and $(1, 0)$ the point B. In general, if a, b are any two elements of a set, (a, b) is called the *ordered pair* formed from a and b. We will say $(a, b) = (c, d)$ if and only if $a = c$ and $b = d$. (a, b) is called an *ordered* pair because the order of a and b matters; (a, b) is not the same as (b, a) if $a \neq b$.

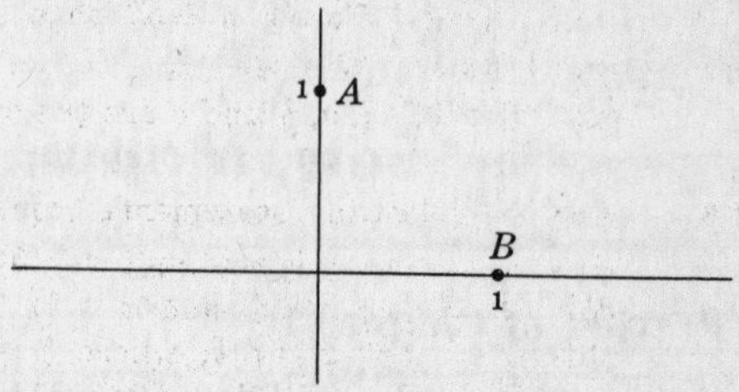

This idea enables us to *define* the Euclidean plane as $\{p \mid p = (x, y) \text{ where } x, y \in R\}$, where the distance between (x, y) and (x_1, y_1) is defined by $\sqrt{(x - x_1)^2 + (y - y_1)^2}$. We write $R^2 = \{p \mid p = (x, y) \text{ with } x, y \in R\}$. R^2 is not the Euclidean plane we normally think of. Rather it can be interpreted as the set of coordinates of the Euclidean plane. Having defined the Euclidean plane it is easy to define, for example, circles and discs.

(iv) *A circle* with radius r and center at the origin is defined by $\{p \mid p = (x, y) \in R^2 \text{ and } x^2 + y^2 = r^2\}$.

(v) The *disc* with radius r and center at the origin is defined by $\{p \mid p = (x, y) \in R^2 \text{ and } x^2 + y^2 \leqq r^2\}$.

We say that two sets S and T are equal, and write $S = T$, if every element of S belongs to T and every element of T belongs to S. Thus $\{2, 2, 3, 3\} = \{2, 3\} = \{3, 2\}$.

If every element of the set S is also an element of the set T, we say *S is a subset of T* and express this briefly by writing $S \subseteq T$. $S \subset T$ means $S \subseteq T$ but $S \neq T$. Thus $P \subseteq N$ and $P \subset N$.

We can use the notion of subset to give a criterion for the equality of sets.

Proposition 1.1: Set S is equal to set T if and only if $S \subseteq T$ and $T \subseteq S$.

Proof: $S \subseteq T$ and $T \subseteq S$ expresses in symbols "every element of S belongs to T and every element of T belongs to S".

Problems

1.1. Are the following statements true?

(i) $2 \in \{2\}$
(ii) $3 \in \{2, 4\}$
(iii) $z \notin \{a, b\}$, $z \neq a$ and $z \neq b$
(iv) $a \notin \{a, b\}$
(v) $5 \in P$
(vi) $\frac{3}{4} \in P$
(vii) $\sqrt{2} \in Q$
(viii) $(\sqrt{-1})^2 \in Z$
(ix) $(\sqrt{-1})^2 \in P$

Solution:

(i) True
(ii) False
(iii) True
(iv) False
(v) True
(vi) False
(vii) False, since $\sqrt{2}$ is not a rational number.
(viii) True, for $(\sqrt{-1})^2 = -1$ and -1 is an integer.
(ix) False, since $-1 \notin P$.

1.2. Check the truth of the following assertions.

(i) $\{2\}$ is a subset of $\{2\}$.
(ii) If S is any set, $S \subseteq S$.
(iii) $\{a\} \subset \{a, b\}$, $a \neq b$
(iv) $\{2, 3\} = \{3, 4\}$
(v) $P \subset Q$
(vi) $Z \subseteq Q$
(vii) $Q \subset R$
(viii) $R \subset C$
(ix) $Q \subset Q$
(x) $\{3, a, b, c\} = \{3. a, b, 3, c, b\}$

Solution:

(i) True. The only element of $\{2\}$ is 2, and $2 \in \{2\}$.
(ii) True. Any element in S is an element in S.

(iii) True. a is the only element of $\{a\}$ and $a \in \{a, b\}$. But $b \in \{a, b\}$ and $b \notin \{a\}$. Consequently $\{a\} \neq \{a, b\}$.

(iv) False. $2 \in \{2, 3\}$ and $2 \notin \{3, 4\}$.

(v) True. Any positive integer is a rational number but not all rational numbers are positive integers.

(vi) True. All integers are rational numbers.

(vii) True. All rational numbers are real numbers but $R \neq Q$.

(viii) True. For if $a \in R$, then $a = a + 0i \in C$. But $\sqrt{-1} \notin R$ implies $R \neq C$.

(ix) False. $Q = Q$.

(x) True. 3, a, b and c are the only elements of $\{3, a, b, 3, c, b\}$. Therefore the sets are equal.

1.3. Are the following statements true?

(i) $Z = \{x \mid x \text{ is real and } x > 0\}$

(ii) $N = \{x \mid x \in Q \text{ and } x \geqq 0\}$

(iii) $Q = \{x \mid x = a/b, \text{ where } b \neq 0 \text{ and } a, b \in Z\}$

(iv) $P = \{x \mid x \in N \text{ and } x^2 \geqq 1\}$

(v) $C = \{x \mid x = u + iv, \text{ where } u, v \in R \text{ and } i^2 = -1\}$

(vi) $Z = \{x \mid x \text{ is real and } x^2 \in P\}$

Solution:

(i) False. $\{x \mid x \text{ is real and } x > 0\}$ has no negative elements.

(ii) False. $\frac{3}{4} \in \{x \mid x \in Q \text{ and } x \geqq 0\}$ but $\frac{3}{4} \notin N$. Hence the sets are not equal.

(iii) True. A rational number is defined as the set of all numbers of the form a/b where $b \neq 0$ and $a, b \in Z$.

(iv) True. For if $x \in P$, then $x \in N$ and $x^2 \geqq 1$. Thus $x \in \{x \mid x \in N \text{ and } x^2 \geqq 1\}$. Now if $x \in N$ and $x^2 \geqq 1$, i.e. $x \in \{x \mid x \in N \text{ and } x^2 \geqq 1\}$, then $x \neq 0$. Hence $x \in P$, as the only element in N which is not in P is zero.

(v) True. The property that $x = u + iv$ where $u, v \in R$ and $i^2 = -1$ is the defining property for complex numbers.

(vi) False. $x^2 \in P$ implies $x \neq 0$. But $0 \in Z$.

b. Union and intersection

Let S and T be sets. Then the *union* of S and T, written $S \cup T$ and read "S union T", is defined as the set whose elements are either in S or in T (or in both S and T). For example, $\{1, 2, 3\} \cup \{2, 5, 6\} = \{1, 2, 3, 5, 6\}$ and $P \cup \{0\} = N$. Clearly, $S \subseteq S \cup T$ and $T \subseteq S \cup T$. Indeed it follows from the definition of $S \cup T$ that any set containing both S and T contains $S \cup T$, so we say $S \cup T$ is the *smallest* set containing S and T.

Similarly if $\{S, T, U \ldots\}$ is any set of sets, we define $S \cup T \cup U \cup \cdots$, the union of S and T and U and $\ldots$, to be the set whose elements are the elements that belong to at least one of the sets $S, T, U, \ldots$.

$S \cup T \cup U \cup \ldots$ is said to be the smallest set containing the sets $S, T, U, \ldots$. To illustrate, $\{1, 2\} \cup \{3, 4\} \cup \{5, 6\} \cup \cdots = P$.

If S and T are sets, we may consider the *common part* or *intersection* of S and T. The intersection is denoted by $S \cap T$ and read as "S intersection T". For example, suppose $S = \{1, 2, 3\}$ and $T = \{2, 5, 6\}$. Then $S \cap T = \{2\}$. Repeating the definition, $S \cap T$ is the set of those elements which belong simultaneously to S and to T. Here the possibility arises that there are *no* elements of S which belong also to T. We shall agree to the convention that there is a *set*, which we denote by $\emptyset$, with no elements. Again we shall agree to the convention that the *empty set* $\emptyset$ is a subset of every set. Two sets are termed *disjoint* if they have an empty intersection. Thus $\{1, 2\}$ and $\{3, 4\}$ are disjoint. This notion of intersection can be generalized to any number of sets in the same way that the notion of union was generalized from two to any number. To be precise, the intersection of sets $S, T, U, \ldots$, written $S \cap T \cap U \cap \cdots$, is the set of all those elements which belong simultaneously to S, to T, to $U, \ldots$. Notice that $S \cap T$ can be thought of as the *largest subset* of S which is also a subset of T. Similarly $S \cap T \cap U \cap \cdots$ is the largest subset of S which is contained in T and in U and in $\ldots$.

If $S \subseteq T$ we define $T - S$, the *difference between T and S*, by

$$T - S = \{x \mid x \in T \text{ and } x \notin S\}$$

Thus if $T = \{1,2,3,4\}$ and $S = \{1,2\}$, then $T - S = \{3,4\}$. For any sets T and S such that $T \supseteq S$,

$$T - (T - S) = S \tag{1.1}$$

We prove equation (*1.1*) by showing that

right side of equation (*1.1*) $\subseteq$ left side of equation (*1.1*)

and left side of equation (*1.1*) $\subseteq$ right side of equation (*1.1*)

To do this suppose $x \in S$. Then clearly $x \in T$ but $x \notin T - S$. So $x \in T - (T - S)$; in other words, the right side of equation (*1.1*) is contained in the left side of equation (*1.1*). The reverse inclusion is obtained similarly. Suppose $x \in T - (T - S)$. Then $x \in T$ and $x \notin T - S$. Therefore $x \in T$ and $x \in S$, i.e. $x \in S$. So the left side is contained in the right side and we have proved equation (*1.1*) by virtue of Proposition 1.1.

Problems

1.4. Check that the following statements are correct:

(i) $\{1,2\} \cup \{1,2,3,4\} = \{1,2,3,4\}$

(ii) $\{a,e\} \cup \{e,f\} \cup \{g,h\} = \{a,e,f,g,h\}$

(iii) $\{\ldots,-2,-1,0\} \cup \{0,1,2,\ldots\} = Z$

(iv) If $a < b$, $a,b \in R$, then $[a,b] = \langle a,b\rangle \cup \{a\} \cup \{b\}$.

(v) $\{p \mid p = (x,y) \in R^2,\ x^2+y^2 = 7^2\} \cup \{p \mid p = (x,y) \in R^2,\ x^2+y^2 < 7^2\} = \{p \mid p = (x,y) \in R^2,\ x^2+y^2 \leqq 7^2\}$. (In other words, the union of the circle of radius 7 and the disc of radius 7 without its boundary, is the disc of radius 7 itself.)

Solution:

(iv) Let $x \in [a,b]$. By definition, $a \leqq x \leqq b$. If $a < x < b$, we have $x \in \langle a,b\rangle$. If $x = a$ or $x = b$, then $x \in \{a\}$ or $x \in \{b\}$ respectively. Therefore $[a,b] \subseteq \langle a,b\rangle \cup \{a\} \cup \{b\}$. Now for any $x \in \langle a,b\rangle \cup \{a\} \cup \{b\}$, either $x \in \langle a,b\rangle$, $x \in \{a\}$, or $x \in \{b\}$; and in each case $a \leqq x \leqq b$. Hence $\langle a,b\rangle \cup \{a\} \cup \{b\} \subseteq [a,b]$. The equality follows from Proposition 1.1.

(v) If $p = (x,y)$ is any element of the disc, then $x^2+y^2 \leqq 7^2$. $x^2+y^2 < 7^2$ implies $p \in$ disc without its boundary; and $x^2+y^2 = 7^2$ implies $p \in$ boundary. Thus the disc $\subseteq$ boundary $\cup$ disc without its boundary. The reverse inclusion can be checked similarly. Proposition 1.1 then implies the sets are equal.

1.5. Check the following statements:

(i) $\{1,2\} \cap \{1,2,3,4\} = \{1,2,3\}$

(ii) $\{a,e\} \cap \{e,f\} \cap \{g,h\} = \emptyset$ where a,e,f,g,h are distinct.

(iii) $\{\ldots,-2,-1,0\} \cap \{0,1,2,\ldots\} = \{0\}$

(iv) If $a < b$, $a,b \in R$, then $[a,b] \cap \{a,b\} = \{a\} \cup \{b\}$.

Solution:

(i) False. $3 \in \{1,2,3,4\}$ but $3 \notin \{1,2\}$.

(ii) True. The three sets $\{a,e\}$, $\{e,f\}$, $\{g,h\}$ have no elements in common.

(iii) True, since 0 is the only element in the intersection and 0 is the only element in $\{0\}$.

(iv) True. a and b are the only elements of $[a,b] \cap \{a,b\}$ and $a,b \in \{a\} \cup \{b\}$. Furthermore, a and b are the only elements of $\{a\} \cup \{b\}$.

1.6. Check that the following statements are correct:

(i) $\{1,2,3,4\} - \{1,2\} = \{3,4\}$

(ii) $\{1,2,3,4\} - \{1,2,3,4\} = \emptyset$

(iii) If $a,b \in R$ and $a < b$, then $[a,b] - \langle a,b\rangle = \{a,b\}$.

1.7. If S, T and U are any three sets, prove the following:

(i) $S \cup T = T \cup S$

(ii) $S \cap T = T \cap S$

(iii) $S \subseteq S \cup T$ and $S \subseteq T \cup S$

(iv) $S \cap T \subseteq S$ and $T \cap S \subseteq S$

(v) $S \cup \emptyset = S$

(vi) $S \cap \emptyset = \emptyset$

(vii) $S \cup S = S$

(viii) $S \cap S = S$

(ix) $S \cup (T \cup U) = S \cup T \cup U$

(x) $S \cap (T \cap U) = (S \cap T) \cap U$

(xi) $S \cup (T \cap U) = (S \cup T) \cap (S \cup U)$

(xii) $S - S = \emptyset$

(xiii) $S - (S - S) = S$

Solution:

(i) Let $x \in S \cup T$. By the definition of union of sets, $x \in S$ or $x \in T$. Hence $x \in T \cup S$ and $S \cup T \subseteq T \cup S$. Similarly if $x \in T \cup S$, it follows that $T \cup S \subseteq S \cup T$. Consequently $S \cup T = T \cup S$ by Proposition 1.1.

(ii) If $x \in S \cap T$, then, by the definition of intersection, $x \in S$ and $x \in T$. x is therefore an element of $T \cap S$ and $S \cap T \subseteq T \cap S$. The reverse inclusion, $T \cap S \subseteq S \cap T$, is established in the same manner. The equality follows from Proposition 1.1.

(iii) By the definition of union, $S \cup T$ contains all elements of S and of T. So $x \in S$ implies $x \in S \cup T$ and $S \subseteq S \cup T$. Using (i) above, we also have $S \subseteq T \cup S$.

(iv) $x \in S \cap T$ implies $x \in S$ and $x \in T$. In particular, $x \in S$. Thus $S \cap T \subseteq S$. Part (ii) allows us to write $T \cap S \subseteq S$.

(v) $S \subseteq S \cup \emptyset$ by (iii). If $x \in S \cup \emptyset$, then $x \in S$, for $x \notin \emptyset$ by definition. Therefore $S \cup \emptyset \subseteq S$. Hence by Proposition 1.1, $S \cup \emptyset = S$.

(vi) By (iv), $S \cap \emptyset \subseteq \emptyset$. But we know that $\emptyset$ is a subset of any set and, in particular, $\emptyset \subseteq S \cap \emptyset$. Hence $S \cap \emptyset = \emptyset$.

(vii) $S \subseteq S \cup S$, using (iii). Now, $x \in S \cup S$ implies $x \in S$. Hence $S \cup S \subseteq S$ and the equality follows.

(viii) By (iv), $S \cap S \subseteq S$. But $x \in S$ implies $x \in S \cap S$. Thus $S \subseteq S \cap S$ and the equality follows.

(ix) Let $x \in S \cup (T \cup U)$. Then $x \in S$ or $x \in T \cup U$. Thus $x \in S$ or $x \in T$ or $x \in U$. Consequently $x \in S \cup T \cup U$. Hence $S \cup (T \cup U) \subseteq S \cup T \cup U$. If $x \in S \cup T \cup U$, then $x \in S$ or $x \in T$ or $x \in U$. If $x \in S$, $x \in S \cup (T \cup U)$. If $x \in T$ or U, $x \in S \cup (T \cup U)$. Hence $S \cup T \cup U \subseteq S \cup (T \cup U)$. The result follows.

(x) $x \in S \cap (T \cap U)$ implies $x \in S$ and $x \in (T \cap U)$, which in turn implies $x \in T$ and $x \in U$. From $x \in S$ and $x \in T$ it follows that $x \in (S \cap T)$ and, as $x \in U$, $x \in (S \cap T) \cap U$. Therefore $S \cap (T \cap U) \subseteq (S \cap T) \cap U$. Similarly $(S \cap T) \cap U \subseteq S \cap (T \cap U)$.

(xi) $S \cup (T \cap U) \subseteq (S \cup T) \cap (S \cup U)$. For, if $x \in S \cup (T \cap U)$, then $x \in S$ or $x \in T \cap U$. Now, $x \in S$ implies $x \in S \cup T$, $x \in S \cup U$, and consequently $x \in (S \cup T) \cap (S \cup U)$. If $x \in T \cap U$, then $x \in T$ and $x \in U$; hence $x \in S \cup T$, $x \in S \cup U$ and, as before, $x \in (S \cup T) \cap (S \cup U)$. $(S \cup T) \cap (S \cup U) \subseteq S \cap (T \cup U)$ is established in a similar manner.

(xii) If $S - S$ contains an element x, then $x \in S$ and $x \notin S$ which is impossible. Hence $S - S$ must be empty.

(xiii) $S - S = \emptyset$ from (xii). Thus $S - (S - S) = S - \emptyset$ and clearly $S - \emptyset = S$.

1.8. Prove the following statements:

(i) If $S \subseteq T$ and U is any set, then $S \cup U \subseteq T \cup U$.

(ii) If $S \subseteq T$ and U is any set, then $S \cap U \subseteq T \cap U$.

(iii) If $S \subseteq T$ and $T \subseteq U$, then $S \subseteq U$.

(iv) $S \subseteq T$ if and only if $S \cap T = S$.

(v) $T \subseteq S$ if and only if $S = T \cup S$.

(vi) If $T \subseteq S$, then $(S - T) \cup T = S$.

Solution:

(i) Let $x \in S \cup U$. Then $x \in S$ or $x \in U$. If $x \in S$ then $x \in T$ since $S \subseteq T$, and consequently $x \in T \cup U$. If $x \in U$, then $x \in T \cup U$. Thus $S \cup U \subseteq T \cup U$.

(ii) $x \in S \cap U$ implies $x \in S$ and $x \in U$. As $S \subseteq T$, we also have $x \in T$. Therefore $x \in T \cap U$ and $S \cap U \subseteq T \cap U$.

(iii) $S \subseteq T$ means that if $x \in S$, then $x \in T$; and since $T \subseteq U$, this in turn implies $x \in U$. Hence $S \subseteq U$.

(iv) First, assume $S \cap T = S$. The equality implies any element x in S belongs to $S \cap T$. But $x \in S \cap T$ implies $x \in T$. Hence $x \in T$ and $S \subseteq T$. Secondly, let $S \subseteq T$. If $x \in S$, then $x \in T$ and so $x \in S \cap T$. Therefore $S \subseteq S \cap T$. The reverse inclusion $S \cap T \subseteq S$ is true by Problem 1.7(iv). By Proposition 1.1, $S = S \cap T$.

(v) Assume $S = T \cup S$. If $x \in T$, then $x \in T \cup S$ and, since $S = T \cup S$, $x \in S$. Hence $T \subseteq S$. On the other hand if we assume $T \subseteq S$, then $x \in T \cup S$ implies that $x \in S$. Consequently $T \cup S \subseteq S$. By Problem 1.7(iii) we have $S \subseteq T \cup S$. Thus $T \cup S = S$.

(vi) It follows from the definition that $(S - T) \subseteq S$. Using (i) above, $(S-T) \cup T \subseteq S \cup T$. But by (v), $T \subseteq S$ implies $S = T \cup S$, and we have $(S-T) \cup T \subseteq S$. To show $S \subseteq (S-T) \cup T$, let $x \in S$. Either $x \in T$ or $x \notin T$. If $x \in T$, then $x \in (S-T) \cup T$. If $x \notin T$, then $x \in S - T$ and we also have $x \in (S-T) \cup T$. Thus $S \subseteq (S-T) \cup T$. The equality follows by Proposition 1.1.

1.2 CARTESIAN PRODUCTS

a. Definition

The plane R^2 (see Section 1.1a) consists of all pairs (x, y) of real numbers x and y. We shall also denote R^2 by $R \times R$; thus $R \times R$ is defined as the set of all ordered pairs (x, y) with $x \in R$ and $y \in R$.

There is a natural extension of this notation to any two sets, S and T:

$$S \times T = \{p \mid p = (s, t),\ s \in S,\ t \in T\}$$

For example,

$$\{1, 2\} \times \{1, 2, 3\} = \{(1,1), (1,2), (1,3), (2,1), (2,2), (2,3)\}$$

In words, $S \times T$ is defined to be the set of all ordered pairs of elements (s, t), the first member of each pair always belonging to S and the second member always belonging to T. We term $S \times T$ the *cartesian product* of S and T. It is worth pointing out that, just as in R^2, two elements of $S \times T$ are equal iff (if and only if) they are identical, i.e. $(s, t) = (s', t')$ if and only if $s = s'$ and $t = t'$. If either $S = \emptyset$ or $T = \emptyset$, we interpret $S \times T$ as $\emptyset$.

One defines similarly the cartesian product $S_1 \times S_2 \times \cdots \times S_n$ of the n sets $S_1, S_2, \ldots, S_n$ $(n < \infty)$ as the set of all n-tuples $(s_1, s_2, \ldots, s_n)$ with $s_1 \in S_1$, $s_2 \in S_2$, $\ldots$, $s_n \in S_n$. As with the cartesian product of two sets, $(s_1, s_2, \ldots, s_n) = (s_1', s_2', \ldots, s_n')$ iff $s_1 = s_1'$, $s_2 = s_2'$, $\ldots$, $s_n = s_n'$. For example,

$$\begin{aligned}\{1,2\} \times \{2,3\} \times \{4,5\} = \{&(1,2,4), (1,2,5), (1,3,4), (1,3,5),\\ &(2,2,4), (2,2,5), (2,3,4), (2,3,5)\}\end{aligned}$$

If any one of the sets $S_1, S_2, \ldots, S_n = \emptyset$, then we shall interpret $S_1 \times S_2 \times \cdots \times S_n$ to be $\emptyset$. If each $S_i = S$, then $S_1 \times S_2 \times \cdots \times S_n$ is denoted by S^n.

We often are interested in certain subsets of $S \times T$. For example, in elementary analytic geometry one investigates lines, circles (see Section 1.1a), ellipses and other figures in the plane; these are subsets of R^2.

Problems

1.9. Let $S = \{1, 2, 3\}$, $T = \{1, 5\}$. Verify the following statements:

(i) $S \times T = \{(1,1), (1,5), (2,1), (2,5), (3,1), (3,5)\}$

(ii) $T \times S = \{(1,1), (5,1), (1,2), (5,2), (1,3), (5,3)\}$

(iii) $S \times T \neq T \times S$

(iv) $S^2 \neq T^2$ ($S^2 = S \times S$ and $T^2 = T \times T$)
(v) $(S \times T) \times S \neq S \times (T \times S)$

Solution:
(i) Clearly these are the only ordered pairs (x, y) with $x \in S$ and $y \in T$.
(ii) As in (i), these are the only ordered pairs (x, y) with $x \in T$ and $y \in S$.
(iii) Looking at (i) and (ii), we find $(5,1) \in T \times S$ and $(5,1) \notin S \times T$. Hence $S \times T \neq T \times S$.
(iv) $(3,3) \in S^2$ but $(3,3) \notin T^2$.
(v) An element (x, y) of $(S \times T) \times S$ has $x \in S \times T$ and $y \in S$. Thus $((1,1),5) \in (S \times T) \times S$. But $((1,1),5) \notin S \times (T \times S)$, since $(1,1) \notin S$.

1.10. Let S, T and U be any three sets. Prove the following:
(i) $S \times T = T \times S$ iff either $S = T$ or at least one of the two is empty.
(ii) If $(x, y) \in S^2$, then $(y, x) \in S^2$.
(iii) $S \times T = S \times U$ iff either $T = U$ or $S = \emptyset$.
(iv) $(S \times T) \times U = S \times (T \times U)$ iff at least one of the sets S, T, U is empty.

Solution:
(i) $S = T$ implies $S \times T = T^2 = T \times S$; and $S = \emptyset$ or $T = \emptyset$ implies, by the definition of the cartesian product of any set and the empty set, $S \times T = \emptyset = T \times S$. Therefore $S \times T = T \times S$ whenever $S = T$, $S = \emptyset$, or $T = \emptyset$. To prove the converse, assume $S \times T = T \times S$. We may also assume $S \neq \emptyset$ and $T \neq \emptyset$. Let $t \in T$ and $s \in S$ (such elements exist since $S \neq \emptyset$ and $T \neq \emptyset$). Then $(s, t) \in S \times T$ and, as $S \times T = T \times S$, $(s, t) \in T \times S$. It follows, from the definition of $T \times S$, that $t \in S$ and $s \in T$. Therefore $T \subseteq S$ and $S \subseteq T$. We conclude, using Proposition 1.1, $S = T$.
(ii) $(x, y) \in S^2$ means $x \in S$ and $y \in S$. Hence $(y, x) \in S^2$.
(iii) Clearly if $T = U$ or $S = \emptyset$, we have $S \times T = S \times U$. Conversely, let $S \times T = S \times U$ and $S \neq \emptyset$ (if $S = \emptyset$ we have nothing to prove). If $T \neq \emptyset$, let $t \in T$. Then $(s, t) \in S \times T$ for any $s \in S$; and, as $S \times T = S \times U$, $(s, t) \in S \times U$. But $(s, t) \in S \times U$ means $t \in U$. Hence $T \subseteq U$. A similar argument gives $U \subseteq T$, and we conclude $T = U$. If $T = \emptyset$, then $S \times T = \emptyset$ and $S \times U = \emptyset$. $U = \emptyset$ follows from $S \times U = \emptyset$; for if $U \neq \emptyset$, then $S \times U$ would not be empty, by virtue of our assumption $S \neq \emptyset$. Thus both $T \neq \emptyset$ and $T = \emptyset$ give $T = U$.
(iv) If either S, T or U is empty, $(S \times T) \times U = \emptyset = S \times (T \times U)$. Conversely, assume $(S \times T) \times U = S \times (T \times U)$. If $S \neq \emptyset$, $T \neq \emptyset$ and $U \neq \emptyset$, there is at least one element (x, y) in $(S \times T) \times U$, $x \in S \times T$ and $y \in U$. But (x, y) must also be an element of $S \times (T \times U)$. Hence $x \in S$. This is a contradiction, for x cannot be both an element of S and an element of $S \times T$. Therefore the assumption that $S \neq \emptyset$, $T \neq \emptyset$ and $U \neq \emptyset$ must be false and so either S, T or U is empty.

1.11. (i) If $A = \{p \mid p = (x, y) \in P^2 \text{ and } x < y\}$ (recall that P is the set of positive integers), prove that:
(*a*) $(x, x) \notin A$ for every $x \in P$.
(*b*) if $(x, y) \in A$ and if $(y, z) \in A$, then $(x, z) \in A$.

(ii) Let $B = \{p \mid p = (x, y) \in P^2, x \leqq y\}$.
(*a*) Show that $(x, x) \in B$ for every $x \in P$.
(*b*) Prove that if $(x, y) \in B$ and if $(y, z) \in B$, then $(x, z) \in B$.
(*c*) If $(x, y) \in B$, when is $(y, x) \in B$?

(iii) Let $A = \{p \mid p = (x, y) \in Z^2 \text{ with } x - y \text{ divisible by } 3\}$. Prove:
(*a*) $(x, x) \in A$ for all $x \in Z$.
(*b*) If $(x, y) \in A$, then $(y, x) \in A$.
(*c*) If $(x, y) \in A$ and $(y, z) \in A$, then $(x, z) \in A$.

(iv) Let U be the points of the plane R^2 on or above the X axis and let L be the points below the X axis. Put $V = U^2 \cup L^2$. Notice that $V \subseteq R^2 \times R^2$. Prove:
(*a*) If $(\alpha, \alpha) \in R^2 \times R^2$, then $(\alpha, \alpha) \in V$.
(*b*) If $(\alpha, \beta) \in V$, then $(\beta, \alpha) \in V$.
(*c*) If $(\alpha, \beta) \in V$ and $(\beta, \gamma) \in V$, then $(\alpha, \gamma) \in V$.
(*d*) $V \neq R^2 \times R^2$.

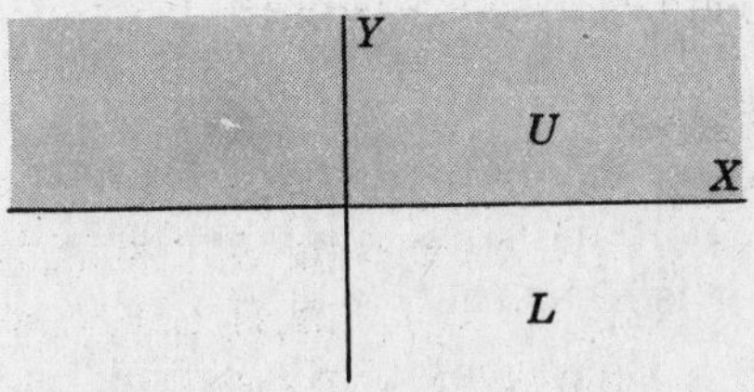

Solution:

(i) (*a*) $(x,x) \notin A$ for every $x \in P$, since x is not less than x.

(*b*) $(x,y) \in A$ implies $x < y$, and $(y,z) \in A$ implies $y < z$. Now $x < y$ and $y < z$ imply $x < z$, since $x, y, z \in P$. By definition of set A, we have $(x,z) \in A$.

(ii) (*a*) $x \leq x$ for all $x \in P$. Hence $(x,x) \in B$.

(*b*) $(x,y) \in B$ and $(y,z) \in B$ imply $x \leq y$ and $y \leq z$ respectively. It follows that $x \leq z$ and, by definition of set B, $(x,z) \in B$.

(*c*) $(y,x) \in B$ iff $x = y$. For if $(x,y) \in B$, $x \leq y$; and if $(y,x) \in B$, $y \leq x$. But $x \leq y$ and $y \leq x$ iff $x = y$.

(iii) (*a*) For any $x \in Z$, $x - x = 0$, and zero is divisible by 3. Consequently $(x,x) \in A$.

(*b*) $(x,y) \in A$ means $x - y$ is divisible by 3, i.e. $x - y = 3q$ where q is some integer. Now $y - x = -(x-y) = -3q = 3(-q)$. Therefore $y - x$ is divisible by 3 and $(y,x) \in A$.

(*c*) $(x,y) \in A$ means $x - y = 3q$ for some integer q, and $(y,z) \in A$ means $y - z = 3r$ for some integer r. Thus

$$x - z \;=\; (x-y) + (y-z) \;=\; 3q + 3r \;=\; 3(q+r)$$

and so $x - z$ is divisible by 3 and $(x,z) \in A$.

(iv) (*a*) $(\alpha,\alpha) \in R^2 \times R^2$ implies $\alpha \in R^2$, which means $\alpha \in U$ or $\alpha \in L$. Hence $(\alpha,\alpha) \in U^2$ or $(\alpha,\alpha) \in L^2$, and $(\alpha,\alpha) \in U^2 \cup L^2 = V$.

(*b*) If $(\alpha,\beta) \in V$, then $(\alpha,\beta) \in U^2$ or $(\alpha,\beta) \in L^2$ and by Problem 1.10(ii), $(\beta,\alpha) \in U^2$ or $(\beta,\alpha) \in L^2$. Hence $(\beta,\alpha) \in V$.

(*c*) $(\alpha,\beta) \in V$ implies either $\alpha,\beta \in U$ or $\alpha,\beta \in L$, but not both, since $U \cap L = \emptyset$. Now $(\beta,\gamma) \in V$ implies $\beta,\gamma \in U$ or L. If $\beta,\gamma \in U$, then $\alpha,\beta \in U$ since β cannot be an element of both U and L, and hence $(\alpha,\gamma) \in U^2$. Similarly if $\beta,\gamma \in L$, we have $(\alpha,\gamma) \in L^2$. Therefore $(\alpha,\gamma) \in U^2 \cup L^2 = V$.

(*d*) Let $\alpha \in U$ and $\beta \in L$. Then $(\alpha,\beta) \in R^2 \times R^2$, but $(\alpha,\beta) \notin V$.

b. Equivalence relations

Similarity of triangles is an example of an equivalence relation. This means that if s, t and u are any triangles, the following three conditions hold:

(i) s is similar to s.

(ii) If s is similar to t, then t is similar to s.

(iii) If s is similar to t, and t is similar to u, then s is similar to u.

Another example of an equivalence relation is *congruence of triangles* since (i), (ii) and (iii) above hold also if "similar" is replaced by "congruent". Continuing in this vein, if X is any non-empty set and R is a "relation on X", i.e. if for any pair of elements $x, y \in X$ either x is related to y by R (written xRy and read "x is related to y by R") or not, then R is termed an *equivalence relation in* X if:

(i) xRx for all $x \in X$.

(ii) If xRy, then yRx.

(iii) If xRy and yRz, then xRz.

One objection to this definition of equivalence relation is that "relation on X" is vaguely defined. We shall therefore define the idea of equivalence relation by means of sets and subsets.

Let us consider again the notion of similarity of triangles. Let T be the set of all triangles. Let S be the subset of $T \times T$ defined by $(s,t) \in S$ iff s is similar to t. If t is a triangle, t is similar to itself, so $(t,t) \in S$. If (s,t) and $(t,u) \in S$, then s and t are similar and t and u are similar. Thus s and u are similar. Hence $(s,u) \in S$. Similarly if $(s,t) \in S$, it is clear that $(t,s) \in S$.

Consequently, discarding our informal approach, we have the following

Definition: Let X be a non-empty set and let R be a subset of X^2. Then R is called an *equivalence* (or an *equivalence relation*) in X if the following conditions are satisfied.
(i) $(x,x) \in R$ for all $x \in X$ (reflexive property).
(ii) If $(x,y) \in R$, then $(y,x) \in R$ (symmetric property).
(iii) If $(x,y) \in R$ and $(y,z) \in R$, then $(x,z) \in R$ (transitive property).

Problem

1.12. Examine Problem 1.11 for equivalence relations.

Solution:

(i) A is not an equivalence relation in P, as $(x,x) \notin A$ for all x.

(ii) B is not an equivalence relation in P, as we know $(x,y) \in B$ and $(y,x) \in B$ occurs only if $x = y$. Thus though $(1,2) \in B$, $(2,1) \notin B$.

(iii) A is an equivalence relation in Z^2 as $A \subseteq Z^2$ and the reflexive, symmetric and transitive properties hold.

(iv) $V \subseteq R^2 \times R^2$ and the reflexive, symmetric and transitive properties hold, so V is an equivalence in R^2.

c. Partitions and equivalence relations

Suppose R is an equivalence relation in X. We say x is R-related to y, or x is related to y by R, if $(x,y) \in R$. If $(x,y) \in R$ we shall sometimes write xRy. To illustrate let $X = \{1,2,3,4\}$ and let

$$R = \{(1,1),(2,2),(3,3),(4,4),(1,3),(3,1),(3,4),(1,4),(4,3),(4,1)\} \tag{1.2}$$

Then it is easy to check that R satisfies the three necessary conditions for it to be an equivalence relation. Now $3R4$ since $(3,4) \in R$, but $2R4$ is an incorrect assertion since $(2,4) \notin R$.

Note that we have used a notation that fits in with the notation of Section 1.2b where we informally introduced an equivalence relation.

An equivalence relation in X is intimately connected with a *partition* of X, i.e. a decomposition of X into disjoint subsets of X such that every element of X belongs to some subset. Examples of partitions of $\{1,2,3,4,5\}$ are

$$\{1\},\ \{2,3\},\ \{4,5\}$$

and

$$\{1,3,4\},\ \{2\},\ \{5\}$$

On the other hand $\{1,2\}, \{2\}, \{3,4,5\}$ is *not* a partition of $\{1,2,3,4,5\}$.

If R is the equivalence relation (*1.2*) in $\{1,2,3,4\}$, then all the elements of $\{1,3,4\}$ are R-related to 1, i.e.,

$$1R1,\ 1R3,\ 1R4$$

This suggests a means of getting a partition of a set X. In order to explain, we need some additional notation. Let R be an equivalence relation in a set X. If $x \in X$, we define $xR = \{y \mid y \in X \text{ and } (x,y) \in R\}$. xR is thus a certain subset of X. This subset xR is called the *R-class of x*, or the *R-equivalence class of x*, or the *R-block of x*. A subset of X will be called an *R-class or R-block* if it is the R-class or R-block of some element $x \in X$. To illustrate these terms, consider the equivalence relation R given by (*1.2*). Here

$$1R = \{1,3,4\},\ 2R = \{2\},\ \text{and}\ 3R = 4R = 1R$$

Thus the R-classes here are simply $\{1,3,4\}$ and $\{2\}$. Notice that these R-classes constitute a partition of $\{1,2,3,4\}$.

More generally, we have the following

Theorem 1.2: Let X be a non-empty set and let R be an equivalence relation in X. Then

(i) if $xR \cap x'R \neq \emptyset$, then $xR = x'R$,

(ii) $x \in xR$ for every $x \in X$.

Thus the R-classes constitute a partition of X, for (i) guarantees that distinct R-classes are disjoint, while (ii) shows that every element of X appears in at least one of the R-classes.

Proof: First, we verify (i). Suppose $xR \cap x'R \neq \emptyset$. Then there is an element $y \in xR$ which lies also in $x'R$, i.e. $(x, y) \in R$ and $(x', y) \in R$. As R is an equivalence, it follows from the symmetric property that $(y, x') \in R$. But $(x, y) \in R$ and $(y, x') \in R$ imply, by the transitive property, $(x, x') \in R$. Now if $z \in x'R$, then $(x', z) \in R$; and hence by the transitive property applied to (x, x') and (x', z), we find $(x, z) \in R$. This means, by the very definition of xR, that $z \in xR$. Since z was any element of $x'R$, we have proved $x'R \subseteq xR$. The reverse inequality follows by a similar argument. Hence $x'R = xR$ as required.

The verification of (ii) is trivial since $(x, x) \in R$ means $x \in xR$. This completes the proof of Theorem 1.2.

Problems

1.13. (i) Prove that $E = \{(0,0), (1,1), (2,2), (3,3), (0,2), (1,3), (2,0), (3,1)\}$ is an equivalence relation in $S = \{0,1,2,3\}$.

(ii) Find the E-equivalence blocks (*a*) $0E$, (*b*) $1E$, (*c*) $2E$, (*d*) $3E$.

Solution:

(i) The reflexive property holds, i.e. $(x, x) \in E$ for all $x \in S$, since $(0,0), (1,1), (2,2), (3,3) \in E$. To show that E is symmetric, let us examine all pairs (x, y) where $x \neq y$. There are only four, namely $(0,2), (1,3), (2,0), (3,1)$. Clearly if (x, y) is any one of the four, so is (y, x). When $x = y$, $(x, y) = (y, x)$. Thus $(x, y) \in E$ implies $(y, x) \in E$. E is also transitive. Let $(x, y) \in E$ and $(y, z) \in E$. Suppose $x \neq y$. Then (x, y) can be $(0,2), (1,3), (2,0)$ or $(3,1)$. If $(x, y) = (0,2)$, then $(y, z) = (2,0)$ or $(2,2)$ and $(x, z) = (0,0)$ or $(0,2)$ respectively. Hence $(x, z) \in E$. Similarly if $(x, y) = (1,3), (2,0)$ or $(3,1)$, it can be shown that $(x, z) \in E$. When $x = y$, $(y, z) \in E$ means $(x, z) \in E$. Therefore for any $(x, y) \in E$ and $(y, z) \in E$, we have $(x, z) \in E$ and E is transitive.

(ii) (*a*) $0E = \{0,2\}$, (*b*) $1E = \{1,3\}$, (*c*) $2E = \{2,0\}$, (*d*) $3E = \{3,1\}$. Observe that $0E = 2E$ and $1E = 3E$.

1.14. Let $A = \{p \mid p = (x, y) \in Z^2$ with $x - y$ divisible by $3\}$. Prove that A is an equivalence relation in Z and find the R-equivalence classes.

Solution:

It was shown in Problem 1.11(iii), page 7, that A satisfies the three conditions of an equivalence relation. The R-equivalance classes are:

(1) $0A = \{3q \mid q \in Z\}$; for if $(0, x) \in A$, $0 - x = -x$ is divisible by 3. Also, $(0, 3q) \in A$.

(2) $1A = \{1 - 3q \mid q \in Z\}$; for if $(1, x) \in A$, $1 - x = 3q$ and hence $x = 1 - 3q$. If $x = 1 - 3q$, $1 - x$ is divisible by 3; hence $(1, x) \in 1A$.

(3) $2A = \{2 - 3q \mid q \in Z\}$; for if $(2, x) \in A$, $2 - x = 3q$ and so $x = 2 - 3q$. If $x = 2 - 3q$, $2 - x$ is divisible by 3; hence $(2, x) \in A$.

$0A, 1A, 2A$ are the only R-blocks, for any integer can be written as $3q$, $1 - 3q$, or $2 - 3q$. Consequently $0A \cup 1A \cup 2A = Z$.

1.15. Let Z^* be the set of nonzero integers and let $S = Z \times Z^*$. (Recall that Z is the set of *all* integers.) Let $E = \{p \mid p = ((r, s), (t, u)) \in S^2$ with $ru = st\}$. Prove that E is an equivalence relation in S.

Solution:

E is reflexive, for $(r,s) \in S$ implies $((r,s),(r,s)) \in S^2$; and since $rs = sr$, $((r,s),(r,s)) \in E$. The symmetric property of E is established by noticing $((r,s),(t,w)) \in E$ means (r,s) and $(t,w) \in S^2$, and $rw = st$. But $rw = st$ can be rewritten as $ts = wr$. Hence $((t,w),(r,s)) \in E$. To show E is transitive, let $((r,s),(t,u)) \in E$ and $((t,u),(v,w)) \in E$. Then $ru = st$ and $tw = vu$. Since $u \neq 0$, $r = \dfrac{st}{u}$ and $rw = \dfrac{stw}{u} = \dfrac{s}{u}tw = \dfrac{s}{u}vu = sv$. Thus $rw = sv$ and so $((r,s),(v,w)) \in E$. Hence E is transitive. Therefore E is an equivalence relation.

1.16. Prove that $S \times S$ is an equivalence relation in S.

Solution:

$S \times S$ is reflexive since $(x,x) \in S \times S$ for all $x \in S$. If $(x,y) \in S \times S$, then x and $y \in S$ and, by definition of $S \times S$, $(y,x) \in S \times S$. Hence $S \times S$ is symmetric. Now $(x,y) \in S$ and $(y,z) \in S$ imply x, y and $z \in S$. But then $(x,z) \in S \times S$ and $S \times S$ is transitive. Thus $S \times S$ is an equivalence relation on S.

1.17. Prove that a set X is infinite if and only if there are infinitely many equivalences in X. (Hard.)

Solution:

Assume there are an infinite number of equivalences in X. If X is finite, then there are at most a finite number of distinct subsets of X^2. Therefore when X is finite there are at most a finite number of equivalences in X, which contradicts our hypothesis. Hence X must be infinite. Conversely, assume X is infinite. We exhibit an infinite number of equivalences in X as follows: For each pair $a, b \in R$, $a \neq b$, we define

$$R_{(a,b)} = \{p \mid p = (x,y) \in X^2, \text{ where either } x = y,\ (x,y) = (a,b) \text{ or } (x,y) = (b,a)\}$$

Now $R_{(a,b)} = R_{(c,d)}$ if and only if $\{a,b\} = \{c,d\}$. Therefore since X is infinite, we can find an infinite number of different pairs $a, b \in X$ each of which gives a distinct set $R_{(a,b)}$. Furthermore, each $R_{(a,b)}$ is an equivalence. To prove that $R_{(a,b)}$ satisfies the three conditions of an equivalence relation, we first notice $(x,x) \in R_{(a,b)}$ for all $x \in X$, by the very definition of $R_{(a,b)}$. Secondly, $R_{(a,b)}$ is symmetric since $(x,y) \in R_{(a,b)}$ means $(x,y) = (a,b)$ or (b,a), in which case $(y,x) = (b,a)$ or (a,b) respectively, or $x = y$ and then $(x,y) = (y,x)$. Thirdly, if (x,y) and $(y,z) \in R_{(a,b)}$, then $(x,z) \in R_{(a,b)}$. To see this, notice that (x,y) can only be (a,b), (b,a) or (x,x). $(x,y) = (a,b)$ implies $(y,z) = (b,a)$ or (b,b); hence $(x,z) = (a,a)$ or (a,b), which are both elements of $R_{(a,b)}$. Similarly, $(x,y) = (b,a)$ implies $(x,z) \in R_{(a,b)}$. Finally, $(x,y) = (x,x)$ means $(x,z) = (y,z) \in R_{(a,b)}$.

d. The division notation

We find it useful to introduce a notation for the R-classes of an equivalence relation R in a set X, namely X/R.

Problems

1.18. What is S/E in Problem 1.13?

Solution: $S/E = \{0E, 1E\}$.

1.19. What is Z/A in Problem 1.14?

Solution: $Z/A = \{0A, 1A, 2A\}$.

1.3 MAPPINGS

a. Definition of mapping

Assign to each even integer the value 1, and to each odd integer the value -1. Let us give the name α to this assignment; thus α assigns to each even element in Z, the set of all integers, the unique element $+1$ in the set $\{1,-1\}$ and to each odd element in Z the unique element -1 in $\{1,-1\}$. In less detailed terms α assigns to each element in Z a unique element in $\{1,-1\}$. Such an assignment is termed *a mapping from* Z *into* $\{1,-1\}$ or *a map from* Z *into* $\{1, -1\}$. More generally, if S and T are any two non-empty sets, a mapping or a map from S into T is an assignment of a unique element of T to each element of S. For the most part we shall denote mappings by lower case Greek letters such as α, β, γ. If α is a mapping from S into T, we shall express this fact more briefly by writing $\alpha: S \to T$; this is read "α is a mapping from S into T". We call S the *domain* and T the *codomain* of α.

We find it useful to provide further notation and definitions. Suppose that $\alpha: S \to T$. If α assigns to s in S the element t in T, we write $\alpha: s \to t$ and read this as "α sends s into t". We call t the *image of* s (*under* α) if $\alpha: s \to t$. It is convenient to have a number of notations for the image of an element s in S under a mapping $\alpha: S \to T$; thus we shall write $s\alpha$ or s^α, or even $\alpha(s)$ for the image of s under α. For the most part we use the first notation. If $t \in T$ and $s\alpha = t$, we call s a *preimage* of t. By $S\alpha$ we mean $\{s\alpha \mid s \in S\}$. We call $S\alpha$ the *range* of α.

Problems

1.20. Suppose $\alpha: P \to P$ is defined by

(i) $\alpha: n \to n^2$ for all $n \in P$
(ii) $\alpha: n \to n+1$ for all $n \in P$
(iii) $\alpha: n \to 2n$ for all $n \in P$
(iv) $\alpha: n \to 1$ for all $n \in P$
(v) $\alpha: 1 \to 2,\ n \to 1$ for all $n \in P,\ n > 1$

Note: In (i)-(v) above, the mapping α is defined by describing its "action" on every element of P. For example, in (i), $1\alpha = 1^2 = 1$, $2\alpha = 4$, $3\alpha = 9$, ... Note that each element of P has a unique element assigned to it.

In each case determine: (*a*) $2\alpha, 5\alpha, 6\alpha$; (*b*) a preimage (under α) of 2, 5, 6, 27. (*c*) Is every element of P an image of some element of P in (i)-(v)? How many preimages (under α) does 2 have in (v)? How many preimages does 1 have in (iv)? in (v)?

Solution:

(i) (*a*) $2\alpha = 4$; $5\alpha = 25$; $6\alpha = 36$.
(*b*) There is no preimage of 2, 5, 6 or 27.
(*c*) Not every element of P is an image, e.g. 2 is not an image.

(ii) (*a*) $2\alpha = 3$; $5\alpha = 6$; $6\alpha = 7$.
(*b*) $1\alpha = 2$; $4\alpha = 5$; $5\alpha = 6$; $26\alpha = 27$. Hence 1, 4, 5, 26 are the required preimages.
(*c*) The only element of P which has no preimage is 1.

(iii) (*a*) $2\alpha = 4$; $5\alpha = 10$; $6\alpha = 12$.
(*b*) 1 is a preimage of 2; 5 has no preimage; 3 is the preimage of 6; 27 has no preimage.
(*c*) 1 has no preimage, so not every element of P is an image.

(iv) (*a*) $2\alpha = 1$; $5\alpha = 1$; $6\alpha = 1$.
(*b*) 2, 5, 6 and 27 have no preimages.
(*c*) 1 has every element of P as a preimage and no other element of P has a preimage.

(v) (*a*) $2\alpha = 1$; $5\alpha = 1$; $6\alpha = 1$.
(*b*) 1 is a preimage of 2; 5, 6 and 7 have no preimages.
(*c*) 2 has one preimage, namely 1. 1 has an infinite number of preimages. In fact any element of P not equal to 1 is mapped onto 1. 1 and 2 are the only elements of P which have preimages.

1.21. Let $S = \{1, 2, 3\}$, $T = \{1, 4, 5\}$.

(i) Write down a mapping of S into T.
(ii) Let $\alpha: S \to T$ be defined by $1\alpha = 4$, $2\alpha = 5$, $3\alpha = 4$. Is every element of T an image of some element in S under α? Is every element of T the image of more than one element in S? Give preimages of 1 and 4.

Solution:

(i) For example, $\alpha: S \to T$ defined by $1\alpha = 1$, $2\alpha = 4$, $3\alpha = 5$.
(ii) Not every element of T is an image, for 1 has no preimage. Only 4 is an image of more than one element in S. 1 has no preimage, and 4 has preimages 1 and 3.

1.22. Suppose $\alpha: P \to C$ is defined by

(i) $\alpha: x \to x^2$

(ii) $\alpha: x \to 2x + 27$

(iii) $\alpha: x \to z$ where $z \in C$ is such that $z^2 = x$.

(iv) $\alpha: x \to z$ where $z \in C$ is such that $z^3 = x - 1$.

(v) $\alpha: x \to ix + 1$

(vi) $\alpha: x \to \dfrac{ix+1}{ix-1}$

(vii) $\alpha: x \to \log_{10} x$

(*a*) Do all of these descriptions really define mappings of P into C?

(*b*) Is every element in C a preimage of some element of P in (i), (ii), (v), (vi), (vii)?

Solution:

(*a*) (i) and (ii) define mappings since every $x \in P$ has a unique image. (iii) does not define a mapping; for example, either 2 or -2 could be taken as 4α. Similarly (iv) is not a mapping, since there are three complex cube roots of $x - 1$, so that each x has three different images. (v), (vi) and (vii) define mappings.

(*b*) In (i), (ii), (v) and (vii), i has no preimage: for (i) $i = x^2$ implies $x = \sqrt{i} \notin P$; (ii) $2x + 27 = i$ implies $x = \dfrac{i-27}{2} \notin P$; (v) $ix + 1 = i$ gives $x = 1 + i \notin P$; (vii) if $\log_{10} x = i$, then $10^i = x$ and thus $x \notin P$. In (vi) 1 has no preimage because $\dfrac{ix+1}{ix-1} = 1$ implies $1 = -1$.

1.23. What is $P\alpha$ in each of the cases in Problem 1.20?

Solution:

(i) $P\alpha$ is the set of squares $\{1, 4, 9, 16, \ldots\}$.

(ii) $P\alpha = \{2, 3, 4, 5, \ldots\}$

(iii) $P\alpha = \{2, 4, 6, 8, \ldots\}$, i.e. all the even integers.

(iv) $P\alpha = \{1\}$

(v) $P\alpha = \{1, 2\}$

b. Formal definition of mapping

The reader may ask whether our definition of mapping is precise. After all, it depends upon an English word, assignment, a word that is used in many different ways.

A comparison with Section 1.2b is valuable. In Section 1.2b we introduced the concept of equivalence relation in X, but as we felt uneasy about it, we redefined it in terms of a subset of X^2. Here too we feel uneasy about our definition of mapping and so we shall redefine it in terms of sets.

A subset α of $S \times T$ is called a mapping of S into T if (s, t_1) and $(s, t_2) \in \alpha$ occurs only if $t_1 = t_2$, and for each $s \in S$ there exists an element $(s, t) \in \alpha$. S is called the *domain* and T the *codomain* of α. If α is a mapping of S into T (written briefly as $\alpha: S \to T$) and $(s, t) \in \alpha$, we call t the *image of* s under α and write $\alpha: s \to t$. We also write $t = s\alpha$.

It is easy to see the relationship between the old definition and the new. In the old definition the elements of S were assigned unique elements of T.

Consider the subset of $S \times T$ consisting of the pairs (s, t) where t is assigned to s. The two conditions of the new definition are satisfied by this subset.

In the sequel we will use the definition of Section 1.3a, being confident that if necessary we could justify our arguments using the definition of a mapping in terms of a subset.

c. Types of mappings

We have talked of mappings without defining what is meant by the equality of two mappings. We will now remedy this. Suppose $\alpha: S \to T$ and $\beta: S' \to T'$. Then we define $\alpha = \beta$ if and only if $S = S'$, $T = T'$, and, for every element $s \in S$, $s\alpha = s\alpha'$. In other words, two mappings are equal if and only if they have the same domain, the same codomain, and the same "action" on each element of S. For example, let $S = \{1, 2, 3, 4\}$, $T = \{4, 5, 6\}$. Let $\alpha: S \to T$ be defined by $1\alpha = 4$, $2\alpha = 5$, $3\alpha = 6$, $4\alpha = 4$; let $\beta: S \to T$ be defined by $1\beta = 4$, $2\beta = 5$, $3\beta = 6$, $4\beta = 5$. Then $\alpha \neq \beta$ since $4\alpha \neq 4\beta$.

It is important to distinguish certain types of mappings. Thus suppose $\alpha: S \to T$. Then we say α is a mapping from S *onto* T (notice that *into* has now been replaced by *onto*) if every element in T has at least one preimage in S, i.e. if for every $t \in T$ there is at least one element $s \in S$ for which $s\alpha = t$; in this case we call α an *onto mapping*.

On the other hand we say α is *one-to-one* if $s\alpha = s'\alpha$ implies $s = s'$, i.e. distinct elements of S have distinct images in T (under α). Finally, we say α is a *matching of S and T* or that α *matches S with T* or α is a *bijection* if α is both onto and one-to-one. Two sets are termed *equipotent* or *of the same cardinality* if there exists a matching of the one with the other. If S is finite and α matches T, then we say S and T have the same number of elements. We denote the number of elements in a set S by $|S|$. If S is infinite this definition no longer makes sense unless one takes from all the sets which match S a single fixed set which we then term $|S|$, the *cardinality* of S†. A set which matches P is called *enumerable, countable,* or *countably infinite.* Important results are that the set of rational numbers is enumerable and the set of real numbers is not†.

We require one further definition. Suppose $\alpha: S \to T$ and suppose $S' \subseteq S$. Then we define a mapping from S' into T by simply restricting the domain of α to S'; this mapping is denoted by $\alpha_{|S'}$ (read α restricted to S') and is called the *restriction of* α *to* S'. To be quite explicit,

$$\alpha_{|S'}: S' \to T \text{ is defined by } \alpha_{|S'}: s \to s\alpha \text{ for all } s \in S'$$

Problems

1.24. Which of the mappings defined in Problems 1.20-1.22 are (*a*) onto, (*b*) one-to-one, (*c*) matchings?

Solution:

(*a*) None of the mappings defined in Problems 1.20 and 1.22 is onto by the solution already given to these problems. The mapping defined in Problem 1.21(ii) is not onto, since 1 has no preimage.

(*b*) Problems 1.20(i), (ii), and (iii) define one-to-one mappings: for in (i), if $n\alpha = m\alpha$, then $n^2 = m^2$ and so $n = m$, since there is only one positive square root of an element in P; in (ii), $n\alpha = m\alpha$ gives $n+1 = m+1$ or $n = m$; and in (iii), $n\alpha = m\alpha$ implies $2n = 2m$ or $n = m$. Clearly Problems 1.20(iv) and (v) do not define one-to-one mappings. The mapping α in Problem 1.21(ii) is not one-to-one since $1\alpha = 3\alpha$. All the mappings defined in Problem 1.22 are one-to-one: for in (i), $x\alpha = x'\alpha$ means $x^2 = x'^2$ and, since $x, x' \in P$, $x = x'$; in (ii), $x\alpha = x'\alpha$ implies $2x + 27 = 2x' + 27$ or $x = x'$; in (v), $x\alpha = x'\alpha$ implies $ix + 1 = ix' + 1$ or $x = x'$; (vi) $x\alpha = x'\alpha$ means $\dfrac{ix+1}{ix-1} = \dfrac{ix'+1}{ix'-1}$ or $2ix = 2ix'$ or $x = x'$; in (vii), $x\alpha = x'\alpha$ gives $\log_{10} x = \log_{10} x' = y$ and hence $10^y = x = x'$. (Note that (iii) and (iv) are not mappings.)

(*c*) None of the mappings defined in Problems 1.20-1.22 is a matching, since none of them is onto.

1.25. Which mappings in Problems 1.20-1.22 are equal?

Solution:

None. Problem 1.20 features $\alpha: P \to P$; Problem 1.21, $\alpha: S \to T$; and Problem 1.22, $\alpha: P \to C$. Hence we need only compare the mappings in each exercise.

1.26. Let $\alpha: P \to Z$ be defined by $n\alpha = -n$ for all $n \in P$. Is α onto? One-to-one? A matching?

Solution:

α is neither onto nor a matching, since 1 has no preimage. α is one-to-one, for if $n\alpha = n'\alpha$, then $-n = -n'$ and hence $n = n'$.

†For more details see, for example, G. Birkhoff and S. MacLane, *A Survey of Modern Algebra*, Macmillan, 1953.

1.27. Suppose m is a fixed positive integer. Every integer n can be written uniquely in the form $n = qm + r$ where the remainder r is an element of the set $\{0, 1, \ldots, m-1\}$. For example, if $m = 4$, then every integer can be expressed in precisely one of the following forms: $4k, 4k+1, 4k+2, 4k+3$. Let $\alpha: Z \to \{0, 1, \ldots, m-1\}$ be defined by $n\alpha = r$, if $n = qm + r$ where $r \in \{0, 1, \ldots, m-1\}$. Prove:

(a) α is onto $\{0, 1, \ldots, m-1\}$.

(b) If n_1, n_2 are any two integers, then $(n_1 n_2)\alpha = (n_1\alpha n_2\alpha)\alpha$.

Solution:

(a) α is onto because $0, 1, \ldots, m-1$ have preimages $0, 1, \ldots, m-1$ respectively.

(b) Let $n_1 = q_1 m + r_1$ and $n_2 = q_2 m + r_2$. Then

$$n_1 n_2 = (q_1 m + r_1)(q_2 m + r_2) = q_1 q_2 m^2 + r_1 q_2 m + r_2 q_1 m + r_1 r_2 = (q_1 q_2 m + r_1 q_2 + r_2 q_1)m + r_1 r_2$$

Now letting $r_1 r_2 = q_3 m + r_3$, we obtain $(n_1 n_2)\alpha = r_3 = (r_1 r_2)\alpha = (n_1\alpha n_2\alpha)\alpha$.

1.28. Check which of the following mappings are onto, one-to-one, bijections:

(i) $\alpha: C \to R$ defined by $\alpha: a + ib \to a^2 + b^2$

(ii) $\alpha: Z \to P$ defined by $\alpha: n \to n^2 + 1$

(iii) $\alpha: P \to Q$ defined by $\alpha: n \to \dfrac{n}{2n+1}$.

Solution:

(i) α is neither onto nor one-to-one, because $a^2 + b^2 \geqq 0$, for any $a, b \in R$, and $-1\alpha = 1\alpha = 1$. Hence α is not a bijection.

(ii) As 3 has no preimage, α is neither onto nor a bijection. Also α is not one-to-one, since $1\alpha = 2$ and $-1\alpha = 2$.

(iii) 1 has no preimage, since $\dfrac{n}{2n+1} = 1$ implies $n = -1 \notin P$. Therefore α is not onto and hence not a bijection. $n\alpha = n'\alpha$ means $\dfrac{n}{2n+1} = \dfrac{n'}{2n'+1}$ or $2n'n + n = 2n'n + n'$; hence $n = n'$. Thus each image has a unique preimage and α is one-to-one.

1.29. Let S be the set of open intervals $\langle a, b \rangle$ on the real line and let T be the set of closed intervals $[a, b]$. Define $\alpha: S \to T$ by $\langle a, b \rangle \alpha = [a, b]$. Is α one-to-one? Onto?

Solution:

The mapping is one-to-one and onto. For, if $\langle a, b \rangle \alpha = \langle a', b' \rangle \alpha$ then $[a, b] = [a', b']$. But this equality holds iff $a = a'$ and $b = b'$. α is therefore one-to-one. α is also onto since a closed interval $[a, b]$ has a preimage $\langle a, b \rangle$.

1.30. How many mappings are there from $\{1, 2\}$ into itself? From $\{1, 2, 3\}$ into itself? In each case, how many of these mappings are one-to-one? Onto?

Solution:

There are four mappings of $\{1, 2\}$ into itself, namely: α_1 defined by $1\alpha_1 = 1$ and $2\alpha_1 = 2$; α_2 defined by $1\alpha_2 = 1$ and $2\alpha_2 = 1$; α_3 defined by $1\alpha_3 = 2$ and $2\alpha_3 = 1$; α_4 defined by $1\alpha_4 = 2$ and $2\alpha_4 = 2$. Only α_1 and α_3 are one-to-one and onto.

To find the number of mappings of $\{1, 2, 3\}$ into itself, we proceed as follows: 1 may have any of three images under such a mapping, i.e. $1 \to 1$ or $1 \to 2$ or $1 \to 3$. Also, 2 may have any of 3 images, either 1, 2 or 3. So we have in all 3×3 possibilities for the actions of mappings on 1 and 2. Then 3 can be sent into 1, 2 or 3, giving $3 \times 3 \times 3 = 27$ possible mappings of $\{1, 2, 3\}$ into itself. There are $3 \times 2 \times 1 = 6$ possible one-to-one and onto mappings; for when we once choose an image for 1 there are only two possible images for 2, and then the image of 3 is uniquely determined.

1.31. Let $S = \{1, 2, 3\}$, $T = \{3, 4, 5\}$, $U = \{4, 5, 6\}$, and let $\alpha: S \to T$ be defined by $\alpha: 1 \to 3, 2 \to 3, 3 \to 5$. Let β and γ be the mappings from T into U given by $\beta: 3 \to 4, 4 \to 6, 5 \to 4$, $\gamma: 3 \to 4, 4 \to 4, 5 \to 4$. Compute $(1\alpha)\beta$, $(2\alpha)\beta$, $(3\alpha)\beta$, $(1\alpha)\gamma$, $(2\alpha)\gamma$, $(3\alpha)\gamma$.

Solution:

$(1\alpha)\beta = 3\beta = 4$; $(2\alpha)\beta = 3\beta = 4$; $(3\alpha)\beta = 5\beta = 4$; $(1\alpha)\gamma = 3\gamma = 4$; $(2\alpha)\gamma = 3\gamma = 4$; $(3\alpha)\gamma = 5\gamma = 4$.

1.32. Let α, β, γ be the mappings of Q into Q defined by $\alpha : x \to \frac{x}{2}$, $\beta : x \to x+1$, $\gamma : x \to x-1$. Prove that if x is any element of Q, then $((x\gamma)\alpha)\beta = \frac{x+1}{2}$. What is $\alpha_{|P}$? $\alpha_{|Z}$? Is $\alpha_{|P} = \alpha_{|Z}$?

Solution:

$((x\gamma)\alpha)\beta = ((x-1)\alpha)\beta = \frac{x-1}{2}\beta = \frac{x-1}{2}+1 = \frac{x-1+2}{2} = \frac{x+1}{2}$. $\alpha_{|P}$ is a mapping of P into Q. $\alpha_{|Z}$ is a mapping of Z into Q. $\alpha_{|P} \neq \alpha_{|Z}$ since the domain of $\alpha_{|P}$ is not the same as the domain of $\alpha_{|Z}$.

1.33. If S is a non-empty set, prove that S is infinite if and only if there are infinitely many mappings of S into S.

Solution:

First we show that if there are infinitely many mappings of S into S, then S is infinite. Assume, on the contrary, that S is finite and $|S| = n$. Now each of the n elements can be mapped onto at most n images. Hence there are at most n^n different mappings of S into S. This contradicts our assumption that there is an infinite number of such mappings. Hence S is infinite. Conversely, let S be infinite. We define, for each $s \in S$, the mapping $\alpha_s : S \to S$ by $\alpha_s : x \to s$ for all $x \in S$. $\{\alpha_s \mid s \in S\}$ is an infinite set since $\alpha_s = \alpha_{s'}$ if and only if $s = s'$, and we assumed S to be infinite. Therefore we have found an infinite number of mappings of S into S.

1.34. If S is any non-empty set, verify that S matches S.

Solution:

Define the mapping $\alpha : S \to S$ by $\alpha : s \to s$ for each $s \in S$. α is clearly one-to-one and onto. Hence α is a matching.

1.35. If S matches T, prove that T matches S.

Solution:

If S matches T, then there is a mapping $\alpha : S \to T$ which is one-to-one and onto.

Define $\bar{\alpha} : T \to S$ as follows. Let $t \in T$. Then there is an $s \in S$ such that $s\alpha = t$. The image of t under $\bar{\alpha}$ is defined to be s.

We now show $\bar{\alpha}$ is a matching. In the first place, $\bar{\alpha}$ is a mapping. For if $t\bar{\alpha} = s$ and $t\bar{\alpha} = s'$, for some $t \in T$, then by definition of $\bar{\alpha}$, $s\alpha = t$ and $s'\alpha = t$. But α is one-to-one, so that $s\alpha = s'\alpha$ implies $s = s'$. Thus the image of an element under $\bar{\alpha}$ is unique. Secondly, $\bar{\alpha}$ is one-to-one, because $t\bar{\alpha} = t'\bar{\alpha} = s$ implies $s\alpha = t$ and $s\alpha = t'$, which in turn implies, since α is a mapping, $t = t'$. Thirdly, if $s \in S$, then $s\alpha = t$ for some $t \in T$. By definition of $\bar{\alpha}$, $t\bar{\alpha} = s$. Hence every element of S has a preimage under $\bar{\alpha}$ and $\bar{\alpha}$ is onto.

1.36. If S matches T and if T matches U, prove that S matches U. (Hard.)

Solution:

Let $\alpha : S \to T$ and $\beta : T \to U$ be matchings. Then $\bar{\alpha} : S \to U$, defined by $s\bar{\alpha} = (s\alpha)\beta$, is a matching. $\bar{\alpha}$ is a mapping of S into U; for $s\bar{\alpha} \in U$, and if $s\bar{\alpha} = u_1$ and $s\bar{\alpha} = u_2$ then $(s\alpha)\beta = u_1$ and $(s\alpha)\beta = u_2$, which implies $u_1 = u_2$ since $s\alpha$ has a unique image under β. $\bar{\alpha}$ is one-to-one, for $s\bar{\alpha} = s'\bar{\alpha}$ implies $(s\alpha)\beta = (s'\alpha)\beta$. But α and β are one-to-one, so that $s\alpha = s'\alpha$ and $s = s'$. $\bar{\alpha}$ is also onto, for if $u \in U$ then there is a $t \in T$ such that $t\beta = u$. Now t has a preimage $s \in S$ under α. Thus $s\bar{\alpha} = (s\alpha)\beta = t\beta = u$.

1.37. Let α be the mapping of $S = \{1,2,3,4,5,6,7,8,9,10\}$ into itself given by $\alpha : 1 \to 3, 2 \to 4, 3 \to 5, 4 \to 7, 5 \to 9, 6 \to 10, 7 \to 1, 8 \to 3, 9 \to 4, 10 \to 5$. Then there is a useful alternative way of describing α: we list the elements of S on one line (in any order) and on the following line we place under each element of S its image under α, enclosing the entire description in parentheses as follows.

$$\begin{pmatrix} 1 & 2 & 3 & 4 & 5 & 6 & 7 & 8 & 9 & 10 \\ 3 & 4 & 5 & 7 & 9 & 10 & 1 & 3 & 4 & 5 \end{pmatrix}$$

Describe the following mappings of S into S, using this notation.

(i) $\beta: 1 \to 2, 2 \to 2, 3 \to 2, 4 \to 2, 5 \to 2, 6 \to 3, 7 \to 4, 8 \to 4, 9 \to 4, 10 \to 5.$

(ii) $\gamma: 1 \to 6, 2 \to 7, 3 \to 8, 4 \to 9, 5 \to 10, 6 \to 1, 7 \to 1, 8 \to 1, 9 \to 1, 10 \to 1.$

Use these descriptions to decide whether (iii) $\beta = \gamma$, (iv) β is one-to-one, (v) γ is onto.

Solution:

(i) $\begin{pmatrix} 1 & 2 & 3 & 4 & 5 & 6 & 7 & 8 & 9 & 10 \\ 2 & 2 & 2 & 2 & 2 & 3 & 4 & 4 & 4 & 5 \end{pmatrix}$

(ii) $\begin{pmatrix} 1 & 2 & 3 & 4 & 5 & 6 & 7 & 8 & 9 & 10 \\ 6 & 7 & 8 & 9 & 10 & 1 & 1 & 1 & 1 & 1 \end{pmatrix}$

(iii) $\beta \neq \gamma$ since $1\beta = 2$ and $1\gamma = 6$. It is only necessary to compare the bottom rows.

(iv) β is not one-to-one, e.g. $1\beta = 2\beta = 2$. It is only necessary to find a repetition on the bottom row.

(v) γ is not onto, e.g. 2 has no preimage. It is only necessary to check whether all the integers $1, 2, \ldots, 10$ appear in the bottom row.

1.4 COMPOSITION OF MAPPINGS

Definition

Let $\alpha: S \to T$, $\beta: T \to U$. Because $s\alpha \in T$, we can compute $(s\alpha)\beta$. This suggests "composing the mappings α and β", i.e. defining a mapping of S into U by performing α and β in succession on each of the elements in S. More precisely we define $\alpha \circ \beta$, the *composition of α with β* (in this order) as a mapping of S into U defined by

$$s(\alpha \circ \beta) = (s\alpha)\beta, \quad \text{for all } s \text{ in } S$$

(Some authors use exactly the opposite order, so that their $\alpha \circ \beta$ is our $\beta \circ \alpha$.) For example, let

$$S = \{1,2\}, \quad T = \{3,4,5\}, \quad U = \{6,7\}$$

and let $\alpha: S \to T$, $\beta: T \to U$ be defined by

$$\alpha: 1 \to 3, 2 \to 5, \quad \beta: 3 \to 6, 4 \to 7, 5 \to 6$$

Then

$$1(\alpha \circ \beta) = (1\alpha)\beta = 3\beta = 6$$
$$2(\alpha \circ \beta) = (2\alpha)\beta = 5\beta = 6$$

Hence $\alpha \circ \beta: \{1,2\} \to \{6,7\}$ is defined by

$$\alpha \circ \beta: 1 \to 6, 2 \to 6$$

This notion of the composition of two mappings is of tremendous importance; hence we give the following drill problems.

Problems

1.38. Let $\alpha: P \to C$ be defined by $n\alpha = in + 1$ and let $\beta: C \to P$ be defined by $\beta: a + ib \to b^2$, where $i^2 = -1$. What do $\alpha \circ \beta$, $(\alpha \circ \beta) \circ \alpha$, and $\alpha \circ (\beta \circ \alpha)$ map $n \in P$ to? Why is $\alpha \circ \beta \neq \beta \circ \alpha$?

Solution:

Let $n \in P$. Then $n(\alpha \circ \beta) = (n\alpha)\beta = (in + 1)\beta = n^2$. Now $n((\alpha \circ \beta) \circ \alpha) = (n(\alpha \circ \beta))\alpha = n^2\alpha = in^2 + 1$ and $n(\alpha \circ (\beta \circ \alpha)) = (n\alpha)(\beta \circ \alpha) = ((in + 1)\beta)\alpha = n^2\alpha = in^2 + 1$. Hence $(\alpha \circ \beta) \circ \alpha = \alpha \circ (\beta \circ \alpha)$. $\alpha \circ \beta: P \to P$ while $\beta \circ \alpha: C \to C$. Hence $\alpha \circ \beta \neq \beta \circ \alpha$.

1.39. Let $\alpha: Q \to Q$ be defined by $\alpha: a \to a^2 + 2$ and let $\beta: Q \to Q$ be defined by $\beta: a \to \frac{1}{2}a - 2$. Compute $\alpha \circ \beta$, $\beta \circ \alpha$. Are these mappings equal? Compute $(\alpha \circ \beta) \circ \alpha$, $\alpha \circ (\beta \circ \alpha)$.

Solution:

Let $a \in Q$. $a(\alpha \circ \beta) = (a\alpha)\beta = (a^2+2)\beta = \frac{1}{2}(a^2+2) - 2 = \frac{1}{2}a^2 - 1$. $a(\beta \circ \alpha) = (a\beta)\alpha = (\frac{1}{2}a - 2)\alpha = (\frac{1}{2}a - 2)^2 + 2$. Clearly $\alpha \circ \beta \neq \beta \circ \alpha$. Furthermore, $a((\alpha \circ \beta) \circ \alpha) = (a(\alpha \circ \beta))\alpha = (\frac{1}{2}a^2 - 1)\alpha = (\frac{1}{2}a^2 - 1)^2 + 2$ and $a(\alpha \circ (\beta \circ \alpha)) = (a\alpha)(\beta \circ \alpha) = (a^2+2)\beta \circ \alpha = ((a^2+2)\beta)\alpha = ((\frac{1}{2}(a^2+2) - 2)\alpha = (\frac{1}{2}a^2 - 1)^2 + 2$. Note then that $(\alpha \circ \beta) \circ \alpha = \alpha \circ (\beta \circ \alpha)$.

1.40. Employing the notation of Problem 1.37, compute the following:

(i) $\begin{pmatrix} 1 & 2 & 3 & 4 & 5 \\ 2 & 3 & 1 & 5 & 4 \end{pmatrix} \circ \begin{pmatrix} 1 & 2 & 3 & 4 & 5 \\ 1 & 2 & 4 & 5 & 3 \end{pmatrix}$ (iii) $\begin{pmatrix} 1 & 2 & 3 & 4 & 5 \\ 2 & 4 & 5 & 3 & 1 \end{pmatrix} \circ \begin{pmatrix} 1 & 2 & 3 & 4 & 5 \\ 1 & 2 & 4 & 3 & 5 \end{pmatrix}$

(ii) $\begin{pmatrix} 1 & 2 & 3 & 4 & 5 \\ 2 & 1 & 4 & 3 & 5 \end{pmatrix} \circ \begin{pmatrix} 1 & 2 & 3 & 4 & 5 \\ 2 & 3 & 4 & 5 & 1 \end{pmatrix}$ (iv) $\begin{pmatrix} 1 & 2 & 3 & 4 & 5 \\ 1 & 2 & 3 & 4 & 5 \end{pmatrix} \circ \begin{pmatrix} 1 & 2 & 3 & 4 & 5 \\ 2 & 5 & 3 & 4 & 1 \end{pmatrix}$

Solution:

(i) $\begin{pmatrix} 1 & 2 & 3 & 4 & 5 \\ 2 & 4 & 1 & 3 & 5 \end{pmatrix}$ (ii) $\begin{pmatrix} 1 & 2 & 3 & 4 & 5 \\ 3 & 2 & 5 & 4 & 1 \end{pmatrix}$ (iii) $\begin{pmatrix} 1 & 2 & 3 & 4 & 5 \\ 2 & 3 & 5 & 4 & 1 \end{pmatrix}$ (iv) $\begin{pmatrix} 1 & 2 & 3 & 4 & 5 \\ 2 & 5 & 3 & 4 & 1 \end{pmatrix}$

1.41. Let $\alpha : S \to T$, $\beta : T \to U$, $\gamma : U \to V$. Prove that $(\alpha \circ \beta) \circ \gamma = \alpha \circ (\beta \circ \gamma)$. (Hard.)

Solution:

If $s \in S$, then $s((\alpha \circ \beta) \circ \gamma) = (s(\alpha \circ \beta))\gamma = ((s\alpha)\beta)\gamma$ and $s(\alpha \circ (\beta \circ \gamma)) = (s\alpha)(\beta \circ \gamma) = ((s\alpha)\beta)\gamma$. Consequently $(\alpha \circ \beta) \circ \gamma = \alpha \circ (\beta \circ \gamma)$.

1.42. Prove that if $\alpha : S \to T$ and $\beta : T \to U$ and $\alpha \circ \beta$ is onto, then β is onto. Is α onto? (Hard.)

Solution:

Let $u \in U$. As $\alpha \circ \beta$ is onto, we can find a preimage of u under $\alpha \circ \beta$. Let $s \in S$ be a preimage of u under $\alpha \circ \beta$, i.e. $s(\alpha \circ \beta) = u$. Thus $s(\alpha \circ \beta) = (s\alpha)\beta = u$ and $s\alpha$ is a preimage of u under β. Hence β is onto. α need not be onto, e.g. let $S = \{1\}$, $T = \{1, 2\}$, $U = \{1\}$. Define $\alpha : S \to T$ by $1\alpha = 1$, and $\beta : T \to U$ by $1\beta = 2\beta = 1$. $\alpha \circ \beta$ is onto but α is not.

1.43. Prove that if $\alpha : S \to T$, $\beta : T \to U$ and $\alpha \circ \beta$ is one-to-one, then α is one-to-one. Is β one-to-one? (Hard.)

Solution:

Let $s_1, s_2 \in S$ and $s_1\alpha = s_2\alpha$. $s_1\alpha = s_2\alpha$ implies $(s_1\alpha)\beta = (s_2\alpha)\beta$ and, by definition of $\alpha \circ \beta$, $s_1\alpha \circ \beta = s_2\alpha \circ \beta$. But $\alpha \circ \beta$ is one-to-one, so that $s_1 = s_2$. Hence α is one-to-one. β is not necessarily one-to-one, e.g. let $S = \{1\}$, $T = \{1, 2\}$ and $U = \{1\}$. Define $\alpha : S \to T$ by $1\alpha = 1$, and $\beta : T \to U$ by $1\beta = 1$ and $2\beta = 1$. $\alpha \circ \beta$ is one-to-one but β is not one-to-one.

1.44. Prove that $\beta : T \to U$ $(T \neq \emptyset)$ is one-to-one if and only if for every set S and every pair of mappings $\alpha : S \to T$ and $\alpha' : S \to T$, $\alpha \circ \beta = \alpha' \circ \beta$ implies $\alpha = \alpha'$. (Hard.)

Solution:

First assume that for every set S and every pair of mappings $\alpha : S \to T$ and $\alpha' : S \to T$, $\alpha \circ \beta = \alpha' \circ \beta$ implies $\alpha = \alpha'$. Under this hypothesis suppose β is not one-to-one. Then we can find $t, t' \in T$ $(t \neq t')$ such that $t\beta = t'\beta = u \in U$. Let $S = \{1, 2\}$, let $\alpha : S \to T$ be defined by $1\alpha = t$ and $2\alpha = t'$, and let $\alpha' : S \to T$ be defined by $1\alpha' = t'$ and $2\alpha' = t$. Now $\alpha \circ \beta = \alpha' \circ \beta$, since $1\alpha \circ \beta = (1\alpha)\beta = t\beta = u$, $2\alpha \circ \beta = (2\alpha)\beta = t'\beta = u$, $1\alpha' \circ \beta = (1\alpha')\beta = t'\beta = u$, and $2\alpha' \circ \beta = (2\alpha')\beta = t\beta = u$. But $\alpha \neq \alpha'$. Hence the assumption that β is not one-to-one is false and β must be one-to-one.

To prove the converse, let β be one-to-one. Say we can find a set S and a pair of mappings α and α' of S into T such that $\alpha \circ \beta = \alpha' \circ \beta$ and $\alpha \neq \alpha'$. So there exists $s \in S$ such that $s\alpha \neq s\alpha'$. $\alpha \circ \beta = \alpha' \circ \beta$ means that $s(\alpha \circ \beta) = s(\alpha' \circ \beta)$. Hence $(s\alpha)\beta = (s\alpha')\beta$. As β is one-to-one, $s\alpha = s\alpha'$. Therefore we have a contradiction and α must be equal to α'.

1.45. Prove that $\alpha : S \to T$ $(T \neq \emptyset)$ is onto iff for every set U and every pair of mappings $\beta : T \to U$ and $\beta' : T \to U$ such that $\alpha \circ \beta = \alpha \circ \beta'$, it follows that $\beta = \beta'$. (Hard.)

Solution:

Let us assume that for every set U and every pair of mappings β and β' of T into U such that $\alpha \circ \beta = \alpha \circ \beta'$, it follows that $\beta = \beta'$. Say α is not onto, and t_1 is an element of T which has no preimage under α. Let $U = \{1,2\}$ and define the mappings β and β' of T into U as follows: $t\beta = 1$ for all $t \in T$, $t\beta' = 1$ for all $t \neq t_1$ in T and $t_1\beta = 2$. Now if $s \in S$, $s\alpha \circ \beta = (s\alpha)\beta = 1$ and $s\alpha \circ \beta' = (s\alpha)\beta' = 1$, since $s\alpha \neq t_1$. Hence $\alpha \circ \beta = \alpha \circ \beta'$ and $\beta \neq \beta'$. This contradicts the assumption that $\alpha \circ \beta = \alpha \circ \beta'$ implies $\beta = \beta'$. Thus α must be onto.

Conversely, assume α is onto and we can find a set U and two mappings, β and β', of T into U such that $\alpha \circ \beta = \alpha \circ \beta'$ and $\beta \neq \beta'$. $\beta \neq \beta'$ means there is a $t_1 \in T$ such that $t_1\beta \neq t_1\beta'$. Furthermore, since α is onto, we can find $s \in S$ such that $s\alpha = t_1$. But $\alpha \circ \beta = \alpha \circ \beta'$ implies $(s\alpha)\beta = (s\alpha)\beta'$ or $t_1\beta = t_1\beta'$. Here we have a contradiction because we chose $t_1\beta \neq t_1\beta'$. We therefore conclude $\beta = \beta'$.

1.5 BINARY OPERATIONS

a. Definition

The idea of a binary operation is illustrated by the usual operation of addition in Z, which may be analyzed in the following way. For every pair of integers (m,n) there is associated a unique integer $m+n$. We may therefore think of addition as being a brief description of a mapping of $Z \times Z$ into Z where the image of $(m,n) \in Z \times Z$ is denoted by $m+n$. Any mapping β of $S \times S$ into S, where S is any non-empty set, is called a *binary operation* in S. We shall sometimes write instead of $(s,t)\beta$ (the image of (s,t) under β) one of $s \circ t$, $s \cdot t$, st, $s+t$ or $s \times t$. We stress that in all these cases the *meaning* of the various expressions $s \circ t$, $s \cdot t$, st, $s+t$ and $s \times t$ is simply the image of (s,t) under the given mapping β of $S \times S$ into S. These notations suppress the binary operation β, so there is danger of confusion. However, we will work with binary operations and the various notations so frequently that the reader will become familiar with the pitfalls. Incidentally, we read

$s \circ t$ as "ess circle tee"

$s \cdot t$ as "ess dot tee" or "ess times tee"

st as "ess tee" or "ess times tee"

$s+t$ as "ess plus tee"

$s \times t$ as "ess times tee".

The notation $s \circ t$ is called the *circle* notation, the notations $s \cdot t$ and st are termed *multiplicative*, and the notation $s+t$ is termed *additive*. We sometimes refer to $s \cdot t$ or st as the *product* of s and t, and $s+t$ as the *sum* of s and t. The following problems will help to make the various notations clear.

Problems

1.46. Convince yourself that the following mappings are binary operations in P.

(i) $\alpha: P \times P \to P$ defined by $\alpha: (i,j) \to i^2$, where $(i,j) \in P$.

(ii) $\alpha: P \times P \to P$ defined by $\alpha: (i,j) \to i+j$, where $(i,j) \in P$.

(iii) $\alpha: P \times P \to P$ defined by $\alpha: (i,j) \to i \times j$ (regular multiplication of integers), where $(i,j) \in P$.

(iv) $\alpha: P \times P \to P$ defined by $\alpha: (i,j) \to 2i+3j$, $(i,j) \in P \times P$.

(v) $\alpha: P \times P \to P$ defined by $\alpha: (i,j) \to i+j+1$, $(i,j) \in P \times P$.

1.47. Which of the following are binary operations in P (throughout $(i,j) \in P^2$)?

(i) $\alpha: (i,j) \to i+j$ (iii) $\alpha: (i,j) \to i \div j$ (v) $\alpha: (i,j) \to j$

(ii) $\alpha: (i,j) \to i-j$ (iv) $\alpha: (i,j) \to i+j+i^2$

Solution:

In (i), α is clearly a mapping from $P \times P$ into P. So α is a binary operation in P. (ii) and (iii) do not define binary operations in P because not every element in $P \times P$ has an image in P: e.g. in (ii), $\alpha: (1,2) \to 1-2 = -1 \notin P$; and in (iii), $\alpha: (1,2) \to 1 \div 2 = \frac{1}{2} \notin P$. (iv) and (v) define mappings from $P \times P$ into P. Hence they define binary operations in P.

1.48. Interpret the following (abbreviated) definitions of binary operations in Z, where i and j denote arbitrary elements of Z.

(i) $i \circ j = (i+j)^2$ (iii) $i \circ j = i + j$ (v) $i \times j = i^j$

(ii) $i + j = i - j - (i \times j)$ (iv) $ij = i - i \times j$ (vi) $i \cdot j = i + 27j$

Solution:

Throughout, $(i,j) \in Z^2$.

(i) $\alpha:\ (i,j) \to (i+j)^2$ (iii) $\alpha:\ (i,j) \to i+j$ (v) $\alpha:\ (i,j) \to i^j$

(ii) $\alpha:\ (i,j) \to i - j - (i \times j)$ (iv) $\alpha:\ (i,j) \to i - i \times j$ (vi) $\alpha:\ (i,j) \to i + 27j$

1.49. Check that the following are binary operations in the plane R^2.

(i) $(x,y) \circ (x',y') =$ midpoint of the line joining the point $(x,y) \in R^2$ to the point $(x',y') \in R^2$ if $(x,y) \neq (x',y')$. If $(x,y) = (x',y')$, define $(x,y) \circ (x',y') = (x,y)$.

(ii) $(x,y) + (x',y') = (x+x', y+y')$, where $(x,y), (x',y') \in R^2$.

(iii) $(x,y) \cdot (x',y') = (xx', yy')$, where $(x,y), (x',y') \in R^2$.

(iv) $(x,y) - (x',y') = (x-x', y-y')$, where $(x,y), (x',y') \in R^2$.

(v) $(x,y) \circ (x',y') = (x+x', x'y+y')$, where $(x,y), (x',y') \in R^2$.

(Notice here how we have abbreviated the definitions of the binary operations considered.)

Solution:

(i) $\circ$ is a binary operation since two points determine a unique line and each line has a unique midpoint.

(ii)-(v) are binary operations because each has a unique image by virtue of the fact that addition, multiplication and subtraction are binary operations in R.

1.50. Let $S = \{1,2,3\}$ and let α and β be the following binary operations in S:

$$\alpha:\ (1,1) \to 1,\ (1,2) \to 1,\ (1,3) \to 2,\ (2,1) \to 2,\ (2,2) \to 3,\ (2,3) \to 3,\ (3,1) \to 3,\ (3,2) \to 2,\ (3,3) \to 1;$$

$$\beta:\ (1,1) \to 1,\ (1,2) \to 1,\ (1,3) \to 2,\ (2,1) \to 3,\ (2,2) \to 3,\ (2,3) \to 3,\ (3,1) \to 3,\ (3,2) \to 1,\ (3,3) \to 2.$$

(i) Is $\alpha = \beta$?

(ii) Compute $((1,1)\alpha, 1)\beta$, $((1,1)\beta, 1)\alpha$.

(iii) Compute $((1,2)\alpha, 3)\alpha$, $(1, (2,3)\alpha)\alpha$, $((1,2)\beta, 3)\beta$, $(1, (2,3)\beta)\beta$.

Solution:

(i) $\alpha \neq \beta$ for $(2,1)\alpha = 2$ and $(2,1)\beta = 3$.

(ii) $((1,1)\alpha, 1)\beta = (1,1)\beta = 1$; $((1,1)\beta, 1)\alpha = (1,1)\alpha = 1$.

(iii) $((1,2)\alpha, 3)\alpha = (1,3)\alpha = 2$; $((1,2)\beta, 3)\beta = (1,3)\beta = 2$;

$(1,(2,3)\alpha)\alpha = (1,3)\alpha = 2$; $(1,(2,3)\beta)\beta = (1,3)\beta = 2$.

1.51. Let $S = \{1,2,3\}$ and let X be the set of all mappings of S into S.

(i) Compute $|X|$.

(ii) Verify that the composition $\circ$ of mappings is a binary operation in X.

(iii) Is $\alpha \circ \beta = \beta \circ \alpha$ for all elements $\alpha, \beta \in X$?

Solution:

(i) $|X| = 27$ (see Problem 1.30, page 15).

(ii) The composition of mappings $\alpha: S \to S$ and $\beta: S \to S$ is again a mapping of S into S. Therefore $\pi: X^2 \to X$, defined by $(\alpha,\beta)\pi = \alpha \circ \beta$, $(\alpha,\beta) \in X^2$, is a binary operation.

(iii) No. For example, if $\alpha: S \to S$ defined by $s\alpha = 1$ for all $s \in S$, and $\beta: S \to S$ defined by $s\beta = 2$ for all $s \in S$, are two mappings of S into S, then $s(\alpha \circ \beta) = (s\alpha)\beta = 1\beta = 2$ and $s\beta \circ \alpha = (s\beta)\alpha = 2\alpha = 1$. Hence $\alpha \circ \beta \neq \beta \circ \alpha$.

1.52. Let Q^* be the set of nonzero rational numbers. Make sense of the remark that *division* (denoted as usual by $\div$) is a binary operation in Q^*. Check whether the following statements hold for all $a, b, c \in Q^*$.

(i) $a \div b = b \div a$.

(ii) $(a \div b) \div c = a \div (b \div c)$.

(iii) $((a \div b) \div c) \div d = a \div (b \div (c \div d))$.

(iv) If $a \div b = a \div c$, then $b = c$.

(v) If $b \div a = c \div a$, then $b = c$.

Solution:

Division is a binary operation in Q^*; for if $w/x \in Q^*$ and $y/z \in Q^*$ where w, x, y, z are integers, then $w/x \div y/z = wz/xy \in Q^*$. Hence $\div$ is a mapping of $Q^* \times Q^* \to Q^*$.

(i) False; e.g. $2 \div 3 \neq 3 \div 2$.

(ii) False; e.g. $(2 \div 1) \div 2 = 1$ and $2 \div (1 \div 2) = 2 \div \frac{1}{2} = 4$.

(iii) False; e.g. $((1 \div 2) \div 2) \div 2 = \frac{1}{8}$ and $1 \div (2 \div (2 \div 2)) = \frac{1}{2}$.

(iv) True. $a \div b = a \div c$ implies $ac = ab$ and, as $a \neq 0$, $c = b$.

(v) True. $b \div a = c \div a$ implies $ab = ca$. Hence $b = c$, since $a \neq 0$.

1.53. Let $\circ$ be the binary operation in R defined by $a \circ b = a + b + ab$. Verify that:

(i) For all $a, b, c \in R$, $(a \circ b) \circ c = a \circ (b \circ c)$.

(ii) For all $a, b \in R$, $a \circ b = b \circ a$.

(iii) Prove that if $a \neq -1$, then $a \circ b = a \circ c$ iff $b = c$.

Solution:

(i)
$$\begin{aligned}(a \circ b) \circ c &= (a + b + ab) \circ c = a + b + ab + c + (a + b + ab)c \\ &= a + b + c + bc + ab + ac + abc\end{aligned}$$
$$\begin{aligned}a \circ (b \circ c) &= a \circ (b + c + bc) = a + b + c + bc + a(b + c + bc) \\ &= a + b + c + bc + ab + ac + abc\end{aligned}$$

(ii) $a \circ b = a + b + ab = b + a + ba = b \circ a$

(iii) If $b = c$, then $a \circ b = a \circ c$ for any a. If $a \circ b = a \circ c$ and $a \neq -1$, then $a + b + ab = a + c + ac$. Therefore $b + ab = c + ac$, $b(1 + a) = c(1 + a)$ and, since $a \neq -1$, $b = c$.

1.54. Let $\circ$ be the binary operation in R^2 defined by $(x, y) \circ (x', y') = (xx' - yy', yx' + xy')$. Verify that for all $(x, y), (x', y'), (x'', y'') \in R^2$:

(i) $(x, y) \circ (x', y') = (x', y') \circ (x, y)$

(ii) $((x, y) \circ (x', y')) \circ (x'', y'') = (x, y) \circ ((x', y') \circ (x'', y''))$

Solution:

(i) $(x, y) \circ (x', y') = (xx' - yy', yx' + xy') = (x'x - y'y, y'x + x'y) = (x', y') \circ (x, y)$

(ii)
$$\begin{aligned}((x, y) \circ (x', y')) \circ (x'', y'') &= (xx' - yy', yx' + xy') \circ (x'', y'') \\ &= ((xx' - yy')x'' - (yx' + xy')y'', (yx' + xy')x'' + (xx' - yy')y'') \\ &= (xx'x'' - yy'x'' - yx'y'' - xy'y'', yx'x'' + xy'x'' + xx'y'' - yy'y'')\end{aligned}$$
$$\begin{aligned}(x, y) \circ ((x', y') \circ (x'', y'')) &= (x, y) \circ (x'x'' - y'y'', y'x'' + y''x') \\ &= (x(x'x'' - y'y'') - y(y'x'' + y''x'), y(x'x'' - y'y'') + x(y'x'' + y''x')) \\ &= (xx'x'' - xy'y'' - yy'x'' - yy''x', yx'x'' - yy'y'' + xy'x'' + xy''x') \\ &= ((x, y) \circ (x', y')) \circ (x'', y'')\end{aligned}$$

1.55. Let $\circ$ be the binary operation in Q defined by
$$(a)\ a \circ b = a - b + ab, \qquad (b)\ a \circ b = \frac{a + b + ab}{2}, \qquad (c)\ a \circ b = \frac{a + b}{3}$$

Determine which of the above binary operations satisfy

(i) $(a \circ b) \circ c = a \circ (b \circ c)$ for all $a, b, c \in Q$

(ii) $a \circ b = b \circ a$ for all $a, b \in Q$

Solution:

(i) (*a*) $(a \circ b) \circ c \neq a \circ (b \circ c)$ for some $a, b, c \in Q$; e.g. $(2 \circ 0) \circ 2 = (2 - 0 + 0) \circ 2 = 2 \circ 2 = 2 - 2 + 4 = 4$ and $2 \circ (0 \circ 2) = 2 \circ (0 - 2 + 0) = 2 \circ -2 = 2 - (-2) - 4 = 0$.

(*b*) $(a \circ b) \circ c \neq a \circ (b \circ c)$ for some $a, b, c \in Q$; e.g. $(1 \circ 0) \circ 0 = \frac{1}{2} \circ 0 = \frac{1}{4}$ and $1 \circ (0 \circ 0) = 1 \circ 0 = \frac{1}{2}$.

(*c*) $(a \circ b) \circ c \neq a \circ (b \circ c)$ for some $a, b, c \in Q$; e.g. $(1 \circ 0) \circ 0 = \frac{1}{3} \circ 0 = \frac{1}{9}$ and $1 \circ (0 \circ 0) = 1 \circ 0 = \frac{1}{3}$.

(ii) (*a*) $a \circ b \neq b \circ a$ since $1 \circ 0 = 1$ and $0 \circ 1 = -1$.

(*b*) $a \circ b = \dfrac{a + b + ab}{2} = \dfrac{b + a + ba}{2} = b \circ a$

(*c*) $a \circ b = \dfrac{a + b}{3} = \dfrac{b + a}{3} = b \circ a$

b. The multiplication table

So far we have introduced a number of definitions and notations and familiarized ourselves with them. The object of this section is to introduce a "table" as a convenient way of either defining a binary operation in a finite set S or tabulating the effect of a binary operation in a set S. To explain this procedure, suppose $S = \{1, 2, 3\}$ and let μ be the binary operation in S defined by

$$\mu: (1,1) \to 1,\ (1,2) \to 1,\ (1,3) \to 2,\ (2,1) \to 2,\ (2,2) \to 3,$$
$$(2,3) \to 3,\ (3,1) \to 1,\ (3,2) \to 3,\ (3,3) \to 2$$

Then a table which sums up this description of μ is

	1	2	3
1	1	1	2
2	2	3	3
3	1	3	2

We put the number $(2,3)\mu = 3$ in the square that is the intersection of the row facing 2 (on the left) and the column below 3 (on the top). More generally, in the (i,j)th square, i.e. the intersection of the ith row (the row labelled or faced by i) and the jth column (the column labelled by j), we put $(i,j)\mu$.

A table of this kind is termed a *multiplication* table because it looks like the usual multiplication tables. One often calls μ a *multiplication in S.* Thus when we talk about a multiplication μ in a set S, we mean that μ is a binary operation in S.

There is a reverse procedure to the one described above. For example, suppose we start out with a table

	1	2	3
1	1	1	2
2	2	3	3
3	1	3	2

Then there is a natural way of associating with this table a binary operation μ in $\{1, 2, 3\}$. We simply define $(i,j)\mu$ to be the entry in the (i,j)th place in the table. For example,

$$(1,1)\mu = 1, \quad (2,3)\mu = 3, \quad (3,2)\mu = 3$$

We shall usually define multiplications in a finite set by means of such tables.

Problems

1.56. Write down the multiplication tables for the following binary operations in $S = \{1, 2, 3\}$.

(i) $\alpha: (1,1) \to 2,\ (1,2) \to 3,\ (1,3) \to 1,\ (2,1) \to 3,\ (2,2) \to 1,\ (2,3) \to 2,\ (3,1) \to 1,\ (3,2) \to 2,\ (3,3) \to 3.$

(ii) $\beta: S^2 \to S$ defined by $(i,j)\beta = 1$ for all $(i,j) \in S^2$.

(iii) $\gamma: (1,1) \to 1,\ (2,2) \to 1,\ (3,3) \to 1,\ (1,3) \to 2,\ (3,1) \to 2,\ (2,3) \to 1,\ (3,2) \to 1,\ (2,1) \to 3,\ (1,2) \to 3.$

Solution:

(i)

	1	2	3
1	2	3	1
2	3	1	2
3	1	2	3

(ii)

	1	2	3
1	1	1	1
2	1	1	1
3	1	1	1

(iii)

	1	2	3
1	1	3	2
2	3	1	1
3	2	1	1

1.57. Does the following table define a binary operation in $\{1,2,3\}$? In $\{1,2,3,4\}$?

	1	2	3
1	1	3	4
2	2	1	3
3	1	3	2

Solution:

The table does not give a binary operation in $\{1,2,3\}$ since $(1,3) \to 4 \notin \{1,2,3\}$. The table also does not define a binary operation in $\{1,2,3,4\}$, because $(1,4)$, $(2,4)$, $(3,4)$, etc., have no images.

1.58. Write down explicitly the binary operations in $\{1,2,3,4\}$ defined by the following tables.

	1	2	3	4
1	1	2	3	4
2	2	3	4	1
3	3	4	1	2
4	4	1	2	3

(i)

	1	2	3	4
1	1	2	3	4
2	2	4	1	3
3	3	1	4	2
4	4	3	2	1

(ii)

	1	2	3	4
1	1	1	1	1
2	2	2	2	2
3	3	3	3	3
4	4	4	4	4

(iii)

Solution:

(i) $(1,j) \to j$ $(j = 1,2,3,4)$; $(j,1) \to j$ $(j = 2,3,4)$; $(2,2) \to 3$; $(2,3) \to 4$; $(3,2) \to 4$; $(3,3) \to 1$; $(3,4) \to 2$; $(4,3) \to 2$; $(4,4) \to 3$; $(4,2) \to 1$; $(2,4) \to 1$.

(ii) $(1,j) \to j$ $(j = 1,2,3,4)$; $(j,1) \to j$ $(j = 2,3,4)$; $(2,2) \to 4$; $(2,3) \to 1$; $(3,2) \to 1$; $(3,3) \to 4$; $(3,4) \to 2$; $(4,3) \to 2$; $(2,4) \to 3$; $(4,2) \to 3$; $(4,4) \to 1$.

(iii) $(i,j) \to i$ $(i = 1,2,3,4$ and $j = 1,2,3,4)$.

1.59. Rewrite the three binary operations in Problem 1.58, using

(*a*) circle notation, i.e. write $(i,j)\beta$ as $i \circ j$,

(*b*) additive notation, i.e. write $(i,j)\beta$ as $i + j$,

(*c*) multiplicative notation, i.e. write $(i,j)\beta$ as $i \cdot j$.

Solution:

(*a*) Problem 1.58(i): $1 \circ j = j$ $(j = 1,2,3,4)$; $j \circ 1 = j$ $(j = 1,2,3,4)$; $2 \circ 2 = 3$; $2 \circ 3 = 3 \circ 2 = 4$; $3 \circ 3 = 1$; $3 \circ 4 = 4 \circ 3 = 2$; $4 \circ 4 = 3$; $4 \circ 2 = 2 \circ 4 = 1$.

Problem 1.58(ii): $1 \circ j = j$ $(j = 1,2,3,4)$; $j \circ 1 = j$ $(j = 2,3,4)$; $2 \circ 2 = 4$; $2 \circ 3 = 3 \circ 2 = 1$; $3 \circ 3 = 4$; $3 \circ 4 = 4 \circ 3 = 2$; $2 \circ 4 = 4 \circ 2 = 3$; $4 \circ 4 = 1$.

Problem 1.58(iii): $i \circ j = i$ $(i = 1,2,3,4$ and $j = 1,2,3,4)$.

(*b*) Problem 1.58(i): $1 + j = j$ $(j = 1,2,3,4)$; $j + 1 = j$ $(j = 1,2,3,4)$; $2 + 2 = 3$; $2 + 3 = 3 + 2 = 4$; $3 + 3 = 1$; $3 + 4 = 4 + 3 = 2$; $4 + 4 = 3$; $4 + 2 = 2 + 4 = 1$.

Problem 1.58(ii): $1 + j = j$ and $j + 1 = j$ $(j = 1,2,3,4)$; $2 + 2 = 4$; $2 + 3 = 3 + 2 = 1$; $3 + 3 = 4$; $3 + 4 = 4 + 3 = 2$; $2 + 4 = 4 + 2 = 3$; $4 + 4 = 1$.

Problem 1.58(iii): $i + j = i$ $(i = 1,2,3,4$ and $j = 1,2,3,4)$.

(*c*) Problem 1.58(i): $1 \cdot j = j$ and $j \cdot 1 = j$ $(j = 1,2,3,4)$; $2 \cdot 2 = 3$; $2 \cdot 3 = 3 \cdot 2 = 4$; $3 \cdot 3 = 1$; $3 \cdot 4 = 4 \cdot 3 = 2$; $4 \cdot 4 = 3$; $4 \cdot 2 = 2 \cdot 4 = 1$.

Problem 1.58(ii): $1 \cdot j = j$ and $j \cdot 1 = j$ $(j = 1,2,3,4)$; $2 \cdot 2 = 4$; $2 \cdot 3 = 3 \cdot 2 = 1$; $3 \cdot 3 = 4$; $3 \cdot 4 = 4 \cdot 3 = 2$; $2 \cdot 4 = 4 \cdot 2 = 3$; $4 \cdot 4 = 1$.

Problem 1.58(iii): $i \cdot j = i$ $(i = 1,2,3,4$ and $j = 1,2,3,4)$.

A look back at Chapter 1

We began with a few remarks about sets. We then introduced the idea of cartesian products. This led to the idea of an equivalence relation on a set. Then the notion of a mapping was defined, followed by the definition of a binary operation.

In this book we are mainly concerned with binary operations in sets. At this stage the reader may wonder what one could possibly say about binary operations in a set. Without some specialization we can say very little. In Chapter 2 we begin to place restrictions on binary operations.

Supplementary Problems

SETS

1.60. The sets S_i $(i = 1, 2, \ldots, n)$ are such that $S_i \subseteq S_{i+1}$ $(i = 1, 2, \ldots, n-1)$. Find $S_1 \cap S_2 \cap \cdots \cap S_n$ and $S_1 \cup S_2 \cup \cdots \cup S_n$.

1.61. Let $E = \{n \mid n \in Z \text{ and } n \text{ even}\}$, $O = \{n \mid n \in Z \text{ and } n \text{ odd}\}$, $T = \{n \mid n \in Z \text{ and } n \text{ divisible by } 3\}$, and $F = \{n \mid n \in Z \text{ and } n \text{ divisible by } 4\}$. Find (i) $E \cap T$, (ii) $E \cup T$, (iii) $T \cap F \cap E \cap O$, (iv) $T \cup F$, (v) $O \cap F$, (vi) $O \cap T$.

1.62. Given $A = \{-3, -2, -1, 0, \{1, 2, 3,\}\}$ and $B = \{-3, -1, \{1, 3\}\}$. Find $A \cup B$ and $A \cap B$.

1.63. Prove $S \cap (T \cup U) = (S \cap T) \cup (S \cap U)$.

1.64. Let $A = \{-5, -4, -3, \ldots, 3, 4, 5\}$, $B = \{-4, -2, 0, 2, 4\}$, $C = \{-5, -3, -1, 1, 3, 5\}$, $D = \{-4, 4\}$, $E = \{-3, -2, -1, 0\}$, $F = \emptyset$. Which, if any, of these sets take the place of X if (i) $X - C = \emptyset$, (ii) $X \cap B = C$, (iii) $X \subseteq C$ but X is not a subset of A, (iv) $X \subseteq B$ and X is not a subset of E, (v) $X \cap C \subseteq A$, (vi) $X \cup (B \cap D) = A$?

1.65. Prove $S \subseteq T$ if and only if $(T \cap C) \cup S = T \cap (C \cup S)$ for every set C.

CARTESIAN PRODUCTS

1.66. Prove $S \times (T \cup W) = (S \times T) \cup (S \times W)$ for any sets S, T and W.

1.67. Let P be the set of positive integers and $S = P^2$. Show that $E = \{p \mid p = ((r, s), (t, w)) \in S^2 \text{ and } r + w = s + t\}$ is an equivalence relation on S. Find the E-class determined by $(4, 7)$.

1.68. Find the equivalence relation, E, on Z: (i) if the equivalence classes of E are $\{n \mid n = 4q \text{ for } q \in Z\}$, $\{n \mid n = 1 + 4q \text{ for } q \in Z\}$, $\{n \mid n = 2 + 4q \text{ for } q \in Z\}$, and $\{n \mid n = 3 + 4q \text{ for } q \in Z\}$; (ii) if every equivalence class consists of a single integer; (iii) if the equivalence classes are $\{q, -q\}$ for each $q \in Z$.

1.69. If E and F are equivalence relations on S, is (i) $E \cap F$, (ii) $E \cup F$ an equivalence relation on S?

1.70. What is wrong with the following argument: E is a non-empty subset of S^2 which has the symmetric and transitive properties. If $(a, b) \in E$, then by the symmetric property, $(b, a) \in E$. But by the transitive property, $(a, b) \in E$ and $(b, a) \in E$ implies $(a, a) \in E$. Therefore E is also reflexive.

1.71. (*a*) If P is the set of positive integers, show that $E = \{(a, b) \mid (a, b) \in P^2 \text{ and } a \text{ divides } b\}$ is reflexive and transitive but not symmetric.

(*b*) Find an example of a subset E of P^2 which is both symmetric and transitive but not reflexive.

(*c*) Find an example of a subset E of P^2 which is reflexive and symmetric but not transitive.

MAPPINGS

1.72. Show that the following define mappings of Z into Z.

(i) $\alpha: x \to \begin{cases} 0 & \text{if } x = 0 \\ x/|x| & \text{if } x \neq 0 \end{cases}$ $\quad |x|$ is the absolute value of x, i.e. $|x| = \begin{cases} x & \text{if } x \geqq 0 \\ -x & \text{if } x < 0 \end{cases}$

(ii) $\beta: x \to \begin{cases} 0 & \text{if } |x| = 0 \text{ or } 1 \\ (-1)^r & \text{if } |x| \neq 0 \text{ or } 1, \end{cases}$ and r is the number of distinct primes dividing x

(iii) $\gamma: x \to \begin{cases} -1 & \text{if } x < 0 \\ 0 & \text{if } x = 0 \\ 1 & \text{if } x > 0 \end{cases}$

(iv) $\delta: x \to \sin^2 x + \cos^2 x$

1.73. Which of the mappings $\alpha, \beta, \gamma, \delta$ of the preceding problem are equal?

1.74. Find subsets $S(\neq \emptyset)$ and $T(\neq \emptyset)$ of the real numbers R such that the mappings (i) $\alpha: x \to \cos x$, (ii) $\beta: x \to \sin x$, and (iii) $\gamma: x \to \tan x$ are one-to-one mappings of S onto T. Do α, β, γ define mappings of R into R?

1.75. Let $E = \{n \mid n \in P$ and n even$\}$. Define $\alpha: E \to E$ by $n\alpha = n$ for all $n \in E$, and $\beta: E \to E$ by $n\beta = 2n$ for all $n \in E$. Find an infinite number of mappings α', β' of P into P such that $\alpha'_{|E} = \alpha$ and $\beta'_{|E} = \beta$.

1.76. Suppose α is a mapping of a set S into a set T and, for any subset W of S, $W\alpha = \{t \mid t \in T$ and $t = s\alpha$ for some $s \in W\}$. If A and B are any subsets of S, show: (i) $(A \cup B)\alpha = A\alpha \cup B\alpha$; (ii) $(A \cap B)\alpha \subseteq A\alpha \cap B\alpha$; and (iii) $A \subseteq B$ implies $A\alpha \subseteq B\alpha$.

1.77. Let S be a subset of a set T, and $\alpha: T \to W$. Prove: (i) α one-to-one implies $\alpha_{|S}$ is one-to-one; (ii) $\alpha_{|S}$ onto implies α is onto.

COMPOSITION OF MAPPINGS

1.78. Suppose $\alpha: S \to T$ is onto and $\beta: T \to U$ is onto. Show $\alpha \circ \beta$ is an onto mapping.

1.79. Given $\alpha: S \to T$ is one-to-one and $\beta: T \to U$ is one-to-one. Prove $\alpha \circ \beta$ is one-to-one.

1.80. (i) $\alpha: x \to \sin(x^2)$, (ii) $\beta: x \to \sin(\sin(x))$, and (iii) $\gamma: x \to \sqrt{1-x^2}$ define mappings of non-empty subsets of the real numbers R into R. First, find an appropriate subset in each case. Secondly, write α, β and γ as the composition of two mappings, giving in each case the domain and codomain of each mapping defined.

BINARY OPERATIONS

1.81. Let S be a set and $\mathcal{S}$ the set of all subsets of S, i.e. $\mathcal{S} = \{A \mid A \subseteq S\}$. Show that intersection and union define binary operations on $\mathcal{S}$.

1.82. How many different binary operations can be defined on a set of 3 elements?

1.83. Consider the set, F, of mappings $f_i\ (i = 1, 2, \ldots, 6)$ of $R - \{0, 1\}$ into R defined for each $x \in R - \{0, 1\}$ by: $f_1: x \to x$; $f_2: x \to \dfrac{1}{1-x}$; $f_3: x \to \dfrac{x-1}{x}$; $f_4: x \to \dfrac{1}{x}$; $f_5: x \to \dfrac{x}{x-1}$; $f_6: x \to 1-x$. Show that composition of mappings is a binary operation on F and write a multiplication table for the operation.

Chapter 2

Groupoids

Preview of Chapter 2

In this chapter we define a set G together with a fixed binary operation $\circ$ to be a groupoid. As we remarked at the end of Chapter 1, there is little one can say about binary operations without making restrictions.

The first restriction is that of associativity. A groupoid with set G and operation $\circ$ is said to be associative if $g_1 \circ (g_2 \circ g_3) = (g_1 \circ g_2) \circ g_3$ for all g_1, g_2, g_3 in G. Such a groupoid is called a semigroup. In order of increasing specialization we have the concepts of groupoid, semigroup, and group.

To define a group we need the concepts of identity and inverse. Hence we discuss these ideas.

We introduce the semigroup M_X of mappings of X into X. The importance of M_X is that, but for the names of the elements, each semigroup is contained in some M_X.

Two other important concepts we deal with are homomorphism and isomorphism. Homomorphism is a more general concept than isomorphism. There is an isomorphism between two groupoids if they are essentially the same but for the names of their elements.

2.1 GROUPOIDS

a. Definition of a groupoid

Consider the set Z of integers. Z has two binary operations, addition $(+)$ and multiplication $(\times)$. The set Z is one thing, a binary operation in Z is another; the two together constitute a *groupoid*. Repeating this definition in general terms.

Definition: A *groupoid* is a pair (G, μ) consisting of a non-empty set G, called the *carrier*, and a binary operation μ in G.

We shall mostly use a multiplicative notation when dealing with groupoids. Thus we write $g \cdot h$ or simply gh for $(g, h)\mu$, $g, h \in G$. This notation has been used in Chapter 1 in our consideration of binary operation. As an example, let $G = \{1, 2, 3\}$ and let μ be the binary operation in G defined by the following table.

	1	2	3
1	1	3	2
2	2	1	1
3	3	2	3

Then the pair (G, μ) is a groupoid. Suppose we use multiplicative notation $\cdot$; then

$$1 \cdot 1 = 1, \quad 1 \cdot 2 = 3, \quad 2 \cdot 2 = 1, \quad 3 \cdot 2 = 2, \quad \text{etc.}$$

These products look bizarre unless we recall that the notation employed is a shorthand version of

$$(1,1)\mu = 1, \quad (1,2)\mu = 3, \quad (2,2)\mu = 1, \quad (3,2)\mu = 2, \quad \text{etc.}$$

If we use the expression "the groupoid G," where G is a set, it is understood that we have already been given a binary operation μ in G, and that we have been talking about the groupoid (G, μ).

Suppose now that (G, μ) is a groupoid. If we use the circle notation for μ, i.e. we write $g \circ h$ instead of $(g, h)\mu$, we shall sometimes write $(G, \circ)$ to refer to the groupoid (G, μ). Similarly we write $(G, \cdot)$, $(G, +)$, $(G, \times)$ if we employ $g \cdot h$, $g + h$, $g \times h$ respectively for $(g, h)\mu$.

Example 1: Let $G = \{1, 2\}$ and let μ be the binary operation in G defined as follows:

$$(1,1)\mu = 1, \quad (1,2)\mu = 2, \quad (2,1)\mu = 1, \quad (2,2)\mu = 2$$

If we use the circle notation, we have

$$1 \circ 1 = 1, \quad 1 \circ 2 = 2, \quad 2 \circ 1 = 1, \quad 2 \circ 2 = 2$$

The pair (G, μ) or $(G, \circ)$ is then a groupoid.

Example 2: Let S be the set of all mappings of $\{1, 2, 3\}$ into $\{1, 2, 3\}$. Then $(S, \circ)$ is a groupoid, where $\circ$ is interpreted as the usual composition of mappings. The composition makes sense, for if $\alpha, \beta \in S$, then

$$\alpha: \{1,2,3\} \to \{1,2,3\} \quad \text{and} \quad \beta: \{1,2,3\} \to \{1,2,3\}$$

Therefore $\alpha \circ \beta$, the composition of α and β, is defined by $a\alpha \circ \beta = (a\alpha)\beta$, $a \in \{1, 2, 3\}$, and is once again a mapping of $\{1, 2, 3\}$ into itself. $(S, \circ)$ is indeed a groupoid.

Example 3: Let $\circ$ be the binary operation in Q, the rational numbers, defined by $a \circ b = a + b + ab$. Then $(Q, \circ)$ is a groupoid, because for every pair of rational numbers a and b, $a \circ b$ defines a unique rational number $a + b + ab$.

Example 4: Let R^2 be the plane. Further, let there be a cartesian coordinate system in R^2 and let C be the disc of radius 1 with center at the point $(2, 0)$ of the coordinate system. Consider the region $R^2 - C$, the unshaded area in the diagram. We term any path beginning and ending at O, which does not meet any point of C (i.e. it is entirely in $R^2 - C$), a loop in $R^2 - C$. By a path we mean any line which can be traced out by a pencil without raising the point from the paper. For example, l and m are loops in $R^2 - C$. Let L be the set of all such loops in $R^2 - C$. Then there is a natural binary operation in L which we denote by $\cdot$; thus if $l_1, l_2 \in L$, then $l_1 \cdot l_2$ is the loop obtained by first tracing out l_1, followed by tracing l_2. This type of groupoid $(L, \cdot)$ is of considerable importance in modern topology.

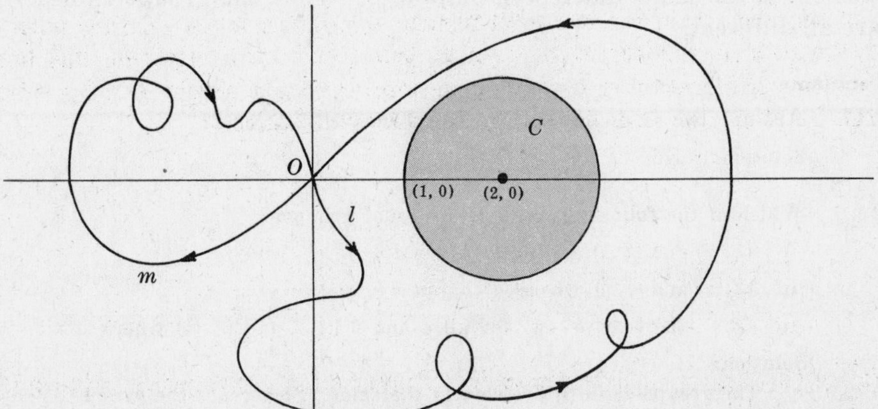

Example 5: Let F be the set of all mappings of R, the real numbers, into R. Consider two elements $\alpha, \beta \in F$. We define the mapping $\alpha + \beta: R \to R$ by $a(\alpha + \beta) = a\alpha + a\beta$, $a \in R$. $\alpha + \beta$ is clearly a unique mapping of R into R and hence is an element in F. Therefore $+$ is a binary operation in F and $(F, +)$ is a groupoid. Notice that $(F, +)$ is *not* the groupoid with F as the carrier and the composition of mappings as the binary operation.

Problems

2.1. Are the following groupoids?

(i) $(S, \circ)$ where $S = \{1, 2, 3, 4\}$ and $i \circ j = 1$ for i and j elements of S.

(ii) $(Z, -)$, the set of integers with the usual subtraction of integers as binary operation.

(iii) $(P, -)$, the set of positive integers with the usual subtraction as binary operation.

(iv) $(Q, \div)$, the set of rational numbers with the usual binary operation of division.

(v) $(Z, \div)$, the set of integers with the usual binary operation of division.

Solution:

(i) $(S, \circ)$ is a groupoid since $\circ$ is clearly a binary operation in S.

(ii) $(Z, -)$ is a groupoid; for if $a, b \in Z$, then $a - b$ is a unique element of Z.

(iii) $(P, -)$ is not a groupoid since $a - b \notin P$ for all $a, b \in P$. Therefore $-$ is not a binary operation in P.

(iv) $(Q, \div)$ is not a groupoid because $a \div 0$ is not defined for any $a \in Q$ and hence $\div$ is not a binary operation in Q.

(v) $(Z, \div)$ is not a groupoid since $a \div b \notin Z$ for all $a, b \in Z$, e.g. $2 \div 3 \notin Z$. Therefore division is not a binary operation in Z.

2.2. Is $(Z, \circ)$ a groupoid if $\circ$ is defined as (i) $a \circ b = \sqrt{a + b}$, (ii) $a \circ b = (a + b)^2$, (iii) $a \circ b = a - b - ab$, (iv) $a \circ b = 0$, (v) $a \circ b = a$?

Solution:

All but (i) define a binary operation in Z. Therefore $(Z, \circ)$ is a groupoid in (ii) through (v). The multiplication $\circ$ in (i) does not define a binary operation in Z since $\sqrt{a + b}$ is not always an integer.

2.3. Let S be any non-empty set and T the set of all subsets of S. Are $(T, \cap)$ and $(T, \cup)$ groupoids?

Solution:

Both intersection $\cap$ and union $\cup$ are binary operations on T, for the intersection or union of two subsets of S is again a unique subset of S. Thus $(T, \cap)$ and $(T, \cup)$ are groupoids.

b. Equality of groupoids

Two groupoids are *equal* if and only if they have the same carriers and the same binary operation. Remember, a binary operation was defined as a mapping and two mappings are equal if and only if they have the same domain and codomain, and the image of each element is the same under both mappings. Thus the groupoids described in Examples 1-5 are all different.

Problems

2.4. Are any two of the groupoids in Problems 2.1-2.3 equal?

Solution: No.

2.5. Which of the following pairs define equal groupoids?

(i) $(Z, +)$ and (Z, μ), where $(a, b)\mu = a + b$.

(ii) $(Z, -)$ and $(Z, \circ)$, where $a \circ b = a - b$.

(iii) $(Z, \circ)$ where $a \circ b = a$ for all a and b in Z, and $(Z, \times)$ where $a \times b = b$ for all a and b in Z.

Solution:

The groupoids in (i) are clearly the same. So too are the groupoids in (ii). In (iii) $(Z, \circ)$ is not the same as $(Z, \times)$; for if $a \neq b$, $a \circ b \neq a \times b$.

2.2 COMMUTATIVE AND ASSOCIATIVE GROUPOIDS

Definition of commutative and associative groupoids

Let $(Z, +)$ be the groupoid of integers under the usual operation of addition. Then

$$a + b = b + a \tag{2.1}$$

and

$$(a+b)+c = a+(b+c) \tag{2.2}$$

for all $a, b, c \in Z$.

Similarly if $(Z, \cdot)$ is the groupoid of integers under multiplication,

$$a \cdot b = b \cdot a \tag{2.3}$$

and

$$(a \cdot b) \cdot c = a \cdot (b \cdot c) \tag{2.4}$$

for all $a, b, c \in Z$.

The analog of (*2.1*) and (*2.3*) in an arbitrary groupoid $(G, \circ)$ is

$$a \circ b = b \circ a \tag{2.5}$$

for all $a, b \in G$. Similarly the analog of (*2.2*) and (*2.4*) in $(G, \circ)$ is

$$(a \circ b) \circ c = a \circ (b \circ c) \tag{2.6}$$

for all $a, b, c \in G$. We term a groupoid satisfying (*2.5*) *commutative* or *abelian,* and a groupoid satisfying (*2.6*) *associative* or a *semigroup.* Thus a semigroup is an associative groupoid. Of course it is not clear that there are non-commutative groupoids, i.e. groupoids which are definitely not commutative, and similarly it is not clear that there are non-associative groupoids. We settle the issue now. Let $G = \{1, 2\}$ and let $\circ$ be the following binary operation in G:

	1	2
1	1	1
2	2	1

Then $(G, \circ)$ is a groupoid. Observe that $1 \circ 2 = 1$ but $2 \circ 1 = 2$, so G is not commutative. Furthermore, $(2 \circ 1) \circ 2 = 2 \circ 2 = 1$ but $2 \circ (1 \circ 2) = (2 \circ 1) = 2$, so G is also non-associative, i.e. G is not a semigroup.

For the most part we shall use the multiplicative notation for a groupoid (G, μ) and simply talk about the groupoid G. If the groupoid is commutative we will use the additive notation instead of the multiplicative notation, since we are accustomed to addition as a commutative binary operation, e.g. in the integers.

The *order* of a groupoid (G, μ) is the number of elements in G and is denoted by $|G|$; (G, μ) is *infinite* if $|G|$ is infinite, and *finite* if $|G|$ is finite.

Problems

2.6. Which of the groupoids in Examples 1, 2, 3 and 5 are commutative and which are associative?

Solution:

The groupoid of Example 1 is not commutative, since $1 \circ 2 = 2$ and $2 \circ 1 = 1$, but is associative. To show $(G, \circ)$ is associative we must examine the following 8 cases:

(*a*) $1 \circ (1 \circ 1) = 1 \circ 1 = 1,\ (1 \circ 1) \circ 1 = 1 \circ 1 = 1$

(*b*) $2 \circ (1 \circ 1) = 2 \circ 1 = 1,\ (2 \circ 1) \circ 1 = 1 \circ 1 = 1$

(*c*) $1 \circ (2 \circ 1) = 1 \circ 1 = 1,\ (1 \circ 2) \circ 1 = 2 \circ 1 = 1$

(*d*) $1 \circ (1 \circ 2) = 1 \circ 2 = 2,\ (1 \circ 1) \circ 2 = 1 \circ 2 = 2$

(*e*) $2 \circ (2 \circ 1) = 2 \circ 1 = 1,\ (2 \circ 2) \circ 1 = 2 \circ 1 = 1$

(*f*) $2 \circ (1 \circ 2) = 2 \circ 2 = 2,\ (2 \circ 1) \circ 2 = 1 \circ 2 = 2$

(*g*) $1 \circ (2 \circ 2) = 1 \circ 2 = 2,\ (1 \circ 2) \circ 2 = 2 \circ 2 = 2$

(*h*) $2 \circ (2 \circ 2) = 2 \circ 2 = 2,\ (2 \circ 2) \circ 2 = 2 \circ 2 = 2$

The groupoid of Example 2 is not commutative; for if α is defined by $1\alpha = 1$, $2\alpha = 3$ and $3\alpha = 2$, and β is defined by $1\beta = 2$, $2\beta = 1$ and $3\beta = 1$, then $1\alpha \circ \beta = (1\alpha)\beta = 1\beta = 2$, $1\beta \circ \alpha = (1\beta)\alpha = 2\alpha = 3$. Hence $\alpha \circ \beta \neq \beta \circ \alpha$. Since the binary composition of mappings is an associative binary operation (Problem 1.41, page 18), $(S, \circ)$ is an associative groupoid.

The groupoid in Example 3 is both commutative and associative. $a \circ b = a + b + ab = b + a + ba = b \circ a$, since addition and multiplication are commutative binary operations in Q. Also, $a \circ (b \circ c) = a \circ (b + c + bc) = (a + (b + c + bc)) + a(b + c + bc) = a + b + c + bc + ab + ac + abc$ and $(a \circ b) \circ c = (a + b + ab) \circ c = a + b + ab + c + (a + b + ab)c = a + b + ab + c + ac + bc + abc$. Using the associative and commutative properties of addition and multiplication in the rationals, we see $a \circ (b \circ c) = (a \circ b) \circ c$.

In Example 5 the groupoid $(F, +)$ is commutative because $a(\alpha + \beta) = a\alpha + a\beta = a\beta + a\alpha = a(\beta + \alpha)$, $\alpha, \beta \in F$ and $a \in R$ (here we use the fact that $a\alpha, a\beta \in R$ and addition is a commutative binary operation in R). $(F, +)$ is also a semigroup, for $a((\alpha + \beta) + \gamma) = a(\alpha + \beta) + a\gamma = a\alpha + (a\beta + a\gamma) = a\alpha + a(\beta + \gamma) = a(\alpha + (\beta + \gamma))$ (here we use the associativity of addition in R).

2.7. Construct an example of a commutative groupoid of order 3.

Solution:

Let $S = \{a, b, c\}$ and the binary operation $\circ$ be defined by the multiplication table

	a	b	c
a	a	b	c
b	b	a	c
c	c	c	a

$(S, \circ)$ is clearly a commutative groupoid.

2.8. Show that the set Q^* of nonzero rational numbers with binary operation the usual division of rational numbers, is a groupoid. Is it commutative? Is it associative?

Solution:

If $\frac{a}{b}$ and $\frac{c}{d}$ are any two elements of Q^*, then $\frac{a}{b} \div \frac{c}{d} = \frac{ad}{bc}$ is a unique element in Q^* $\left(\frac{ad}{bc} \neq 0 \text{ since } a, b, c, d \neq 0\right)$. Therefore division is a binary operation in Q^*. However, division is neither commutative (e.g. $\frac{1}{2} \div \frac{1}{4} = 2 \neq \frac{1}{2} = \frac{1}{4} \div \frac{1}{2}$) nor associative (e.g. $\frac{1}{2} \div (\frac{1}{4} \div \frac{1}{3}) = \frac{1}{2} \div \frac{3}{4} = \frac{2}{3} \neq 6 = (\frac{1}{2} \div \frac{1}{4}) \div \frac{1}{3}$).

2.9. Which of the groupoids in Examples 1-3, 5 and in Problems 2.7 and 2.8 are finite?

Solution:

The groupoid of Example 1 is clearly finite of order 3.

In Example 2 the set S of all mappings of $\{1, 2, 3\}$ into itself contains 27 elements, and so is finite.

In Example 3, $(Q, \circ)$ is not finite as there are an infinite number of rational numbers.

In Example 5 the set F is infinite. To show that F is not finite we construct an infinite number of mappings of R into R as follows. Let $\rho_i : R \to R$ $(i = 1, 2, 3, \ldots)$ be defined by $r\rho_i = r$ for all $r \neq i \in R$ and $i\rho_i = 0$. Clearly $\rho_i = \rho_j$ iff $i = j$. Therefore we have found an infinite number of different elements in F. Notice the ρ_i are not all the elements of F.

In Problem 2.7 the groupoid has only three elements and is therefore finite.

In Problem 2.8, since there is an infinite number of nonzero rational numbers, Q^* is not finite.

2.3 IDENTITIES AND INVERSES IN GROUPOIDS

a. The identity of a groupoid

Let G be a groupoid written multiplicatively. An element e in G is called an *identity element* of G if

$$eg = ge = g$$

for every $g \in G$. For example in the multiplicative groupoid Z of integers, 1 is an identity element. A less natural example is the groupoid $\{1,2,3\}$ with multiplication table given by

	1	2	3
1	2	3	1
2	3	1	2
3	1	2	3

The element 3 is an identity element of G since

$$3 \cdot 1 = 1 = 1 \cdot 3, \quad 3 \cdot 2 = 2 = 2 \cdot 3, \quad 3 \cdot 3 = 3 = 3 \cdot 3$$

One might ask whether or not a groupoid can have more than one identity element. The following theorem settles this question.

Theorem 2.1: If a groupoid G has an identity element, it has precisely one identity element. In other words if e and e' are identity elements of G, then $e = e'$.

Proof: Since e is an identity element of G, $ee' = e'$. But e' is also an identity element of G. Hence $ee' = e$ and so $e = e'$.

It is instructive to reformulate the proof of Theorem 2.1 in a different notation. Thus we revert to the notation (G, μ) and instead of the multiplicative notation gh we write $(g, h)\mu$. Suppose e and e' are identity elements. Then as e is an identity element, $(e, e')\mu = e'$. But e' is also an identity element. Hence $(e, e')\mu = e$; and because the image of any element under the mapping is unique, $e = e'$.

Problem

2.10. Which of the following groupoids have identity elements?

(i) The groupoid $(Z, +)$ under the usual operation of addition.

(ii) The groupoid of nonzero rational numbers under division.

(iii) The groupoid of complex numbers under multiplication.

(iv) The groupoid of all mappings of $\{1,2,3,4\}$ into itself under the composition of mappings.

(v) The groupoid with carrier $\{1,2\}$ and multiplication table

(*a*)

	1	2
1	1	1
2	1	1

(*b*)

	1	2
1	1	2
2	1	2

(*c*)

	1	2
1	2	2
2	1	1

(*d*)

	1	2
1	1	2
2	2	1

Solution:

(i) $(Z, +)$ has the identity element 0, since $0 + z = z + 0 = z$ for all $z \in Z$.

(ii) If this groupoid had an identity element e, then $e \div q = q$ for all q in the groupoid. In particular, $e \div e = e$ and so $e = 1$. But $1 \div 2 = 1/2 \neq 2$. Hence there is no identity element.

(iii) Recall that a complex number is any number of the form $a + bi$ where $a, b \in R$ and $i = \sqrt{-1}$. $1 + 0i$ is the identity element of this groupoid, since $(a+bi)(1+0i) = a + 0ai + 0bi^2 + bi = a + bi = (1+0i)(a+bi)$.

(iv) The identity mapping ι, defined by $j\iota = j$, $j \in \{1,2,3,4\}$, is the identity element of the groupoid since $j(\sigma \circ \iota) = (j\sigma)\iota = j\sigma = (j\iota)\sigma = j(\iota \circ \sigma)$.

(v) Only (*d*) has an identity element, namely 1.

b. Inverses in a groupoid

If we use multiplicative notation for groupoids, we shall *for the most part* reserve the symbol 1 for the identity element.

In the multiplicative groupoid of nonzero rational numbers, one talks about inverses; for example, $\frac{1}{2}$ is the inverse of 2, the determining factor being $2 \times \frac{1}{2} = 1 = \frac{1}{2} \times 2$. In general if G is any groupoid with an identity 1, we term h an *inverse* of g $(h, g \in G)$ if

$$gh = 1 = hg$$

Clearly if h is an inverse of g, then g is an inverse of h. Examples follow.

Example 6: Let $G = \{a, b, c\}$ be the groupoid with multiplication table

	a	b	c
a	a	b	c
b	b	c	a
c	c	a	b

Then a is the identity element of G, as inspection of the table shows. Furthermore, $bc = a = cb$ and $aa = a = aa$, so that c is an inverse of b, b is an inverse of c, and a is its own inverse. a, b and c have no other inverses.

Example 7: Let G be the groupoid of mappings of $\{1, 2, 3\}$ into itself. Then the identity mapping ι defined by $j\iota = j$ $(j = 1, 2, 3)$ is the identity element of the groupoid (see Problem 2.10(iv)). Now let $\sigma \in G$ be defined by $1\sigma = 2$, $2\sigma = 3$, $3\sigma = 1$. Then $\tau \in G$, defined by $1\tau = 3$, $2\tau = 1$, $3\tau = 2$, is an inverse of σ because

$$1\sigma \circ \tau = (1\sigma)\tau = 2\tau = 1, \quad 2\sigma \circ \tau = (2\sigma)\tau = 3\tau = 2, \quad 3\sigma \circ \tau = (3\sigma)\tau = 1\tau = 3$$

which implies $\sigma\tau = \iota$. Similarly, $\tau\sigma = \iota$.

Unlike identities, inverses in groupoids are *not* always unique. For example, let $G = \{1, 2, 3, 4\}$ be the groupoid with multiplication table

	1	2	3	4
1	1	2	3	4
2	2	1	1	1
3	3	1	1	4
4	4	2	3	4

Here 1 is the identity element in G. Moreover, $2 \cdot 2 = 1 = 2 \cdot 2$ and $2 \cdot 3 = 1 = 3 \cdot 2$; thus 2 has two inverses, 2 and 3. Notice that in this groupoid, 4 is *not* an inverse of 2 even though $2 \cdot 4 = 1$. The definition of an inverse requires both $2 \cdot 4$ *and* $4 \cdot 2$ to equal 1.

Problems

2.11. Consider the groupoids in Problem 2.10 which have identity elements. What elements in each of the groupoids have inverses?

Solution:

(i) Any integer z has an additive inverse, namely $-z$; for $z + (-z) = 0 = (-z) + z$.

(iii) All nonzero elements in the groupoid of complex numbers under multiplication have inverses; i.e. if $a + bi$ (a and b not both zero) is any element in the groupoid, then

$$\frac{1}{a+bi} = \left(\frac{1}{a+bi}\right)\left(\frac{a-bi}{a-bi}\right) = \frac{a-bi}{a^2+b^2} = \frac{a}{a^2+b^2} - \frac{b}{a^2+b^2}i$$

is an element in the groupoid and $\frac{1}{a+bi}(a+bi) = 1 = (a+bi)\frac{1}{a+bi}$. $(0+0i)$ has no inverse, since $(0+0i)(a+bi) = (0+0i)$.

(iv) Only mappings τ which are one-to-one and onto have inverses. Say τ is a one-to-one mapping of $\{1,2,3,4\}$ onto itself, defined by $1\tau = a$, $2\tau = b$, $3\tau = c$ and $4\tau = d$, where a, b, c, d are the elements of $\{1,2,3,4\}$. We define the mapping σ which will be an inverse of τ, by $a\sigma = 1$, $b\sigma = 2$, $c\sigma = 3$ and $d\sigma = 4$. σ is a mapping of $\{1,2,3,4\}$ onto $\{1,2,3,4\}$, since $\{a,b,c,d\} = \{1,2,3,4\}$. $1\tau \circ \sigma = (1\tau)\sigma = a\sigma = 1$, $2\tau \circ \sigma = b\sigma = 2$, $3\tau \circ \sigma = c\sigma = 3$, and $4\tau \circ \sigma = d\sigma = 4$. Hence $\tau \circ \sigma = \iota$. We must also show $\sigma \circ \tau = \iota$. Now $a\sigma \circ \tau = 1\tau = a$, $b\sigma \circ \tau = 2\tau = b$, $c\sigma \circ \tau = 3\tau = c$, $d\sigma \circ \tau = 4\tau = d$. As $\{a,b,c,d\}$ is $\{1,2,3,4\}$, $\sigma \circ \tau = \iota$. If τ is not one-to-one then there are at least two elements, a and b $(a \neq b)$ in $\{1,2,3,4\}$ which are mapped onto the same element, c, by τ, i.e. $a\tau = c$ and $b\tau = c$. Now if σ is an inverse of τ, then $a(\tau \circ \sigma) = a$ and $b(\tau \circ \sigma) = b$ or $(a\tau)\sigma = c\sigma = a$ and $(b\tau)\sigma = c\sigma = b$; thus $c\sigma = a$ and $c\sigma = b$. But under a mapping each element has a unique image. Hence we have a contradiction. So τ has no inverse.

(v) Both 1 and 2 have inverses since $1 \cdot 1 = 1 = 1 \cdot 1$ and $2 \cdot 2 = 1 = 2 \cdot 2$.

2.12. Find the identity of the groupoid $(Q, \circ)$ where $a \circ b = a + b + ab$ (see Example 3, page 27). What elements have inverses?

Solution:

0 is the identity of $(Q, \circ)$, since $0 \circ a = 0 + a + 0a = a$ and $a \circ 0 = a + 0 + 0a = a$. To find an inverse for an element $a \in Q$, we must find an $x \in Q$ such that $a \circ x = 0 = x \circ a$. Now $a \circ x = a + x + ax$, so that x must satisfy the equation $a + x + ax = 0$. If $a = -1$, we obtain $-1 + x - x = -1 = 0$; thus if $a = -1$, there is no x such that $a \circ x = 0$. If $a \neq -1$, the equation $a + x + ax = 0$ can be solved, giving $x = -\dfrac{a}{1+a}$; thus $a \circ \dfrac{-a}{1+a} = 0$. Since $(Q, \circ)$ is commutative, $\dfrac{-a}{1+a} \circ a = 0$ so that $\dfrac{-a}{1+a}$ is an inverse of a. Hence all elements of Q, except -1, have inverses.

2.13. Find the identity of the groupoid $(F, +)$ in Example 5 and show that every element has an inverse.

Solution:

The mapping $\omega: R \to R$ defined by $r\omega = 0$ for all $r \in R$, is the identity of $(F, +)$. For $a(\alpha + \omega) = a\alpha + a\omega = a\alpha + 0 = a\alpha = 0 + a\alpha = a\omega + a\alpha = a(\omega + \alpha)$ for all $a \in R$ implies $\alpha + \omega = \alpha = \omega + \alpha$. If β is an inverse of α, then $\alpha + \beta = \omega$ and $a(\alpha + \beta) = a\omega$ for all $a \in R$. But $a(\alpha + \beta) = a\alpha + a\beta$ and $a\omega = 0$. Thus $a\alpha + a\beta = 0$ or $-(a\alpha) = a\beta$. Consequently if β is an inverse of α, the image of $a \in R$ under β must be the negative of the image of a under α. We therefore define an inverse β for the mapping α by $a\beta = -(a\alpha)$, $a \in R$. β is a mapping of R into R, since $-(a\alpha)$ is a unique element of R. Furthermore, $a(\alpha + \beta) = a\alpha + a\beta = a\alpha + (-(a\alpha)) = 0 = a\omega$ implies $\alpha + \beta = \omega$. $(F, +)$ is a commutative groupoid. Hence $\alpha + \beta = \omega = \beta + \alpha$ and β is an inverse of α.

2.4 SEMIGROUPS WITH AN IDENTITY ELEMENT

a. Uniqueness of inverses

Suppose G is a groupoid with an identity 1 and suppose h is the inverse of g: $gh = 1 = hg$. It is tempting to employ the notation used when dealing with real numbers and write g^{-1} for an inverse of g. The trouble with this notation is that in a groupoid an element may have more than one inverse, as we have already seen in Section 2.3b. However, the associative law on a groupoid prohibits this as we see from

Theorem 2.2: Let G be a semigroup with an identity element 1. If $g \in G$ has an inverse, it has precisely one; i.e. if h and h' are inverses of g, then $h = h'$.

Proof:

$$\begin{aligned} h &= h1 && \text{(since } h1 = h \text{ for all } h \in G) \\ &= h(gh') && (gh' = 1, \text{ since } h' \text{ is an inverse of } g) \\ &= (hg)h' && \text{(by associativity)} \\ &= 1h' && (hg = 1, \text{ since } h \text{ is an inverse of } g) \\ &= h' \end{aligned}$$

Theorem 2.2 entitles us to denote *the* inverse of an element g in a semigroup, written multiplicatively, by g^{-1}. Note that if g and h have inverses, then gh has an inverse, namely $h^{-1}g^{-1}$. (Notice the reversed order.) For,

$$(gh)(h^{-1}g^{-1}) = ((gh)h^{-1})g^{-1} = (g(hh^{-1}))g^{-1} = (g1)g^{-1} = gg^{-1} = 1$$

Similarly, $(h^{-1}g^{-1})(gh) = 1$.

Problems

2.14. Let $G = \{1, 2, 3, 4\}$. The binary operations on G given by the following tables make G into a groupoid.

(*a*)

	1	2	3	4
1	1	2	3	4
2	2	4	2	4
3	3	2	1	4
4	4	4	4	4

(*b*)

	1	2	3	4
1	1	2	3	4
2	2	1	4	3
3	3	4	1	2
4	4	2	3	1

(*c*)

	1	2	3	4
1	1	1	1	1
2	2	2	2	2
3	3	3	3	3
4	4	4	4	4

Which of the groupoids are semigroups? Which have an identity? Which elements have inverses?

Solution:

(*a*) In order to see that in this case G is a semigroup, we must check associativity, i.e. we must show $a(bc) = (ab)c$ for every $a, b, c \in G$. Notice that when either a, b or c is 1, $a(bc)$ is clearly equal to $(ab)c$; e.g. if $c = 1$, $(ab)1 = ab = a(b1)$. Since $4g = 4 = g4$ for any $g \in G$, then $(ab)c = 4 = a(bc)$ if either a, b or c is 4. Therefore we need only check the products when a, b and c have values 2 or 3. If two of the three elements a, b, c are equal to 2 and the other is equal to 2 or 3, then $(ab)c = 4 = a(bc)$ because $2 \cdot 2 = 4$ and $2 \cdot 3 = 2 = 3 \cdot 2$. The following calculations take care of the remaining cases:

$$3(3 \cdot 3) = 3 \cdot 1 = 3 = 1 \cdot 3 = (3 \cdot 3)3 \qquad 3(2 \cdot 3) = 3 \cdot 2 = 2 = 2 \cdot 3 = (3 \cdot 2)3$$

$$2(3 \cdot 3) = 2 \cdot 1 = 2 = 2 \cdot 3 = (2 \cdot 3)3 \qquad 3(3 \cdot 2) = 3 \cdot 2 = 2 = (1 \cdot 2) = (3 \cdot 3)2$$

1 is the identity element of G. The only elements which have inverses are 1 and 3; these inverses are unique.

(*b*) The associative law does not hold, since $4(2 \cdot 3) = 4 \cdot 4 = 1$ and $(4 \cdot 2)3 = 2 \cdot 3 = 4$. 1 is the identity, and 1, 2, 3 and 4 have inverses. Notice that the inverses are unique even though G is not a semigroup.

(*c*) G is a semigroup. Associativity follows from the fact that $ab = a$ for all $a, b \in G$; hence $(ab)c = a = a(bc)$ for any a, b, c in G. G has no identity element; therefore no element has an inverse.

2.15. Let G be the groupoid with carrier Q, the set of rational numbers, and binary operation $\circ$ defined by $a \circ b = a + b - ab$. Is the groupoid $(Q, \circ)$ a semigroup? Is there an identity element in $(Q, \circ)$? Which elements of the groupoid have inverses?

Solution:

Notice $a + b - ab$ is a unique rational number, so that $(Q, \circ)$ is a groupoid. Now if a, b, c are any elements in $(Q, \circ)$,

$$(a \circ b) \circ c = (a + b - ab) \circ c = a + b - ab + c - (a + b - ab)c$$
$$= a + b + c - ab - ac - bc + abc$$

and

$$a \circ (b \circ c) = a \circ (b + c - bc) = a + b + c - bc - a(b + c - bc)$$
$$= a + b + c - bc - ab - ac + abc$$

Hence $(Q, \circ)$ is a semigroup. The identity of $(Q, \circ)$ is 0 since $a \circ 0 = a + 0 - a0 = a$ and $0 \circ a = 0 + a - 0a = a$. Using an analysis similar to that in Problem 2.12, we find that for $a \neq 1$ the inverse of a is $\dfrac{a}{a-1}$ and that $a = 1$ has no inverse. To check that $\dfrac{a}{a-1}$ is the inverse of a,

$$a \circ \frac{a}{a-1} \;=\; a + \frac{a}{a-1} - a\left(\frac{a}{a-1}\right) \;=\; a + \frac{-a(a-1)}{a-1} \;=\; 0$$

Similarly, $\frac{a}{a-1} \circ a = 0$.

2.16. Let G be the groupoid with carrier Q and binary operation $\circ$ defined by $a \circ b = a - b + ab$. Is the groupoid $(Q, \circ)$ a semigroup? Is there an identity element in $(Q, \circ)$? Which elements of the groupoid have inverses?

Solution:

$(Q, \circ)$ is clearly a groupoid.

$$a \circ (b \circ c) \;=\; a \circ (b - c + bc) \;=\; a - (b - c + bc) + a(b - c + bc) \;=\; a - b + c - bc + ab - ac + abc$$

and

$$(a \circ b) \circ c \;=\; (a - b + ab) \circ c \;=\; a - b + ab - c + (a - b + ab)c \;=\; a - b + ab - c + ac - bc + abc$$

These two expressions are not the same for all values of a, b and c. For example, $(0 \circ 0) \circ 1 = 0 \circ 1 = -1$ while $0 \circ (0 \circ 1) = 0 \circ -1 = 1$. Hence $(Q, \circ)$ is not a semigroup. Furthermore, $(Q, \circ)$ has no identity element; for if e were an identity, then $e \circ 1 = 1$ and $e \circ 0 = 0$, since $1, 0 \in Q$. But $e \circ 1 = e - 1 + e = 1$ implies $2e = 2$ or $e = 1$, and $0 = e \circ 0 = e - 0 + 0$ implies $e = 0$. This is clearly impossible. Therefore $(Q, \circ)$ has no identity element.

2.17. If Q is replaced by Z, the integers, as the carrier for the groupoids in Problem 2.15, are the solutions the same? (Hard.)

Solution:

$(Z, \circ)$ with $\circ$ defined as in Problem 2.15 is a semigroup with an identity, since the argument for the associativity depended only upon the associativity and commutativity of addition and multiplication in Q. These laws also hold in Z. The same is true for the proof that 0 is the identity of $(Z, \circ)$. However, the inverse of an element $a(\neq 1) \in Z$ would be $\frac{a}{a-1} \in Z$. Our problem then is: for which integers $a \neq 1$ is $\frac{a}{a-1}$ an integer?

Let $\frac{a}{a-1} = r$, an integer. Then $a = r(a-1)$. Clearly $r = 1$ is impossible. (*a*) Assume first that $a > 1$. Then r must be positive, and so $r \geqq 2$. Hence $a \geqq 2(a-1)$ and thus $0 \geqq a - 2$. Therefore $a = 2$. If $a = 2$, then $\frac{a}{a-1} = 2$ is an integer. (*b*) Now assume $a \leqq 0$. If $a = 0$, then $\frac{a}{a-1} = 0$ is an integer. If $a < 0$, r must be positive and $\geqq 2$. Then $a = r(a-1) \leqq 2(a-1)$ and $0 \leqq a - 2$, which is impossible. Thus 0 and 2 are the only elements with inverses.

2.18. Let G be the mappings of P, the positive integers, into P. Determine whether G is a semigroup with an identity element if the binary operation in P is (i) the composition of mappings, (ii) the addition of mappings, where $\alpha + \beta$ is defined by $a(\alpha + \beta) = a\alpha + a\beta$, $\alpha, \beta \in G$ and $a \in P$.

Solution:

(i) The composition of mappings is an associative binary operation (see Problem 1.41, Page 18). Then G with the binary operation of composition of mappings is a semigroup. The mapping ι defined by $j \circ \iota = j$ for all $j \in P$ is the identity of G; for if $\tau \in G$, then $j(\iota \circ \tau) = (j\iota)\tau = j\tau = (j\tau)\iota = j(\tau \circ \iota)$.

(ii) In Problem 2.6 we showed that the addition of mappings in the set F of all mappings of R into R is an associative binary operation. The argument here is similar. Thus G is a semigroup.

If α were an identity element in G and if $\beta \in G$, then $\alpha + \beta = \beta$. Thus if $j \in P$, then $j(\alpha + \beta) = j\alpha + j\beta = j\beta$; hence $j\alpha = 0$. But $\alpha \in G$, $\alpha : P \to P$ and, since $0 \notin P$, this is a contradiction. Thus $(G, +)$ has no identity. (Compare with Problem 2.13.)

2.19. Let a, b, c be three elements in a semigroup which have inverses. Prove that $a(bc)$ has an inverse and that this inverse is $(c^{-1}b^{-1})a^{-1}$.

Solution:

We need only verify that

$$\{a(bc)\}\{(c^{-1}b^{-1})a^{-1}\} \;=\; a\{(bc)[(c^{-1}b^{-1})a^{-1}]\} \;=\; a\{[(bc)(c^{-1}b^{-1})]a^{-1}\} \;=\; a\{1a^{-1}\} \;=\; aa^{-1} \;=\; 1$$

and similarly that $\{(c^{-1}b^{-1})a^{-1}\}\{a(bc)\} = 1$ to prove the result. Note our use of the associative law.

b. The semigroup of mappings of a set into itself

A particularly important semigroup is the set M_X of all mappings of a given non-empty set X into itself, where the binary operation is composition of mappings. We repeat the definition of the composition of mappings in this special case. Suppose $\mu: X \to X$ and $\gamma: X \to X$, i.e. $\mu, \gamma \in M_X$. We define $\mu \circ \gamma$ to be the mapping of X into X given by

$$x(\mu \circ \gamma) = (x\mu)\gamma \quad \text{for all } x \in X$$

It is clear that $\circ$ is a binary operation in M_X. We shall use the multiplicative notation $\mu \cdot \gamma$, or simply $\mu\gamma$, instead of $\mu \circ \gamma$. We now show M_X is a semigroup with an identity.

Theorem 2.3: If X is any non-empty set, M_X is a semigroup with an identity element.

Proof: We begin by proving M_X is a semigroup. Let $\mu, \gamma, \rho \in M_X$ and let $x \in X$. Then using the definition of composition of mappings,

$$x(\mu(\gamma\rho)) = (x\mu)(\gamma\rho) = ((x\mu)\gamma)\rho = (x(\mu\gamma))\rho = x((\mu\gamma)\rho)$$

Since x is any element of X, $(\mu\gamma)\rho$ and $\mu(\gamma\rho)$ have the same effect on every element of X. Thus $(\mu\gamma)\rho = \mu(\gamma\rho)$, and so M_X is a semigroup.

To show that M_X has an identity element, let $\iota: X \to X$ be defined by $x\iota = x$ for all $x \in X$. Then if $\mu \in M_X$,

$$x(\iota\mu) = (x\iota)\mu = x\mu = (x\mu)\iota = x(\mu\iota)$$

Thus $\iota\mu = \mu = \mu\iota$ for all $\mu \in M_X$; hence ι is an identity element of M_X. The proof of the theorem is complete.

Not every element in M_X necessarily has an inverse. For example, if $X = \{1,2,3\}$ and $\sigma \in M_X$ is defined by $1\sigma = 1$, $2\sigma = 1$, $3\sigma = 1$, then σ has no inverse; for if $\sigma\gamma = \iota$, we have $1 = 1\iota = 1(\sigma\gamma) = (1\sigma)\gamma = 1\gamma$ and $2 = 2\iota = 2(\sigma\gamma) = (2\sigma)\gamma = 1\gamma$, so that 1 would have two distinct images under γ, which contradicts the assumption that γ is a mapping. The subset of M_X consisting of all those elements which have inverses is very important. We characterize these elements in the following theorem.

Theorem 2.4: An element in M_X has an inverse if and only if it is one-to-one and onto.

Proof: Suppose μ has an inverse γ; then μ is onto. For if $x \in X$, then $x = x\iota = x(\gamma\mu) = (x\gamma)\mu$ and $x\gamma \in X$ is a preimage of x. Moreover, μ is one-to-one. For if $x\mu = y\mu$, then

$$(x\mu)\gamma = x(\mu\gamma) = x\iota = x \quad \text{and} \quad (y\mu)\gamma = y(\mu\gamma) = y\iota = y$$

Therefore, since γ is a mapping, $x\mu = y\mu$ implies $x = y$; in other words, μ is one-to-one.

To prove the converse, we assume μ is one-to-one and onto. Define γ, which will be shown to be an inverse of μ, as follows: if $x \in X$, we define $x\gamma = y$ where y is that element of X which is the preimage of x under μ. To check that the definition of γ is meaningful, observe that as μ is onto there certainly is at least one element $y \in X$ such that $y\mu = x$. But μ is one-to-one, i.e. distinct elements have distinct images. So y is the *unique* element such that $y\mu = x$. To conclude we show γ is the inverse of μ. Let $x \in X$ and $x\mu = y$. Then $x(\mu\gamma) = (x\mu)\gamma = y\gamma = x$ by the definition of γ, and so $\mu\gamma = \iota$.

Since μ is onto, each $x \in X$ must have a preimage $\bar{y} \in X$, i.e. $\bar{y}\mu = x$. Then

$$x(\gamma\mu) = (\bar{y}\mu)(\gamma\mu) = ((\bar{y}\mu)\gamma)\mu = (\bar{y}(\mu\gamma))\mu = (\bar{y}\iota)\mu = \bar{y}\mu = x$$

and so $\gamma\mu = \iota$. Thus γ is the inverse of μ and the proof of Theorem 2.4 is complete.

c. Notation for a mapping

When X is finite, say $X = \{a_1, \ldots, a_n\}$, there is a convenient way of denoting any $\sigma \in M_X$, namely

$$\sigma = \begin{pmatrix} a_1 & \ldots & a_n \\ a_1\sigma & \ldots & a_n\sigma \end{pmatrix}$$

That is, we place below each element a_i of X $(i = 1, 2, \ldots, n)$ its image under σ; e.g. if $X = \{1, 2, 3\}$ and $\sigma \in X$ is defined by $1\sigma = 1$, $2\sigma = 1$ and $3\sigma = 2$, then σ is represented by

$$\begin{pmatrix} 1 & 2 & 3 \\ 1 & 1 & 2 \end{pmatrix}$$

Notice that every element of X has a unique image under an element σ in M_X; therefore the top row $\begin{pmatrix} a_1 & a_2 & a_3 & \ldots & a_n \\ & & & & \end{pmatrix}$ will contain all the elements of X and under each element will appear its unique image under σ. All elements of X need not appear in the bottom row, as we see from the example above.

Problems

2.20. (*a*) Write the element σ in M_X, $X = \{1, 2, 3, 4, 5, 6\}$, in the notation introduced above, when σ is defined by

(i) $1\sigma = 1,\ 2\sigma = 4,\ 3\sigma = 5,\ 4\sigma = 6,\ 5\sigma = 2,\ 6\sigma = 6$

(ii) $1\sigma = 1,\ 2\sigma = 1,\ 3\sigma = 1,\ 4\sigma = 1,\ 5\sigma = 1,\ 6\sigma = 1$

(iii) $1\sigma = 6,\ 2\sigma = 5,\ 3\sigma = 4,\ 4\sigma = 3,\ 5\sigma = 2,\ 6\sigma = 1$

(*b*) What elements of M_X, $X = \{a, b, c, d, e\}$, are represented by the following?

(i) $\begin{pmatrix} a & b & c & d & e \\ b & a & c & d & e \end{pmatrix}$ (ii) $\begin{pmatrix} a & b & c & d & e \\ a & e & d & c & b \end{pmatrix}$ (iii) $\begin{pmatrix} a & b & c & d & e \\ e & a & c & b & a \end{pmatrix}$

Solution:

(*a*) (i) $\begin{pmatrix} 1 & 2 & 3 & 4 & 5 & 6 \\ 1 & 4 & 5 & 6 & 2 & 6 \end{pmatrix}$ (ii) $\begin{pmatrix} 1 & 2 & 3 & 4 & 5 & 6 \\ 1 & 1 & 1 & 1 & 1 & 1 \end{pmatrix}$ (iii) $\begin{pmatrix} 1 & 2 & 3 & 4 & 5 & 6 \\ 6 & 5 & 4 & 3 & 2 & 1 \end{pmatrix}$

(*b*) The mappings defined by

(i) $a \to b,\ b \to a,\ c \to c,\ d \to d,\ e \to e$

(ii) $a \to a,\ b \to e,\ c \to d,\ d \to c,\ e \to b$

(iii) $a \to e,\ b \to a,\ c \to c,\ d \to b,\ e \to a$

2.21. Exhibit all elements of M_X when (i) $X = \{1\}$, (ii) $X = \{1, 2\}$, (iii) $X = \{a_1, a_2\}$.

Solution:

(i) There is only one element in $M_{\{1\}}$, namely $\iota: 1 \to 1$, the identity mapping.

(ii) The following are the elements of $M_{\{1,2\}}$: $\begin{pmatrix} 1 & 2 \\ 1 & 2 \end{pmatrix}, \begin{pmatrix} 1 & 2 \\ 1 & 1 \end{pmatrix}, \begin{pmatrix} 1 & 2 \\ 2 & 1 \end{pmatrix}, \begin{pmatrix} 1 & 2 \\ 2 & 2 \end{pmatrix}$.

(iii) $\begin{pmatrix} a_1 & a_2 \\ a_1 & a_2 \end{pmatrix}, \begin{pmatrix} a_1 & a_2 \\ a_1 & a_1 \end{pmatrix}, \begin{pmatrix} a_1 & a_2 \\ a_2 & a_1 \end{pmatrix}, \begin{pmatrix} a_1 & a_2 \\ a_2 & a_2 \end{pmatrix}$ are the only elements of $M_{\{a_1, a_2\}}$.

2.22. Write out the multiplication tables of the three semigroups in Problem 2.21(i) and (ii).

(i)

	ι
ι	ι

(ii) The identity of $M_{\{1,2\}}$ is $\iota = \begin{pmatrix} 1 & 2 \\ 1 & 2 \end{pmatrix}$.

Let $\sigma_1 = \begin{pmatrix} 1 & 2 \\ 1 & 1 \end{pmatrix}$, $\sigma_2 = \begin{pmatrix} 1 & 2 \\ 2 & 1 \end{pmatrix}$ and $\sigma_3 = \begin{pmatrix} 1 & 2 \\ 2 & 2 \end{pmatrix}$. Then the multiplication table is

	ι	σ_1	σ_2	σ_3
ι	ι	σ_1	σ_2	σ_3
σ_1	σ_1	σ_1	σ_3	σ_3
σ_2	σ_2	σ_1	ι	σ_3
σ_3	σ_3	σ_1	σ_1	σ_3

The multiplication is calculated as follows. Since ι is the identity, ι leaves every element of $M_{\{1,2\}}$ unchanged; hence the first row and first column are easily written down. Since $j\sigma_1 = 1$, $j = 1, 2$, and if $\sigma \in M_{\{1,2\}}$, then $k(\sigma\sigma_1) = (k\sigma)\sigma_1 = 1$; thus $\sigma\sigma_1 = \sigma_1$. Hence the second column consists of σ_1.

Similarly $\sigma\sigma_3 = \sigma_3$, so the last column consists of σ_3. We must still calculate $\sigma_1\sigma_2$, $\sigma_2\sigma_2$ and $\sigma_3\sigma_2$. Now each element of $\{1,2\}$ is taken to 1 by σ_1, and 1 is taken to 2 by σ_2, so $\sigma_1\sigma_2$ takes every element to 2; hence $\sigma_1\sigma_2 = \sigma_3$. Now $1\sigma_2\sigma_2 = (1\sigma_2)\sigma_2 = 2\sigma_2 = 1$ and $2\sigma_2\sigma_2 = (2\sigma_2)\sigma_2 = 1\sigma_2 = 2$; hence $\sigma_2\sigma_2 = \iota$. If $j \in \{1,2\}$, then $j\sigma_3 = 2$ and so $j\sigma_3\sigma_2 = 2\sigma_2 = 1$. Thus $\sigma_3\sigma_2 = \sigma_1$.

2.23. Let $X = \{1, 2, \ldots, n\}$ and $\sigma \in M_X$. Show that σ has an inverse if and only if the bottom row in the representation of σ given on page 37, viz. $\begin{pmatrix} 1 & 2 & \ldots & n \\ 1\sigma & 2\sigma & \ldots & n\sigma \end{pmatrix}$, contains every element of X once and only once.

Solution:

By Theorem 2.4, if σ has an inverse then σ is one-to-one and onto. Hence $\{1\sigma, 2\sigma, 3\sigma, \ldots, n\sigma\} = \{1, 2, \ldots, n\}$ and the bottom row of $\begin{pmatrix} 1 & 2 & \ldots & n \\ 1\sigma & 2\sigma & \ldots & n\sigma \end{pmatrix}$ contains all elements of X once and only once. Conversely if the bottom row of the representation has all the elements of X once and only once, then each element in X has a unique image under σ, namely the entry under that particular element. Therefore σ is one-to-one. σ is also onto, for if $j \in X$ it must be one of the elements in the bottom row and j is then an image of the element of X appearing above it in the representation. As σ is a one-to-one and onto mapping it has, by Theorem 2.4, an inverse.

2.24. List all elements of $M_{\{1,2\}}$ which have inverses.

Solution:

Problem 2.23 shows us that we must find all representations of elements of $M_{\{1,2\}}$ with bottom row containing 1 and 2. Using the result of Problem 2.21(ii), we have $\begin{pmatrix} 1 & 2 \\ 1 & 2 \end{pmatrix}$ and $\begin{pmatrix} 1 & 2 \\ 2 & 1 \end{pmatrix}$ as the only elements of $M_{\{1,2\}}$ which have inverses. In the notation of Problem 2.22, ι and σ_2 are the only elements which have inverses.

2.25. List all elements of $M_{\{1,2,3\}}$ which have inverses.

Solution:

The possible representations of elements of $M_{\{1,2,3\}}$ which have inverses are

$$\sigma_1 = \begin{pmatrix} 1 & 2 & 3 \\ 1 & 2 & 3 \end{pmatrix} \quad \sigma_2 = \begin{pmatrix} 1 & 2 & 3 \\ 1 & 3 & 2 \end{pmatrix} \quad \sigma_3 = \begin{pmatrix} 1 & 2 & 3 \\ 2 & 1 & 3 \end{pmatrix}$$

$$\sigma_4 = \begin{pmatrix} 1 & 2 & 3 \\ 2 & 3 & 1 \end{pmatrix} \quad \sigma_5 = \begin{pmatrix} 1 & 2 & 3 \\ 3 & 1 & 2 \end{pmatrix} \quad \sigma_6 = \begin{pmatrix} 1 & 2 & 3 \\ 3 & 2 & 1 \end{pmatrix}$$

2.26. What are the inverses of the elements in Problem 2.25?

Solution:

Theorem 2.4 explains how to find the inverse of a mapping which is one-to-one and onto. For example, to find the inverse of σ_3, we note σ_3 takes $1 \to 2$. Hence $\sigma_3^{-1}: 2 \to 1$. $\sigma_3: 3 \to 3$, hence $\sigma_3^{-1}: 3 \to 3$. But then σ_3 is its own inverse. Similarly $\sigma_1^{-1} = \sigma_1$, $\sigma_2^{-1} = \sigma_2$, $\sigma_4^{-1} = \sigma_5$, $\sigma_5^{-1} = \sigma_4$, $\sigma_6^{-1} = \sigma_6$.

2.27. Prove that the subset of elements of M_X which have inverses is a semigroup with identity under the usual compositions of mappings.

Solution:

Let S be the elements of M_X which have inverses. Is $\circ$, the usual composition of mappings, a binary operation in S? In other words, is $(\alpha, \beta) \to \alpha \circ \beta$ a mapping of $S \times S \to S$? If $\alpha \circ \beta \in S$, the answer is yes. So we ask: if α, β have inverses, does $\alpha \circ \beta$? Note that $(\alpha \circ \beta) \circ (\beta^{-1} \circ \alpha^{-1}) = \alpha \circ (\beta \circ \beta^{-1}) \circ \alpha^{-1} = \alpha \circ \iota \circ \alpha^{-1} = \iota$. Hence $\alpha \circ \beta$ has the inverse $\beta^{-1} \circ \alpha^{-1}$. As ι is its own inverse, $\iota \in S$. As M_X satisfies the associative law, so does S. Hence S is a semigroup.

d. The order in a product

There is one way in which the associative law makes it easier to work in a semigroup than in a non-associative groupoid. Suppose S is a semigroup and let $a_1, a_2, a_3 \in S$. There are two ways in which one can multiply a_1, a_2 and a_3 together (in this order): $(a_1a_2)a_3$ and $a_1(a_2a_3)$. The point of the associative law is that these products coincide. Suppose now $a_1, a_2, a_3, a_4 \in S$. Then we can multiply a_1, a_2, a_3, a_4 together (in this order) in the following five ways:

$$a_1(a_2(a_3a_4)),\quad (a_1a_2)(a_3a_4),\quad ((a_1a_2)a_3)a_4,\quad (a_1(a_2a_3))a_4,\quad a_1((a_2a_3)a_4)$$

It is conceivable that some of these products give rise to different elements of S. However, the associative law, which of course involves products of only three elements, prohibits this. To see this consider first $(a_1a_2)(a_3a_4)$. By the associative law, $(a_1a_2)(a_3a_4) = a_1(a_2(a_3a_4))$; hence the second product coincides with the first. In fact all of the products equal the first. As a second illustration consider $((a_1a_2)a_3)a_4$. Here we have, as desired,

$$((a_1a_2)a_3)a_4 = (a_1(a_2a_3))a_4 = a_1((a_2a_3)a_4) = a_1(a_2(a_3a_4))$$

In general, we have

Theorem 2.5: Let S be a semigroup and let $a_1, a_2, \ldots, a_n \in S$. Then any two products of $a_1, a_2, \ldots, a_n$ coincide, when the a_1 appears first in each product, a_2 second, ..., and a_n last.

Proof: Assume the contrary, that not all possible products of $a_1, \ldots, a_n$ in that order are equal. We may assume that n is the first integer for which two different products give rise to different elements. Let x and y be these two different products. Now $x = uv$ and $y = u_1v_1$ for some u, v and some u_1, v_1. Suppose u is the product of the elements $a_1, \ldots, a_r$ and v the product of the elements $a_{r+1}, \ldots, a_n$, while u_1 is the product of $a_1, \ldots, a_s$ and v_1 is the product of $a_{s+1}, \ldots, a_n$. Without loss of generality we may suppose that $s \leq r$. If $s = r$ then $u = u_1$ and $v = v_1$, since n is the first integer for which there exist two unequal products of the same elements. If $s < r$, then $u = (a_1 \ldots a_s)(a_{s+1} \ldots a_r)$ while $v_1 = (a_{s+1} \ldots a_r)(a_{r+1} \ldots a_n)$. Hence $x = uv = \{(a_1 \ldots a_s)(a_{s+1} \ldots a_r)\}(a_{r+1} \ldots a_n)$ while $y = u_1v_1 = (a_1 \ldots a_s)\{(a_{s+1} \ldots a_r)(a_{r+1} \ldots a_n)\}$. By the associative law for the three elements $(a_1 \ldots a_s)$, $(a_{s+1} \ldots a_r)$ and $(a_{r+1} \ldots a_n)$, we have $x = y$, contradicting the assumption that not all possible products are equal. Hence the result follows.

It follows from Theorem 2.5 that in a semigroup, if we are given the order of a product, the bracketing is immaterial. Thus we write simply $a_1a_2 \ldots a_n$, without brackets, for the product of $a_1, a_2, \ldots, a_n$ in this order. For example, if m is any positive integer, we write $a \cdot a \cdot \cdots \cdot a$ for the product of m a's; a useful abbreviation for such a product is a^m. If n is a second positive integer, then $a^m \cdot a^n = a^{m+n}$ since $a^m \cdot a^n$ is simply the product of $m + n$ a's. Similarly, $(a^m)^n = a^{mn}$.

For example, if $S = \{1,2,3\}$ is the semigroup with binary operation given by the following table, then $1^3 = 3,\ 2^3 = 3,\ 3^3 = 3$.

	1	2	3
1	2	3	1
2	3	1	2
3	1	2	3

2.5 HOMOMORPHISMS OF GROUPOIDS AND CAYLEY'S THEOREM

a. Definition of a homomorphism

Before we give the formal definition of a homomorphism of one groupoid into another, we will give an example. Let $(\bar{R},\cdot)$ be the groupoid of positive real numbers with the binary operation of ordinary multiplication. Let $(R,+)$ be the groupoid of real numbers with binary operation the usual addition inside R. Let us define a mapping $\theta : \bar{R} \to R$ by $x\theta = \log_{10} x$. Recall that $\log_{10}(xy) = \log_{10} x + \log_{10} y$; hence $(xy)\theta = x\theta + y\theta$. θ is an example of a homomorphism. The formal definition is

Definition: A homomorphism from a groupoid (G,α) into a groupoid (H,β) is a mapping $\theta : G \to H$ which satisfies the condition

$$((g_1, g_2)\alpha)\theta = (g_1\theta, g_2\theta)\beta$$

for all g_1, g_2 in G.

Usually groupoids are written in multiplicative notation. The definition then takes the form: A homomorphism of $(G,\cdot)$ into $(H,\cdot)$ is a mapping $\theta : G \to H$ such that

$$(g_1 g_2)\theta = g_1\theta g_2\theta \tag{2.7}$$

for all g_1, g_2 in G. (*2.7*) is often expressed as: "θ preserves multiplication."

Problems

2.28. Let G be the semigroup of integers under the usual addition of integers and let H be the semigroup of even integers under the usual addition. Verify that the mapping $\theta : G \to H$ defined by $\theta : g \to 2g$ for all $g \in G$ is a homomorphism of G into H.

Solution:

First we must check to see if θ is a mapping. Since $2g$ is a unique integer, θ is clearly a mapping. Let $g_1, g_2 \in G$. Then $g_1\theta + g_2\theta = 2g_1 + 2g_2 = 2(g_1 + g_2) = (g_1 + g_2)\theta$. Therefore θ preserves multiplication and is a homomorphism.

2.29. Let G be the semigroup of integers under the usual multiplication of integers, and let H be the semigroup of even integers under the usual multiplication. Is $\sigma : G \to H$ defined by $\sigma : g \to 2g$ for all $g \in G$ a homomorphism of G into H?

Solution:

As in the preceding problem, σ is a mapping. But σ is not a homomorphism; for if $g_1, g_2 \in G$, then $(g_1g_2)\sigma = 2g_1g_2$ and $g_1\sigma g_2\sigma = 2g_1 2g_2 = 4g_1g_2$. Hence $(g_1g_2)\sigma \neq (g_1\sigma)(g_2\sigma)$ for all $g_1, g_2 \in G$.

2.30. Let $(G,+)$ and $(H,+)$ be the semigroups of Problem 2.28. Find a homomorphism of $(G,+)$ into $(H,+)$ which is not equal to θ.

Solution:

Define $\tau_n : G \to H$ by $\tau_n : g \to ng$ where n is a fixed even integer. τ_n is a mapping of G into H, since $ng \in H$ for any even integer n and ng is unique. For each n, τ_n is a homomorphism because

$g_1, g_2 \in G$ implies $(g_1+g_2)\tau_n = n(g_1+g_2) = ng_1 + ng_2 = g_1\tau_n + g_2\tau_n$. $\tau_n \neq \theta$, since $1\theta \neq 1\tau_n$ if $n \neq 2$. Therefore there is an infinite number of different homomorphisms between $(G,+)$ and $(H,+)$. Notice τ_n is an onto mapping if $n = 2$, because any element in H is of the form $2q$, q an integer, and $q\tau_2 = 2q$. But when $n \neq 2$, τ_n is not onto since 2 has no preimage under τ. For if $q\tau_n = 2$, $q \in G$, then $nq = 2$ or $q = 2/n$ (if $n \neq 0$) and $2/n \notin G$.

2.31. Let G and H be as in Problem 2.28. Verify that the mapping $\sigma: G \to H$ defined by $\sigma: g \to g^2$, for all $g \in G$, is not a homomorphism.

Solution:

Let $g_1, g_2 \in G$. Then $(g_1+g_2)\sigma = (g_1+g_2)^2 = g_1^2 + g_2^2 + 2g_1g_2$ and $g_1\sigma + g_2\sigma = g_1^2 + g_2^2$. Hence σ is not a homomorphism for $(g_1+g_2)\sigma \neq g_1\sigma + g_2\sigma$ for all $g_1, g_2 \in G$.

2.32. Is the mapping σ of G into H, G and H as in Problem 2.28, defined by $g\sigma = 0$ for all $g \in G$, a homomorphism?

Solution:

If $g_1, g_2 \in G$, then $(g_1+g_2)\sigma = 0$ and $g_1\sigma + g_2\sigma = 0+0 = 0$. Hence σ is a homomorphism.

2.33. Let G be the semigroup of positive integers P under the usual addition, and let H be the semigroup of positive integers P under the usual multiplication. Show that the mapping $\eta: G \to H$ defined by $g\eta = 2^g$ for all $g \in G$ is a homomorphism.

Solution:

η is clearly a mapping. Let $g_1, g_2 \in G$. Then $(g_1+g_2)\eta = 2^{g_1+g_2} = 2^{g_1}2^{g_2} = g_1\eta g_2\eta$. Hence η is a homomorphism.

2.34. Let $G = (\{1,2,3\}, \alpha)$ and $H = (\{a,b,c\}, \beta)$ be the groupoids with binary operations α and β defined by the multiplication tables

α:

	1	2	3
1	1	2	3
2	2	3	1
3	3	1	2

β:

	a	b	c
a	c	a	b
b	b	b	a
c	c	a	b

Which of the following mappings are homomorphisms?

(a) $1 \to a$, $2 \to b$, $3 \to c$ (d) $1 \to b$, $2 \to c$, $3 \to c$

(b) $1 \to a$, $2 \to a$, $3 \to a$ (e) $1 \to b$, $2 \to b$, $3 \to b$

(c) $1 \to a$, $2 \to b$, $3 \to b$ (f) $1 \to c$, $2 \to a$, $3 \to b$

Solution:

We use σ to indicate the mapping in each case.

(a) $1\sigma 2\sigma = ab = a$ and $(1\cdot 2)\sigma = 2\sigma = b$. σ is not a homomorphism. (b) $1\sigma 2\sigma = aa = c$ and $(1\cdot 2)\sigma = 2\sigma = a$. σ is not a homomorphism. (c) $1\sigma 2\sigma = ab = a$ and $(1\cdot 2)\sigma = b$. σ is not a homomorphism. (d) $1\sigma 3\sigma = bc = a$ and $(1\cdot 3)\sigma = 3\sigma = c$. σ is not a homomorphism. (e) σ is a homomorphism since the image of 1, 2 and 3 is b, so that $(ij)\sigma = b$ for any $i, j \in \{1,2,3\}$, and $i\sigma j\sigma = bb = b$. (f) $1\sigma 2\sigma = ca = c$ and $(1\cdot 2)\sigma = 2\sigma = a$. σ is not a homomorphism.

2.35. Let G be the semigroup of positive integers under the usual addition. Determine which mappings of G into G are homomorphisms:

(i) $\sigma: n \to 2n+1$, (ii) $\sigma: n \to 2n^2$, (iii) $\sigma: n \to 1$.

Solution:

(i) $(n_1+n_2)\sigma = 2(n_1+n_2)+1$ and $n_1\sigma + n_2\sigma = 2n_1+1+2n_2+1 = 2(n_1+n_2)+2$. Hence $(n_1+n_2)\sigma \neq n_1\sigma + n_2\sigma$, and so σ is not a homomorphism.

(ii) $(n_1+n_2)\sigma = 2(n_1+n_2)^2 = 2n_1^2 + 2n_2^2 + 4n_1n_2$. $n_1\sigma + n_2\sigma = 2n_1^2 + 2n_2^2$. Then $n_1\sigma + n_2\sigma \neq (n_1+n_2)\sigma$, and so σ is not a homomorphism.

(iii) $(n_1+n_2)\sigma = 1$. $n_1\sigma + n_2\sigma = 1+1 = 2$. Thus $n_1\sigma + n_2\sigma \neq (n_1+n_2)\sigma$ and hence σ is not a homomorphism.

b. Epimorphism, monomorphism, and isomorphism

Three special types of homomorphism arise naturally.

1. A homomorphism of groupoid G into groupoid H may be an onto mapping.
2. A homomorphism of a groupoid G into a groupoid H may be a one-to-one mapping.
3. A homomorphism of a groupoid G into a groupoid H may be both onto and one-to-one.

We give these three types of homomorphisms special names.

Definition: Let θ be a homomorphism of a groupoid G into a groupoid H. Then

1. θ is called an *epimorphism* if θ maps the carrier of G onto the carrier of H, i.e. $G\theta = H$. (See Section 1.3a, page 12, for the definition of $G\theta$.)
2. θ is called a *monomorphism* if θ is a one-to-one mapping of the carrier of G into the carrier of H.
3. θ is called a *isomorphism* if θ is both an epimorphism and a monomorphism, i.e. θ is one-to-one and onto.

If there is an isomorphism from groupoid G onto the groupoid H, then we say G and H are isomorphic, or G is isomorphic to H, and write $G \cong H$.

Problems

2.36. Let G be the groupoid of integers with addition as binary operation, and H the even integers with addition as the binary operation. Let τ_n for n an even integer be the homomorphism (Problem 2.30) defined by $g\tau_n = ng$, for $g \in G$. When is τ_n an isomorphism, monomorphism or epimorphism?

Solution:

If $n \neq 0$, τ_n is one-to-one since $g\tau_n = g'\tau_n$ implies $ng = ng'$ and so $g = g'$. τ_0 is not one-to-one, so it is not a monomorphism. If τ_n is onto, there exists $g \in G$ such that $g\tau_n = ng = 2$. Then $g = 2/n$ and $n = \pm 2$. Hence $\tau_{\pm 2}$ are the only epimorphisms. Thus $\tau_{\pm 2}$ are isomorphisms and τ_n is a monomorphism when $n \neq 0$.

2.37. G and H are finite groupoids and $|G| \neq |H|$. Show that G cannot be isomorphic to H.

Solution:

Let $\theta: G \to H$ be an isomorphism and let $g_1, \ldots, g_n$ be the (distinct) elements of G. Then $g_1\theta, g_2\theta, \ldots, g_n\theta$ are distinct (since θ is one-to-one) and are all the elements of H (since θ is onto). Hence $|H| = n$, which contradicts $|G| \neq |H|$. Thus there exists no isomorphism $\theta: G \to H$.

2.38. Prove that if G, H, K are groupoids, then: (i) $G \cong G$; (ii) if $G \cong H$, then $H \cong G$; (iii) if $G \cong H$ and $H \cong K$, then $G \cong K$. In other words "$\cong$" is an equivalence relation. (Hard.)

Solution:

(i) Let $\iota: G \to G$ be the mapping defined by $g\iota = g$ for all $g \in G$. ι is a one-to-one epimorphism. Hence it is an isomorphism, and so $G \cong G$.

(ii) Let $\alpha = G \to H$ be an isomorphism. Then we define a mapping $\beta: H \to G$ as follows: Let $h \in H$. As α is one-to-one and onto, there exists a unique $g \in G$ such that $g\alpha = h$. Put $h\beta = g$. Note that $h\beta\alpha = h$.

Now β is onto G, for if $g \in G$, $g\alpha = h \in H$ and $h\beta = g$ by definition. Also β is one-to-one; for if $h_1\beta = h_2\beta$, then $h_1\beta\alpha = h_2\beta\alpha$ and so $h_1 = h_2$.

Finally β is a homomorphism. Let $h_1, h_2 \in H$. Suppose $g_1\alpha = h_1$, $g_2\alpha = h_2$. Then $h_1\beta = g_1$, $h_2\beta = g_2$. Note that $(g_1g_2)\alpha = g_1\alpha g_2\alpha = h_1h_2$. Hence $(h_1h_2)\beta = g_1g_2 = h_1\beta h_2\beta$.

(iii) Let $\alpha: G \to H$ and $\beta: H \to K$ be isomorphisms. Let $\gamma = \alpha\beta$. We shall prove that γ is an isomorphism. First, $\gamma: G \to K$. Secondly, γ is onto; because if $k \in K$, there exists $h \in H$ such that $h\beta = k$, and there exists $g \in G$ such that $g\alpha = h$, so $g(\alpha\beta) = h\beta = k$. Next γ is one-to-one; for if $g_1\gamma = g_2\gamma$, $(g_1\alpha)\beta = (g_2\alpha)\beta$, and as β is one-to-one, $g_1\alpha = g_2\alpha$, which implies, since α is one-to-one, $g_1 = g_2$. Finally, we must show that γ is a homomorphism. $(g_1g_2)\gamma = ((g_1g_2)\alpha)\beta = (g_1\alpha g_2\alpha)\beta = (g_1\alpha)\beta(g_2\alpha)\beta = g_1(\alpha\beta)g_2(\alpha\beta) = g_1\gamma g_2\gamma$. Hence $G \cong K$.

2.39. (*a*) Is the semigroup of integers under the usual addition isomorphic to the multiplicative groupoid of nonzero real numbers? (*Hint*: The integers and the reals are not equipotent.)

(*b*) Is the groupoid of nonzero rational numbers under division isomorphic to the groupoid of nonzero rational numbers under multiplication?

(*c*) Is the semigroup of integers under the usual addition isomorphic to the semigroup of rational numbers under the usual addition? (*Hint*: Show that under any homomorphism θ of the integers under addition into the rationals under addition, $r\theta = r(1\theta)$.)

Solution:

(*a*) No. Because if the integers were isomorphic to the reals, the isomorphism between them would constitute a matching and hence the reals and the integers would be equipotent.

(*b*) Let $\alpha: (Q^*, \div) \to (Q^*, \cdot)$ be a homomorphism between the nonzero rationals under division and the nonzero rationals under multiplication. Now $1 = 1 \div 1$. Thus $1\alpha = (1 \div 1)\alpha = 1\alpha \cdot 1\alpha$ and so $1\alpha = \pm 1$. Now $\pm 1 = 1\alpha = (2 \div 2)\alpha = 2\alpha 2\alpha$ and so $2\alpha = \pm 1$. Similarly, $3\alpha = \pm 1$. Hence α is not one-to-one. In particular, α is not an isomorphism. Thus there is no isomorphism between the two groupoids.

(*c*) Let $\theta: (Z, +) \to (Q, +)$ be any homomorphism. Let $1\theta = q$. We shall show by induction on r that $r\theta = rq$ for all $r \in N$. Now $q = 1\theta = (0+1)\theta = 0\theta + 1\theta = 0\theta + q$ and so $0\theta = 0$. Suppose $r\theta = rq$ for $r = n$. Consider $r = n+1$. $(n+1)\theta = n\theta + 1\theta = nq + q = (n+1)q$. Hence $r\theta = rq$ for all $r \in N$, by induction. If $r \in Z$ and r is negative, then $-r \in N$. Then since $0 = (r + -r)\theta = r\theta + (-r)\theta = r\theta + (-r)q$, $r\theta = rq$. Hence $r\theta = rq$ for all $r \in Z$.

If $q = 0$, θ is not onto. If $q = m/n$, with m, n integers and $m, n \neq 0$, then $1/2n \in Q$. Has $1/2n$ a pre-image? If r were an integer such that $r\theta = 1/2n$, then $rq = 1/2n$ and so $r = 1/2m$. But $1/2m$ is not an integer. Thus θ is not an epimorphism and there is no isomorphism.

2.40. Let $(Z, \circ)$ be the groupoid with binary operation $\circ$ defined by

$$a \circ b = a + b + ab$$

and let $(Z, *)$ be the groupoid with binary operation $*$ defined by

$$a * b = a + b - ab$$

Is $(Z, \circ) \cong (Z, *)$?

Solution:

Let $\sigma: (Z, \circ) \to (Z, *)$ be defined by $a\sigma = -a$ for $a \in Z$. σ is clearly a mapping. $(a \circ b)\sigma = (a + b + ab)\sigma = -(a + b + ab)$ and $a\sigma * b\sigma = -a * -b = (-a) + (-b) - (-a)(-b) = -(a + b + ab)$. Hence σ is a homomorphism. σ is onto, for if $a \in (Z, *)$, then $-a \in (Z, \circ)$ and $(-a)\sigma = a$. Now $a\sigma = b\sigma$ implies $-a = -b$ and $a = b$. Therefore σ is one-to-one and $(Z, \circ) \cong (Z, *)$.

2.41. Is $M_{\{1\}} \cong M_{\{1,2\}}$? Is $M_{\{1,2\}} \cong M_{\{1,2,3\}}$?

Solution:

$|M_{\{1\}}| = 1$ and $|M_{\{1,2\}}| = 4$ (see Problem 2.21, page 37). Therefore, by Problem 2.37, $M_{\{1\}}$ cannot be isomorphic to $M_{\{1,2\}}$. From Problem 2.25, the subset of elements in $M_{\{1,2,3\}}$ which have inverses has 6 elements. Thus $|M_{\{1,2\}}|$ is less then the order of $M_{\{1,2,3\}}$, and so $M_{\{1,2\}}$ is not isomorphic to $M_{\{1,2,3\}}$.

2.42. Give an example of two groupoids of order two which are not isomorphic.

Solution:

	a	*b*
a	*a*	*a*
b	*a*	*a*

	c	*d*
c	*c*	*d*
d	*d*	*c*

These two groupoids are not isomorphic, since there are only two one-to-one mappings, namely $\theta: a \to c, b \to d$ and $\psi: a \to d, b \to c$. Now θ is not an isomorphism, for $(ab)\theta = a\theta = c$ while $a\theta b\theta = cd = d$. ψ is not an isomorphism, for $(aa)\psi = a\psi = d$ while $a\psi a\psi = dd = c$.

2.43. Prove that the mapping $\theta: a + ib \to a - ib$ is an isomorphism from the groupoid C of complex numbers under addition with itself.

Solution:

θ is onto; for if $x \in C$, $x = a + ib$. Now $(a + i(-b))\theta = a + ib = x$. Also θ is one-to-one, since $(a_1 + ib_1)\theta = (a_2 + ib_2)\theta$ implies $a_1 - ib_1 = a_2 - ib_2$ and hence $a_1 = a_2$ and $b_1 = b_2$.

Finally θ is a homomorphism, for

$$[(a_1 + ib_1) + (a_2 + ib_2)]\theta \;=\; (a_1 + a_2) - i(b_1 + b_2) \;=\; (a_1 + ib_1)\theta + (a_2 + ib_2)\theta$$

2.44. Is the homomorphism θ of the groupoid C of complex numbers under the usual multiplication of complex numbers to the groupoid of real numbers under the usual multiplication defined by

$$\theta: \; a + ib \rightarrow |a + ib| \;=\; +\sqrt{a^2 + b^2}$$

an epimorphism or monomorphism?

Solution:

It is well known that if x_1, x_2 are two complex numbers, then $|x_1x_2| = |x_1|\,|x_2|$. The calculations are

$$\begin{aligned} |a_1 + ib_1|\,|a_2 + ib_2| &= \sqrt{(a_1^2 + b_1^2)(a_2^2 + b_2^2)} \\ &= \sqrt{(a_1a_2 - b_1b_2)^2 + (b_1a_2 + b_2a_1)^2} \\ &= |a_1a_2 - b_1b_2 + i(b_1a_2 + b_2a_1)| \\ &= |(a_1 + ib_1)(a_2 + ib_2)| \end{aligned}$$

Then $(x_1x_2)\theta = x_1\theta x_2\theta$ and so θ is a homomorphism. On the other hand $(a + ib)\theta = (-a - ib)\theta$, so θ is not one-to-one. Finally θ is not onto, for it is always the case that $|x| \geqq 0$, and thus there exists no x such that $x\theta = -1$.

2.45. Let $(P, \cdot)$ be the groupoid P under the usual multiplication of positive integers and $(R, +)$ the groupoid R under the usual addition of real numbers. Is the mapping θ of $(P, \cdot)$ into $(R, +)$ defined by $\theta: a \rightarrow \log_{10} a$ an epimorphism, monomorphism or isomorphism?

Solution:

As in Section 2.5a, θ is a homomorphism. If $a\theta = b\theta$, then $\log_{10} a = \log_{10} b$ and hence $a = b$. Thus θ is a monomorphism. Since $0 = \log_{10} 1 < \log_{10} 2 < \log_{10} 3 < \cdots$, there is no integer such that $\log_{10} x = -1$. Hence θ is not onto. Therefore θ is a monomorphism but not an epimorphism nor an isomorphism.

c. Properties of epimorphisms

We will show in this section that if θ is an epimorphism from the groupoid G to the groupoid H, then H shares some of the properties of G.

Theorem 2.6: Let θ be an epimorphism from the groupoid G to the groupoid H. Then

(*a*) if G is a groupoid with an identity 1, so is H and 1θ is the identity of H. Furthermore if f is an inverse of g in G, then $f\theta$ is an inverse of $g\theta$ in H.

(*b*) if G is commutative, so is H.

(*c*) if G is a semigroup, so is H.

Proof:

(*a*) Let $h \in H$. We shall prove 1θ is the identity of H, i.e. $h \cdot 1\theta = h = 1\theta \cdot h$. As θ is an epimorphism, θ is onto and we can find an element g in G such that $g\theta = h$. Then

$$h \cdot 1\theta \;=\; g\theta \cdot 1\theta \;=\; (g \cdot 1)\theta \;=\; g\theta \;=\; h$$

and

$$1\theta \cdot h \;=\; 1\theta \cdot g\theta \;=\; (1 \cdot g)\theta \;=\; g\theta \;=\; h$$

Thus 1θ is the identity of H and H is a groupoid with an identity.

Now suppose $g \in G$ has an inverse f. Then $gf = 1 = fg$. Therefore

$$g\theta \cdot f\theta \;=\; (gf)\theta \;=\; 1\theta \;=\; (fg)\theta \;=\; f\theta \cdot g\theta$$

which means $f\theta$ is the inverse of $g\theta$ in H.

(*b*) Suppose G is commutative. We show H is commutative. To this end let $h, h' \in H$. Because θ is onto, we can find $g, g' \in G$ for which $g\theta = h$ and $g'\theta = h'$. Hence

$$hh' = g\theta \cdot g'\theta = (g \cdot g')\theta = (g' \cdot g)\theta = g'\theta \cdot g\theta = h'h$$

and so H is commutative as claimed.

(*c*) To show H is a semigroup, we must prove that multiplication is associative in H. Let $h, h', h'' \in H$. Then we can find $g, g', g'' \in G$ such that $g\theta = h$, $g'\theta = h'$ and $g''\theta = h''$. Since G is associative we have, as required,

$$\begin{aligned}(hh')h'' &= (g\theta \cdot g'\theta)g''\theta = [(gg')\theta]g''\theta = ((gg')g'')\theta \\ &= (g(g'g''))\theta = g\theta[(g'g'')\theta] = g\theta(g'\theta \cdot g''\theta) = h(h'h'')\end{aligned}$$

d. Naming and isomorphisms

In our study of groupoids we will take isomorphic groupoids to be essentially the same. To explain why, we will describe a "naming process," beginning with an example.

Let G be the groupoid with binary operation $\cdot$, elements 1 and a, and multiplication table

	1	a
1	1	a
a	a	1

We define a new groupoid $\bar{G}$ by relabeling the elements of G. Let $\bar{G}$ consist of the elements α, β and have multiplication table

	α	β
α	α	β
β	β	α

What we have done is to call the elements of G by different names.

In general if G is any groupoid, we can form a new groupoid $\bar{G}$ by renaming the elements of G. Thus for each $g \in G$ we take a new element $\bar{g}$, ensuring only that $\bar{f} \neq \bar{g}$ if $f \neq g$, i.e. don't use the same name twice. If $fg = h$, then we define multiplication of elements of $\bar{G}$ by $\bar{f} \circ \bar{g} = \bar{h}$. It is easy to prove that $\bar{G}$ is a groupoid with this multiplication.

We are not interested in distinguishing between groupoids which differ only because their elements have different names. Considering groupoids to be the same if they are isomorphic overcomes this snag. To see this we will show that the $\bar{G}$ constructed from G above by renaming is isomorphic with G. We must find a one-to-one onto homomorphism θ. Define $g\theta = \bar{g}$, i.e. the image of g under θ is the new name of g. θ is one-to-one onto, as one and only one $\bar{g}$ corresponds to each g. Also $(fg)\theta = \bar{h}$ where $fg = h$. But $f\theta \circ g\theta = \bar{f} \circ \bar{g} = \bar{h}$, by definition of the multiplication of $\bar{G}$. Hence $(fg)\theta = f\theta \circ g\theta$, and G is isomorphic with $\bar{G}$. Thus isomorphism gets rid of the difficulty of obtaining a new groupoid on simply renaming.

We look at the problem from another point of view. Suppose F and G are two isomorphic groupoids, and that θ is an isomorphism between F and G. Then we will apply our renaming process to show that G and $\bar{F}$, F suitably renamed, cannot be distinguished either as regards their elements or the way they multiply.

Let us as before rename each element $f \in F$, $\bar{f}$. But we shall choose $\bar{f}$ to be $f\theta$. This is a proper renaming since $\bar{f} = \bar{g}$ means that $f\theta = g\theta$, and, as θ is one-to-one, $f = g$. So we have not used the same name twice. As before, if $fg = h$ we define $\bar{f} \circ \bar{g}$ to be $\bar{h}$. Thus $\bar{F}$ becomes a groupoid with respect to the binary operation $\circ$.

How does $\bar{F}$ compare with G? $\bar{F}$ has the same elements as G. But do these elements multiply the same way? Suppose $\bar{f}_1$ and $\bar{f}_2$ are two elements of $\bar{F}$. Then $\bar{f}_1$ was previously called f_1, and $\bar{f}_2$ was called f_2. If $f_1 f_2 = f_3$, then we defined $\bar{f}_1 \circ \bar{f}_2 = \bar{f}_3$.

Now $\bar{f}_1, \bar{f}_2$ are also elements of G; in fact, $\bar{f}_1 = f_1\theta$, $\bar{f}_2 = f_2\theta$. The product $\bar{f}_1\bar{f}_2$ in G is therefore $f_1\theta f_2\theta = (f_1 f_2)\theta = f_3\theta = \bar{f}_3$, since θ is a homomorphism. Hence the product $\bar{f}_1 \circ \bar{f}_2$ of two elements in $\bar{F}$ is the same element as the product $\bar{f}_1\bar{f}_2$ inside G.

Thus a groupoid isomorphic with a groupoid F is indistinguishable from a suitable renaming of F as far as the elements and the way they multiply are concerned. For this reason we do not distinguish between groupoids that are isomorphic.

e. M_X and semigroups

The importance of M_X is explained by the following theorem, which says that an isomorphic copy of any semigroup S is contained in some M_X.

Theorem 2.7: (*Cayley's Theorem*): Let S be a semigroup with identity. Then there is a monomorphism of S into M_S. (The semigroup S is an abbreviation for the semigroup (S, μ) where μ is a binary operation. M_S is the semigroup of all mappings of the set S into itself, with binary operation the composition of mappings.)

Proof: Let $s \in S$ and let $\rho_s : S \to S$ be defined by $x\rho_s = xs$, $x \in S$. Here xs is the product of x and s in S, i.e. $(x, s)\mu$. It is clear that ρ_s is a mapping of S into S, i.e. $\rho_s \in M_S$.

Let $\theta : S \to M_S$ be defined by $s\theta = \rho_s$. We shall show that θ is a monomorphism. First we have to check that θ is a homomorphism. Let $s, s' \in S$.

$$(ss')\theta = \rho_{ss'} \quad \text{and} \quad s\theta s'\theta = \rho_s \rho_{s'}$$

Now if $x \in S$, $x\rho_s\rho_{s'} = (x\rho_s)\rho_{s'} = (xs)\rho_{s'} = (xs)s' = x\rho_{ss'}$. As this is true for all $x \in S$, $\rho_{ss'} = \rho_s\rho_{s'}$. Hence $(ss')\theta = s\theta s'\theta$.

Secondly we must show that θ is one-to-one. Suppose $s\theta = s'\theta$; then $\rho_s = \rho_{s'}$. In particular, if 1 is the identity of S, $1\rho_s = 1\rho_{s'}$. But then $1\rho_s = 1 \cdot s = s = 1 \cdot \rho_{s'} = 1 \cdot s' = s'$ and so $s = s'$ and θ is a monomorphism. This completes the proof of Theorem 2.7.

As an illustration of the proof, let $S = \{a_1, a_2, a_3\}$ be the semigroup given by the following multiplication table:

	a_1	a_2	a_3
a_1	a_1	a_2	a_3
a_2	a_2	a_3	a_1
a_3	a_3	a_1	a_2

Notice that a_1 is the identity element of S and that S is a semigroup. Now

$$\rho_{a_1} = \begin{pmatrix} a_1 & a_2 & a_3 \\ a_1 & a_2 & a_3 \end{pmatrix}, \quad \rho_{a_2} = \begin{pmatrix} a_1 & a_2 & a_3 \\ a_2 & a_3 & a_1 \end{pmatrix}, \quad \rho_{a_3} = \begin{pmatrix} a_1 & a_2 & a_3 \\ a_3 & a_1 & a_2 \end{pmatrix}$$

The mapping θ is defined by $a_1\theta = \rho_{a_1}$, $a_2\theta = \rho_{a_2}$, $a_3\theta = \rho_{a_3}$. It can be checked directly that θ is a monomorphism.

Let $S\theta = \{s\theta \mid s \in S\}$ where θ is the monomorphism of Theorem 2.7; then $S \cong S\theta$ and hence M_S contains an isomorphic copy of S. In Section 2.5d we pointed out that, but for naming, S and an isomorphic copy were the same. Thus we see that in a rough way every semigroup with identity appears in some M_X. Hence the importance of M_X.

A look back at Chapter 2

We defined a groupoid, associative groupoid (called a semigroup), and commutative groupoid.

We showed that in a groupoid the identity is unique, while inverses are unique in a semigroup.

We defined a mapping $\theta: (G, \alpha) \to (G, \beta)$ to be a homomorphism if $[(g_1, g_2)\alpha]\theta = [(g_1\theta, g_2\theta)]\beta$. If θ is one-to-one and onto, we called it an isomorphism of (G, α) onto (G, β). We proved Cayley's theorem, that each semigroup has an isomorphic copy in M_X for some suitable X.

Supplementary Problems

GROUPOIDS

2.46. Let $G = \{1, -1\}$. Is $(G, \cdot)$ a groupoid if $\cdot$ is the usual multiplication of integers?

2.47. Show that $(G, \cdot)$ is a groupoid when $G = \{1, -1, i, -i\}$, $i = \sqrt{-1}$, and $\cdot$ is the usual multiplication of complex numbers.

2.48. Suppose G_n is the set of all integers divisible by the integer n. For which n is $(G_n, \cdot)$, with the usual multiplication of integers, a groupoid?

2.49. Let $f_i, i = 1, 2, 3, 4, 5, 6,$ be the set G of mappings of $R - \{0, 1\}$ into R defined for each $x \in R - \{0, 1\}$ by $f_1 : x \to x$; $f_2 : x \to \dfrac{1}{1-x}$; $f_3 : x \to \dfrac{x-1}{x}$; $f_4 : x \to \dfrac{1}{x}$; $f_5 : x \to \dfrac{x}{x-1}$; $f_6 : x \to 1 - x$. Suppose $G_{ij} = \{f_i, f_j\}$ and that $\circ$ is the composition of mappings. Determine which of the following are groupoids: (i) $(G_{1,2}, \circ)$, (ii) $(G_{1,3}, \circ)$, (iii) $(G_{1,4}, \circ)$, (iv) $(G_{1,5}, \circ)$, (v) $(G_{1,6}, \circ)$, (vi) $(G_{5,6}, \circ)$.

2.50. Let $F = \{f_1, f_2, f_3\}$ and $H = \{f_1, f_2, f_4, f_6\}$ where f_i are the mappings defined in Problem 2.49. Prove $(F, \circ)$ is a groupoid while $(H, \circ)$ is not a groupoid ($\circ$ is the composition of mappings).

COMMUTATIVE AND ASSOCIATIVE GROUPOIDS

2.51. Let R^* be the set of nonzero real numbers. Define the binary operation $\circ$ on R^* by $a \circ b = |a|\, b$ for $a, b \in R^*$. Prove $(R^*, \circ)$ is an associative groupoid but not a commutative groupoid. *Hint:* $|a|\,|b| = |ab|$.

2.52. Define the binary operation $\cdot$ on $G = R \times R$ as $(a, b) \cdot (c, d) = (ac, bc + d)$. Is $(G, \cdot)$ a commutative or an associative groupoid?

2.53. The binary operation α on R is defined by $\alpha: (a,b) \to |a-b|$ for $a, b \in R$. Show that (R, α) is commutative but not associative.

2.54. Suppose we define a binary operation $+$ on R by $a+b =$ the minimum of a and b $(a, b \in R)$. Show that $(R, +)$ is both associative and commutative.

2.55. Let $G = \{\alpha \mid \alpha: R \to R\}$. For $\alpha, \beta \in G$ define the mapping $\alpha * \beta = \alpha \circ \beta - \beta \circ \alpha$ where $\circ$ is the usual composition of mappings and $x(\alpha \circ \beta - \beta \circ \alpha) = x(\alpha \circ \beta) - x(\beta \circ \alpha)$ for all $x \in R$. Prove: (i) $(G, *)$ is a groupoid; (ii) $(G, *)$ is neither associative nor commutative; (iii) $(\alpha * \beta) * \alpha = \alpha * (\beta * \alpha)$ for all $\alpha, \beta \in G$; (iv) $\alpha * \beta = (-\beta) * \alpha$ where $-\beta$ is the element of G defined by $-\beta: x \to -(x\beta)$ for all $x \in R$. (Hard.)

2.56. Let G be the set in Problem 2.55. For $\alpha, \beta \in G$ define $\alpha \cdot \beta = \dfrac{\alpha \circ \beta + \beta \circ \alpha}{2}$ where $\circ$ is the usual composition of mappings and $x\dfrac{\alpha \circ \beta + \beta \circ \alpha}{2} = \dfrac{x(\alpha \circ \beta) + x(\beta \circ \alpha)}{2}$ for all $x \in R$. Prove: (i) $(G, \cdot)$ is a groupoid; (ii) $(G, \cdot)$ is commutative but not associative; (iii) $(\alpha \cdot \beta) \cdot \alpha = \alpha \cdot (\beta \cdot \alpha)$ for all $\alpha, \beta \in G$.

INVERSES IN GROUPOIDS

2.57. Let R^+ be the set of all non-negative real numbers. Define $a * b = \sqrt{a^2 + b^2}$ for all $a, b \in R^+$ ($\sqrt{a^2+b^2}$ is the positive square root). Find an identity in $(R^+, *)$. What elements have inverses?

2.58. For all $\alpha, \beta \in G = \{\alpha \mid \alpha: Z \to Z\}$, let $\alpha \times \beta$ be the mapping defined by $x(\alpha \times \beta) = x\alpha \cdot x\beta$ where $x \in Z$ and $\cdot$ is the usual multiplication of integers. Is $(G, \times)$ a groupoid? Does it have an identity? What elements have inverses?

2.59. Let $G = \{\alpha \mid \alpha: Z \to Q,\ Q \text{ the rational numbers}\}$. Define $\alpha \times \beta$ as in the preceding problem. Does $(G, \times)$ have an identity? What elements have inverses?

2.60. Which of the groupoids in Problems 2.57, 2.58 and 2.59 are commutative and which are associative?

2.61. Define the following binary operation $+$ in R^+, the non-negative real numbers: $a + b =$ the maximum of a and b, $a, b \in R^+$. Does $(R^+, +)$ have an identity? What elements have inverses?

2.62. Let $(G, *)$ be the groupoid of Problem 2.56. What is the identity of $(G, *)$? Find an infinite number of elements which have inverses.

2.63. Show that $(G, *)$, the groupoid of Problem 2.55 has no identity.

SEMIGROUPS WITH AN IDENTITY

2.64. Which elements of $(G, \cdot)$, where $G = \{1, -1, i, -i\}$, $i = \sqrt{-1}$, and $\cdot$ the usual multiplication of complex numbers, have inverses?

2.65. Let $G = \{f_1, f_2, f_3, f_4, f_5, f_6\}$ of Problem 2.49. Prove that G with the binary operation $\circ$ of composition of mappings, is a semigroup with an identity. Find the inverse of each element in $(G, \circ)$.

2.66. Let $G = \{(a, b, c, d) \mid a, b, c, d \in Z\}$. Define

$$(a, b, c, d) \cdot (a', b', c', d') = (aa' + cb',\ ba' + db',\ ac' + cd',\ bc' + dd')$$

Show that $(G, \cdot)$ is a semigroup with an identity. Show that the subsets $H = \{(1, 0, 0, 1), (-1, 0, 0, -1)\}$ and $F = \{(1, 0, c, 1) \mid c \in Z\}$, with the binary operation of $(G, \cdot)$ restricted to H and F respectively, are semigroups with an identity. Find the inverses of the elements of H and F.

2.67. Let $G = \{\alpha \mid \alpha \text{ a mapping of } \{1, 2, 3, 4, 5\} \text{ into } \{1, 2\}\}$. For $\alpha, \beta \in G$, let $\alpha \circ \beta$ be the usual composition of mappings. Is $(G, \circ)$ a semigroup with an identity?

2.68. For $\alpha, \beta \in G = \{\alpha \mid \alpha: \{1,2,3,4,5\} \to \{1,-1\}\}$, define the mapping $\alpha \times \beta$ by $n(\alpha \times \beta) = n\alpha \cdot n\beta$, where $n \in \{1,2,3,4,5\}$ and $\cdot$ the usual multiplication of integers. Is $(G, \times)$ a semigroup with an identity? If so, what elements have inverses?

HOMOMORPHISMS OF GROUPOIDS AND CAYLEY'S THEOREM

2.69. Let $H = \{(1,0,0,1), (-1,0,0,-1)\}$ be the groupoid defined in Problem 2.66. Show that H is isomorphic to the groupoid $(G, \cdot)$ where $G = \{1,-1\}$ and $\cdot$ is the usual multiplication of integers.

2.70. If $F = \{(1,0,c,1) \mid c \in Z\}$ is the groupoid defined in Problem 2.66 and $(Z, +)$ the groupoid of integers under addition, prove the two groupoids are isomorphic.

2.71. Which of the groupoids of Problem 2.49 are isomorphic?

2.72. Let $G = \{f_1, f_2, f_3, f_4, f_5, f_6\}$ be the groupoid of Problem 2.65, $(H, \cdot)$ the groupoid with $H = \{1,-1\}$ and $\cdot$ the usual multiplication of integers. Find all possible homomorphisms of G into H. Show that there is no homomorphism of $(G, \circ)$ into $(F, \circ)$ where $F = \{f_1, f_2, f_3\}$ and $\circ$ the usual composition of mappings. (Hard.)

2.73. Can the groupoid $(G, *)$ of Problem 2.55 be a homomorphic image of the groupoid $(M_R, \circ)$ where $\circ$ is the usual composition of mappings?

2.74. Suppose $G = \{\alpha \mid \alpha: Z \to Z\}$, $\times$ the binary operation defined in Problem 2.58 and $\circ$ the usual composition of mappings. Show that $\Psi: (G, \times) \to (G, \circ)$ defined by $\Psi: \alpha \to \alpha$ is not a homomorphism. (See Theorem 2.6, page 44.)

Chapter 3

Groups and Subgroups

Preview of Chapter 3

We define a *group* as a *semigroup with an identity in which every element has an inverse.* The object of this chapter is to show that the concept of a group is natural. This is done by providing illustrations of groups which arise in various branches of mathematics. The most important concepts of this chapter are group and subgroup.

3.1 GROUPS

Definition

As we remarked in the preview, a *semigroup with an identity in which every element has an inverse is termed a group.* We repeat the definition in more detail:

Definition: A non-empty set S together with a binary operation in S is called a group if

(i) there exists an identity element (usually denoted by) $1 \in S$; in other words,

$$a \cdot 1 = a = 1 \cdot a \quad \text{for all } a \in S$$

Recall that the identity is unique by Theorem 2.1, page 31.

(ii) for every choice of the elements $a, b, c \in S$,

$$(a \cdot b) \cdot c = a \cdot (b \cdot c)$$

Thus (i) and (ii) are the conditions for $(S, \cdot)$ to be a semigroup with identity.

(iii) every element $a \in S$ has an inverse in S, i.e. there is an element $b \in S$ such that

$$a \cdot b = 1 = b \cdot a$$

This element b is often denoted by a^{-1}. The inverse is unique by Theorem 2.2, page 33.

Whenever we define a group we shall follow the pattern:

I. Define a set S $(\neq \emptyset)$.

II. Define a binary operation in S.

III. Verify that the groupoid $(S, \cdot)$ contains an identity element.

IV. Verify that the groupoid $(S, \cdot)$ is a semigroup, i.e. is associative.

V. Verify that every element of S has an inverse.

The number of elements of S, $|S|$, is called the *order* of the group. (Compare Section 2.2, page 29.) We will exhibit groups of infinite and finite order. (See following examples and problems.)

Examples of groups of numbers

Example 1: *The additive group of integers.*

I. Let Z be the set of integers.

II. Let $+$ be the binary operation of addition in Z.

III. $n+0 = n = 0+n$ for every $n \in Z$. Thus $(Z,+)$ has an identity element.

IV. If l, m, n are integers,
$$(l+m)+n \;=\; l+(m+n)$$
i.e. $(Z,+)$ is a semigroup.

V. If $n \in Z$, then $-n$ in Z has the property
$$n+(-n) \;=\; 0 \;=\; (-n)+n$$
i.e. $-n$ is an inverse of n in $(Z,+)$.

Thus we have shown that the groupoid $(Z,+)$ is a group. This group is usually referred to as the *additive group of integers.*

Example 2: *The additive group of rationals.*

I. Let Q be the set of rational numbers.

II. Let $+$ be the binary operation of addition in Q.

III. $a+0 = a = 0+a$ for every $a \in Q$, so 0 is an identity element for $(Q,+)$.

IV. If $a, b, c \in Q$, then $(a+b)+c \;=\; a+(b+c)$.

V. If $a \in Q$, then $-a$ in Q has the property $a+(-a) \;=\; 0 \;=\; (-a)+a$.

Example 3: *The additive group of complex numbers.*

The description of this group is left to the reader.

Example 4: *The multiplicative group of nonzero rationals.*

I. Let Q^* be the set of nonzero rational numbers.

II. Let $\cdot$ be the binary operation of multiplication, i.e. the usual multiplication of rational numbers.

III. The rational number 1 is clearly an identity in the groupoid $(Q^*,\cdot)$.

IV. If $a, b, c \in Q^*$, then
$$(a\cdot b)\cdot c \;=\; a\cdot(b\cdot c)$$

V. If $z \in Q^*$, so is $1/a$ and
$$a\cdot\frac{1}{a} \;=\; 1 \;=\; \frac{1}{a}\cdot a$$
Thus every element of Q^* has an inverse.

Example 5: *The multiplicative group of nonzero complex numbers.*

This group is very similar to that in Example 4. We shall go through the usual five stages in setting up and describing the group.

I. Let C^* be the set of all nonzero complex numbers. Thus
$$C^* \;=\; \{x \mid x = a+ib \text{ where } x \neq 0+i0 \text{ and } a, b \in R\}$$
Recall that $i^2 = -1$.

II. We define multiplication of complex numbers as follows:
$$(a+ib)(c+id) \;=\; (ac-bd)+i(ad+bc)$$
This is a binary operation in C^* since $(ac-bd)+i(ad+bc)$ is a unique element in C^* (not both $ac-bd$ and $ad+bc$ can be zero).

III. $1+i\cdot 0 = 1 \in C^*$ and it is clearly an identity in $(C^*,\cdot)$.

IV. Suppose $a+ib,\ c+id,\ e+if \in C^*$. Then

$$[(a+ib)(c+id)](e+if) = [(ac-bd)+i(bc+ad)](e+if)$$
$$= [(ac-bd)e-(bc+ad)f] + i[(bc+ad)e+(ac-bd)f]$$

On the other hand,

$$(a+ib)[(c+id)(e+if)] = (a+ib)[(ce-df)+i(de+cf)]$$
$$= [a(ce-df)-b(de+cf)] + i[b(ce-df)+a(de+cf)]$$

It follows from these two computations that

$$(a+ib)[(c+id)(e+if)] = [(a+ib)(c+id)](e+if)$$

V. We have to check the existence of inverses. Thus suppose $a+ib \in C^*$; then not both a and b are zero. Hence $a^2+b^2 \neq 0$ and so

$$\frac{a}{a^2+b^2} - i\frac{b}{a^2+b^2} \in C^*$$

Moreover,

$$\left(\frac{a}{a^2+b^2} - i\frac{b}{a^2+b^2}\right)(a+ib) = 1 = (a+ib)\left(\frac{a}{a^2+b^2} - i\frac{b}{a^2+b^2}\right)$$

Thus we have proved $(C^*, \cdot)$ is a group and we term this group the multiplicative group of nonzero complex numbers.

Problems

3.1. Is $(S, \circ)$ a group if

(i) $S = Z$ and $\circ$ is the usual multiplication of integers?

(ii) $S = Q$ and $\circ$ is the usual multiplication in Q?

(iii) $S = \{q \mid q \in Q \text{ and } q > 0\}$ and $\cdot$ is the usual multiplication of rational numbers?

(iv) $S = \{z \mid z \in Z \text{ and } z = \sqrt{2}\}$ and $\circ$ is the usual multiplication in Z?

(v) $S = R$ and $\circ$ is the usual addition of real numbers?

(vi) $S = Z$ and $\circ$ is defined by $a \circ b = 0$ for all a, b in Z?

Solutions:

(i) The identity element is the integer 1. $(S, \circ)$ is not a group because $5 \in Z$ but there is no *integer* z in Z such that $z \circ 5 = 5 \circ z = 1$.

(ii) Again the identity is the number 1. There is no $q \in Q$ such that $q \circ 0 = 1$. Hence $(S, \circ)$ is not a group.

(iii) $(S, \cdot)$ is a group. Clearly $S \neq \emptyset$ and $\cdot$ is a binary operation on S. $q \cdot 1 = 1 \cdot q = q$ for all $q \in S$; hence 1 is an identity. Multiplication of rational numbers is associative and every element in S has an inverse; for if $q \in S$, then $\frac{1}{q} \in S$ and $\frac{1}{q} \cdot q = 1 = q \cdot \frac{1}{q}$.

(iv) $S = \emptyset$ since $\sqrt{2} \notin Z$. Therefore $(S, \circ)$ is not a group.

(v) $(S, \circ)$ is a group. $S \neq \emptyset$ and addition is an associative binary operation on S. $r + 0 = 0 + r = r$ and $r + (-r) = 0 = (-r) + r$ for all $r \in S$.

(vi) $(S, \circ)$ is not a group because there is no identity element in S.

3.2. Let S be the set of even integers. Show that S is a group under addition of integers.

Solution:

Let $a = 2a_1$ and $b = 2b_1$ be any two elements in S. $a + b = 2(a_1 + b_1)$ is a unique element in S; thus addition is a binary operation on S. Associativity of addition in S follows from the associativity of addition in Z. $0 = 2 \cdot 0$ is an identity element in S. If $a \in S$, then $-a \in S$ since $a = 2a_1$ implies $-a = 2(-a_1)$. Hence a has an inverse in S, as $a + (-a) = (-a) + a = 0$.

3.3. Let S be the set of real numbers of the form $a + b\sqrt{2}$ where $a, b \in Q$ and are not simultaneously zero. Show that S becomes a group under the usual multiplication of real numbers.

Solution:

$(a+b\sqrt{2})(c+d\sqrt{2}) = (ac+2bd)+(cb+ad)\sqrt{2}$. If neither $a+b\sqrt{2}$ nor $c+d\sqrt{2}$ is zero, i.e. $(0+0\sqrt{2})$, then their product cannot be zero. Hence the product of elements in S belongs to S. $1 = 1+0\sqrt{2}$ is an identity for S. Multiplying $\dfrac{1}{a+b\sqrt{2}}$ by $\dfrac{a-b\sqrt{2}}{a-b\sqrt{2}}$, we obtain

$$\frac{1}{a+b\sqrt{2}} = \frac{a-b\sqrt{2}}{a^2-2b^2} = \frac{a}{a^2-2b^2} - \frac{b}{(a^2-2b^2)}\sqrt{2}$$

Hence $\dfrac{1}{a+b\sqrt{2}} \in S$. The associativity holds because it is true for multiplication of real numbers.

3.4. Let S be the set of complex numbers of the form $a+b\sqrt{-5}$ where $a,b \in Q$ and are not both simultaneously zero. Show that S becomes a group under the usual multiplication of complex numbers.

Solution:

$(a+b\sqrt{-5})(c+d\sqrt{-5}) = (ac-5bd)+(bc+ad)\sqrt{-5}$ and cannot be zero if its factors are not zero; hence the product of two elements of S belongs to S. $1 = 1+0\sqrt{-5}$ is an identity for S. $\dfrac{1}{a+b\sqrt{-5}}$ is certainly an inverse for $a+b\sqrt{-5}$. Multiplying top and bottom by $a-b\sqrt{-5}$, we get

$$\frac{1}{a+b\sqrt{-5}} = \frac{a-b\sqrt{-5}}{a^2+5b^2} = \frac{a}{a^2+5b^2} + \frac{-b}{a^2+5b^2}\sqrt{-5}$$

and so $\dfrac{1}{a+b\sqrt{-5}} \in S$. The associativity of multiplication in S follows from associativity of multiplication of complex numbers.

3.5. Let m be any fixed positive integer and let $S = \{0,1,2,\ldots,m-1\}$. Define a binary operation in S by

$$a \circ b = a+b \quad \text{if} \quad a+b < m$$

$$a \circ b = r \quad \text{if} \quad a+b = m+r, \; 0 \leqq r < m$$

Prove that $(S,\circ)$ is a group of order m. (Hard.)

Solution:

If $a,b \in S$, then $a \circ b$ is uniquely defined and belongs to S. $a \circ 0 = 0 \circ a = a$, so 0 is an identity. Note that $a \circ b = a+b-\delta m$ where δ is 0 or 1, for any $a,b \in S$. So $b \circ c = b+c-\delta_1 m$ where δ_1 is 0 or 1. $a \circ (b \circ c) = a+(b \circ c)-\delta_2 m$ where δ_2 is 0 or 1. Then $a \circ (b \circ c) = a+b+c-(\delta_1+\delta_2)m$ where both δ_1 and δ_2 could be 0 or 1. Hence

$$a \circ (b \circ c) = a+b+c-\eta_1 m \quad \text{where} \quad \eta_1 \text{ is 0 or 1 or 2}$$

Similarly

$$(a \circ b) \circ c = a+b+c-\eta_2 m \quad \text{where} \quad \eta_2 \text{ is 0 or 1 or 2}$$

Now $0 \leqq a \circ (b \circ c) < m$ and $0 \leqq (a \circ b) \circ c < m$. Suppose $\eta_1 > \eta_2$; then

$$a \circ (b \circ c) \leqq a+b+c-(\eta_2+1)m = a+b+c-\eta_2 m - m$$

because η_1 is at least η_2+1. But $0 \leqq a+b+c-\eta_2 m < m$ and the above equation implies that $a \circ (b \circ c) < 0$; this contradicts $0 \leqq a \circ (b \circ c)$. Hence $\eta_1 \leqq \eta_2$. $\eta_2 > \eta_1$ leads in a similar way to a contradiction. Thus $\eta_1 = \eta_2$ and $a \circ (b \circ c) = (a \circ b) \circ c$. If $a \in S$, then $m-a \in S$ and $a \circ (m-a) = (m-a) \circ a = 0$; hence $m-a$ is an inverse to a. Thus S is a group.

3.6. Let $S \subseteq C$ (C the set of complex numbers) be the set of all mth roots of unity, where m is a fixed positive integer. Prove that under the usual multiplication of complex numbers, S becomes a group of order m.

Solution:

Recall that a complex number x is an mth root of unity if $x^m = 1$ and that there are exactly m distinct roots of unity, viz. $e^{i2\pi\tau/m}$, $\tau = 1,2,\ldots,m$; also, $e^{i\pi x} = \cos x + i \sin x$. If $a,b \in S$, then ab is uniquely defined. Since $(ab)^m = a^m b^m = 1$, ab is an mth root of unity and hence $ab \in S$. $1 \cdot a = a \cdot 1 = a$, so 1 is an identity. Associativity is true, since it is true for complex numbers in general. If $a \in S$, then $(1/a)^m = 1/a^m = 1$; thus $1/a \in S$ and $1/a$ is the inverse of a.

3.7. Let $S \subseteq C$ (C the set of complex numbers) be the set of *all* roots of unity. Describe one way of making S into a group.

Solution:

Use as binary operation the usual multiplication of complex numbers. If $a, b \in S$ and a is an mth root of unity, b an nth root of unity, then ab is an mnth root of unity because $(ab)^{mn} = a^{mn}b^{mn} = (a^m)^n(b^n)^m = 1^n 1^m = 1$. Hence $ab \in S$ and is, of course, uniquely defined. $1 \in S$ and acts as an identity. $1/a \in S$ and is the inverse of a. Associativity holds for multiplication of complex numbers. Thus S is a group with respect to the usual multiplication of complex numbers.

3.8. The following table defines a binary operation. Is the resultant groupoid a group?

	1	2
1	1	2
2	2	1

Solution:

We need to check only (*a*) associativity, (*b*) existence of identity and inverse.

(*a*) To check associativity, we have the following possible questions:

(*a*) Does $1 \cdot (1 \cdot 1)$ equal $(1 \cdot 1) \cdot 1$?
(*b*) Does $1 \cdot (1 \cdot 2)$ equal $(1 \cdot 1) \cdot 2$?
(*c*) Does $1 \cdot (2 \cdot 1)$ equal $(1 \cdot 2) \cdot 1$?
(*d*) Does $1 \cdot (2 \cdot 2)$ equal $(1 \cdot 2) \cdot 2$?
(*e*) Does $2 \cdot (1 \cdot 1)$ equal $(2 \cdot 1) \cdot 1$?
(*f*) Does $2 \cdot (1 \cdot 2)$ equal $(2 \cdot 1) \cdot 2$?
(*g*) Does $2 \cdot (2 \cdot 2)$ equal $(2 \cdot 2) \cdot 2$?
(*h*) Does $2 \cdot (2 \cdot 1)$ equal $(2 \cdot 2) \cdot 1$?

Checking all these products, we see that associativity holds.

(*b*) 1 acts as an identity. The inverse of 1 is 1, the inverse of 2 is 2. Hence the table defines a group.

3.9. Write the multiplication table for the group of Problem 3.5 with $m = 3$ and $m = 4$.

Solution:

If $m = 3$, the multiplication table is

	0	1	2
0	0	1	2
1	1	2	0
2	2	0	1

If $m = 4$, the multiplication table is

	0	1	2	3
0	0	1	2	3
1	1	2	3	0
2	2	3	0	1
3	3	0	1	2

3.2 SUBGROUPS

Definition

Let $(G, \cdot)$ be a group with binary operation $\cdot$ and let H be a non-empty subset of G. Then we say H is a *subgroup* of G if the operation $\cdot$ restricted to H is a binary operation in H which makes H into a group.

For example, if G is the group with $m = 4$ of Problem 3.9, then the subset $H = \{0, 2\}$ is a subgroup of G. For when the operation $\circ$ in G, as defined in the multiplication table for G, is restricted to H, it is a binary operation in H, i.e. $0 \circ 0 = 0 \in H$, $0 \circ 2 = 2 \in H$, $2 \circ 0 = 2 \in H$, and $2 \circ 2 = 0 \in H$. H is a group because: $H \neq \emptyset$; 0, the identity, is in H; the operation $\circ$ restricted to H is an associative binary operation (since the operation in G is associative); and every element in H has an inverse in H.

The following lemma facilitates proving a subset of a group is a subgroup.

Lemma 3.1: Let $(G, \cdot)$ be a group. Then a subset H of G is a subgroup of G iff

$$\text{(i)}\ \ H \neq \emptyset \quad \text{and} \quad \text{(ii)}\ \ \text{if } a, b \in H, \text{ then } ab^{-1} \in H$$

Proof: If H satisfies these conditions, then H is a group with respect to the binary operation. For if $H \neq \emptyset$, then there exists $a \in H$. Hence $aa^{-1} = 1 \in H$. Also, if $b \in H$ then $1b^{-1} = b^{-1} \in H$. Hence $a, b \in H$ implies $a(b^{-1})^{-1} = ab \in H$. Associativity is true in H, as it is true in G. Thus $\cdot$ is an associative binary operation on H, $1 \in H$, and the inverse of every element of H is an element of H. Therefore $(H, \cdot)$ is a subgroup.

Conversely if H is a group with respect to $\cdot$, then clearly H satisfies conditions (i) and (ii) above.

Problems

3.10. Let $(Q, +)$ be the additive group of rationals. Is Z a subgroup of Q? Is P a subgroup of Q?

Solution:

Clearly $Z \neq \emptyset$, and $Z \subseteq Q$. If $a, b \in (Q, +)$ then the binary operation is $+$, and the inverse of b is $-b$. So, if $a, b \in Z$, we ask in accordance with Lemma 3.1, whether $a + (-b) = a - b \in Z$. It is, and so Z is a subgroup of $(Q, +)$. Clearly $P \neq \emptyset$ and $P \subseteq Q$. If $a, b \in P$, is $a + (-b) = a - b \in P$? No, for P does not contain negative numbers; and if $a = 1$ and $b = 2$, then $a - b$ is negative.

3.11. Is Q a subgroup of $(C, +)$, the additive group of complex numbers?

Solution:

$Q \neq \emptyset$. $Q \subseteq C$; for if $a \in Q$, $a = a + 0i \in C$. If $a, b \in Q$, then $a + (-b) = a - b \in Q$. Hence Q is a subgroup of $(C, +)$.

3.12. Is $Z - \{0\}$ a subgroup of $(Q^*, \cdot)$, the multiplicative group of nonzero rational numbers?

Solution:

1 is the identity. $3 \in Z - \{0\}$, but 3 has no inverse in $Z - \{0\}$. Therefore $Z - \{0\}$ is not a subgroup of $(Q^*, \cdot)$.

3.13. Is Q^*, as above, a subgroup of $(C^*, \cdot)$, the multiplicative group of nonzero complex numbers?

Solution:

$Q^* \neq \emptyset$. $a, b \in Q^*$ implies ab^{-1} is a nonzero rational. Thus Q^* is a subgroup of $(C^*, \cdot)$.

3.14. Is Q^* a subgroup of the group of real numbers of the form $a + b\sqrt{2}$, $a, b \in Q$ and a, b not simultaneously zero, under multiplication?

Solution:

$Q^* \neq \emptyset$. If $a \in Q^*$, then $a = a + 0\sqrt{2} \in \{a + b\sqrt{2} \mid a, b \in Q \text{ and not both zero}\}$, since $a \neq 0$. Thus $Q^* \subseteq \{a + b\sqrt{2} \mid a, b \in Q \text{ and not both } 0\}$. $a, b \in Q^*$ implies ab^{-1} is a nonzero rational. Hence Q^* is a subgroup.

3.15. Prove that the intersection of two subgroups H and K of a group G is a subgroup.

Solution:

$1 \in H$, for as H is not empty, there is an element $h \in H$. But then H contains $hh^{-1} = 1$, the identity of G. Similarly, $1 \in K$. Hence $1 \in H \cap K$ and $H \cap K \neq \emptyset$. If $g, h \in H \cap K$, then $g, h \in H$ and $gh^{-1} \in H$. Also, $gh^{-1} \in K$. Thus $gh^{-1} \in H \cap K$ and $H \cap K$ is a subgroup of G.

3.16. By considering the group of Problem 3.5 with $m = 6$, show that the union of two subgroups is not necessarily a subgroup.

Solution:

That $\{0,3\}$ and $\{0,2,4\}$ are subgroups is easily verified. But $U = \{0,2,3,4\} = \{0,3\} \cup \{0,2,4\}$ is not a subgroup as $3 \circ 4 = 1 \notin U$. Therefore $\circ$ is not a binary operation in U.

3.3 THE SYMMETRIC AND ALTERNATING GROUPS

a. The symmetric group on X

Let X be any non-empty set. A very important group arises from the set S_X of all one-to-one mappings of X onto X, called the *symmetric group on X*. We describe this group in the usual five steps.

I. S_X is the set of all matchings of the non-empty set X with itself. Clearly, $S_X \subseteq M_X$ (see Section 2.4b, page 36).

II. If $\sigma, \tau \in S_X$, then we define $\sigma\tau$ to be the composition of the mappings σ and τ. Here we must verify that $\sigma\tau$ is a matching of X with itself. Suppose $x \in X$; then as τ is onto, we can find $x'' \in X$ such that $x''\tau = x$. But σ is also onto, so we can find $x' \in X$ such that $x'\sigma = x''$. Consequently

$$x'(\sigma\tau) = (x'\sigma)\tau = x''\tau = x$$

and hence $\sigma\tau$ is onto. If $x(\sigma\tau) = y(\sigma\tau)$, then $(x\sigma)\tau = (y\sigma)\tau$ by the definition of the composition of mappings; this means, since τ is one-to-one, that $x\sigma = y\sigma$. This in turn implies, since σ is one-to-one, that $x = y$. Therefore $\sigma\tau$ is also one-to-one. Thus composition of mappings is a binary operation in S_X.

III. Clearly the identity mapping $\iota: x \to x$ is in S_X and is an identity element of S_X.

IV. The groupoid $(S_X, \cdot)$ is a semigroup, since composition of mappings is associative. (Theorem 2.3, page 36.)

V. Let $\sigma \in S_X$. Since σ is one-to-one and onto, Theorem 2.4, page 36, implies σ has an inverse, τ, in M_X. Now, $\sigma\tau = \iota = \tau\sigma$ means σ is an inverse of τ. By Theorem 2.4 the only elements in M_X which have inverses are those which are one-to-one onto mappings. Therefore $\tau \in S_X$ and τ is the required inverse of σ. The proof that S_X is a group is complete.

We will call an element of S_X a *permutation of X*, or, for short, a *permutation*.

In the particular case where $X = \{1, 2, \ldots, n\}$, we write $S_X = S_n$. S_n is called the *symmetric group of degree n*.

$|S_n|$ is calculated as follows. If $\sigma \in S_n$, then 1σ can be one of n elements. 2σ can be one of $n-1$ elements, as 1σ has been chosen and σ must be one-to-one; so $2\sigma \neq 1\sigma$. 3σ can be one of $n-2$ elements, as 1σ and 2σ have been chosen and σ must be one-to-one; so 3σ is not equal to 2σ or 1σ. Continuing in this way, we conclude that there are

$$n \cdot (n-1)(n-2) \cdots \cdot 2 \cdot 1 = n!$$

elements of S_n, i.e. $|S_n| = n!$

The elements of S_3, for example, are

$$\iota = \begin{pmatrix} 1 & 2 & 3 \\ 1 & 2 & 3 \end{pmatrix} \qquad \sigma_2 = \begin{pmatrix} 1 & 2 & 3 \\ 3 & 1 & 2 \end{pmatrix} \qquad \tau_2 = \begin{pmatrix} 1 & 2 & 3 \\ 3 & 2 & 1 \end{pmatrix}$$

$$\sigma_1 = \begin{pmatrix} 1 & 2 & 3 \\ 2 & 3 & 1 \end{pmatrix} \qquad \tau_1 = \begin{pmatrix} 1 & 2 & 3 \\ 1 & 3 & 2 \end{pmatrix} \qquad \tau_3 = \begin{pmatrix} 1 & 2 & 3 \\ 2 & 1 & 3 \end{pmatrix}$$

Here we are using the notation of Section 2.4c, page 37. To find the multiplication table, we must compute the products. As an example, we calculate $\sigma_1\tau_1$.

$$\sigma_1\tau_1 = \begin{pmatrix} 1 & 2 & 3 \\ 1\sigma_1\tau_1 & 2\sigma_1\tau_1 & 3\sigma_1\tau_1 \end{pmatrix} = \begin{pmatrix} 1 & 2 & 3 \\ 2\tau_1 & 3\tau_1 & 1\tau_1 \end{pmatrix} = \begin{pmatrix} 1 & 2 & 3 \\ 3 & 2 & 1 \end{pmatrix} = \tau_2$$

To do this calculation mentally, we think as follows:

$$1 \to 2 \text{ (in } \sigma_1), \quad 2 \to 3 \text{ (in } \tau_1)$$

Write down
$$\begin{pmatrix} 1 & 2 & 3 \\ 3 & & \end{pmatrix}$$

$2 \to 3$ (in σ_1), $3 \to 2$ (in τ_1). Put 2 beneath 2 to get

$$\begin{pmatrix} 1 & 2 & 3 \\ 3 & 2 & \end{pmatrix}$$

Only one other number can appear, namely 1. Thus

$$\sigma_1\tau_1 = \begin{pmatrix} 1 & 2 & 3 \\ 3 & 2 & 1 \end{pmatrix} = \tau_2$$

The multiplication table for S_3 is

	ι	σ_1	σ_2	τ_1	τ_2	τ_3
ι	ι	σ_1	σ_2	τ_1	τ_2	τ_3
σ_1	σ_1	σ_2	ι	τ_2	τ_3	τ_1
σ_2	σ_2	ι	σ_1	τ_3	τ_1	τ_2
τ_1	τ_1	τ_3	τ_2	ι	σ_2	σ_1
τ_2	τ_2	τ_1	τ_3	σ_1	ι	σ_2
τ_3	τ_3	τ_2	τ_1	σ_2	σ_1	ι

The reader should check some of the entries. Note that $\sigma_1\tau_1 = \tau_2$ and $\tau_1\sigma_1 = \tau_3$, so that $\sigma_1\tau_1 \neq \tau_1\sigma_1$. Hence S_3 is not commutative.

Problems

3.17. Calculate $\alpha\beta$, $\beta\alpha$, α^{-1}, β^{-1}, $(\alpha\beta)^{-1}$, and $(\beta\alpha)^{-1}$ if

$$\alpha = \begin{pmatrix} 1 & 2 & 3 & 4 & 5 & 6 \\ 2 & 3 & 6 & 5 & 4 & 1 \end{pmatrix} \quad \text{and} \quad \beta = \begin{pmatrix} 1 & 2 & 3 & 4 & 5 & 6 \\ 1 & 3 & 5 & 6 & 2 & 4 \end{pmatrix}$$

Solution:
$$\alpha\beta = \begin{pmatrix} 1 & 2 & 3 & 4 & 5 & 6 \\ 3 & 5 & 4 & 2 & 6 & 1 \end{pmatrix}, \quad \beta\alpha = \begin{pmatrix} 1 & 2 & 3 & 4 & 5 & 6 \\ 2 & 6 & 4 & 1 & 3 & 5 \end{pmatrix}$$

To find α^{-1}, we note that $x(\alpha\alpha^{-1}) = x$ and hence α^{-1} must carry $x\alpha$ to x. Now we determine which x is taken onto 1. $6\alpha = 1$, so we must have $1\alpha^{-1} = 6$. Next, since $1\alpha = 2$, $2\alpha^{-1} = 1$. Proceeding in this way, we obtain

$$\alpha^{-1} = \begin{pmatrix} 1 & 2 & 3 & 4 & 5 & 6 \\ 6 & 1 & 2 & 5 & 4 & 3 \end{pmatrix}$$

An easy method of calculating the answer mechanically follows. Take

$$\alpha = \begin{pmatrix} 1 & 2 & 3 & 4 & 5 & 6 \\ 2 & 3 & 6 & 5 & 4 & 1 \end{pmatrix}$$

Interchange the rows,

$$\begin{pmatrix} 2 & 3 & 6 & 5 & 4 & 1 \\ 1 & 2 & 3 & 4 & 5 & 6 \end{pmatrix}$$

Rearrange the columns so that the top row reads 1 2 3 4 5 6,

$$\alpha^{-1} = \begin{pmatrix} 1 & 2 & 3 & 4 & 5 & 6 \\ 6 & 1 & 2 & 5 & 4 & 3 \end{pmatrix}$$

This method is conceptually the same as the first. To find β^{-1}, interchange the rows to obtain

$$\begin{pmatrix} 1 & 3 & 5 & 6 & 2 & 4 \\ 1 & 2 & 3 & 4 & 5 & 6 \end{pmatrix}$$

and rearrange to get

$$\beta^{-1} = \begin{pmatrix} 1 & 2 & 3 & 4 & 5 & 6 \\ 1 & 5 & 2 & 6 & 3 & 4 \end{pmatrix}. \quad (\alpha\beta)^{-1} = \begin{pmatrix} 1 & 2 & 3 & 4 & 5 & 6 \\ 6 & 4 & 1 & 3 & 2 & 5 \end{pmatrix}, \quad (\beta\alpha)^{-1} = \begin{pmatrix} 1 & 2 & 3 & 4 & 5 & 6 \\ 4 & 1 & 5 & 3 & 6 & 2 \end{pmatrix}$$

3.18. Verify that $\alpha(\beta\gamma) = (\alpha\beta)\gamma$ where

$$\alpha = \begin{pmatrix} 1 & 2 & 3 & 4 & 5 \\ 3 & 1 & 2 & 4 & 5 \end{pmatrix} \quad \beta = \begin{pmatrix} 1 & 2 & 3 & 4 & 5 \\ 3 & 2 & 1 & 5 & 4 \end{pmatrix} \quad \gamma = \begin{pmatrix} 1 & 2 & 3 & 4 & 5 \\ 4 & 3 & 1 & 5 & 2 \end{pmatrix}$$

Solution:

$$(\alpha\beta)\gamma = \begin{pmatrix} 1 & 2 & 3 & 4 & 5 \\ 1 & 3 & 2 & 5 & 4 \end{pmatrix}\begin{pmatrix} 1 & 2 & 3 & 4 & 5 \\ 4 & 3 & 1 & 5 & 2 \end{pmatrix} = \begin{pmatrix} 1 & 2 & 3 & 4 & 5 \\ 4 & 1 & 3 & 2 & 5 \end{pmatrix}$$

$$\alpha(\beta\gamma) = \begin{pmatrix} 1 & 2 & 3 & 4 & 5 \\ 3 & 1 & 2 & 4 & 5 \end{pmatrix}\begin{pmatrix} 1 & 2 & 3 & 4 & 5 \\ 1 & 3 & 4 & 2 & 5 \end{pmatrix} = \begin{pmatrix} 1 & 2 & 3 & 4 & 5 \\ 4 & 1 & 3 & 2 & 5 \end{pmatrix}$$

3.19. Is the subset $R = \{\iota, \sigma_1, \sigma_2\}$ a subgroup of S_3? (For this notation see above.)

Solution:

$R \neq \emptyset$. From the multiplication table of S_3, page 57, the product of any two elements in R is again in R. Since $\sigma_1^{-1} = \sigma_2$ and $\sigma_2^{-1} = \sigma_1$, it follows that $xy^{-1} \in R$ for any $x, y \in R$. Hence R is a subgroup of S_3.

3.20. Find all elements of S_1 and S_2 and exhibit multiplication tables for these groups.

Solution:

S_1 contains one element $\iota = \begin{pmatrix} 1 \\ 1 \end{pmatrix}$.

	ι
ι	ι

is a multiplication table for S_1. There are two elements in S_2: $\iota = \begin{pmatrix} 1 & 2 \\ 1 & 2 \end{pmatrix}$ and $\beta = \begin{pmatrix} 1 & 2 \\ 2 & 1 \end{pmatrix}$. The multiplication table for S_2 is

	ι	β
ι	ι	β
β	β	ι

3.21. Find the elements of S_4.

Solution:

$$\iota = \begin{pmatrix} 1 & 2 & 3 & 4 \\ 1 & 2 & 3 & 4 \end{pmatrix} \qquad \sigma_6 = \begin{pmatrix} 1 & 2 & 3 & 4 \\ 3 & 1 & 4 & 2 \end{pmatrix} \qquad \tau_3 = \begin{pmatrix} 1 & 2 & 3 & 4 \\ 3 & 2 & 4 & 1 \end{pmatrix} \qquad \alpha_1 = \begin{pmatrix} 1 & 2 & 3 & 4 \\ 2 & 1 & 3 & 4 \end{pmatrix}$$

$$\sigma_1 = \begin{pmatrix} 1 & 2 & 3 & 4 \\ 2 & 3 & 4 & 1 \end{pmatrix} \qquad \sigma_7 = \begin{pmatrix} 1 & 2 & 3 & 4 \\ 3 & 4 & 2 & 1 \end{pmatrix} \qquad \tau_4 = \begin{pmatrix} 1 & 2 & 3 & 4 \\ 4 & 2 & 1 & 3 \end{pmatrix} \qquad \alpha_2 = \begin{pmatrix} 1 & 2 & 3 & 4 \\ 3 & 2 & 1 & 4 \end{pmatrix}$$

$$\sigma_2 = \begin{pmatrix} 1 & 2 & 3 & 4 \\ 3 & 4 & 1 & 2 \end{pmatrix} \qquad \sigma_8 = \begin{pmatrix} 1 & 2 & 3 & 4 \\ 2 & 1 & 4 & 3 \end{pmatrix} \qquad \tau_5 = \begin{pmatrix} 1 & 2 & 3 & 4 \\ 2 & 4 & 3 & 1 \end{pmatrix} \qquad \alpha_3 = \begin{pmatrix} 1 & 2 & 3 & 4 \\ 4 & 2 & 3 & 1 \end{pmatrix}$$

$$\sigma_3 = \begin{pmatrix} 1 & 2 & 3 & 4 \\ 4 & 1 & 2 & 3 \end{pmatrix} \qquad \sigma_9 = \begin{pmatrix} 1 & 2 & 3 & 4 \\ 4 & 3 & 1 & 2 \end{pmatrix} \qquad \tau_6 = \begin{pmatrix} 1 & 2 & 3 & 4 \\ 4 & 1 & 3 & 2 \end{pmatrix} \qquad \alpha_4 = \begin{pmatrix} 1 & 2 & 3 & 4 \\ 1 & 3 & 2 & 4 \end{pmatrix}$$

$$\sigma_4 = \begin{pmatrix} 1 & 2 & 3 & 4 \\ 2 & 4 & 1 & 3 \end{pmatrix} \qquad \tau_1 = \begin{pmatrix} 1 & 2 & 3 & 4 \\ 1 & 3 & 4 & 2 \end{pmatrix} \qquad \tau_7 = \begin{pmatrix} 1 & 2 & 3 & 4 \\ 2 & 3 & 1 & 4 \end{pmatrix} \qquad \alpha_5 = \begin{pmatrix} 1 & 2 & 3 & 4 \\ 1 & 4 & 3 & 2 \end{pmatrix}$$

$$\sigma_5 = \begin{pmatrix} 1 & 2 & 3 & 4 \\ 4 & 3 & 2 & 1 \end{pmatrix} \qquad \tau_2 = \begin{pmatrix} 1 & 2 & 3 & 4 \\ 1 & 4 & 2 & 3 \end{pmatrix} \qquad \tau_8 = \begin{pmatrix} 1 & 2 & 3 & 4 \\ 3 & 1 & 2 & 4 \end{pmatrix} \qquad \alpha_6 = \begin{pmatrix} 1 & 2 & 3 & 4 \\ 1 & 2 & 4 & 3 \end{pmatrix}$$

We give the multiplication table for future reference.

THE SYMMETRIC GROUP OF DEGREE 4

	ι	σ_1	σ_2	σ_3	σ_4	σ_5	σ_6	σ_7	σ_8	σ_9	τ_1	τ_2	τ_3	τ_4	τ_5	τ_6	τ_7	τ_8	α_1	α_2	α_3	α_4	α_5	α_6
ι	ι	σ_1	σ_2	σ_3	σ_4	σ_5	σ_6	σ_7	σ_8	σ_9	τ_1	τ_2	τ_3	τ_4	τ_5	τ_6	τ_7	τ_8	α_1	α_2	α_3	α_4	α_5	α_6
σ_1	σ_1	σ_2	σ_3	ι	τ_6	α_2	τ_2	τ_4	α_5	τ_8	σ_7	α_3	σ_4	α_1	σ_9	α_4	σ_6	α_6	τ_1	σ_8	τ_7	τ_3	σ_5	τ_5
σ_2	σ_2	σ_3	ι	σ_1	α_4	σ_8	α_3	α_1	σ_5	α_6	τ_4	τ_7	τ_6	τ_1	τ_8	τ_3	τ_2	τ_5	σ_7	α_5	σ_6	σ_4	α_2	σ_9
σ_3	σ_3	ι	σ_1	σ_2	τ_3	α_5	τ_7	τ_1	α_2	τ_5	α_1	σ_6	α_4	σ_7	α_6	σ_4	α_3	σ_9	τ_4	σ_5	τ_2	τ_6	σ_8	τ_8
σ_4	σ_4	τ_8	α_3	τ_1	σ_5	σ_6	ι	τ_6	α_4	τ_3	α_2	σ_9	α_1	σ_1	σ_3	α_6	σ_7	α_5	τ_2	τ_5	σ_8	σ_2	τ_4	τ_7
σ_5	σ_5	α_5	σ_8	α_2	σ_6	ι	σ_4	α_6	σ_2	α_1	τ_5	τ_3	τ_2	τ_8	τ_1	τ_7	τ_6	τ_4	σ_9	σ_3	α_4	α_3	σ_1	σ_7
σ_6	σ_6	τ_4	α_4	τ_5	ι	σ_4	σ_5	τ_7	α_3	τ_2	σ_3	α_1	σ_9	α_5	α_2	σ_7	α_6	σ_1	τ_3	τ_1	σ_2	σ_8	τ_8	τ_6
σ_7	σ_7	τ_6	α_6	τ_7	τ_1	α_1	τ_4	σ_8	σ_9	ι	α_3	σ_1	σ_3	α_4	σ_6	α_2	α_5	σ_4	σ_2	τ_2	τ_8	τ_5	τ_3	σ_5
σ_8	σ_8	α_2	σ_5	α_5	α_3	σ_2	α_4	σ_9	ι	σ_7	τ_8	τ_6	τ_7	τ_5	τ_4	τ_2	τ_3	τ_1	α_6	σ_1	σ_4	σ_6	σ_3	α_1
σ_9	σ_9	τ_2	α_1	τ_3	τ_8	α_6	τ_5	ι	σ_7	σ_8	σ_4	α_2	α_5	σ_6	α_4	σ_1	σ_3	α_3	σ_5	τ_6	τ_1	τ_4	τ_7	σ_2
τ_1	τ_1	σ_4	τ_8	α_3	α_1	τ_4	σ_7	α_2	τ_5	σ_3	τ_2	ι	σ_2	τ_6	τ_7	σ_5	σ_8	τ_3	σ_1	σ_6	σ_9	α_6	α_4	α_5
τ_2	τ_2	α_1	τ_3	σ_9	σ_1	τ_6	α_2	σ_6	τ_7	α_3	ι	τ_1	τ_8	σ_5	σ_8	τ_4	τ_5	σ_2	σ_4	σ_7	σ_3	α_5	α_6	α_4
τ_3	τ_3	σ_9	τ_2	α_1	α_5	τ_7	σ_3	σ_4	τ_6	α_4	σ_5	τ_5	τ_4	ι	σ_2	τ_8	τ_1	σ_8	σ_6	α_6	α_2	σ_1	σ_7	α_3
τ_4	τ_4	α_4	τ_5	σ_6	σ_7	τ_1	α_1	α_5	τ_8	σ_1	τ_7	σ_2	ι	τ_3	τ_2	σ_8	σ_5	τ_6	σ_3	α_3	α_6	σ_9	σ_4	α_2
τ_5	τ_5	σ_6	τ_4	α_4	σ_9	τ_8	α_6	σ_3	τ_1	α_2	τ_3	σ_5	σ_8	τ_7	τ_6	ι	σ_2	τ_2	α_5	σ_4	α_1	σ_7	α_3	σ_1
τ_6	τ_6	α_6	τ_7	σ_7	α_2	τ_2	σ_1	α_4	τ_3	σ_4	σ_8	τ_8	τ_1	σ_2	ι	τ_5	τ_4	σ_5	α_3	σ_9	α_5	σ_3	α_1	σ_6
τ_7	τ_7	σ_7	τ_6	α_6	σ_3	τ_3	α_5	α_3	τ_2	σ_6	σ_2	τ_4	τ_5	σ_8	σ_5	τ_1	τ_8	ι	α_4	α_1	σ_1	α_2	σ_9	σ_4
τ_8	τ_8	α_3	τ_1	σ_4	α_6	τ_5	σ_9	σ_1	τ_4	α_5	τ_6	σ_8	σ_5	τ_2	τ_3	σ_2	ι	τ_7	α_2	α_4	σ_7	α_1	σ_6	σ_3
α_1	α_1	τ_3	σ_9	τ_2	τ_4	σ_7	τ_1	σ_5	α_6	σ_2	σ_6	σ_3	σ_1	σ_4	α_3	α_5	α_2	α_4	ι	τ_7	τ_5	τ_8	τ_6	σ_8
α_2	α_2	σ_5	α_5	σ_8	τ_2	σ_1	τ_6	τ_5	σ_3	τ_1	σ_9	σ_4	α_3	α_6	σ_7	σ_6	α_4	α_1	τ_8	ι	τ_3	τ_7	σ_2	τ_4
α_3	α_3	τ_1	σ_4	τ_8	σ_2	α_4	σ_8	τ_2	σ_6	τ_7	σ_1	σ_7	α_6	α_2	α_5	α_1	σ_9	σ_3	τ_6	τ_4	ι	σ_5	τ_5	τ_3
α_4	α_4	τ_5	σ_6	τ_4	σ_8	α_3	σ_2	τ_3	σ_4	τ_6	α_5	α_6	σ_7	σ_3	σ_1	σ_9	α_1	α_2	τ_7	τ_8	σ_5	ι	τ_1	τ_2
α_5	α_5	σ_8	α_2	σ_5	τ_7	σ_3	τ_3	τ_8	σ_1	τ_4	α_6	α_4	σ_6	σ_9	α_1	α_3	σ_4	σ_7	τ_5	σ_2	τ_6	τ_2	ι	τ_1
α_6	α_6	τ_7	σ_7	τ_6	τ_5	σ_9	τ_8	σ_2	α_1	σ_5	α_4	α_5	α_2	α_3	σ_4	σ_3	σ_1	σ_6	σ_8	τ_3	τ_4	τ_1	τ_2	ι

b. Even and odd permutations

We are interested in a special subgroup of S_n, the *alternating group of degree n*, usually denoted by A_n. This subgroup A_n is obtained from S_n by singling out certain elements. To begin with, consider S_3. Let $\sigma = \begin{pmatrix} 1 & 2 & 3 \\ 2 & 3 & 1 \end{pmatrix}$. Then

$$\frac{2\sigma - 1\sigma}{2-1}\cdot\frac{3\sigma - 1\sigma}{3-1}\cdot\frac{3\sigma - 2\sigma}{3-2} = \frac{3-2}{2-1}\cdot\frac{1-2}{3-1}\cdot\frac{1-3}{3-2} = 1$$

If $\tau = \begin{pmatrix} 1 & 2 & 3 \\ 3 & 2 & 1 \end{pmatrix}$, then

$$\frac{2\tau - 1\tau}{2-1}\cdot\frac{3\tau - 1\tau}{3-1}\cdot\frac{3\tau - 2\tau}{3-2} = \frac{2-3}{2-1}\cdot\frac{1-3}{3-1}\cdot\frac{1-2}{3-2} = -1$$

We say σ is even and τ is odd.

More generally, let us call $\sigma \in S_n$ *even* (or an *even permutation*) if

$$\frac{2\sigma - 1\sigma}{2-1}\cdot\frac{3\sigma - 1\sigma}{3-1}\cdot\frac{3\sigma - 2\sigma}{3-2}\cdot\,\cdots\,\cdot\frac{n\sigma - 1\sigma}{n-1}\cdot\frac{n\sigma - 2\sigma}{n-2}\cdot\,\cdots\,\cdot\frac{n\sigma - (n-1)\sigma}{n-(n-1)} = 1$$

On the other hand, we call $\sigma \in S_n$ *odd* (or an *odd permutation*) if

$$\frac{2\sigma - 1\sigma}{2-1}\cdot\frac{3\sigma - 1\sigma}{3-1}\cdot\frac{3\sigma - 2\sigma}{3-2}\cdot\,\cdots\,\cdot\frac{n\sigma - 1\sigma}{n-1}\cdot\frac{n\sigma - 2\sigma}{n-2}\cdot\,\cdots\,\cdot\frac{n\sigma - (n-1)\sigma}{n-(n-1)} = -1$$

The definition of even or odd is written more briefly as

$$\sigma \text{ is even if } \prod_{i<k}\frac{k\sigma - i\sigma}{k-i} = 1$$

$$\sigma \text{ is odd if } \prod_{i<k}\frac{k\sigma - i\sigma}{k-i} = -1$$

We shall show that an element in S_n is either even or odd, i.e. $\prod_{i<k}\frac{k\sigma - i\sigma}{k-i} = \pm 1$.

Corresponding to each factor $k-i$ in the denominator, we will find a factor in the numerator which is either $k-i$ or $-(k-i)$. As σ is a permutation, there exist unique integers l, m such that $l\sigma = k$, $m\sigma = i$. If $l > m$, the factor $l\sigma - m\sigma = k - i$ appears in the numerator. If $l < m$, the factor $m\sigma - l\sigma = -(k-i)$ appears in the numerator. The quotient $\frac{l\sigma - m\sigma}{k-i}$ or $\frac{m\sigma - l\sigma}{k-i}$ is thus ± 1. Note that distinct factors $k-i$, $k'-i'$ in the denominator give rise to distinct associated factors $\pm(l\sigma - m\sigma)$, $\pm(l'\sigma - m'\sigma)$ in the numerator. For if $l = l'$, $m = m'$, we have $k = k'$ and $i = i'$.

Thus regrouping the factors in the numerator, the product becomes the product of factors ± 1 and hence is itself ± 1. Therefore every permutation of S_n is either even or odd.

There is an easy way of determining whether a permutation σ is even or odd. If we are given a row of integers, we call the number of integers in the row smaller than the first integer, the number of inversions. Thus for example the number of inversions in the row $7, 4, 3, 2, 1, 6, 8$ is 5. We will use this concept of inversion to find out if a given permutation

$$\sigma = \begin{pmatrix} 1 & 2 & \ldots & n \\ 1\sigma & 2\sigma & \ldots & n\sigma \end{pmatrix}$$

is even or odd.

To do this we must calculate $\prod_{i<k}\left(\frac{k\sigma - i\sigma}{k-i}\right)$. Much of the calculation is redundant as we have proved that the result is always 1 or -1. We must only determine the sign. In the denominator we always have positive numbers. In the numerator a negative number $k\sigma - i\sigma$ arises if $i\sigma > k\sigma$. For fixed i and varying k, the number of negative factors that arise is the number of $k > i$ for which $i\sigma > k\sigma$. But this is the number of inversions in the row $i\sigma, (i+1)\sigma, \ldots, n\sigma$. The total number of negative factors is the number of inversions in $1\sigma, 2\sigma, \ldots, n\sigma$ + the number of inversions in $2\sigma, 3\sigma, \ldots, n\sigma$ + the number of inversions in $(n-1)\sigma, n\sigma$. Let this total be I. The product of I negative numbers is positive if I is even, and negative if I is odd. So σ is even or odd according as I is even or odd.

Example 6: Is $\sigma = \begin{pmatrix} 1 & 2 & 3 \\ 3 & 1 & 2 \end{pmatrix}$ even or odd?

The number of inversions in 3, 1, 2 is 2. The number of inversions in 1, 2 is 0. Thus the total number of inversions is 2, and hence σ is even.

Problem

3.22. Is σ even or odd?

(i) $\sigma = \begin{pmatrix} 1 & 2 & 3 & 4 \\ 2 & 1 & 4 & 3 \end{pmatrix}$

(ii) $\sigma = \begin{pmatrix} 1 & 2 & 3 & 4 & 5 \\ 5 & 3 & 2 & 4 & 1 \end{pmatrix}$

(iii) $\sigma = \begin{pmatrix} 1 & 2 & 3 & 4 & 5 & 6 \\ 3 & 5 & 2 & 1 & 6 & 4 \end{pmatrix}$

(iv) $\sigma = \begin{pmatrix} 1 & 2 & 3 & 4 & 5 & 6 & 7 \\ 4 & 3 & 1 & 2 & 6 & 7 & 5 \end{pmatrix}$

(v) $\sigma = \begin{pmatrix} 1 & 2 & 3 & 4 & 5 & 6 \\ 3 & 1 & 4 & 6 & 2 & 5 \end{pmatrix}\begin{pmatrix} 1 & 2 & 3 & 4 & 5 & 6 \\ 4 & 1 & 6 & 3 & 2 & 5 \end{pmatrix}$

Solution:

(i) Number of inversions in

$$\begin{aligned} 2\,1\,4\,3 &= 1 \\ 1\,4\,3 &= 0 \\ 4\,3 &= 1 \\ \hline & \;\;\, 2 \end{aligned}$$

Total number of inversions 2

Hence σ is even.

(ii) Number of inversions in

$$\begin{aligned} 5\,3\,2\,4\,1 &= 4 \\ 3\,2\,4\,1 &= 2 \\ 2\,4\,1 &= 1 \\ 4\,1 &= 1 \\ \hline & \;\;\, 8 \end{aligned}$$

Total number of inversions 8

Hence σ is even.

(iii) Odd. (iv) Odd. (v) Even.

c. The alternating groups

We shall show that the set of even permutations forms a subgroup of S_n.

1. Let A_n be the set of even permutations in S_n. Then $A_n \neq \emptyset$, since the identity permutation $\iota \in A_n$:

$$\prod_{i<k}\frac{k\iota - i\iota}{k-i} = \prod_{i<k}\frac{k-i}{k-i} = 1$$

2. If σ and $\tau \in S_n$, then

$$\begin{aligned} \prod_{i<k}\frac{k(\sigma\tau) - i(\sigma\tau)}{k-i} &= \prod_{i<k}\frac{k(\sigma\tau) - i(\sigma\tau)}{k-i}\cdot\frac{k\sigma - i\sigma}{k\sigma - i\sigma} \\ &= \prod_{i<k}\frac{(k\sigma)\tau - (i\sigma)\tau}{k\sigma - i\sigma}\cdot\prod_{i<k}\frac{k\sigma - i\sigma}{k-i} \end{aligned} \qquad (3.1)$$

We will prove that

$$\prod_{i<k} \frac{(k\sigma)\tau - (i\sigma)\tau}{k\sigma - i\sigma} = \prod_{m<l} \frac{l\tau - m\tau}{l - m} \qquad (3.2)$$

To do this we will show that each of the factors $\frac{l\tau - m\tau}{l-m}$ corresponds to one and only one of the factors $\frac{(k\sigma)\tau - (i\sigma)\tau}{k\sigma - i\sigma}$. Corresponding to each of the factors $\frac{l\tau - m\tau}{l-m}$ in the right-hand side of (*3.2*), there exists, since σ is a permutation, unique integers p, q such that $p\sigma = l$ and $q\sigma = m$. If $p > q$, then a factor

$$\frac{(p\sigma)\tau - (q\sigma)\tau}{p\sigma - q\sigma} = \frac{l\tau - m\tau}{l - m}$$

appears on the left-hand side of (*3.2*). If $p < q$, then a factor

$$\frac{(q\sigma)\tau - (p\sigma)\tau}{q\sigma - p\sigma} = \frac{(p\sigma)\tau - (q\sigma)\tau}{p\sigma - q\sigma} = \frac{l\tau - m\tau}{l - m}$$

appears on the left-hand side of (*3.2*). We associate with the factor $\frac{l\tau - m\tau}{l-m}$ the (equal) factor $\frac{(p\sigma)\tau - (q\sigma)\tau}{p\sigma - q\sigma}$ if $p > q$, and the (equal) factor $\frac{(q\sigma)\tau - (p\sigma)\tau}{q\sigma - p\sigma}$ if $p < q$. It is easy to see that each of the factors of the right-hand side of (*3.2*) corresponds in this way to one and only one of the (equal) factors of the left-hand side of (*3.2*). Hence (*3.2*) holds. From (*3.1*) we therefore obtain

$$\prod_{i<k} \frac{k(\sigma\tau) - i(\sigma\tau)}{k - i} = \prod_{m<l} \frac{l\tau - m\tau}{l - m} \cdot \prod_{i<k} \frac{k\sigma - i\sigma}{k - i} \qquad (3.3)$$

It follows from (*3.3*) and the rules for multiplying +1 and −1, that we have

Lemma 3.2: The product of

(i) two even permutations is even

(ii) two odd permutations is even

(iii) an odd permutation and an even permutation is odd

(iv) an even permutation and an odd permutation is odd.

Accordingly if σ is even, then σ^{-1} is even too, since $\sigma\sigma^{-1} = \iota$ is even. Thus if $\sigma, \tau \in A_n$, $\sigma\tau^{-1} \in A_n$. A_n is therefore a *subgroup* of S_n. It is called the *alternating group of degree n.*

As an illustration let us find the multiplication table for A_4. From the list of elements of S_4 given in Problem 3.21 we determine that the even permutations are

$$\iota = \begin{pmatrix} 1 & 2 & 3 & 4 \\ 1 & 2 & 3 & 4 \end{pmatrix} \quad \sigma_8 = \begin{pmatrix} 1 & 2 & 3 & 4 \\ 2 & 1 & 4 & 3 \end{pmatrix} \quad \tau_3 = \begin{pmatrix} 1 & 2 & 3 & 4 \\ 3 & 2 & 4 & 1 \end{pmatrix} \quad \tau_6 = \begin{pmatrix} 1 & 2 & 3 & 4 \\ 4 & 1 & 3 & 2 \end{pmatrix}$$

$$\sigma_2 = \begin{pmatrix} 1 & 2 & 3 & 4 \\ 3 & 4 & 1 & 2 \end{pmatrix} \quad \tau_1 = \begin{pmatrix} 1 & 2 & 3 & 4 \\ 1 & 3 & 4 & 2 \end{pmatrix} \quad \tau_4 = \begin{pmatrix} 1 & 2 & 3 & 4 \\ 4 & 2 & 1 & 3 \end{pmatrix} \quad \tau_7 = \begin{pmatrix} 1 & 2 & 3 & 4 \\ 2 & 3 & 1 & 4 \end{pmatrix}$$

$$\sigma_5 = \begin{pmatrix} 1 & 2 & 3 & 4 \\ 4 & 3 & 2 & 1 \end{pmatrix} \quad \tau_2 = \begin{pmatrix} 1 & 2 & 3 & 4 \\ 1 & 4 & 2 & 3 \end{pmatrix} \quad \tau_5 = \begin{pmatrix} 1 & 2 & 3 & 4 \\ 2 & 4 & 3 & 1 \end{pmatrix} \quad \tau_8 = \begin{pmatrix} 1 & 2 & 3 & 4 \\ 3 & 1 & 2 & 4 \end{pmatrix}$$

The multiplication table is easily written down from the multiplication table of Problem 3.21.

	ι	σ_2	σ_5	σ_8	τ_1	τ_2	τ_3	τ_4	τ_5	τ_6	τ_7	τ_8
ι	ι	σ_2	σ_5	σ_8	τ_1	τ_2	τ_3	τ_4	τ_5	τ_6	τ_7	τ_8
σ_2	σ_2	ι	σ_8	σ_5	τ_4	τ_7	τ_6	τ_1	τ_8	τ_3	τ_2	τ_5
σ_5	σ_5	σ_8	ι	σ_2	τ_5	τ_3	τ_2	τ_8	τ_1	τ_7	τ_6	τ_4
σ_8	σ_8	σ_5	σ_2	ι	τ_8	τ_6	τ_7	τ_5	τ_4	τ_2	τ_3	τ_1
τ_1	τ_1	τ_8	τ_4	τ_5	τ_2	ι	σ_2	τ_6	τ_7	σ_5	σ_8	τ_3
τ_2	τ_2	τ_3	τ_6	τ_7	ι	τ_1	τ_8	σ_5	σ_8	τ_4	τ_5	σ_2
τ_3	τ_3	τ_2	τ_7	τ_6	σ_5	τ_5	τ_4	ι	σ_2	τ_8	τ_1	σ_8
τ_4	τ_4	τ_5	τ_1	τ_8	τ_7	σ_2	ι	τ_3	τ_2	σ_8	σ_5	τ_6
τ_5	τ_5	τ_4	τ_8	τ_1	τ_3	σ_5	σ_8	τ_7	τ_6	ι	σ_2	τ_2
τ_6	τ_6	τ_7	τ_2	τ_3	σ_8	τ_8	τ_1	σ_2	ι	τ_5	τ_4	σ_5
τ_7	τ_7	τ_6	τ_3	τ_2	σ_2	τ_4	τ_5	σ_8	σ_5	τ_1	τ_8	ι
τ_8	τ_8	τ_1	τ_5	τ_4	τ_6	σ_8	σ_5	τ_2	τ_3	σ_2	ι	τ_7

Problems

3.23. Write out a multiplication table for (i) A_1, (ii) A_2, (iii) A_3.

Solution:

(i) There is only one element in S_1, namely $\iota = \begin{pmatrix} 1 \\ 1 \end{pmatrix}$, and it is an even permutation. Hence a multiplication table for A_1 is $\begin{array}{c|c|} & \iota \\ \hline \iota & \iota \\ \hline \end{array}$. Notice A_1 is the same group as S_1.

(ii) $S_2 = \left\{\begin{pmatrix} 1 & 2 \\ 1 & 2 \end{pmatrix}, \begin{pmatrix} 1 & 2 \\ 2 & 1 \end{pmatrix}\right\}$. $\begin{pmatrix} 1 & 2 \\ 1 & 2 \end{pmatrix}$ is an even permutation and $\begin{pmatrix} 1 & 2 \\ 2 & 1 \end{pmatrix}$ is an odd permutation. Therefore $A_2 = \left\{\begin{pmatrix} 1 & 2 \\ 1 & 2 \end{pmatrix}\right\}$ and $\begin{array}{c|c|} & \iota \\ \hline \iota & \iota \\ \hline \end{array}$ where $\iota = \begin{pmatrix} 1 & 2 \\ 1 & 2 \end{pmatrix}$ is a multiplication table for A_2.

(iii) S_3 contains six elements (see example in Section 3.3a). The elements ι, σ_1 and σ_2 are the even permutations, and a multiplication table for A_3 is

	ι	σ_1	σ_2
ι	ι	σ_1	σ_2
σ_1	σ_1	σ_2	ι
σ_2	σ_2	ι	σ_1

3.24. Prove $A_n = S_n$ implies $n = 1$.

Solution:

If $n > 1$, S_n must contain a permutation which interchanges 1 and 2 and leaves everything else fixed, i.e. $1\tau = 2$, $2\tau = 1$, and $i\tau = i$ $(i = 3, \ldots, n)$. $\tau \notin A_n$, since τ is an odd permutation, and therefore $A_n \neq S_n$. By Problem 3.23(i), $A_1 = S_1$. Hence $A_n = S_n$ implies $n = 1$.

3.25. Show that the set $S = \{\iota, \sigma_2, \sigma_5, \sigma_8\}$ is a subgroup of A_4. (For notation see page 62.)

Solution:

If we look at the multiplication table for A_4, we see that the product of any two elements in S is again in S. It is clear that $\sigma_j, \sigma_k \in S$ implies $\sigma_j\sigma_k^{-1} \in S$ because $\sigma_j = \sigma_j^{-1}$ $(j = 2, 5, 8)$. Hence S is a subgroup of A_4.

d. The order of A_n

We now count the elements of A_n. Suppose that $n \geqq 2$. Let τ be the following element of S_n:

$$\tau: \; 1 \to 2,\; 2 \to 1,\; 3 \to 3,\; \ldots,\; n \to n$$

We claim τ is odd because

$$\frac{2\tau - 1\tau}{2-1} \cdot \frac{3\tau - 1\tau}{3-1} \cdot \frac{3\tau - 2\tau}{3-2} \cdot \cdots \cdot \frac{n\tau - 1\tau}{n-1} \cdot \cdots \cdot \frac{n\tau - (n-1)\tau}{n-(n-1)} = -1$$

Now let $\epsilon_1, \epsilon_2, \ldots, \epsilon_k$ be the *even* permutations (i.e. elements of A_n). Then

$$\epsilon_1\tau,\; \epsilon_2\tau,\; \ldots,\; \epsilon_k\tau \qquad (3.4)$$

are all odd; moreover, if $\epsilon_i\tau = \epsilon_j\tau$, then

$$\epsilon_i = \epsilon_i\iota = \epsilon_i(\tau\tau^{-1}) = (\epsilon_i\tau)\tau^{-1} = (\epsilon_j\tau)\tau^{-1} = \epsilon_j(\tau\tau^{-1}) = \epsilon_j\iota = \epsilon_j$$

Notice that this means there are at least as many odd permutations as even ones. Indeed there are exactly the same number of each; for if ω is odd, then

$$\omega = (\omega\tau)\tau$$

since $\tau^2 = \iota$. Since $\omega\tau$ is even by Lemma 3.2, every odd permutation is listed in (*3.4*). Thus there are precisely the same number of even permutations as odd ones.

Consequently, as k is the number of even permutations, the number of odd permutations is also k, and the number of odd and even permutations is $2k$. But every permutation of S_n is either even or odd, and $|S_n| = n!$ Therefore $k = n!/2$.

Putting our results together, we have proved

Theorem 3.3: If n is any positive integer > 1, then S_n is of order $n!$, and A_n is of order $n!/2$.

Problem

3.26. Check to see whether $|A_j|$ $(j = 2, 3, 4)$ agrees with Theorem 3.3.

Solution:

$|A_2| = 1$ by Problem 3.23, and $2!/2 = 1$.

$|A_3| = 3$ by Problem 3.23, and $3!/2 = 3$.

$|A_4| = 12$ by Section 3.3c, and $4!/2 = 12$.

3.4 GROUPS OF ISOMETRIES

a. Isometries of the line

We begin with certain subgroups of S_R, the symmetric group on R, the set of real numbers. We think of the elements of R arranged as points on the real line. Then if $a, b \in R$ it is clear what we mean by the *distance* between a and b, namely the absolute value, $|a - b|$, of $a - b$. We denote the distance between a and b by $d(a, b)$.

We define a group $I(R)$, the *group of isometries of R*, as a subgroup of S_R in the following way.

1. Let $I(R)$ be the set of all elements of S_R which preserve distance. The elements of this set will be called isometries of R. To put this definition more explicitly, an element $\sigma \in S_R$ is termed an *isometry* if and only if

$$d(a, b) = d(a\sigma, b\sigma)$$

for every pair of elements $a, b \in R$. Since the identity mapping $\iota \in I(R)$, $I(R) \neq \emptyset$.

2. Suppose $\sigma \in I(R)$. Then of course $\sigma^{-1} \in S_R$. We claim $\sigma^{-1} \in I(R)$. To see this, suppose $a, b \in R$. Then

$$d(a\sigma^{-1}, b\sigma^{-1}) = d((a\sigma^{-1})\sigma, (b\sigma^{-1})\sigma) = d(a, b)$$

as σ is an isometry. Hence

$$d(a, b) = d(a\sigma^{-1}, b\sigma^{-1})$$

which is precisely what we need. Thus $\sigma^{-1} \in I(R)$. Suppose $\sigma, \tau \in I(R)$; then $\tau^{-1} \in I(R)$ and

$$d(a(\sigma\tau^{-1}), b(\sigma\tau^{-1})) = d((a\sigma)\tau^{-1}, (b\sigma)\tau^{-1}) = d(a\sigma, b\sigma) = d(a, b)$$

Thus $\sigma\tau^{-1} \in I(R)$ and $I(R)$ is a subgroup of S_R. Considered as a group in its own right, $I(R)$ is called the group of isometries of R.

Problems

3.27. Is σ an isometry? (In the following, $x \in R$.)

(i) $\sigma: x \to x+2$. (ii) $\sigma: x \to nx$, n an integer $\neq 1$ or -1. (iii) $\sigma: x \to x^2$. (iv) $\sigma: x \to -x$.

Solution:

(i) First note that $\sigma \in S_R$. σ is an isometry, since $d(x, x') = |x - x'|$ and

$$d(x\sigma, x'\sigma) = d(x+2, x'+2) = |(x+2) - (x'+2)| = |x - x'|$$

(ii) σ is not an isometry, since $d(x, x') = |x - x'|$ and

$$d(x\sigma, x'\sigma) = d(nx, nx') = |n(x - x')|$$

so that, for $x \neq x'$, $d(x, x') \neq d(x\sigma, x'\sigma)$, e.g. $x = 2$, $x' = 1$ implies $d(x, x') = 1$ and $d(x\sigma, x'\sigma) = |n|$.

(iii) σ is not an isometry, since $d(x, x') = |x - x'|$ and

$$d(x\sigma, x'\sigma) = d(x^2, x'^2) = |x^2 - x'^2|$$

so that, for $x \neq x'$, $d(x, x') \neq d(x\sigma, x'\sigma)$, e.g. $x = 2$, $x' = 0$.

(iv) σ is an isometry, since $\sigma \in S_R$ and

$$d(x\sigma, x'\sigma) = d(-x, -x') = |-x - (-x')| = |-x + x'| = |x - x'| = d(x, x')$$

3.28. Set $1\sigma = 2$, $2\sigma = 1$ and $x\sigma = x$ for all $x \in R$ excepting 1 and 2. Is σ an isometry?

Solution:

No, since $d(1, 3) = |3 - 1| = 2$ and $d(1\sigma, 3\sigma) = d(2, 3) = 1$.

b. Two points determine an isometry

The following lemma gives us a method of determining whether the two isometries are the same.

Lemma 3.4: If $\sigma, \tau \in I(R)$ have the same effect on two distinct real numbers a and b, i.e. $a\sigma = a\tau$ and $b\sigma = b\tau$, then $\sigma = \tau$.

Proof: Let c be any element of R. Then

$$d(c, a) = d(c\sigma, a\sigma) = |c\sigma - a\sigma| = d(c\tau, a\tau) = |c\tau - a\tau|$$

and hence

$$c\sigma - a\sigma = \pm(c\tau - a\tau)$$

Assume $c\sigma \neq c\tau$. Since $a\sigma = a\tau$, $c\sigma - a\sigma = +(c\tau - a\tau)$ implies $c\sigma - c\tau = a\sigma - a\tau = 0$, i.e. $c\sigma = c\tau$, contrary to our assumption. Therefore

$$c\sigma - a\sigma = -(c\tau - a\tau) = -c\tau + a\tau, \quad \text{i.e.} \quad c\sigma + c\tau = 2(a\sigma)$$

Similarly, $c\sigma + c\tau = 2(b\sigma)$. Hence $2a\sigma = 2b\sigma$ and $a\sigma = b\sigma$. But σ is a permutation. Consequently $a = b$, contrary to the hypothesis that a and b are distinct real numbers. We conclude that $c\sigma = c\tau$ and, since c is any element in R, $\sigma = \tau$.

Using this lemma it is possible to describe the elements of $I(R)$. Let $\sigma \in I(R)$ and let $0\sigma = a$. Now $d(0\sigma, 1\sigma) = d(a, 1\sigma) = |a - 1\sigma| = d(0,1) = 1$. Hence

$$a - 1\sigma = \pm 1 \quad \text{or} \quad 1\sigma = a \pm 1$$

(i) $1\sigma = a+1$ and $0\sigma = a$. Let σ^* be the member of $I(R)$ defined by mapping $r \in R$ to $r+a$. σ^* is clearly an isometry. Then σ^* and σ agree on 0 and 1. Hence $\sigma = \sigma^*$.

(ii) $1\sigma = a-1$ and $0\sigma = a$. Let σ^* be the member of $I(R)$ defined by mapping $r \in R$ to $-r+a$. Then σ and σ^* agree on 0 and 1. Hence $\sigma = \sigma^*$.

Thus if $\sigma \in I(R)$, $r\sigma = \epsilon r + 0\sigma$ where ϵ is either 1 or -1.

Geometrically, if $r\sigma = r + a$, it "moves" the real line a units to the right. If $r\sigma = -r + a$, the real line is inverted about the origin and then moved a units to the right.

We come now to an interesting subset of $I = I(R)$ which we will prove is a subgroup. Let $(I : Z) = \{\sigma \mid \sigma \in I,\ n\sigma \in Z \text{ whenever } n \in Z\}$. Let $\sigma, \tau \in (I : Z)$. Let $0\sigma = a$, $0\tau = b$. a and b must be integers. The effects of σ and τ are

$$r\sigma = \epsilon r + a \quad \text{where} \quad \epsilon = 1 \text{ or } -1$$

$$r\tau = \eta r + b \quad \text{where} \quad \eta = 1 \text{ or } -1$$

Let $\mu : r \to \eta r - \eta b$. Then $\mu = \tau^{-1}$, for $r(\tau\mu) = (\eta r + b)\mu = \eta(\eta r + b) - \eta b = r$ and similarly $r(\mu\tau) = r$. Hence $r(\sigma\tau^{-1}) = (\epsilon\eta)r + \eta(a-b)$. Clearly $\epsilon\eta$ is ± 1, and $\eta(a-b) \in Z$. Thus if n is an integer, $n(\sigma\tau^{-1}) \in Z$. Therefore $\sigma\tau^{-1} \in (I : Z)$ and $(I : Z)$ is a subgroup of I.

Problem

3.29. Determine whether the following sets of mappings, with composition as the binary operation, are groups.

(i) The set of all mappings of the plane of the form

$$\tau_a : (x, y) \to (x + a, y + a) \quad \text{with} \quad (x, y) \in R^2,\ a \in R$$

Notice that τ_a is defined for each real number a.

(ii) The set of all τ_a with $a \in Z$.

(iii) The set of all τ_a with $a \in Q$.

(iv) The set of all mappings of real numbers of the form

$$\mu_a : x \to ax \quad \text{with} \quad a \neq 0,\ a \in Q$$

(v) The set of all mappings of the plane R^2 of the form

$$\mu_a : (x, y) \to (ax, ay) \quad \text{with} \quad a \in Q$$

(vi) The set of mappings of (v) but with $a \in Q^*$, i.e. $a \neq 0$.

Solution:

(i) We show first that τ_a is a permutation of R^2. τ_a is a matching of $R^2 \to R^2$, for if $(x, y)\tau_a = (x_1, y_1)\tau_a$, then $(x, y) = (x_1, y_1)$. If $(x, y) \in R^2$, there exists $(x - a, y - a) \in R^2$ and $(x - a, y - a)\tau_a = (x, y)$. Hence τ_a is a one-to-one onto mapping and so τ_a is a permutation of R^2. $\tau_{-a} = \tau_a^{-1}$. The set of all τ_a is not empty, and $\tau_a\tau_b^{-1} = \tau_a\tau_{-b} = \tau_{a-b}$ since $(x, y)\tau_a\tau_b^{-1} = (x, y)\tau_a\tau_{-b} = (x + a, y + a)\tau_{-b} = (x + a - b, y + a - b) = (x, y)\tau_{a-b}$. Thus the set of all τ_a is a subgroup of S_{R^2} and so it constitutes a group.

(ii) Using (i), we need only consider whether $\tau_a\tau_b^{-1} = \tau_{a-b}$ belongs to the given set, which it clearly does.

(iii) This follows by arguing as in (ii).

(iv) It is easy to verify that each μ_a is a permutation of R. Also, $\mu_{(1/a)} = (\mu_a)^{-1}$. Now $\mu_a\mu_b^{-1} = \mu_a\mu_{1/b} = \mu_{a/b}$. Hence the set of all μ_a, $a \neq 0, a \in Q$, constitutes a subgroup of the group of all permutations of R and itself forms a group.

(v) The set of μ_a is not a group, as μ_0 has no inverse. For μ_0 maps all points onto the point $(0, 0)$, and by Theorem 2.4, page 36, the only elements of M_{R^2} which have an inverse are the one-to-one and onto mappings.

(vi) The set of all μ_a in this case forms a group. The set is non-empty. If μ_a is an element, then $\mu_{1/a} = \mu_a^{-1}$ because $(x, y)(\mu_a \mu_{1/a}) = (ax, ay)\mu_{1/a} = (x, y)$ and $(x, y)(\mu_{1/a}\mu_a) = (x, y)$. By Theorem 2.4 μ_a is a one-to-one onto mapping and thus μ_a is a permutation of R^2. Now $\mu_a \mu_b^{-1} = \mu_a \mu_{1/b} = \mu_{a/b}$ since $(x, y)\mu_a \mu_b^{-1} = (ax, ay)\mu_b^{-1} = \left(\frac{a}{b}x, \frac{a}{b}y\right) = (x, y)\mu_{a/b}$. Hence the set of all μ_a is a subgroup of S_{R^2}.

c. Isometries of the plane

Let E be the set $R^2 = R \times R$. If $(x_A, y_A) = A$, $(x_B, y_B) = B$ are two elements of E, we define the distance between A and B as

$$\sqrt{(x_A - x_B)^2 + (y_A - y_B)^2}$$

and denote it by $d(A, B)$. Recall that if we interpret A and B as points of the Euclidean plane with coordinates (x_A, y_A) and (x_B, y_B) relative to a given cartesian coordinate system, then the formula for $d(A, B)$ is the ordinary distance between A and B. The interpretation of E as the Euclidean plane is not an essential part of our argument. Logically we do without it and work abstractly with the ordered pairs (x, y).

$\sigma \in S_E$, the symmetric group on E, is called an isometry if for any A, B points of E, $d(A, B) = d(A\sigma, B\sigma)$.

Theorem 3.5: The set I of all isometries of E forms a subgroup of S_E.

Proof: I is not empty, since the identity mapping is an isometry. We need to show only that $\sigma\tau^{-1} \in I$ whenever $\sigma, \tau \in I$. Consider the effect of τ^{-1}. As $\tau \in S_E$, there exist, for each pair of points $A, B \in E$, $A', B' \in E$ such that $A'\tau = A$, $B'\tau = B$. Then

$$d(A', B') = d(A'\tau, B'\tau) = d(A, B)$$

since τ is an isometry. But $A\tau^{-1} = A'$, $B\tau^{-1} = B'$. Hence

$$d(A\tau^{-1}, B\tau^{-1}) = d(A, B)$$

and τ^{-1} is an isometry. Consequently

$$d(A\sigma\tau^{-1}, B\sigma\tau^{-1}) = d(A\sigma, B\sigma) = d(A, B)$$

and so $\sigma\tau^{-1} \in I$. Hence I is a subgroup of S_E.

In the next four paragraphs we shall argue informally.

Let us imagine that the Euclidean plane E is covered by an infinite rigid metal lamina S. If we move S so that it still covers E, we can define an isometry induced by that movement. Let the points of E be denoted $P, Q, R, \ldots$ and the points of S be denoted $A, B, C, \ldots$.

In the initial position suppose that A lies on top of P, B on top of Q, C on top of $R, \ldots$. After the sheet S is moved, the point A is on top of another point of E, say P_1; the point B is on top of, say, the point Q_1; C is on top of, say, $R_1; \ldots$.

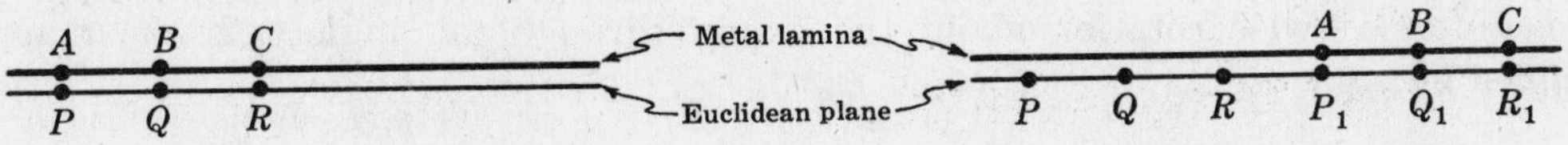

(a) Initial position. (b) After a movement.

Fig. 1. The isometry induced by a movement.

Let $\theta: E \to E$ be defined by $P\theta = P_1$, $Q\theta = Q_1$, $R\theta = R_1$, ... Then we assert θ is an isometry. For if P, Q are any two points of E, with, in the initial position, A of S on top of P, and B of S on top of Q, then we have $d(A,B) = d(P,Q)$. After S is moved, A lies on top of, say, P_1, and B on top of Q_1. Hence $d(A,B) = d(P_1,Q_1)$ and so $d(P_1,Q_1) = d(P,Q)$.

Using this informal approach, we now describe three particular isometries:

1. Rotation about a point. Let O be any point of S. Rotate S about O through an angle Ψ. Then the isometry induced by this movement of S is called the rotation about O through an angle Ψ.

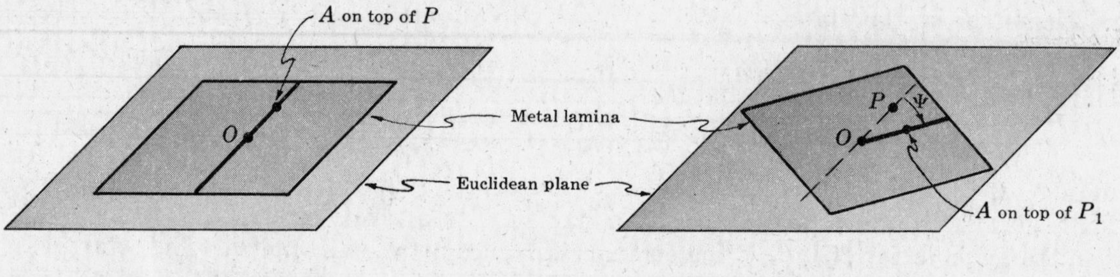

(a) Initial position. (b) A rotation through an angle Ψ.

Fig. 2

2. Reflection in a line. Let us choose a line in E and turn S over this line and back to E. The isometry obtained in this way is called the reflection in XY.

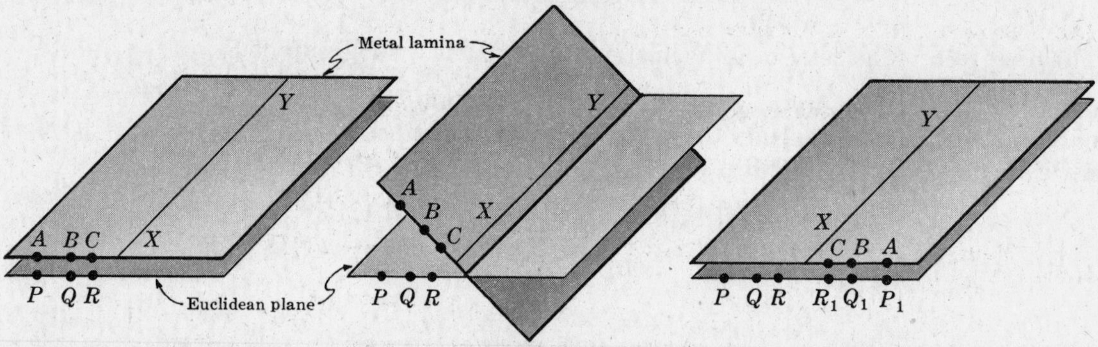

(a) Initial position. (b) Rotating about XY. (c) Final position.

Fig. 3. A reflection in line XY.

3. Let us choose a line XY. Let θ be an isometry corresponding to a movement of S for which $X\theta Y\theta$, the line joining $X\theta$ and $Y\theta$, is parallel to XY. Then θ is called a translation.

We now describe formally these three types of isometries. As a simplification we describe only rotations about the origin and reflections about the axis OX.

1. A translation $\tau_{a,b}$ is the mapping defined by

$$(x, y)\tau_{a,b} = (x+a, y+b)$$

It can be shown that for each a, b, $\tau_{a,b}$ is an isometry and that $(\tau_{a,b})^{-1} = \tau_{-a,-b}$. (See Problem 3.30 for details.)

2. A counterclockwise rotation about the origin through an angle θ is the mapping ρ_θ defined by

$$(x, y)\rho_\theta = (x \cos\theta - y \sin\theta,\ x \sin\theta + y \cos\theta)$$

For each θ, ρ_θ is an isometry and $(\rho_\theta)^{-1} = \rho_{-\theta}$. (See Problem 3.31 for proof. Problem 3.32 shows why ρ_θ is called a rotation through θ.)

3. Define a reflection in OX to be the map σ_y where

$$(x, y)\sigma_y = (x, -y)$$

This is readily seen to be an isometry and it is easy to show that $(\sigma_y)^{-1} = \sigma_y$ (Problem 3.33).

Since ι is the product of two reflections, we call it a reflection. This is a convenience which will simplify the statement of Theorem 3.8.

Problems

3.30. Show that $\tau_{a,b}$ is an isometry and that $(\tau_{a,b})^{-1} = \tau_{-a,-b}$.

Solution:

First we must show that $\tau_{a,b} \in S_E$, so we must show that it is a one-to-one onto mapping. $(x,y)\tau_{a,b} = (x',y')\tau_{a,b}$ clearly implies $(x,y) = (x',y')$. If $(x,y) \in E$, then $(x-a, y-b)\tau_{a,b} = (x,y)$ and so $\tau_{a,b}$ is onto. Hence $\tau_{a,b} \in S_E$. Is $\tau_{a,b}$ an isometry? If $A = (x_A, y_A)$ and $B = (x_B, y_B)$, then

$$d(A,B) = \sqrt{(x_A - x_B)^2 + (y_A - y_B)^2} = d(A\tau_{a,b}, B\tau_{a,b})$$

and $\tau_{a,b}$ is an isometry. $(x,y)\tau_{a,b}\tau_{-a,-b} = (x+a, y+b)\tau_{-a,-b} = (x,y)$. Hence $\tau_{a,b}\tau_{-a,-b} = \iota$. Similarly, $\tau_{-a,-b}\tau_{a,b} = \iota$ and so $(\tau_{a,b})^{-1} = \tau_{-a,-b}$.

3.31. Show that ρ_θ is an isometry and that $(\rho_\theta)^{-1} = \rho_{-\theta}$. (Hard.)

Solution:

We must show first that ρ_θ is an element of S_E. $(x,y)\rho_\theta = (x',y')\rho_\theta$ implies

$$\begin{aligned} x\cos\theta - y\sin\theta &= x'\cos\theta - y'\sin\theta \\ x\sin\theta + y\cos\theta &= x'\sin\theta + y'\cos\theta \end{aligned} \tag{3.5}$$

Multiply the first equation by $\cos\theta$ and the second by $\sin\theta$ and add to obtain

$$\begin{aligned} &x(\cos^2\theta + \sin^2\theta) + (-y\cos\theta\sin\theta + y\cos\theta\sin\theta) \\ &\quad = x'(\cos^2\theta + \sin^2\theta) + (-y'\cos\theta\sin\theta + y'\cos\theta\sin\theta) \end{aligned}$$

Since $\cos^2\theta + \sin^2\theta = 1$, $x = x'$. Similarly by multiplying the first equation of (3.5) by $\sin\theta$ and the second by $\cos\theta$ and subtracting the first from the second, we find $y = y'$. Hence ρ_θ is one-one. Is it onto? In other words, can we find (x,y) such that $(x,y)\rho_\theta = (a,b)$ for any $(a,b) \in R^2$? That is, does there exist a solution to

$$\begin{aligned} x\cos\theta - y\sin\theta &= a \\ x\sin\theta + y\cos\theta &= b \end{aligned} \tag{3.6}$$

for $x, y \in R$? We solve these equations for x and y using the same stratagem as above, i.e. multiply the top equation by $\cos\theta$ and the bottom by $\sin\theta$ and add to obtain

$$x(\cos^2\theta + \sin^2\theta) = a\cos\theta + b\sin\theta \quad \text{or} \quad x = a\cos\theta + b\sin\theta$$

Multiply the first equation of (3.6) by $\sin\theta$ and the second by $\cos\theta$ and subtract the first from the second to obtain

$$y = y(\sin^2\theta + \cos^2\theta) = b\cos\theta - a\sin\theta$$

On substituting these values of x and y in (3.6), it is easily seen that they satisfy the equation. (The reader who knows that the condition for existence of a solution to (3.6) is $\begin{vmatrix} \cos\theta & -\sin\theta \\ \sin\theta & \cos\theta \end{vmatrix} \neq 0$, can verify the existence of a solution to (3.6) immediately.)

Finally, is ρ_θ an isometry? If $A = (x_A, y_A)$ and $B = (x_B, y_B)$, then

$d(A\rho_\theta, B\rho_\theta)$

$$= \sqrt{[(x_A \cos\theta - y_A \sin\theta) - (x_B \cos\theta - y_B \sin\theta)]^2 + [(x_A \sin\theta + y_A \cos\theta) - (x_B \sin\theta + y_B \cos\theta)]^2}$$
$$= \sqrt{[(x_A - x_B)\cos\theta - (y_A - y_B)\sin\theta]^2 + [(x_A - x_B)\sin\theta + (y_A - y_B)\cos\theta]^2}$$
$$= \sqrt{(x_A - x_B)^2[\cos^2\theta + \sin^2\theta] + (y_A - y_B)^2[\cos^2\theta + \sin^2\theta]}$$
$$= \sqrt{(x_A - x_B)^2 + (y_A - y_B)^2} = d(A, B)$$

and thus ρ_θ is an isometry.

$$\begin{aligned}(x, y)\rho_\theta\rho_{-\theta} &= (x\cos\theta - y\sin\theta,\ x\sin\theta + y\cos\theta)\rho_{-\theta} \\ &= [(x\cos\theta - y\sin\theta)\cos(-\theta) - (x\sin\theta + y\cos\theta)\sin(-\theta), \\ &\qquad (x\cos\theta - y\sin\theta)\sin(-\theta) + (x\sin\theta + y\cos\theta)\cos(-\theta)] \\ &= [x(\sin^2\theta + \cos^2\theta),\ y(\sin^2\theta + \cos^2\theta)] = (x, y)\end{aligned}$$

and so $\rho_\theta\rho_{-\theta} = \iota$. Similarly $\rho_{-\theta}\rho_\theta = \iota$.

3.32. Show that $(0,0)$ is equidistant from (x,y) and $(x,y)\rho_\theta$, and that the smallest angle between the line L_1 joining $(x,y)\rho_\theta$ and $(0,0)$, and the line L_2 joining (x,y) and $(0,0)$, is θ when $0 \leq \theta \leq \pi$, or $2\pi - \theta$ when $\pi \leq \theta \leq 2\pi$.

Solution:

As we have shown in Problem 3.31, $d((0,0),(x,y)) = d((0,0)\rho_\theta, (x,y)\rho_\theta) = d((0,0),(x,y)\rho_\theta)$. We remind the reader of the formula for calculating the angle between two lines meeting at the origin. If the one line L_1 has end point (a_1, b_1) and the other L_2 has end point (a_2, b_2), then the cosine of the angle between L_1 and L_2 is given by

$$\frac{a_1a_2 + b_1b_2}{\sqrt{(a_1^2 + b_1^2)(a_2^2 + b_2^2)}}$$

Put $(a_1, b_1) = (x, y)$, $(a_2, b_2) = (x, y)\rho_\theta$. Then if μ is the smallest angle between the lines L_1 and L_2,

$$\cos\mu = \frac{(x^2 + y^2)\cos\theta}{\sqrt{(x^2 + y^2)(x^2 + y^2)}} = \cos\theta$$

Hence $\mu = \theta$ if $0 \leq \theta \leq \pi$, and $\mu = 2\pi - \theta$ if $\pi \leq \theta \leq 2\pi$.

3.33. Show that σ_y is an isometry and that $\sigma_y^2 = \iota$.

Solution:

If $A = (x_A, y_A)$ and $B = (x_B, y_B)$,

$$d(A\sigma_y, B\sigma_y) = \sqrt{(x_A - x_B)^2 + (-y_A - (-y_B))^2} = d(A, B)$$

and so σ_y preserves distance. Obviously $(x,y)\sigma_y = (x', y')\sigma_y$ implies $(x,y) = (x', y')$. And σ_y is clearly onto, for $(x, -y)\sigma_y = (x, y)$. Hence σ_y is an isometry. $(x,y)\sigma_y\sigma_y = (x,-y)\sigma_y = (x,y)$. Thus $\sigma_y^2 = \iota$.

3.34. Show that rotations about the origin form a subgroup of I.

Solution:

ρ_0, the rotation through zero degrees, is the identity; hence the set of rotations is not empty. If ρ_θ, ρ_ϕ are two rotations, $(\rho_\phi)^{-1} = \rho_{(-\phi)}$. Put $-\phi = \Psi$, as it is annoying to carry the minus sign. Is $\rho_\theta(\rho_\phi)^{-1} = \rho_\theta\rho_\Psi$ a rotation?

$$\begin{aligned}(x, y)\rho_\theta\rho_\Psi &= (x\cos\theta - y\sin\theta,\ x\sin\theta + y\cos\theta)\rho_\Psi \\ &= ((x\cos\theta - y\sin\theta)\cos\Psi - (x\sin\theta + y\cos\theta)\sin\Psi, \\ &\qquad (x\cos\theta - y\sin\theta)\sin\Psi + (x\sin\theta + y\cos\theta)\cos\Psi)) \\ &= (x(\cos\theta\cos\Psi - \sin\theta\sin\Psi) - y(\sin\theta\cos\Psi + \cos\theta\sin\Psi), \\ &\qquad x(\cos\theta\sin\Psi + \sin\theta\cos\Psi) + y(\cos\theta\cos\Psi - \sin\theta\sin\Psi)) \\ &= (x\cos(\theta + \Psi) - y\sin(\theta + \Psi),\ x\sin(\theta + \Psi) + y\cos(\theta + \Psi)) \\ &= (x, y)\rho_{\theta + \Psi}\end{aligned}$$

Hence $\rho_\theta\rho_\Psi = \rho_{(\theta + \Psi)}$, and $\rho_\theta(\rho_\phi)^{-1}$ is a rotation. Thus the rotations form a subgroup of I.

3.35. Show that the reflections about OX form a subgroup of I.

Solution:

We have only two elements, for there are only the two reflections ι and σ_y. Since we have shown that $\sigma_y = \sigma_y^{-1}$, we quickly verify that the two possible products of a reflection and the inverse of a reflection is again a reflection. Hence the set of reflections forms a subgroup of I and is of order 2.

3.36. Show that the set of translations forms a subgroup of I.

Solution:

$\tau_{a,b}(\tau_{c,d})^{-1} = \tau_{a,b}\tau_{-c,-d} = \tau_{a-c,b-d}$ is easily verified as

$$(x, y)\tau_{a,b}\tau_{-c,-d} = (x+a, y+b)\tau_{-c,-d} = (x+a-c, y+b-d) = (x, y)\tau_{a-c,b-d}$$

and thus the set of translations forms a subgroup of I.

3.37. Find an element of S_E that is not an isometry.

Solution:

Let $\sigma \in S_E$ be defined by $(x,y)\sigma = (x,y)$ except that $(0,0)\sigma = (1,0)$ and $(1,0)\sigma = (0,0)$. It is easy to verify that $\sigma \in S_E$. Now if $A = (0,0)$, $B = (1,0)$ and $C = (0,1)$, then $1 = d(A,C)$. $d(A\sigma, C\sigma) = d(B,C) = \sqrt{2} \neq d(A,C)$. Hence σ is not an isometry.

d. Isometries are products of reflections, translations and rotations

We will prove in this section that an isometry is determined uniquely by its action on any three points not all on a straight line. This enables us to prove that every isometry is expressible as the product of a reflection, a translation and a rotation. (Note, however, that there are isometries that are neither reflections, rotations, nor translations.)

Lemma 3.6: Let $\sigma \in I$. Let A, B and C be three noncollinear points in E. Let $A\sigma = A'$, $B\sigma = B'$ and $C\sigma = C'$. Then $A'B'C'$ is a triangle congruent to ABC.

Proof: $d(A',B') = d(A,B)$, $d(A',C') = d(A,C)$ and $d(B',C') = d(B,C)$. Thus we have two triangles with corresponding sides equal, and so $\triangle ABC$ is congruent to $\triangle A'B'C'$.

Lemma 3.7: If σ and τ, elements of I, have the same effect on three points A, B, C which do not lie on a straight line, then $\sigma = \tau$.

Proof: Let D be any point of E. Let $d(D,A) = a$, $d(D,B) = b$ and $d(D,C) = c$. Let $A' = A\sigma = A\tau$, $B' = B\sigma = B\tau$ and $C' = C\sigma = C\tau$. Then the distance of $D\sigma$ from A' is a, from B' is b, and from C' is c. Similarly $D\tau$ is a distance a from A', b from B', and c from C'. Geometrically it is possible to see that $D\sigma = D\tau$, for both $D\sigma$ and $D\tau$ lie on the intersection of three circles: O_1 with center A' and radius a; O_2 with center B' and radius b; and O_3 with center C' and radius c.

Two circles intersect in two points at most. Hence O_1 and O_2 determine two points. It is possible for $D\sigma$ and $D\tau$ to be these two points. But we shall show that as $D\sigma$ and $D\tau$ must lie on O_3, they must be the same point. If this is not so, then O_1, O_2 and O_3 have two points in common. We shall prove geometrically then that A, B, C must lie on the same straight line.

Let three circles with centers A', B' and C' intersect in two points P and Q. Without loss of generality we may assume that B' lies between A' and C'. $A'P = A'Q$ and $B'P = B'Q$; hence $A'B'$ lies on the perpendicular bisector of PQ. Similarly $B'C'$ lies on the perpendicular bisector of PQ. Therefore $A'B'C'$ is a straight line. But $\triangle ABC$ is congruent to $\triangle A'B'C'$ by Lemma 3.6. Hence ABC lies in a straight line. But we assumed this was not so. This contradiction proves Lemma 3.7.

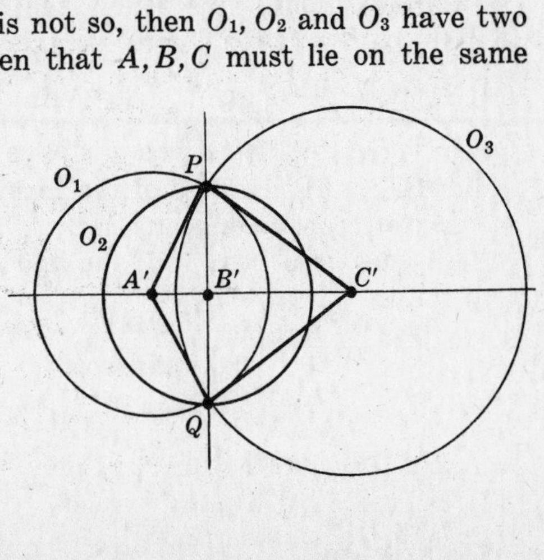

Theorem 3.8: Every isometry of E is expressible as the product of a reflection, a rotation and a translation.

We present an intuitive proof first. The formal proof below follows exactly the same steps.

Intuitive proof: Let $\sigma \in I$. Let $A = (0,0)$, $B = (1,0)$ and $C = (0,1)$ as shown in *Fig. 1*. Our idea is first to find σ^{-1} as a product, Ψ, of a translation, a rotation and a reflection. It is easy then to prove σ is the product of a reflection, rotation and translation. So we must find a Ψ which is the product of a translation, rotation and reflection such that $\sigma\Psi = \iota$. To check that $\sigma\Psi = \iota$, we need only prove that for the three noncollinear points A, B, C the effect of ι and $\sigma\Psi$ is the same (Lemma 3.7). We build Ψ in three stages so that we bring (*a*) $A\sigma$ to A; (*b*) $B\sigma$ to B; (*c*) $C\sigma$ to C. Let $A\sigma = (a, b)$.

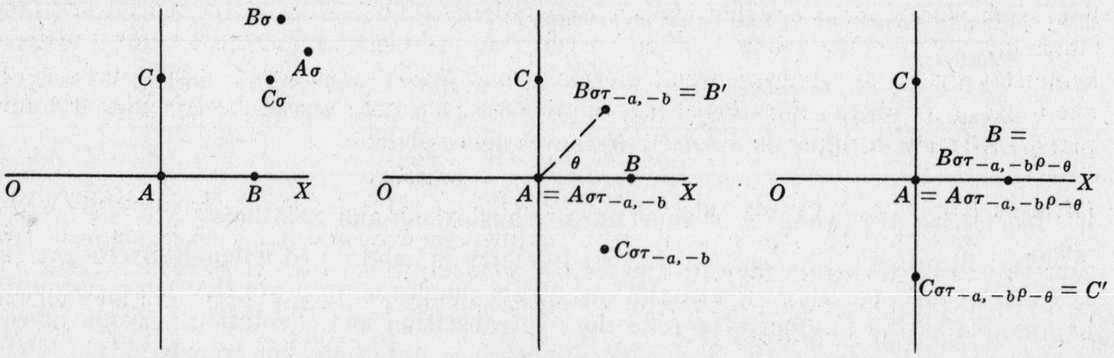

Fig. 1 *Fig. 2* *Fig. 3*

Apply the translation $\tau_{-a,-b}$ which moves each point a distance $d(A, A\sigma)$ parallel to the line joining $A\sigma$ and A, so that $A\sigma\tau_{-a,-b} = A$. Let $B' = B\sigma\tau_{-a,-b}$. Then $B'A$ is at an angle θ, say, to OX and, since $\tau_{-a,-b}$ and σ are isometries, is of length 1. See *Fig. 2*.

Apply the rotation through an angle of $-\theta$ that takes B' onto B. Let $C' = C\sigma\tau_{-a,-b}\rho_{-\theta}$. Then since σ, $\tau_{-a,-b}$ and $\rho_{-\theta}$ are isometries, C' is at a distance 1 from A and a distance $\sqrt{2}$ from B. Hence C' is either as in *Fig. 3*, in which case let μ be the reflection in OX, or else $C' = C$, in which case let μ be the identity reflection. Let $\Psi = \mu_{-a,-b}\rho_{-\theta}\mu$. Since A, B and C are mapped to A, B and C by $\sigma\Psi$, $\sigma\Psi = \iota$ by Lemma 3.7. Hence, using our remark on the inverse of a product in Section 2.4, page 34,

$$\sigma = \Psi^{-1} = \mu^{-1}(\rho_{-\theta})^{-1}(\tau_{-a,-b})^{-1} = \mu\rho_{\theta}\tau_{a,b}$$

and the theorem is true.

Formal proof: We follow the same three steps and use the same notation. Thus $A = (0,0)$, $B = (1,0)$, $C = (0,1)$ and $\sigma \in I$.

(*a*) Suppose $A\sigma = (a, b)$. Then $A\sigma\tau_{-a,-b} = A$.

(*b*) $B' = B\sigma\tau_{-a,-b}$ must be a distance 1 from A, since $1 = d(A,B) = d(A\sigma\tau_{-a,-b}, B\sigma\tau_{-a,-b}) = d(A, B')$. Hence B' is of the form $(\cos\theta, \sin\theta)$ for some angle θ. [This is a well known fact of coordinate geometry. All we must check is that if $B' = (d, e)$ with $d^2 + e^2 = 1$, the equations $\cos\theta = d$ and $\sin\theta = e$ can be solved for θ.] Then

$$\begin{aligned} B'\rho_{-\theta} &= (\cos\theta, \sin\theta)\rho_{-\theta} \\ &= (\cos\theta\,(\cos -\theta) - \sin\theta\,(\sin -\theta),\ \cos\theta\,\sin(-\theta) + \sin\theta\,\cos(-\theta)) \\ &= (\cos^2\theta + \sin^2\theta,\ 0) = (1, 0) = B \end{aligned}$$

Also, $A\rho_{-\theta} = A$.

(c) $C' = C\sigma\tau_{-a,-b}\rho_{-\theta}$ is a distance 1 from A and $\sqrt{2}$ from B, since $A\sigma\tau_{-a,-b}\rho_{-\theta} = A$, $B\sigma\tau_{-a,-b}\rho_{-\theta} = B$, and $\sigma\tau_{-a,-b}\rho_{-\theta}$ is an isometry. Thus $C' = (0,1)$ or $(0,-1)$. Let μ be the identity if $C' = (0,1)$, and let μ be the reflection σ_y if $C' = (0,-1)$. Put $\Psi = \tau_{-a,-b}\rho_{-\theta}\mu$. Then $A\sigma\Psi = A$, $B\sigma\Psi = B$ and $C\sigma\Psi = C$. Hence $\sigma\Psi = \iota$ by Lemma 3.7, and $\sigma = \Psi^{-1} = \mu\tau_{a,b}\rho_{\theta}$.

e. Symmetry groups

Given a figure in the Euclidean plane, we shall define a group which we will call the symmetry group of the figure.

The word symmetry is not sufficiently precise for the needs of Chemistry and Physics; so instead of talking about symmetry, we talk about symmetry groups. In comparing two figures, it usually turns out that what one would normally think of as the more symmetrical figure has a symmetry group of greater order than the less symmetrical figure. Also, the symmetry group of what we would normally call a non-symmetrical figure usually turns out to be of order 1. More generally, we will be concerned with subsets of the Euclidean plane. (Clearly, a figure is a subset of the Euclidean plane.)

Theorem 3.9: Let S be any subset of the Euclidean plane. The set, denoted by I_S, of all $\sigma \in I$ such that (i) $s \in S$ implies $s\sigma \in S$ and (ii) $t\sigma \in S$ implies $t \in S$, forms a subgroup of I, called the *symmetry group of* S. (An element of I_S, therefore, is characterized by its mapping elements of S, and only elements of S, into S.)

Proof: $I_S \neq \emptyset$, as the identity mapping of the Euclidean plane into itself is in I_S. If $\sigma, \tau \in I_S$, is $\sigma\tau^{-1} \in I_S$? We first prove $\tau^{-1} \in I_S$. If $s \in S$, $(s\tau^{-1})\tau = s \in S$. Because $\tau \in I_S$, (ii) implies $s\tau^{-1} \in S$. Thus τ^{-1} satisfies (i), i.e. $s \in S$ implies $s\tau^{-1} \in S$.

To show τ^{-1} also satisfies (ii), let $t\tau^{-1} \in S$. Then $(t\tau^{-1})\tau = t \in S$, since τ satisfies (i). Hence τ^{-1} satisfies (ii), and $\tau^{-1} \in I_S$.

Now we show $\sigma\tau^{-1} \in I_S$. Let $s \in S$. Then $s\sigma \in S$ and $s\sigma\tau^{-1} \in S$, since σ and τ^{-1} are in I_S. Therefore $\sigma\tau^{-1}$ satisfies (i). If $t\sigma\tau^{-1} \in S$, since τ^{-1} satisfies (ii), we have $t\sigma \in S$. Furthermore, since $\sigma \in I_S$, (ii) implies $t \in S$ and consequently $\sigma\tau^{-1}$ also satisfies (ii). Hence $\sigma\tau^{-1} \in I_S$ and I_S forms a subgroup of I.

Problems

3.38. Find a plane figure S such that

$$U = \{\sigma \mid \sigma \in I \text{ and for all } s \in S,\ s\sigma \in S\}$$

does not form a subgroup of I. (In other words, in Theorem 3.9 we cannot drop condition (ii).)

Solution:

Let S be the infinite half-line starting at $(1,0)$, i.e. $S = \{(x,0) \mid x \geq 1\}$. Then $\tau_{1,0} \in U$. Now $\tau_{1,0}^{-1} = \tau_{-1,0}$. But $\tau_{-1,0}$ moves $(1,0) \in S$ to $(0,0) \notin S$. Hence U is not a subgroup.

3.39. Find the orders of the symmetry group of

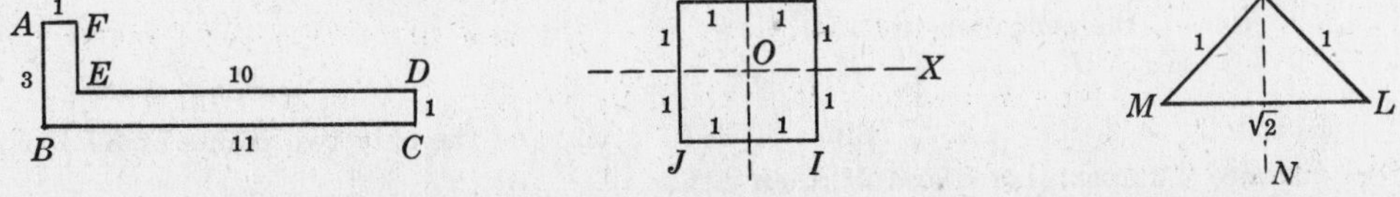

(*Hint*: Use Theorem 3.8 and argue intuitively. A useful intuitive approach for this problem is to cut a cardboard figure corresponding to each figure, label its vertices on both sides, and draw its perimeter onto a sheet of paper. The isometries are obtained on moving each cardboard figure so that it lies on the drawn perimeter.)

Solution:

(i) Let $S = ABCDEF$. Let the images of A, B, C, D, E and F under $\sigma \in I_S$ be denoted by A', B', C', D', E' and F'. By Theorem 3.8 it is easy to see intuitively that the image of the plane figure S will be the congruent figure $A'B'C'D'E'F'$, since Theorem 3.8 states that every element of I is a product of a reflection, rotation and translation. But if a rigid body is rotated, translated or reflected it retains the same shape. Now $A'B'$ must lie along AB, as all other sides of S are either smaller or larger than $d(A', B') = d(A, B)$, which means F' must lie on F and hence E' on E, etc. Thus σ must be the identity. Accordingly $I_s = \{\iota\}$ and is of order 1.

(ii) Let $S = IJGH$. Now $|I_S|$ is at least 8, for we have as members of I_S:

s_1: A reflection in the diagonal GI,
$Gs_1 = G,\ Hs_1 = J,\ Is_1 = I.$

s_2: A reflection in the diagonal HJ,
$Gs_2 = I,\ Hs_2 = H,\ Is_2 = G.$

s_3: A reflection in OX,
$Gs_3 = J,\ Hs_3 = I,\ Is_3 = H.$

s_4: A reflection in OY,
$Gs_4 = H,\ Hs_4 = G,\ Is_4 = J.$

s_5: A clockwise rotation about O of $0°$,
$Gs_5 = G,\ Hs_5 = H,\ Is_5 = I.$

s_6: A clockwise rotation about O of $90°$,
$Gs_6 = H,\ Hs_6 = I,\ Is_6 = J.$

s_7: A clockwise rotation about O of $180°$,
$Gs_7 = I,\ Hs_7 = J,\ Is_7 = G.$

s_8: A clockwise rotation about O of $270°$,
$Gs_8 = J,\ Hs_8 = G,\ Is_8 = H.$

That all of these are distinct is clear. Could $|I_S|$ be greater than 8?

Let $s \in I_S$. $d(G, I) = d(Gs, Is) = \sqrt{2}$. Only two pairs of points of S are a distance $\sqrt{2}$ apart, namely G, I and H, J. Therefore the line $GsIs$ is one of the diagonals of S. Similarly the line $HsJs$ is a diagonal of S. As s is a permutation of S, distinct points of S are mapped by s to distinct points. This means that at most the following possibilities arise:

(1) $GsIs$ is GI, $HsJs$ is HJ or JH

(2) $GsIs$ is IG, $HsJs$ is HJ or JH

(3) $GsIs$ is JH, $HsJs$ is GI or IG

(4) $GsIs$ is HJ, $HsJs$ is GI or IG

Since these represent eight distinct cases and each involves the movement of three points not in a straight line, by Lemma 3.7 at most one isometry could correspond to any one case. Then $|I_S| \leqq 8$. But we have already exhibited eight elements of I_S. Hence $|I_S| = 8$.

(iii) Let S be the triangle KLM. Then I_S contains the following elements:

(*a*) σ_1, the identity mapping of I,
$K\sigma_1 = K,\ L\sigma_1 = L,\ M\sigma_1 = M.$

(*b*) σ_2, the reflection in KN,
$K\sigma_2 = K,\ L\sigma_2 = M,\ M\sigma_2 = L.$

Hence $|I_S| \geqq 2$. Let $\sigma \in I_S$. Since $d(M, L) = \sqrt{2}$ and the only two points of KLM which are a distance $\sqrt{2}$ apart are L and M, then either

(a) $M\sigma = M$, $L\sigma = L$. Then as K must be a distance 1 from both M and L, $K\sigma = K$.

or

(b) $M\sigma = L$, $L\sigma = M$. Then as K must be a distance 1 from both $M\sigma$ and $L\sigma$, i.e. from M and L, and K is the only point of S which is a distance 1 from both L and M, $K\sigma = K$. Hence $|I_S| \leqq 2$, by Lemma 3.7. Therefore $|I_S| = 2$.

f. The dihedral groups

Let S be a regular n-gon, $n > 2$, e.g. one of the figures below. We will show that in any isometry of S, vertices are taken to vertices. This will make it easy to determine the order of the symmetry group of a regular n-gon, $n > 2$.

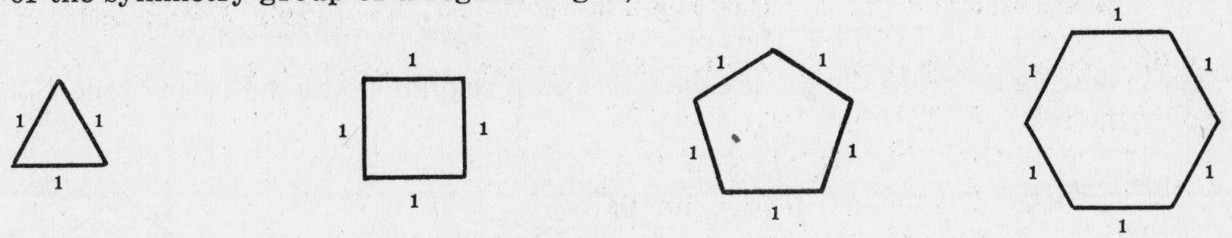

We will take the following geometrical lemma for granted.

Lemma 3.10: Every regular n-gon can be circumscribed by one and only one circle.

We call the center of the circumscribing circle of an n-gon its center.

Lemma 3.11: The center of a regular n-gon S is taken onto itself by any element of I_S.

Proof: Since every point of S is within a distance r, say, from the center O, and σ is an isometry, then every element of $S\sigma$ is within a distance r from $O\sigma$. Also, there are points of $S\sigma$ which are exactly a distance r from $O\sigma$, as there are points of S which are exactly a distance r from O. But $S\sigma = S$. Hence the circle with radius r and center $O\sigma$ is a circumscribing circle of S. But by the previous lemma there is only one circumscribing circle of S. Thus $O\sigma = O$.

Lemma 3.12: If S is a regular n-gon and $\sigma \in I_S$, then vertices of S are taken onto vertices of S by σ.

Proof: If A is a vertex of S and O is the center of S, $O\sigma = O$ by Lemma 3.11. $d(O\sigma, A\sigma) = d(O, A\sigma) = d(O, A)$. Hence $A\sigma$ is a distance r from O, where r is the radius of the circumscribing circle C. The only points of S on the circumference of C are vertices. But $A\sigma$ is an element of S on the circumference of C. Thus $A\sigma$ is a vertex.

The symmetry group of the regular n-gon is called *the dihedral group of degree n*. We can now calculate the orders of the dihedral groups.

Let the vertices of a regular n-gon S with center O be $A_1, \ldots, A_n$ (in a clockwise direction).

Let σ_j, $1 \leqq j \leqq n$, rotate S about O in a clockwise direction through an angle $\frac{2\pi(j-1)}{n}$ radians $\left(= \frac{360}{n}(j-1) \text{ degrees}\right)$ so that $A_1\sigma_j = A_j$. As an example, the effect of σ_3 on the regular pentagon is shown below.

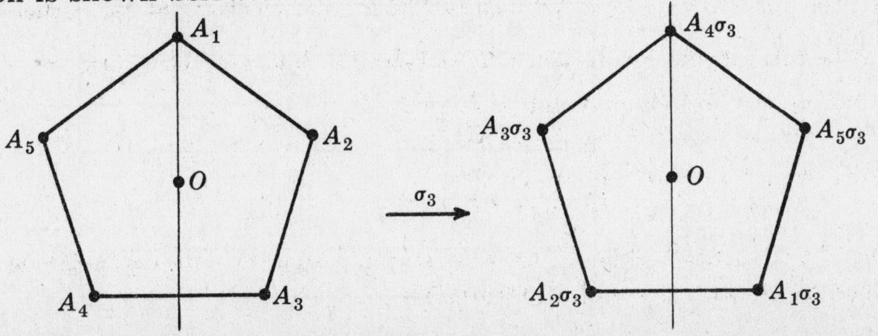

Let τ be the reflection about the line through A_1 and O, so that $A_1\tau = A_1$, $A_2\tau = A_n$. The effect of τ on the regular pentagon is shown.

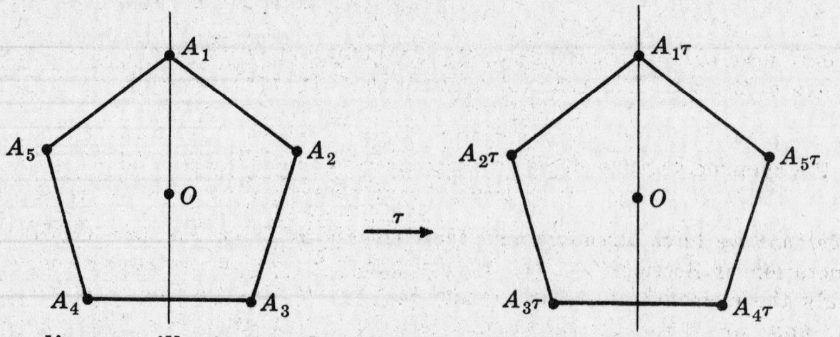

The following diagram illustrates the effect on a regular pentagon of the reflection τ followed by the rotation σ_3.

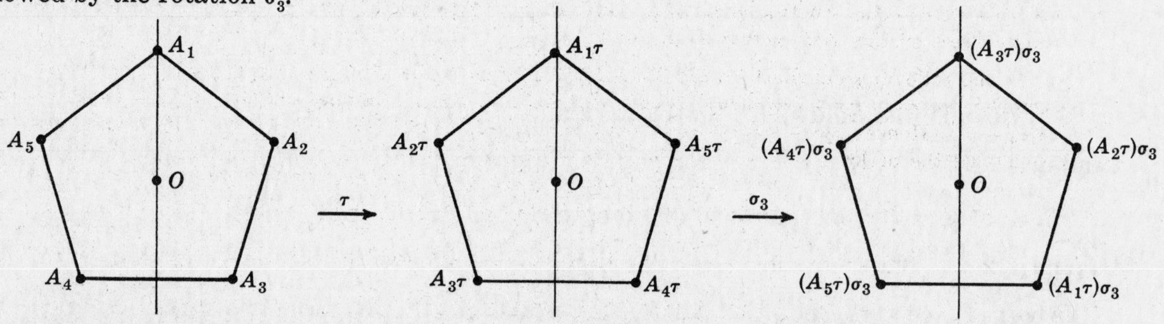

The elements $\sigma_1, \ldots, \sigma_n, \tau\sigma_1, \ldots, \tau\sigma_n$ are all distinct. For certainly $\sigma_j \neq \sigma_k$, $j \neq k$, as $A_1\sigma_j \neq A_1\sigma_k$, $j \neq k$. If $\tau\sigma_j = \sigma_k$, then $A_1\tau\sigma_j = A_1\sigma_j = A_1\sigma_k$. Thus $\tau\sigma_j = \sigma_k$ implies $j = k$. But $\tau\sigma_j = \sigma_j$ implies $\tau = \sigma_1$, the identity, contrary to assumption. Finally, $\tau\sigma_j = \tau\sigma_k$ implies $\sigma_j = \sigma_k$.

So there are at least $2n$ possible elements of the dihedral group of degree n. But we can easily show that there are no more than $2n$. For if $\sigma \in I_S$, S the regular n-gon, then there are n possibilities for $A_1\sigma$. As vertices are taken to vertices, $A_1\sigma$ is one of $A_1, \ldots, A_n$. $A_2\sigma$ has only two possibilities once $A_1\sigma$ is determined as $d(A_1\sigma, A_2\sigma) = d(A_1, A_2)$, and $A_2\sigma$ must also be a vertex. Once $A_1\sigma$ and $A_2\sigma$ are determined, $A_i\sigma$, $i = 3, 4, \ldots, n$ are also determined. Hence there are at most two elements $\sigma \in I_S$ which map $A_1\sigma$ to A_j. Thus there are at most $2n$ elements of I_S, and so $|I_S| = 2n$.

Let D_n denote the dihedral group of degree n.

Problems

3.40. Find D_3 and its multiplication table.

Solution:

The elements of D_3 are the $\sigma_j, \tau\sigma_j$ above. Note that $\sigma_j\sigma_2 = \sigma_{j+1}$ if $1 \leq j \leq 2$, and $\sigma_3\sigma_2 = \sigma_1$. Also note that $\tau^{-1} = \tau$ and $\sigma_i\tau = \tau^2\sigma_i\tau = \tau\tau\sigma_i\tau$. Now $\tau\sigma_1\tau = \sigma_1$, since σ_1 is the identity; $\tau\sigma_2\tau = \sigma_3$, as $A_1\tau\sigma_2\tau = A_1\sigma_2\tau = A_2\tau = A_3$; and $A_2\tau\sigma_2\tau = A_3\sigma_2\tau = A_1\tau = A_1$. So $\sigma_3\tau = \tau\sigma_2$ and $\tau\sigma_3 = \sigma_2\tau$. Accordingly the multiplication table is as follows:

	σ_1	σ_2	σ_3	τ	$\tau\sigma_2$	$\tau\sigma_3$
σ_1	σ_1	σ_2	σ_3	τ	$\tau\sigma_2$	$\tau\sigma_3$
σ_2	σ_2	σ_3	σ_1	$\tau\sigma_3$	τ	$\tau\sigma_2$
σ_3	σ_3	σ_1	σ_2	$\tau\sigma_2$	$\tau\sigma_3$	τ
τ	τ	$\tau\sigma_2$	$\tau\sigma_3$	σ_1	σ_2	σ_3
$\tau\sigma_2$	$\tau\sigma_2$	$\tau\sigma_3$	τ	σ_3	σ_1	σ_2
$\tau\sigma_3$	$\tau\sigma_3$	τ	$\tau\sigma_2$	σ_2	σ_3	σ_1

3.41. Show that the following are subgroups of D_3: (i) $\{\sigma_1\}$, (ii) $\{\sigma_1, \sigma_2, \sigma_3\}$, (iii) $\{\tau\sigma_2, \sigma_1\}$. (Notation is the same as in the preceding problem.)

Solution:

It is only necessary to check in each case that the set is not empty and if g, h belong to the set, gh^{-1} belongs to the set. It is easy to calculate gh^{-1} from the multiplication table of Problem 3.40.

3.42. Find D_4, the symmetry group of the square, and its multiplication table.

Solution:

Notice that we have already found the elements of D_4 in Problem 3.39(ii). We will, however, use the notation of Section 3.4f, i.e. σ_j, $j = 1, 2, 3, 4$, for the rotations and τ for the reflection. Accordingly the elements of D_4 are σ_j and $\tau\sigma_j$, $j = 1, 2, 3, 4$. Now $\sigma_j\sigma_2 = \sigma_{j+1}$ for $1 \leq j \leq 3$, $\sigma_4\sigma_2 = \sigma_1$, and $\sigma_i\sigma_j = \sigma_j\sigma_i$ for all i and j. Also $\tau^{-1} = \tau$, $\sigma_i\tau = \tau\tau\sigma_i\tau$, and $\tau\sigma_1\tau = \sigma_1$.

We show $\tau\sigma_2\tau = \sigma_4$. $A_1\tau\sigma_2\tau = A_1\sigma_2\tau = A_2\tau = A_4$, $A_2\tau\sigma_2\tau = A_4\sigma_2\tau = A_1\tau = A_1$, and $A_3\tau\sigma_2\tau = A_3\sigma_2\tau = A_4\tau = A_2$. Furthermore $A_1\sigma_4 = A_4$, $A_2\sigma_4 = A_1$, and $A_3\sigma_4 = A_2$. Since $\tau\sigma_2\tau$ and σ_4 have the same effect on the three points A_1, A_2 and A_3, $\tau\sigma_2\tau = \sigma_4$ by Lemma 3.7.

The following calculations facilitate the construction of a multiplication table for D_4. $\sigma_2^2 = \sigma_3$ implies $\tau\sigma_3\tau = \tau\sigma_2^2\tau = (\tau\sigma_2\tau)(\tau\sigma_2\tau) = \sigma_4^2 = \sigma_3$, and $\sigma_4 = \sigma_3\sigma_2$ implies $\tau\sigma_4\tau = \tau\sigma_3\sigma_2\tau = (\tau\sigma_3\tau)(\tau\sigma_2\tau) = \sigma_3\sigma_4 = \sigma_2$. Hence $\sigma_2\tau = \tau\tau\sigma_2\tau = \tau\sigma_4$, $\sigma_3\tau = \tau\tau\sigma_3\tau = \tau\sigma_3$, and $\sigma_4\tau = \tau\tau\sigma_4\tau = \tau\sigma_2$. It is now easy to construct the table:

	σ_1	σ_2	σ_3	σ_4	τ	$\tau\sigma_2$	$\tau\sigma_3$	$\tau\sigma_4$
σ_1	σ_1	σ_2	σ_3	σ_4	τ	$\tau\sigma_2$	$\tau\sigma_3$	$\tau\sigma_4$
σ_2	σ_2	σ_3	σ_4	σ_1	$\tau\sigma_4$	τ	$\tau\sigma_2$	$\tau\sigma_3$
σ_3	σ_3	σ_4	σ_1	σ_2	$\tau\sigma_3$	$\tau\sigma_4$	τ	$\tau\sigma_2$
σ_4	σ_4	σ_1	σ_2	σ_3	$\tau\sigma_2$	$\tau\sigma_3$	$\tau\sigma_4$	τ
τ	τ	$\tau\sigma_2$	$\tau\sigma_3$	$\tau\sigma_4$	σ_1	σ_2	σ_3	σ_4
$\tau\sigma_2$	$\tau\sigma_2$	$\tau\sigma_3$	$\tau\sigma_4$	τ	σ_4	σ_1	σ_2	σ_3
$\tau\sigma_3$	$\tau\sigma_3$	$\tau\sigma_4$	τ	$\tau\sigma_2$	σ_3	σ_4	σ_1	σ_2
$\tau\sigma_4$	$\tau\sigma_4$	τ	$\tau\sigma_2$	$\tau\sigma_3$	σ_2	σ_3	σ_4	σ_1

3.5 THE GROUP OF MÖBIUS TRANSFORMATIONS

a. Defining the group

The complex numbers can be represented as points of the Euclidean plane E, the complex number $z = x + iy$ corresponds to the point with coordinates (x, y). Instead of inquiring (as we did in Section 3.4c) what are the permutations of E that preserve distance, we inquire what are the permutations of E that preserve both angles and their orientation. These are called conformal mappings. It can be shown (see Ford, L. R., *Automorphic Functions*, Chelsea, 1951) that the mappings

$$\sigma = \sigma(a, b, c, d) : z \to \frac{az + b}{cz + d}$$

where a, b, c, d are fixed complex numbers such that $ad - bc \neq 0$, preserve angles and their orientation. But $\sigma(a, b, c, d)$ is not always a mapping of E to E. Two things can go wrong. If $c \neq 0$, then:

(i) $z\sigma$ is not defined if $z = -d/c$, as then the denominator becomes 0.

(ii) There is no complex number that maps to a/c. For suppose $z\sigma = a/c$, then $az + b = (cz + d)a/c$ and $b - ad/c = 0$ and hence $bc - ad = 0$, the very condition we assumed did not hold.

It seems as if we have been cheated in our efforts to argue analogously to Section 3.4c in order to prove the σ for various a, b, c, d form a group, because not all the σ are permutations of E.

However, by adding an extra element ∞ to E and forming $E \cup \{\infty\} = \bar{E}$ we can overcome these difficulties. ∞ is any object outside E. It is customary to write ∞ for historical reasons. The reader is cautioned that just as the symbol x can have different meanings (e.g. x is sometimes a number and sometimes an element of a group or a groupoid, etc.), so ∞ has different meanings. The ∞ we introduce should not be confused with the ∞ in such expressions as $\lim_{x \to \infty} \frac{1}{x} = 0$; it is logically distinct.

Our idea is to extend σ to $\bar{E}$ in order to patch up the difficulties (i) and (ii) above so that $\sigma \in S_{\bar{E}}$, the symmetric group on $\bar{E}$.

$$(a)\ \text{If}\ c = 0,\ \text{define}\ z\sigma = \frac{az+b}{cz+d}\ \text{for any complex number}\ z,\ \text{and put}\ \infty\sigma = \infty.$$

$$(b)\ \text{If}\ c \neq 0,\ \text{put}\ z\sigma = \frac{az+b}{cz+d}\ \text{for}\ z \neq -d/c,\ z \in C,\ \text{the complex numbers. Put} \qquad (3.7)$$

$$(-d/c)\sigma = \infty \quad \text{and} \quad \infty\sigma = a/c$$

In (a) we have no real problem. Having had to add an extra element, we just let it map to itself. In (b) we neatly get rid of both difficulties (i) and (ii) above, for we have both defined $(-d/c)\sigma$ and found an element to map to a/c.

M will denote the set of all mappings of $\bar{E}$ to $\bar{E}$ defined by (3.7). We will show that M is a subgroup of $S_{\bar{E}}$, leaving most of the checking of details to the problems.

First, each of the $\sigma(a, b, c, d)$ is a member of $S_{\bar{E}}$ (Problem 3.45). Next, the inverse of $\sigma(a, b, c, d)$ is given by $\sigma(-d, b, c, -a)$ (Problem 3.44). Finally,

$$\sigma(a_1, b_1, c_1, d_1)\,\sigma(a_2, b_2, c_2, d_2) = \sigma(a_3, b_3, c_3, d_3)$$

for some choice of a_3, b_3, c_3, d_3 (Problem 3.46). Note that $\sigma(1, 0, 0, 1)$ is the identity mapping (denoted by ι). Hence the product of an element of M and the inverse of an element in M belongs to M. Thus M is a subgroup of $S_{\bar{E}}$. It is called the group of Möbius transformations.

Problems

3.43. Determine the image of (i) i, (ii) $1+2i$, (iii) ∞, and (iv) $-1/3$ under $\sigma(2, 1, 3, 1)$.

Solutions:

(i) $i\sigma = \frac{2i+1}{3i+1} = \frac{7}{10} - \frac{1}{10}i$ (iii) $\infty\sigma = 2/3$

(ii) $(1+2i)\sigma = \frac{2(1+2i)+1}{3(1+2i)+1} = \frac{9}{13} - \frac{1}{26}i$ (iv) $-1/3\sigma = \infty$

3.44. Show that $\sigma(-d, b, c, -a)$ is the inverse of $\sigma(a, b, c, d)$, given $ad - bc \neq 0$.

Solution:

Case (a): $c = 0$.

$$z\,\sigma(a, b, 0, d)\,\sigma(-d, b, 0, -a) = \left[\left(\frac{az+b}{d}\right)(-d) + b\right] \Big/ (-a) = z$$

$$\infty\,\sigma(a, b, 0, d)\,\sigma(-d, b, 0, -a) = \infty\,\sigma(-d, b, 0, -a) = \infty$$

Case (b): $c \neq 0$.

(i) $z = -d/c$.

$$z\,\sigma(a, b, c, d)\,\sigma(-d, b, c, -a) = \infty\,\sigma(-d, b, c, -a) = -d/c = z$$

(ii) $z \neq -d/c$ or ∞. Then

$$z\,\sigma(a,b,c,d)\,\sigma(-d,b,c,-a) = \left[\left(\frac{az+b}{cz+d}\right)(-d)+b\right]\Big/\left[\left(\frac{az+b}{cz+d}\right)c-a\right]$$

$$= \frac{bcz+bd-adz-bd}{azc+bc-acz-ad} = \frac{(bc-ad)z}{bc-ad} = z$$

$$\infty\,\sigma(a,b,c,d) = a/c \quad\text{and}\quad (a/c)\,\sigma(-d,b,c,-a) = \infty$$

Hence $\sigma(a,b,c,d)\,\sigma(-d,b,c,-a) = \iota$.

Similarly we can show that $\sigma(-d,b,c,-a)\,\sigma(a,b,c,d) = \iota$.

Hence $\sigma(-d,b,c,-a) = \sigma(a,b,c,d)^{-1}$.

3.45. Prove $\sigma(a,b,c,d) \in S_{\bar{E}}$ for any choice of a,b,c,d such that $ad-bc \neq 0$.

Solution:

In Problem 3.44 we have seen that each $\sigma(a,b,c,d)$ has an inverse. By Theorem 2.4, page 36, any mapping of a set into itself which has an inverse is a one-to-one onto mapping. Therefore $\sigma(a,b,c,d)$ is one-to-one and onto, and so $\sigma(a,b,c,d)$ is an element of $S_{\bar{E}}$.

3.46. Show that

$$\sigma(a_1,b_1,c_1,d_1)\,\sigma(a_2,b_2,c_2,d_2) = \sigma(a_3,b_3,c_3,d_3)$$

where $a_3 = a_1a_2+b_2c_1$, $b_3 = a_2b_1+b_2d_1$, $c_3 = a_1c_2+d_2c_1$, and $d_3 = b_1c_2+d_1d_2$. Prove also that $a_3d_3-b_3c_3 \neq 0$. (*Hint*: This is very much an endurance test.)

Solution:

Let $\sigma(a_1,b_1,c_1,d_1) = \sigma_1$, $\sigma(a_2,b_2,c_2,d_2) = \sigma_2$, and $\sigma(a_3,b_3,c_3,d_3) = \sigma_3$. Note that

$$a_3d_3-b_3c_3 = (a_1a_2+b_2c_1)(b_1c_2+d_1d_2)-(b_1a_2+b_2d_1)(a_1c_2+d_2c_1)$$

$$= (a_1d_1-b_1c_1)a_2d_2-(a_1d_1-b_1c_1)b_2c_2 = (a_1d_1-b_1c_1)(a_2d_2-b_2c_2) \neq 0$$

Hence σ_3 is a Möbius transformation.

Now if $z \in \bar{E}$ satisfies

(A) if $c_1 \neq 0$, $z \neq -d_1/c_1$

(B) if $c_2 \neq 0$, $z\sigma_1 \neq -d_2/c_2$

(C) if $c_3 \neq 0$, $z \neq -d_3/c_3$

(D) $z \neq \infty$

then
$$z\sigma_1\sigma_2 = \frac{\dfrac{a_1z+b_1}{c_1z+d_1}a_2+b_2}{\dfrac{a_1z+b_1}{c_1z+d_1}c_2+d_2} = \frac{(a_1a_2+b_2c_1)z+(a_2b_1+b_2d_1)}{(a_1c_2+d_2c_1)z+(b_1c_2+d_1d_2)} = z\sigma_3$$

Thus except for restrictions (A), (B), (C) and (D), there is nothing more to prove. Since $\sigma_1\sigma_2$ and σ_3 are permutations, we may ignore one of these cases, say (D). This obtains because if for all complex numbers z, $z\sigma_1\sigma_2 = z\sigma_3$, then ∞ can only be mapped to one element of $\bar{E}$ by $\sigma_1\sigma_2$ and σ_3. Accordingly we shall not consider $z = \infty$ in the following case by case analysis.

Case (a), $c_1 = 0$. We have two possibilities: (i) $c_2 = 0$, (ii) $c_2 \neq 0$.

(i) $c_2 = 0$. Then $c_3 = a_1c_2+c_1d_2 = 0$. Thus (A), (B) and (C) do not restrict z, and we can conclude $\sigma_1\sigma_2 = \sigma_3$.

(ii) $c_2 \neq 0$. We first show $z = -d_3/c_3$ if and only if $z\sigma_1 = -d_2/c_2$. $z = -d_3/c_3$ implies

$$z\sigma_1 = \frac{a_1(-d_3/c_3)+b_1}{d_1} = \frac{-a_1d_3+c_3b_1}{c_3d_1} = \frac{-a_1(b_1c_2+d_1d_2)+a_1c_2b_1}{a_1c_2d_1} = -\frac{d_2}{c_2}$$

A simple computation shows that $z\sigma_1 = (a_1z+b_1)/d_1 = -d_2/c_2$ implies $z = -d_3/c_3$. Now if $z\sigma_1 = -d_2/c_2$, then $z\sigma_1\sigma_2 = (-d_2/c_2)\sigma_2 = \infty$ and, as $z = -d_3/c_3$, $z\sigma_3 = \infty$. $z = -d_3/c_3$ implies $z\sigma_1 = -d_2/c_2$. Thus $z\sigma_1\sigma_2 = \infty = z\sigma_3$ in this case, and so $\sigma_1\sigma_2 = \sigma_3$.

Case (b), $c_1 \neq 0$. Again there are two possibilities: (i) $c_2 = 0$, (ii) $c_2 \neq 0$.

(i) $c_2 = 0$. Then $c_3 = c_1 d_2$ and, as $a_2 d_2 - c_2 b_2 \neq 0$ implies $d_2 \neq 0$, it follows that $c_3 \neq 0$. Note also that $-d_3/c_3 = -d_1 d_2/c_1 d_2 = -d_1/c_1$. Hence we need consider only the possibility $z = -d_1/c_1$. If $z = -d_1/c_1$, then, as $c_2 = 0$, $z\sigma_1\sigma_2 = \infty\sigma_2 = \infty$ while $z\sigma_3 = \infty$. Hence $\sigma_1\sigma_2 = \sigma_3$.

(ii) $c_2 \neq 0$. $c_3 = a_1 c_2 + c_1 d_2 = 0$ if and only if $a_1/c_1 = -d_2/c_2$.

(α) $c_3 = 0$. If $z = -d_1/c_1$, then $z\sigma_1\sigma_2 = \infty\sigma_2 = a_2/c_2$ while

$$z\sigma_3 = \frac{(-d_1/c_1)(a_1a_2 + c_1b_2) + (b_1a_2 + d_1b_2)}{b_1c_2 + d_1d_2} = \frac{a_1(-a_1d_1 + b_1c_1)}{c_1(b_1c_2 + d_1d_2)}$$

From $a_1/c_1 = -d_2/c_2$ we have $d_2 = (-a_1/c_1)c_2$ and

$$z\sigma_3 = \frac{a_2(-a_1d_1 + b_1c_1)}{c_2(b_1c_1 - a_1d_1)} = \frac{a_2}{c_2}$$

Now $z\sigma_1 = -d_2/c_2 = a_1/c_1$ only if $z = \infty$, and we need not consider this case. Hence $\sigma_1\sigma_2 = \sigma_3$.

(β) $c_3 \neq 0$. Then $a_1/c_1 \neq -d_2/c_2$. If $z = -d_1/c_2$, $z\sigma_1\sigma_2 = \infty\sigma_2 = a_2/c_2$.

$$-\frac{d_1}{c_1}\sigma_3 = \frac{(a_1a_2 + b_2c_1)(-d_1/c_1) + (a_2b_1 + b_2d_1)}{(a_1c_2 + d_2c_1)(-d_1/c_1) + (b_1c_2 + d_1d_2)} = \frac{a_2(-a_1d_1 + b_1c_1)}{c_2(-a_1d_1 + b_1c_1)} = \frac{a_2}{c_2}$$

Finally if $z\sigma_1 = -d_2/c_2$, then $z = -d_3/c_3$; for $(a_1z + b_1)/(c_1z + d_1) = -d_2/c_2$ implies $a_1c_2z + b_1c_2 = -c_1d_2z - d_1d_2$. Hence $(a_1c_2 + c_1d_2)z = -(b_1c_2 + d_1d_2)$ and, as $c_3 = a_1c_2 + d_2c_1 \neq 0$, $z = -(b_1c_2 + d_1d_2)/(a_1c_2 + d_2c_1) = -d_3/c_3$. Therefore

$$-\frac{d_3}{c_3}\sigma_1\sigma_2 = -\frac{d_2}{c_2}\sigma_2 = \infty = -\frac{d_3}{c_3}\sigma_3$$

3.47. Let $\sigma(a, b, c, d) = \sigma_1$ be a Möbius transformation and k a nonzero complex number. Show that $\sigma(a, b, c, d) = \sigma(ka, kb, kc, kd)$.

Solution:

Denote $\sigma(ka, kb, kc, kd)$ by $\bar{\sigma}$. If $z \neq \infty$ or $-d/c$ (if $c \neq 0$), then, as $k \neq 0$, $z\sigma_1 = \dfrac{az + b}{cz + d} = \dfrac{kaz + kb}{kcz + kd} = z\bar{\sigma}$. To treat the special cases $z = \infty$ or $-d/c$ (when $c \neq 0$), we first assume $c = 0$. Then $kc = 0$, and $\infty\sigma_1 = \infty = \infty\bar{\sigma}$ by definition. Secondly, if $c \neq 0$, then $kc \neq 0$. Therefore $\infty\sigma_1 = \dfrac{a}{c} = \dfrac{ka}{kc} = \infty\,\bar{\sigma}$ and $-\dfrac{d}{c}\sigma_1 = \infty = (-kd/kc)\,\bar{\sigma}$. Thus for all possible choices of z, we have $z\sigma_1 = z\bar{\sigma}$ and hence $\sigma_1 = \bar{\sigma}$.

3.48. Show that the set of all Möbius transformations $= \{\sigma(a, b, c, d) \mid ad - bc = 1\}$.

Solution:

Let $\bar{M} = \{\sigma(a, b, c, d) \mid ad - bc = 1\}$. If $\sigma = \sigma(a, b, c, d)$ is any Möbius transformation, then, by definition, $D = ad - bc \neq 0$. From Problem 3.47 above we know that $\sigma = \sigma\left(\dfrac{a}{\sqrt{D}}, \dfrac{b}{\sqrt{D}}, \dfrac{c}{\sqrt{D}}, \dfrac{d}{\sqrt{D}}\right)$. But $\dfrac{a}{\sqrt{D}}\dfrac{d}{\sqrt{D}} - \dfrac{b}{\sqrt{D}}\dfrac{c}{\sqrt{D}} = \dfrac{ad - bc}{ad - bc} = 1$. Hence $\sigma \in \bar{M}$. Furthermore, any element of $\bar{M}$ is obviously a Möbius transformation. Thus $\bar{M}$ is the set of all Möbius transformations.

3.49. Suppose $ad - bc \neq 0$. Prove $\sigma(a, b, c, d) = \iota$ iff $a = d$ and $b = c = 0$.

Solution:

If $\sigma(a, b, c, d) = \iota$, then $z = z\sigma(a, b, c, d)$. Hence $\infty = \infty\sigma(a, b, c, d)$ implies $c = 0$.

$0 = 0\sigma(a, b, c, d) = b/d$ implies $b = 0$.

$1 = 1\sigma(a, b, c, d) = 1 \cdot \dfrac{a}{d}$ implies $a = d$.

If $a = d$ and $c = b = 0$, then, using the results of Problem 3.47, $\sigma(a, b, c, d) = \sigma(a, 0, 0, a) = \sigma(1, 0, 0, 1) = \iota$.

b. 2×2 matrices

In this section we will define the group of two by two matrices and indicate its relationship to the group of Möbius transformations.

An array

$$\begin{pmatrix} a & c \\ b & d \end{pmatrix} \tag{3.8}$$

of complex numbers a, b, c, d is called a two by two (2×2) matrix. (Since we will only deal with 2×2 matrices, we usually omit the adjective 2×2.) a, b, c and d are called the *entries* of matrix *(3.8)*. Two matrices are equal if and only if their entries are the same, i.e. $\begin{pmatrix} a & c \\ b & d \end{pmatrix} = \begin{pmatrix} a' & c' \\ b' & d' \end{pmatrix}$ if and only if $a = a'$, $b = b'$, $c = c'$ and $d = d'$. We define the product of two matrices as follows:

$$\begin{pmatrix} a & c \\ b & d \end{pmatrix}\begin{pmatrix} a' & c' \\ b' & d' \end{pmatrix} = \begin{pmatrix} aa' + cb' & ac' + cd' \\ ba' + db' & bc' + dd' \end{pmatrix}$$

The product of two matrices is clearly a matrix. A calculation shows (Problem 3.51) matrix multiplication is an associative binary operation.

The matrix $I = \begin{pmatrix} 1 & 0 \\ 0 & 1 \end{pmatrix}$ is the identity matrix, since

$$\begin{pmatrix} a & c \\ b & d \end{pmatrix}\begin{pmatrix} 1 & 0 \\ 0 & 1 \end{pmatrix} = \begin{pmatrix} a+0 & 0+c \\ b+0 & 0+d \end{pmatrix} = \begin{pmatrix} a & c \\ b & d \end{pmatrix} = \begin{pmatrix} 1 & 0 \\ 0 & 1 \end{pmatrix}\begin{pmatrix} a & c \\ b & d \end{pmatrix}$$

In order to determine which matrices have inverses, we define the *determinant* of the matrix $A = \begin{pmatrix} a & c \\ b & d \end{pmatrix}$ to be the complex number $D(A) = ad - bc$. It is easy to show $D(A)\,D(B) = D(AB)$ for any two matrices A and B (see Problem 3.52). If $A = \begin{pmatrix} a & c \\ b & d \end{pmatrix}$ and $D(A) \neq 0$, then

$$\begin{pmatrix} \dfrac{d}{D(A)} & \dfrac{-c}{D(A)} \\[2ex] \dfrac{-b}{D(A)} & \dfrac{a}{D(A)} \end{pmatrix}$$

is the inverse of A (see Problem 3.53(i)). If $D(A) = 0$, then A has no inverse (see Problem 3.53(ii)).

We claim that the set

$$\mathcal{M} = \left\{\begin{pmatrix} a & c \\ b & d \end{pmatrix} \middle| \; a, b, c, d \text{ complex numbers, } ad - bc \neq 0\right\}$$

with the operation of matrix multiplication is a group. For if $A, B \in \mathcal{M}$, then, as $D(A)\,D(B) = D(AB)$, $D(AB) \neq 0$ and $AB \in \mathcal{M}$. The determinant of $I = \begin{pmatrix} 1 & 0 \\ 0 & 1 \end{pmatrix}$ is 1. Hence the identity I is in $\mathcal{M}$. Furthermore if $A \in \mathcal{M}$, then $A^{-1} \in \mathcal{M}$ since $D(A)\,D(A^{-1}) = D(I) = 1$ implies $D(A^{-1}) \neq 0$. Therefore $\mathcal{M}$ is a group. We call $\mathcal{M}$ the *group of* 2×2 *matrices* over the complex numbers.

The relationship between the group of Möbius transformations and the group of 2×2 matrices is now evident. For in Problem 3.46 we found

$$\sigma(a_1, b_1, c_1, d_1)\,\sigma(a_2, b_2, c_2, d_2) = \sigma(a_3, b_3, c_3, d_3)$$

where $a_3 = a_1a_2 + b_2c_1$, $b_3 = a_2b_1 + b_2d_1$, $c_3 = a_1c_2 + d_2c_1$, and $d_3 = b_1c_2 + d_1d_2$. But by the definition of multiplication,

$$\begin{pmatrix} a_1 & c_1 \\ b_1 & d_1 \end{pmatrix}\begin{pmatrix} a_2 & c_2 \\ b_2 & d_2 \end{pmatrix} = \begin{pmatrix} a_1a_2 + c_1b_2 & a_1c_2 + c_1d_2 \\ b_1a_2 + d_1b_2 & b_1c_2 + d_1d_2 \end{pmatrix} = \begin{pmatrix} a_3 & c_3 \\ b_3 & d_3 \end{pmatrix}$$

This does not mean the group of 2×2 matrices is identical with the group of Möbius transformations. For we have seen in Problem 3.47 that $\sigma(a, b, c, d) = \sigma(ka, kb, kc, kd)$ for any complex number $k \neq 0$. But $\begin{pmatrix} a & c \\ b & d \end{pmatrix} \neq \begin{pmatrix} ka & kc \\ kb & kd \end{pmatrix}$, e.g. if $k = -1$. The precise relationship between the two groups will be given in Problem 4.81, page 120.

Problems

3.50. (i) Multiply (a) $\begin{pmatrix} 3 & i \\ -2 & 4 \end{pmatrix}\begin{pmatrix} i & -1 \\ 5 & 2 \end{pmatrix}$, (b) $\begin{pmatrix} 1 & c \\ 0 & 1 \end{pmatrix}\begin{pmatrix} 1 & c \\ 0 & 1 \end{pmatrix}$, (c) $\begin{pmatrix} 0 & 1 \\ 1 & 0 \end{pmatrix}\begin{pmatrix} a & c \\ b & d \end{pmatrix}$, (d) $\begin{pmatrix} a & c \\ b & d \end{pmatrix}\begin{pmatrix} 0 & 1 \\ 1 & 0 \end{pmatrix}$.

(ii) Find the inverse of (a) $\begin{pmatrix} 7 & i \\ 3 & -2 \end{pmatrix}$, (b) $\begin{pmatrix} 0 & 1 \\ 1 & 0 \end{pmatrix}$, (c) $\begin{pmatrix} i & 0 \\ 0 & i \end{pmatrix}$, (d) $\begin{pmatrix} a & c \\ 0 & d \end{pmatrix}$.

Solution:

(i) (a) $\begin{pmatrix} 8i & -3 + 2i \\ 20 - 2i & 10 \end{pmatrix}$ (b) $\begin{pmatrix} 1 & 2c \\ 0 & 1 \end{pmatrix}$ (c) $\begin{pmatrix} b & d \\ a & c \end{pmatrix}$ (d) $\begin{pmatrix} c & a \\ d & b \end{pmatrix}$

(ii) (a) $\begin{pmatrix} \dfrac{2}{14 + 3i} & \dfrac{i}{14 + 3i} \\ \dfrac{3}{14 + 3i} & \dfrac{-7}{14 + 3i} \end{pmatrix}$ (b) $\begin{pmatrix} 0 & 1 \\ 1 & 0 \end{pmatrix}$ (c) $\begin{pmatrix} -i & 0 \\ 0 & -i \end{pmatrix}$ (d) $\begin{pmatrix} 1/a & -c/ad \\ 0 & 1/d \end{pmatrix}$

3.51. Show that matrix multiplication is an associative binary operation.

Solution:

Let $A = \begin{pmatrix} a_1 & c_1 \\ b_1 & d_1 \end{pmatrix}$, $B = \begin{pmatrix} a_2 & c_2 \\ b_2 & d_2 \end{pmatrix}$, $C = \begin{pmatrix} a_3 & c_3 \\ b_3 & d_3 \end{pmatrix}$. Then

$$(AB)C = \left[\begin{pmatrix} a_1 & c_1 \\ b_1 & d_1 \end{pmatrix}\begin{pmatrix} a_2 & c_2 \\ b_2 & d_2 \end{pmatrix}\right]\begin{pmatrix} a_3 & c_3 \\ b_3 & d_3 \end{pmatrix} = \begin{pmatrix} a_1a_2 + c_1b_2 & a_1c_2 + c_1d_2 \\ b_1a_2 + d_1b_2 & b_1c_2 + d_1d_2 \end{pmatrix}\begin{pmatrix} a_3 & c_3 \\ b_3 & d_3 \end{pmatrix}$$

$$= \begin{pmatrix} (a_1a_2 + c_1b_2)a_3 + (a_1c_2 + c_1d_2)b_3 & (a_1a_2 + c_1b_2)c_3 + (a_1c_2 + c_1d_2)d_3 \\ (b_1a_2 + d_1b_2)a_3 + (b_1c_2 + d_1d_2)b_3 & (b_1a_2 + d_1b_2)c_3 + (b_1c_2 + d_1d_2)d_3 \end{pmatrix}$$

$$A(BC) = \begin{pmatrix} a_1 & c_1 \\ b_1 & d_1 \end{pmatrix}\begin{pmatrix} a_2a_3 + c_2b_3 & a_2c_3 + c_2d_3 \\ b_2a_3 + d_2b_3 & b_2c_3 + d_2d_3 \end{pmatrix}$$

$$= \begin{pmatrix} a_1(a_2a_3 + c_2b_3) + c_1(b_2a_3 + d_2b_3) & a_1(a_2c_3 + c_2d_3) + c_1(b_2c_3 + d_2d_3) \\ b_1(a_2a_3 + c_2b_3) + d_1(b_2a_3 + d_2b_3) & b_1(a_2c_3 + c_2d_3) + d_1(b_2c_3 + d_2d_3) \end{pmatrix}$$

A check of entries shows $(AB)C = A(BC)$.

3.52. Prove $D(A)\,D(B) = D(AB) = D(B)\,D(A)$, for any two matrices A and B.

Solution:

Let $A = \begin{pmatrix} a & c \\ b & d \end{pmatrix}$, $B = \begin{pmatrix} a' & c' \\ b' & d' \end{pmatrix}$. Then $AB = \begin{pmatrix} aa' + cb' & ac' + cd' \\ ba' + db' & bc' + dd' \end{pmatrix}$ and

$$\begin{aligned} D(A)\,D(B) &= (ad-bc)(a'd'-b'c') = aa'dd' + bb'cc' - adb'c' - bca'd' \\ &= (aa'+cb')(bc'+dd') - (ba'+db')(ac'+cd') \\ &= D(AB) \end{aligned}$$

Because multiplication of complex numbers is commutative, $D(A)\,D(B) = D(B)\,D(A)$.

3.53. Let $A = \begin{pmatrix} a & c \\ b & d \end{pmatrix}$ be a matrix. Prove:

(i) If $D(A) \neq 0$, $\begin{pmatrix} \frac{d}{D(A)} & \frac{-c}{D(A)} \\ \frac{-b}{D(A)} & \frac{a}{D(A)} \end{pmatrix}$ is the inverse of A.

(ii) If $D(A) = 0$, A has no inverse.

Solution:

(i) $$\begin{pmatrix} a & c \\ b & d \end{pmatrix}\begin{pmatrix} \frac{d}{D(A)} & \frac{-c}{D(A)} \\ \frac{-b}{D(A)} & \frac{a}{D(A)} \end{pmatrix} = \begin{pmatrix} \frac{ad-bc}{D(A)} & \frac{-ac+ac}{D(A)} \\ \frac{bd-db}{D(A)} & \frac{-bc+da}{D(A)} \end{pmatrix} = \begin{pmatrix} 1 & 0 \\ 0 & 1 \end{pmatrix} = \begin{pmatrix} \frac{d}{D(A)} & \frac{-c}{D(A)} \\ \frac{-b}{D(A)} & \frac{a}{D(A)} \end{pmatrix}\begin{pmatrix} a & c \\ b & d \end{pmatrix}$$

(ii) If A' is an inverse of A, then $D(A')\,D(A) = D(I)$ where I is the identity matrix. But $D(I) = 1$ and $D(A) = 0$. Since zero times any number is zero, A cannot have an inverse.

3.54. Show that the following sets of matrices are subgroups of the group $\mathcal{M}$ of 2×2 matrices.

(i) $\mathcal{R} = \left\{\begin{pmatrix} a & c \\ b & d \end{pmatrix} \middle|\ a, b, c, d \text{ real numbers, } ad - bc \neq 0\right\}$

(ii) $\mathcal{U} = \left\{\begin{pmatrix} a & c \\ b & d \end{pmatrix} \middle|\ a, b, c, d \text{ complex numbers, } ad - bc = 1\right\}$

(iii) $\mathcal{P} = \left\{\begin{pmatrix} a & c \\ 0 & d \end{pmatrix} \middle|\ a, d, c \text{ complex numbers, } ad \neq 0\right\}$

Solution:

(i) $\mathcal{R} \subseteq \mathcal{M}$ and $I = \begin{pmatrix} 1 & 0 \\ 0 & 1 \end{pmatrix} \in \mathcal{R}$. If $A = \begin{pmatrix} a & c \\ b & d \end{pmatrix} \in \mathcal{R}$ then, as $\frac{a}{D(A)}, \frac{b}{D(A)}, \frac{c}{D(A)}, \frac{d}{D(A)}$ are all real numbers, $A^{-1} \in \mathcal{R}$ and $\mathcal{R}$ is a subgroup of $\mathcal{M}$, as it is easy to check that $\mathcal{R}$ is closed with respect to products.

(ii) $\mathcal{U} \subseteq \mathcal{M}$ and $D(I) = 1$, so $I \in \mathcal{U}$. Let $A \in \mathcal{U}$. Then $D(A)\,D(A^{-1}) = D(I) = 1$ implies $D(A^{-1}) = 1$. Hence $\mathcal{U}$ is a subgroup of $\mathcal{M}$, as it is easy to check that $\mathcal{U}$ is closed with respect to products.

(iii) $\mathcal{P} \subseteq \mathcal{M}$ and $I \in \mathcal{P}$. The inverse of $A = \begin{pmatrix} a & c \\ 0 & d \end{pmatrix}$ is $\begin{pmatrix} 1/a & -c/ad \\ 0 & 1/d \end{pmatrix}$. Hence $A^{-1} \in \mathcal{P}$ if $A \in \mathcal{P}$. Thus $\mathcal{P}$ is a subgroup of $\mathcal{M}$, as it is easily checked that $\mathcal{P}$ is closed with respect to products.

3.6 SYMMETRIES OF AN ALGEBRAIC STRUCTURE

a. Automorphisms of groupoids

We have discussed isometries of plane figures. The corresponding one-to-one onto mapping of a groupoid is defined as follows.

Definition: Let G be a groupoid. Then an automorphism α of G is a one-to-one mapping of G onto G such that $(ab)\alpha = a\alpha b\alpha$ for all $a, b \in G$.

Note that isometries preserve distance whereas automorphisms of groupoids preserve groupoid multiplication. As the analog to Theorem 3.5, page 67, we have

Theorem 3.13: The set A of all automorphisms of a groupoid G is a subgroup of S_G, the symmetric group on G.

Proof:

I. ι, the identity mapping, belongs to A; hence $A \neq \emptyset$.

II. If $\alpha, \beta \in A$ and $a, b \in G$ then

$$(ab)(\alpha\beta) = ((ab)\alpha)\beta = [(a\alpha)(b\alpha)]\beta = [a(\alpha\beta)][b(\alpha\beta)]$$

Thus the composition of mappings is a binary operation in A.

III. The identity mapping is an automorphism, and so A contains an identity.

IV. $(A, \cdot)$ is associative.

V. If $\alpha \in A$, let α^{-1} be the inverse of α; since $\alpha \in S_G$, α^{-1} makes sense. Let $a, b \in G$ and choose $a', b' \in A$ so that $a'\alpha = a$, $b'\alpha = b$. Then $(a'b')\alpha = ab$. Hence $(ab)\alpha^{-1} = a'b' = (a\alpha^{-1})(b\alpha^{-1})$. Thus A is a group, and hence a subgroup of S_G.

We call A the automorphism group of the groupoid G and sometimes denote it by aut(G).

Problems

3.55. Find the automorphism group of $(G, \cdot)$ where $G = \{a, b\}$ and $\cdot$ is defined by the multiplication table

(a)

	a	b
a	a	b
b	b	a

(b)

	a	b
a	a	b
b	a	b

Solution:

(a) ι, the identity mapping, is the only automorphism for the only other possibility is the mapping α defined by $a\alpha = b$ and $b\alpha = a$. But $(bb)\alpha = a\alpha = b$ and $b\alpha b\alpha = aa = a$; hence $(bb)\alpha \neq (b\alpha) \cdot (b\alpha)$. Thus α is not an automorphism.

(b) Define α by $a\alpha = b$ and $b\alpha = a$. Note that $xy = y$. Hence $(xy)\alpha = y\alpha = (x\alpha)(y\alpha)$. The automorphism group therefore contains the two elements ι and α. Notice $\alpha\alpha = \iota$.

3.56. Find the automorphism group of A_3. (For table of A_3 see Problem 3.23(iii).)

Solution:

Theorem 2.6, page 44, showed that for any homomorphism α of a groupoid G into a groupoid G', i.e. a mapping of G into G' such that $(g_1 g_2)\alpha = g_1\alpha g_2\alpha$ for all $g_1, g_2 \in G$, the image of an identity in G is an identity in G' and the image of an inverse of $g \in G$ is an inverse of $g\alpha$, i.e. $1\alpha = 1'$ (1 an identity of G and $1'$ an identity of G') and if $gh = 1 = hg$, $g\alpha h\alpha = 1' = h\alpha g\alpha$.

Now an automorphism of a group G is a one-to-one homomorphism of G onto G. Therefore if α is an automorphism of A_3, $\iota\alpha = \iota$. Also $\sigma_1\alpha$ is either σ_2 or σ_1, as α is a one-to-one onto mapping. Hence there are at most two automorphisms of A_3.

Let I be the identity mapping, i.e. $\iota I = \iota$, $\sigma_1 I = \sigma_1$, $\sigma_2 I = \sigma_2$. Let A be the mapping $\iota A = \iota$, $\sigma_1 A = \sigma_2$, $\sigma_2 A = \sigma_1$. On checking the homomorphism property, we see that I and A are automorphisms. Note that $A^2 = I$. Thus the multiplication table is

	I	A
I	I	A
A	A	I

3.57. Let a be any element of a group G. Define the mapping ρ_a of G into G by $\rho_a: g \to a^{-1}ga$. Prove that ρ_a is an automorphism of G and that $\rho_a\rho_b = \rho_{ab}$ where $\rho_a\rho_b$ is the usual multiplication of mappings.

Solution:

ρ_a is clearly a mapping. If $g_1\rho_a = g_2\rho_a$, then $a^{-1}g_1a = a^{-1}g_2a$ and hence $g_1 = g_2$. Therefore ρ_a is one-to-one. Also ρ_a is onto, for if $g \in G$, g has as pre-image aga^{-1}.

$$(g_1g_2)\rho_a \;=\; a^{-1}(g_1g_2)a \;=\; a^{-1}g_1aa^{-1}g_2a \;=\; g_1\rho_a g_2\rho_a$$

and so ρ_a is a homomorphism. Hence ρ_a is an automorphism. Note that we have used the associative law, so that our argument would not apply to a groupoid which is not a semigroup. If $g \in G$,

$$g(\rho_a\rho_b) \;=\; (g\rho_a)\rho_b \;=\; (a^{-1}ga)\rho_b \;=\; b^{-1}a^{-1}gab \;=\; (ab)^{-1}g(ab) \;=\; g\rho_{ab}$$

and thus $\rho_{ab} = \rho_a\rho_b$.

3.58. Find the automorphism group of S_3. (*Hint:* Use Problem 3.57 to find six automorphisms. Then prove that there are no other automorphisms. This problem is difficult.)

Solution:

Refer to the multiplication table of S_3 given in Section 3.3(a). By Problem 3.57, $\rho_\iota, \rho_{\sigma_1}, \rho_{\sigma_2}, \rho_{\tau_1}, \rho_{\tau_2}, \rho_{\tau_3}$ are all automorphisms of S_3. We use the notation of Section 2.4(c), page 37, to denote the effect of these mappings. We use the multiplication table of Section 3.3(a) to calculate the images under the automorphisms.

$$\rho_\iota = \begin{pmatrix} \iota & \sigma_1 & \sigma_2 & \tau_1 & \tau_2 & \tau_3 \\ \iota & \sigma_1 & \sigma_2 & \tau_1 & \tau_2 & \tau_3 \end{pmatrix} \qquad \rho_{\sigma_2} = \begin{pmatrix} \iota & \sigma_1 & \sigma_2 & \tau_1 & \tau_2 & \tau_3 \\ \iota & \sigma_1 & \sigma_2 & \tau_3 & \tau_1 & \tau_2 \end{pmatrix} \qquad \rho_{\tau_2} = \begin{pmatrix} \iota & \sigma_1 & \sigma_2 & \tau_1 & \tau_2 & \tau_3 \\ \iota & \sigma_2 & \sigma_1 & \tau_3 & \tau_2 & \tau_1 \end{pmatrix}$$

$$\rho_{\sigma_1} = \begin{pmatrix} \iota & \sigma_1 & \sigma_2 & \tau_1 & \tau_2 & \tau_3 \\ \iota & \sigma_1 & \sigma_2 & \tau_2 & \tau_3 & \tau_1 \end{pmatrix} \qquad \rho_{\tau_1} = \begin{pmatrix} \iota & \sigma_1 & \sigma_2 & \tau_1 & \tau_2 & \tau_3 \\ \iota & \sigma_2 & \sigma_1 & \tau_1 & \tau_3 & \tau_2 \end{pmatrix} \qquad \rho_{\tau_3} = \begin{pmatrix} \iota & \sigma_1 & \sigma_2 & \tau_1 & \tau_2 & \tau_3 \\ \iota & \sigma_2 & \sigma_1 & \tau_2 & \tau_1 & \tau_3 \end{pmatrix}$$

If ρ were another automorphism, then $\iota\rho = \iota$. Once $\sigma_1\rho$ is given, $\sigma_2\rho$ is known as $\sigma_2\rho = (\sigma_1\sigma_1)\rho = \sigma_1\rho\sigma_1\rho$. Now $\sigma_1\rho$ must be either σ_1 or σ_2, for if say $\sigma_1\rho = \tau_1$, then $\sigma_2\rho = \sigma_1\rho\sigma_1\rho = \tau_1\tau_1 = \iota$; but this contradicts ρ a one-to-one mapping, since $\iota\rho = \iota$. Hence there are two possible choices for $\sigma_1\rho$.

Now $\tau_1\rho$ must be one of τ_1, τ_2 or τ_3 for if for example, $\tau_1\rho = \sigma_1$, then $\iota\rho = (\tau_1\tau_1)\rho = \sigma_1\sigma_1 = \sigma_2 = \iota$, a contradiction. Hence there are 3 possible choices for $\tau_1\rho$. But once $\sigma_1\rho$ and $\tau_1\rho$ are known, the effect of ρ on all the elements of S_3 is known, since

$$\tau_1\sigma_1 \;=\; \tau_3 \qquad \text{and} \qquad \tau_1\sigma_2 \;=\; \tau_2$$

So this means that there are at most six possible automorphisms.

To find the multiplication table we use the result of Problem 3.57, that $\rho_a\rho_b = \rho_{ab}$.

	ρ_ι	ρ_{σ_1}	ρ_{σ_2}	ρ_{τ_1}	ρ_{τ_2}	ρ_{τ_3}
ρ_ι	ρ_ι	ρ_{σ_1}	ρ_{σ_2}	ρ_{τ_1}	ρ_{τ_2}	ρ_{τ_3}
ρ_{σ_1}	ρ_{σ_1}	ρ_{σ_2}	ρ_ι	ρ_{τ_2}	ρ_{τ_3}	ρ_{τ_1}
ρ_{σ_2}	ρ_{σ_2}	ρ_ι	ρ_{σ_1}	ρ_{τ_3}	ρ_{τ_1}	ρ_{τ_2}
ρ_{τ_1}	ρ_{τ_1}	ρ_{τ_3}	ρ_{τ_2}	ρ_ι	ρ_{σ_2}	ρ_{σ_1}
ρ_{τ_2}	ρ_{τ_2}	ρ_{τ_1}	ρ_{τ_3}	ρ_{σ_1}	ρ_ι	ρ_{σ_2}
ρ_{τ_3}	ρ_{τ_3}	ρ_{τ_2}	ρ_{τ_1}	ρ_{σ_2}	ρ_{σ_1}	ρ_ι

b. Fields of complex numbers

The complex numbers C have customarily two binary operations, addition and multiplication. Notice that if $a, b \in C$, then $a - b \in C$; and if $b \neq 0$, $ab^{-1} \in C$. We are often interested in a subset of C that satisfies the same conditions. This leads us to a *field* of complex numbers.

Definition: A subset F of C is called a *field* of complex numbers if

(*a*) $1 \in F$.

(*b*) Whenever $a, b \in F$, then also $a - b \in F$.

(*c*) Whenever $a, b \in F$ and $b \neq 0$, then $ab^{-1} \in F$.

(The definition of field can be extended to sets which are not contained in the complex numbers. See, for example, Birkhoff and MacLane, *A Survey of Modern Algebra*, Macmillan, 1953.)

Of course the complex numbers themselves form a field. Let F be a field.

Recall that the set of complex numbers is a group under the usual binary operation of addition, denoted by $(C, +)$, and that $C^* = C - \{0\}$ is a group $(C^*, \times)$ under the usual multiplication of complex numbers (see Example 5, page 51). Therefore, using Lemma 3.1, page 55, for a subset of a group to be a subgroup, parts (*a*) and (*b*) of the definition of a field imply $(F, +)$ is a subgroup of $(C, +)$, and parts (*a*) and (*c*) imply $(F^*, \times)$, $F^* = F - \{0\}$, is a subgroup of $(C^*, \times)$. In view of these remarks the definition of a field is equivalent to:

Lemma 3.14: A subset F of C is a field of complex numbers if

(1) $(F, +)$ is a subgroup of $(C, +)$,

(2) $(F^*, \times)$ is a subgroup of $(C^*, \times)$ where $F^* = F - \{0\}$, $C^* = C - \{0\}$.

Problems

3.59. Show that R and Q are fields of complex numbers.

Solution:

$1 \in R$. If $a, b \in R$, then $a - b \in R$ and $ab^{-1} \in R$ whenever $b \neq 0$. The same argument applies for Q.

3.60. Which of the following sets are fields?

(i) $F = \{a + b\sqrt{2} \mid a, b \in Q\}$

(ii) $F = \{a + bi \mid a, b \in Q\}$, $i = \sqrt{-1}$

(iii) $F = \{a + bi \mid a, b \in Z\}$, $i = \sqrt{-1}$

Solution:

(i) $1 = 1 + 0\sqrt{2} \in F$. Let $a + b\sqrt{2}$ and $a' + b'\sqrt{2}$ be two elements in F.

$$(a + b\sqrt{2}) - (a' + b'\sqrt{2}) \quad = \quad (a - a') + (b - b')\sqrt{2} \quad \in \quad F$$

and if $a' + b'\sqrt{2} \neq 0$,

$$(a + b\sqrt{2})(a' + b'\sqrt{2})^{-1} \quad = \quad \frac{aa' - 2bb'}{a'^2 - 2b'^2} + \frac{a'b - ab'}{a'^2 - 2b'^2}\sqrt{2} \quad \in \quad F$$

Therefore F is a field.

(ii) F is a field. $1 + 0i = 1 \in F$.

$$(a + bi) - (a' + b'i) \quad = \quad (a - a') + (b - b')i \quad \in \quad F$$

and, if $a' + b'i \neq 0$,

$$(a + bi)(a' + b'i)^{-1} \quad = \quad \frac{aa' + bb'}{a'^2 + b'^2} + \frac{a'b - ab'}{a'^2 + b'^2}i \quad \in \quad F$$

(iii) F is not a field, since $1 + i \neq 0$, $1 + i \in F$ but $(1 + i)^{-1} = 1/2 - 1/2i \notin F$.

c. Automorphisms of fields

We have discussed isometries of the plane and automorphisms of groupoids. The corresponding one-to-one onto mapping of a field is defined as follows.

Definition: A one-to-one mapping α of a field F onto itself is termed an *automorphism* if

(i) $(a+b)\alpha = a\alpha + b\alpha$ for all $a, b \in F$.

(ii) $(ab)\alpha = (a\alpha)(b\alpha)$ for all $a, b \in F$.

Note that the automorphisms of fields preserve both the operations of addition and multiplication.

Theorem 3.15: The set A of automorphisms of a field F forms a subgroup of the symmetric group S_F.

Proof: We must prove

I. $A \neq \emptyset$; this is true since the identity mapping $\iota \in A$.

II. If $\alpha, \beta \in A$, then

$$(a+b)(\alpha\beta) = ((a+b)\alpha)\beta = (a\alpha + b\alpha)\beta = (a\alpha)\beta + (b\alpha)\beta = a(\alpha\beta) + b(\alpha\beta)$$

and
$$(ab)(\alpha\beta) = ((ab)\alpha)\beta = [(a\alpha)(b\alpha)]\beta = [(a\alpha)\beta][(b\alpha)\beta] = [a(\alpha\beta)][b(\alpha\beta)]$$

for all $a, b \in F$. Thus composition of mappings is a binary operation in A.

III. The identity mapping is in A and is an identity element.

IV. $(A, \cdot)$ is a semigroup, since composition of mappings is associative.

V. If $\alpha \in A$, then $\alpha \in S_F$ the symmetric group on F. Let α^{-1} be the inverse of α. We claim $\alpha^{-1} \in A$ and so α will have an inverse in A as desired. Let $a, b \in F$. Then as α is onto, we can find $a', b' \in F$ such that $a = a'\alpha$, $b = b'\alpha$. Then $ab = (a'b')\alpha$ and $a + b = (a' + b')\alpha$. Consequently $(ab)\alpha^{-1} = a'b' = (a\alpha^{-1})(b\alpha^{-1})$ and $(a+b)\alpha^{-1} = a' + b' = a\alpha^{-1} + b\alpha^{-1}$. Thus $\alpha^{-1} \in A$ as desired.

We have proved that the automorphisms of a field form a group. This group is extremely useful. For additional pertinent remarks and references, see Section 5.4a, page 158.

Problems

3.61. Find the automorphism group of Q.

Solution:

We will use the fact that $(Q, +)$ is a group and $(Q^*, \times)$, $Q^* = Q - \{0\}$, is a group. Notice that $(Q, \times)$ is a groupoid (not a group, since 0 has no inverse) and, because of part (ii) of the definition, the automorphism of the field Q is also an epimorphism (see Section 2.5b, page 42) of groupoid $(Q, \times)$ onto $(Q, \times)$. Hence by Theorem 2.6, page 44, 1, the multiplicative identity of $(Q, \times)$, is mapped onto 1 by any automorohism of Q.

Let α be an auotmorphism of Q; then $1\alpha = 1$. Using mathematical induction we show $n\alpha = n$ for all positive integers n. $1\alpha = 1$. Assume $k\alpha = k$ for some integer $k \geqq 1$. Then $(k+1)\alpha = k\alpha + 1\alpha = k + 1$, by the automorphism property of α. We conclude that $n\alpha = n$ for all positive integers n.

Now $(Q, +)$ is a group and, by definition, any automorphism of Q is an epimorphism of the group $(Q, +)$. Hence by Theorem 2.6, inverses are mapped onto inverses and the identity, 0, of $(Q, +)$ is mapped onto 0. Therefore $(-n)\alpha = -n$ for all positive integers n, since $n\alpha = n$ and $-n$ is the additive inverse of n. Furthermore, $0\alpha = 0$. Hence $r\alpha = r$ for all integers. But the automorphism α is also an epimorphism of the group $(Q^*, \times)$ onto itself so that $(\pm r)^{-1}\alpha = \frac{1}{\pm r}\alpha = \frac{1}{\pm r}$ for all positive integers r, because $\frac{1}{r}$ is the inverse of r. Collecting these facts we see that if $\frac{m}{n}$ $(n \neq 0)$ is any element in Q, then $\frac{m}{n}\alpha = \left(m \cdot \frac{1}{n}\right)\alpha = m\alpha\frac{1}{n}\alpha = m \cdot \frac{1}{n} = \frac{m}{n}$. Therefore α is the identity mapping and is the only possible automorphism of Q. The automorphism group of Q is of order one.

3.62. Find the automorphism group of $F = \{a + b\sqrt{2} \mid a, b \in Q\}$.

Solution:

Any rational number q is an element of F, since $q + 0\sqrt{2} = q$. If α is an automorphism of F, then for any $q \in Q$, $q\alpha = q$ arguing as in Problem 3.61. Now $\sqrt{2} \in F$ and $(\sqrt{2}\sqrt{2})\alpha = 2\alpha = 2$, since 2 is an element of Q. But $\sqrt{2}\alpha\sqrt{2}\alpha = (\sqrt{2}\sqrt{2})\alpha = 2\alpha = 2$, so that $(\sqrt{2}\alpha)^2 = 2$ or $\sqrt{2}\alpha = \pm\sqrt{2}$. We conclude that $\sqrt{2}$ has only two possible images under an automorphism of F. Hence $(a + b\sqrt{2})\alpha = a\alpha + (b\sqrt{2})\alpha = a\alpha + b\alpha\sqrt{2}\alpha = a + b(\sqrt{2}\alpha)$. There are two possibilities: (1) $(a + b\sqrt{2})\alpha = a + b\sqrt{2}$, in which case α is the identity automorphism ι; (2) $(a + b\sqrt{2})\alpha = a + b(-\sqrt{2})$, in which case we must check to see whether α is an automorphism. If $\alpha : a + b\sqrt{2} \to a - b\sqrt{2}$, then

$$\begin{aligned}\{(a + b\sqrt{2})(a' + b'\sqrt{2})\}\alpha &= \{aa' + 2bb' + (a'b + b'a)\sqrt{2}\}\alpha \\ &= aa' + 2bb' - (a'b + b'a)\sqrt{2}\end{aligned}$$

and

$$\begin{aligned}(a + b\sqrt{2})\alpha(a' + b'\sqrt{2})\alpha &= (a - b\sqrt{2})(a' - b'\sqrt{2}) \\ &= aa' + 2bb' - (a'b + b'a)\sqrt{2}\end{aligned}$$

Hence

$$\{(a + b\sqrt{2})(a' + b'\sqrt{2})\}\alpha = (a + b\sqrt{2})\alpha(a' + b'\sqrt{2})\alpha$$

Also,

$$\begin{aligned}\{(a + b\sqrt{2}) + (a' + b'\sqrt{2})\}\alpha &= \{(a + a') + (b + b')\sqrt{2}\}\alpha \\ &= a + a' - (b + b')\sqrt{2}\end{aligned}$$

and

$$\begin{aligned}(a + b\sqrt{2})\alpha + (a' + b'\sqrt{2})\alpha &= (a - b\sqrt{2}) + (a' - b'\sqrt{2}) \\ &= (a + a') - (b + b')\sqrt{2}\end{aligned}$$

Hence

$$\{(a + b\sqrt{2}) + (a' + b'\sqrt{2})\}\alpha = (a + b\sqrt{2})\alpha + (a' + b'\sqrt{2})\alpha$$

Thus α is an automorphism of F. The automorphism group of F has two elements ι and $\alpha : a + b\sqrt{2} \to a - b\sqrt{2}$. Notice $\alpha\alpha = \iota$.

d. Vector spaces

In Physics we represent a force x by a straight line pointing in the direction the force is acting and of length proportional to the magnitude of the force. We shall assume for the moment that all the forces act on a fixed point O and act in the Euclidean plane E. It is then possible to represent a force by its endpoint, as we know it begins at O. Any point of course can be represented by its coordinates, so a force can be represented by the coordinates of its endpoint.

We can talk of increasing the force x in magnitude by a factor 3, say. The resultant force is written as $3x$. If $x = (f_1, f_2)$, then $3x = (3f_1, 3f_2)$. Similarly if f is any real number, we define fx to be the force x increased in magnitude by a factor f and we can prove $fx = (ff_1, ff_2)$.

The sum of two forces $x = (f_1, f_2)$ and $y = (g_1, g_2)$ is a third force z computed by the parallelogram law. Again it can be shown that $z = (f_1 + g_1, f_2 + g_2)$. We write $z = x + y$. The set of all 2-tuples (f_1, f_2) is called a vector space of dimension 2 over the field of real numbers (because we can multiply the 2-tuples by real numbers).

We shall generalize the concept of two dimensional physical forces in two ways:

(i) We shall deal with arbitrary dimensions and not only 2 or 3. (We must therefore relinquish our contact with the real world.)

(ii) We shall consider vectors that involve fields other than the real numbers.

Let F be any field. Let $V = F^n$ be the cartesian product of n copies of F. Then V consists of the n-tuples $(f_1, f_2, \ldots, f_n)$ where $f_i \in F$.

If $x = (f_1, \ldots, f_n)$ and $y = (g_1, \ldots, g_n)$ are two elements of V, and $f \in F$, we define $\mu : V \times V \to V$ (i.e. μ is a binary operation in V) and $\omega : F \times V \to V$ by

$$(x, y)\mu = (f_1 + g_1, f_2 + g_2, \ldots, f_n + g_n)$$
$$(f, x)\omega = (ff_1, \ldots, ff_n)$$

We denote $(x, y)\mu$ by $x + y$, and $(f, x)\omega$ by fx.

V together with μ and ω is called the *vector space of dimension n over the field F.*

The elements of V are called vectors.

Problems

3.63. Find (i) $(1,2,3) + (6,7,8)$, (ii) $4(6,-2,0,3)$.

Solution:

(i) $(1+6, 2+7, 3+8) = (7, 9, 11)$

(ii) $(4 \cdot 6, 4 \cdot (-2), 4 \cdot 0, 4 \cdot 3) = (24, -8, 0, 12)$

3.64. Prove that if x, y and z are elements of a vector space of dimension n, then $x+y = y+x$ and $(x+y)+z = x+(y+z)$.

Solution:

If $x = (f_1, \ldots, f_n)$, $y = (g_1, \ldots, g_n)$ and $z = (h_1, \ldots, h_n)$, then $x+y = y+x = (f_1+g_1, \ldots, f_n+g_n)$ as addition is commutative in any field.

$(x+y)+z = ((f_1+g_1)+h_1, \ldots, (f_n+g_n)+h_n) = x+(y+z)$ by associativity of addition.

3.65. Prove that if V is a vector space of dimension n, then the elements of V form an abelian group under the operation μ.

Solution:

V is an abelian groupoid by the preceding problem. $(0, 0, \ldots, 0)$ is the identity element. The inverse of $(f_1, f_2, \ldots, f_n)$ is $(-f_1, -f_2, \ldots, -f_n)$.

3.66. Prove that if $e_1 = (1, 0, \ldots, 0)$, $e_2 = (0, 1, 0, \ldots, 0)$, $\ldots$, $e_n = (0, 0, \ldots, 1)$, then every element x of V can be represented uniquely in the form

$$x = f_1e_1 + f_2e_2 + \cdots + f_ne_n$$

Solution:

Suppose $x = (f_1, \ldots, f_n)$. Then indeed $x = f_1e_1 + f_2e_2 + \cdots + f_ne_n$.

If $x = g_1e_1 + g_2e_2 + \cdots + g_ne_n$, then $(f_1, \ldots, f_n) = (g_1, \ldots, g_n)$. Hence $f_1 = g_1$, $f_2 = g_2$, $\ldots$, $f_n = g_n$ and the representation is unique.

e. Linear transformations. The full linear group

Let V be a vector space and $\alpha: V \to V$. Then α is said to be a *linear transformation* of V if

(i) $(x+y)\alpha = x\alpha + y\alpha$ (ii) $(fx)\alpha = f(x\alpha)$, for all $x, y \in V$ and $f \in F$

For example, let $(f_1, f_2)\alpha = (f_2, f_1)$. Then

$$\{(f_1, f_2) + (g_1, g_2)\}\alpha = (f_2 + g_2, f_1 + g_1) = (f_2, f_1) + (g_2, g_1) = (f_1, f_2)\alpha + (g_1, g_2)\alpha$$

Also, $$\{f(f_1, f_2)\}\alpha = (ff_2, ff_1) = f(f_2, f_1) = f((f_1, f_2)\alpha)$$

Note that linear transformations preserve both the additive and the multiplicative structures of V.

Now we have the analog of Theorems 3.13 and 3.15. First let us define $L_n(V, F)$ to consist of all one-to-one linear transformations of V, the vector space of dimension n over F.

$L_n(V, F) \subseteq S_V$, the symmetric group of V, clearly.

Theorem 3.16: $L_n(V,F)$ is a subgroup of S_V.

Proof: $\iota \in L_n(V,F)$ as ι preserves both addition and multiplication. Hence $L_n(V,F) \neq \emptyset$.

If $\alpha \in L_n(V,F)$, we ask whether $\alpha^{-1} \in L_n(V,F)$. α^{-1} is one-to-one onto. Is it a linear transformation? Let $x, y \in V$ and $f \in F$. Since α is onto, there exists x_1 and y_1 such that $x_1\alpha = x$ and $y_1\alpha = y$. Of course $x_1 = x\alpha^{-1}$, $y_1 = y\alpha^{-1}$; and $(x_1 + y_1)\alpha = x_1\alpha + y_1\alpha = x + y$. Hence

$$(x_1 + y_1) = (x_1 + y_1)\alpha\alpha^{-1} = (x + y)\alpha^{-1}$$

and so $x\alpha^{-1} + y\alpha^{-1} = (x+y)\alpha^{-1}$. Also, $(fx_1)\alpha = f(x_1\alpha) = fx$; so $((fx_1)\alpha)\alpha^{-1} = fx_1 = (fx)\alpha^{-1}$, i.e. $f(x\alpha^{-1}) = (fx)\alpha^{-1}$. Accordingly $\alpha^{-1} \in L_n(V,F)$. Thus if $\alpha, \beta \in L_n(V,F)$, $\beta^{-1} \in L_n(V,F)$ and we ask whether $\alpha\beta^{-1} \in L_n(V,F)$. We have

$$(x+y)\alpha\beta^{-1} = ((x+y)\alpha)\beta^{-1} = (x\alpha + y\alpha)\beta^{-1}$$
$$= (x\alpha)\beta^{-1} + (y\alpha)\beta^{-1} = x(\alpha\beta^{-1}) + y(\alpha\beta^{-1})$$

and
$$(fx)\alpha\beta^{-1} = ((fx)\alpha)\beta^{-1} = (f(x\alpha))\beta^{-1} = f((x\alpha)\beta^{-1}) = f(x(\alpha\beta^{-1}))$$

and thus $L_n(V,F)$ is a subgroup of S_V. $L_n(V,F)$ is called the full linear group of dimension n.

Problems

3.67. Show that if α is a linear transformation of V, a vector space of dimension n, then the effect of α is uniquely determined by its effect on the elements $e_1, \ldots, e_n$ of Problem 3.66.

Solution:

By Problem 3.66 each element of V is of the form $x = f_1e_1 + \cdots + f_ne_n$. Then $x\alpha = f_1(e_1\alpha) + \cdots + f_n(e_n\alpha)$. Hence the effect of α is known once its effect on the elements $e_1, \ldots, e_n$ is known.

3.68. Show that if α is any mapping of $\{e_1, \ldots, e_n\} \to V$, then there exists a linear transformation $\bar{\alpha}: V \to V$ such that $e_j\alpha = e_j\bar{\alpha}$, $j = 1, \ldots, n$.

Solution:

Each element of V is uniquely of the form $f_1e_1 + \cdots + f_ne_n$. Define $\bar{\alpha}: V \to V$ by

$$(f_1e_1 + \cdots + f_ne_n)\bar{\alpha} = f_1(e_1\alpha) + \cdots + f_n(e_n\alpha)$$

Then $\bar{\alpha}$ is a linear transformation, since

$$\{(f_1e_1 + \cdots + f_ne_n) + (g_1e_1 + \cdots + g_ne_n)\}\bar{\alpha} = (f_1 + g_1)(e_1\alpha) + \cdots + (f_n + g_n)(e_n\alpha)$$
$$= (f_1e_1 + \cdots + f_ne_n)\bar{\alpha} + (g_1e_1 + \cdots + g_ne_n)\bar{\alpha}$$

and
$$\{f(f_1e_1 + \cdots + f_ne_n)\}\bar{\alpha} = (ff_1)(e_1\alpha) + \cdots + (ff_n)(e_n\alpha)$$
$$= f\{(f_1e_1 + \cdots + f_ne_n)\bar{\alpha}\}$$

3.69. Is $\alpha \in L_n(V,F)$ if α is a linear transformation and $e_1\alpha = e_2$, $e_2\alpha = e_3, \ldots, e_{n-1}\alpha = e_n$ and $e_n\alpha = e_1$?

Solution:

Yes. All we must prove is that α is one-to-one and onto. An arbitrary element $f_1e_1 + \cdots + f_ne_n$ has $f_2e_1 + f_3e_2 + \cdots + f_ne_{n-1} + f_1e_n$ as a pre-image. Also,

$$(f_1e_1 + \cdots + f_ne_n)\alpha = (g_1e_1 + \cdots + g_ne_n)\alpha$$

implies $f_1 = g_1, f_2 = g_2, \ldots, f_n = g_n$ by Problem 3.66. Hence α is one-to-one. Thus $\alpha \in L_n(V,F)$.

A look back at Chapter 3

We have met many important groups, including groups of real and complex numbers, the symmetric group S_n, symmetry groups, the dihedral groups, the automorphism groups of groupoids and fields, and the full linear group.

Groups thus arise in many different branches of mathematics, and hence general theorems about groups can be useful in apparently unrelated topics.

In subsequent chapters we will derive general theorems for groups.

Supplementary Problems

GROUPS

3.70. Let n be any positive integer and let $G_n = \{a + b\sqrt{n} \mid a, b \in Z\}$ where Z is the set of integers. Prove that with respect to addition G_n is a group. When does $G_n = Z$?

3.71. Let n be any positive integer. Let $G_n = \{a + ib\sqrt{n} \mid a, b \in Z\}$ where $i = \sqrt{-1}$ and Z is the set of integers. Is G_n a group with respect to addition? Is G_n a group with respect to multiplication of complex numbers?

3.72. Let $D = Z \times Z$, Z the set of integers. Define $(a, b) \circ (c, d) = (a + c, (-1)^c b + d)$. Prove that D is a group with respect to this operation $\circ$.

3.73. Prove that the group D of Problem 3.72 is not abelian.

3.74. Let $G = Z \times Q$, where Z is the set of integers and Q the set of rationals. Define $(a, b) * (c, d) = (a + c, 2^c b + d)$. Prove G is a group with respect to this operation $*$.

3.75. Is the group of Problem 3.74 abelian?

3.76. If we define $(a, b) \circ (c, d) = (a + c, 2^{-c} b + d)$, is G (of Problem 3.74) a group with respect to $\circ$? Is G a group with respect to the operation $\cdot$ defined by $(a, b) \cdot (c, d) = (a + c, 2^c b - d)$?

3.77. Let $B = \{\theta \mid \theta : Z \to Z\}$. Let $W = Z \times B$. We define a multiplication on W by $(m, \theta)(n, \phi) = (m + n, \psi)$ where for each $z \in Z$, $z\psi = (z - n)\theta + z\phi$. Prove that W is a group. (Hard.)

SUBGROUPS

3.78. Let G be a group and $G_1 \subseteq G_2 \subseteq \cdots$ be subgroups of G. Show that $G_1 \cup G_2 \cup \cdots$ is a subgroup of G. Find a group G and two subgroups G_1 and G_2 of G such that $G_1 \cup G_2$ is not a subgroup of G.

3.79. Let $G_1, G_2, \ldots$ be subgroups of G. Prove $G_1 \cap G_2 \cap \cdots$ is a subgroup of G.

3.80. Let G be an abelian group. Let H be a subgroup of G. Let $S(H) = \{x \mid x \in G$ and $xx \in H\}$. Prove that $S(H)$ is a subgroup of G.

3.81. Let D be the group of Problem 3.72. Determine whether $H = \{(a, 0) \mid a \in Z\}$ and $K = \{(0, a) \mid a \in Z\}$ are subgroups of D.

3.82. Let G be the group of Problem 3.74. Determine whether $H = \{(a, 0) \mid a \in Z\}$ and $K = \{(0, q) \mid q \in Q\}$ are subgroups of G.

3.83. Let B be as in Problem 3.77. Let $C = \{\theta \mid \theta : Z \to Z,\ z\theta = z$ for all but a finite number of integers $z\}$. Let $B' = \{x \mid x = (0, b),\ b \in B\}$, $C' = \{x \mid x = (0, c),\ c \in C\}$. Prove that B' is a subgroup of W and C' is a subgroup of B'. (Hard.)

3.84. Using the notation of the preceding problem, let $\hat{W} = \{x \mid x = (m, c)$, where $m \in Z$ and $c \in C\}$. Prove $\hat{W}$ is a subgroup of W. (Hard.)

SYMMETRIC GROUPS AND ALTERNATING GROUPS

3.85. Let $\alpha : Z \to Z$ be defined by $z\alpha = z + 1$ for all $z \in Z$. Let $\beta : Z \to Z$ be defined by $2n\beta = 2n$, $(2n+1)\beta = 2n+3$ for all integers n. Let $\gamma : Z \to Z$ be defined by $2n\gamma = 2(n+1)$, $(2n+1)\gamma = 2n+1$ for all integers n. Prove that $\alpha, \beta, \gamma \in S_Z$ and show that $\alpha\alpha = \beta\gamma = \gamma\beta$.

3.86. Let $G = S_P$, where P is the set of positive integers. Let $\hat{S}_P = \{\theta \mid \theta \in S_P$ and $z\theta = z$ for all but a finite number of $z \in P\}$. Prove that $\hat{S}_P$ is a subgroup of S_P.

3.87. Let $G = \hat{S}_P$. Let $G_n = \{\theta \mid \theta \in S_P$ and $z\theta = z$ for all $z \in P$ such that $z > n\}$. Prove that $G = G_1 \cup G_2 \cup \cdots$.

3.88. Let $H = \{\theta \mid \theta \in S_5,\ 1\theta = 1\}$. Prove that H is a subgroup of S_5. What is its order? Let $K = \{\theta \mid \theta \in S_5,\ 1\theta = 1$ or $1\theta = 2\}$. Prove that K is not a subgroup of S_5.

3.89. Let n and r be positive integers. Let

$$H = \{\theta \mid i\theta \in \{1, 2, \ldots, r\} \text{ for all } i \in \{1, 2, \ldots, r\} \text{ and } \theta \in S_n\}$$

Prove that H is a subgroup of S_n and find $|H|$.

3.90. Let X be a set and Y a proper subset of X. Let $H = \{\theta \mid \theta \in S_X$ and $y\theta = y$ for all $y \in Y\}$. Let $K = \{\theta \mid \theta \in S_X$ and $y\theta \in Y$ for all $y \in Y\}$. Prove that H and K are subgroups of S_X and that $K \supseteq H$. Prove that if $|Y| \geqq 2$, $H \neq K$.

3.91. Let $H = \{\theta \mid \theta \in A_5,\ 1\theta = 1\}$. Prove that H is a subgroup of A_5 and find its order.

3.92. Let A, B be sets with $|A| = 1$. Prove that $S_{A \times B} \cong S_B$. (Hard.)

3.93. Prove that if $|X| = |Y|$, $S_X \cong S_Y$.

GROUPS OF ISOMETRIES

3.94. Let S be the even integers and $(I : S) = \{\theta \mid \theta \in I(R),\ s\theta \in S$ for all $s \in S\}$. Prove $(I : S)$ is a subgroup of $I(R)$.

3.95. Let $(I : Q) = \{\theta \mid \theta \in I(R)$ and $q\theta \in Q$ for all $q \in Q\}$. Prove $(I : Q)$ is a subgroup of $I(R)$.

3.96. Find the symmetry group of the figure W.

3.97. Find the symmetry group of the figure 8.

3.98. What is the symmetry group of the graph of $y = \sin x$?

3.99. Determine the symmetry group of the circle.

3.100. Prove that if S is any subspace of the plane and S' is a congruent figure, i.e. there is an isometry θ such that $S\theta = S'$, then $I_S \cong I_{S'}$. (Hard.)

THE GROUP OF MÖBIUS TRANSFORMATIONS

3.101. Prove that if M is the group of Möbius transformations, then the only element $m \in M$ for which $mn = nm$ for all $n \in M$ is $m = \iota$.

3.102. Let $\sigma(a, b, c, d)$ be the Möbius transformation defined by $\sigma(a, b, c, d) : z \to \dfrac{az + b}{cz + d}$. Prove that $\sigma(a, b, c, d) = \sigma(a', b', c', d')$ if and only if either $a = a'$, $b = b'$, $c = c'$, $d = d'$ or $a = -a'$, $b = -b'$, $c = -c'$, $d = -d'$, given $ad - bc = a'd' - b'c' = 1$. (Hard.)

3.103. Let N be the set of Möbius transformations $\sigma(a, b, c, d)$ with $b = 0$. Prove that N is a subgroup of M, the group of all Möbius transformations.

3.104. Prove that if $m \in N$ (N defined as in Problem 3.103), then there exists $s \in N$ such that $ss = m$.

3.105. Let $\mathcal{U}$ be the set of all matrices $\begin{pmatrix} a & b \\ c & d \end{pmatrix}$, where a, b, c, d are integers such that $ad - bc = 1$. Find the set of all matrices such that

$$\begin{pmatrix} a & b \\ c & d \end{pmatrix}\begin{pmatrix} a & b \\ c & d \end{pmatrix} = \begin{pmatrix} 1 & 0 \\ 0 & 1 \end{pmatrix}$$

SYMMETRIES OF AN ALGEBRAIC STRUCTURE

3.106. Prove that the automorphism group of a finite group is finite.

3.107. Find a finite groupoid G with $|G| > 2$, whose automorphism groupoid is of order 1.

3.108. Let G be a non-abelian group. Prove that the automorphism group of G is not of order 1.

3.109. Prove that the subset K of the symmetric group S_4 defined by

$$K = \left\{\begin{pmatrix} 1 & 2 & 3 & 4 \\ 2 & 1 & 3 & 4 \end{pmatrix}, \begin{pmatrix} 1 & 2 & 3 & 4 \\ 1 & 2 & 4 & 3 \end{pmatrix}, \begin{pmatrix} 1 & 2 & 3 & 4 \\ 2 & 1 & 4 & 3 \end{pmatrix}, \iota\right\}$$

is a subgroup of S_4. Find the automorphism group of K. (Hard.)

3.110. Let $F = \{a + ib\sqrt{17} \mid a, b$ rational numbers, where $i = \sqrt{-1}\}$. Verify that F is a field under the usual operations of addition and multiplication of complex numbers. Determine the automorphism group of F.

3.111. Let V be the vector space over the rationals of dimension $n + 1$. Let

$$S = \{\alpha \mid \alpha \in L_{n+1}(V, F),\ (1, 0, \ldots, 0)\alpha = (1, 0, \ldots, 0)\}$$

Prove that S is a subgroup of $L_{n+1}(V, F)$ and that $S \cong L_n(V, F)$.

Chapter 4

Isomorphism Theorems

Preview of Chapter 4

We say that two groups are isomorphic if they are isomorphic groupoids. Here we shall prove three theorems which provide a means of determining whether two groups are isomorphic. The main concepts that arise are those of subgroups generated by a set, cosets, and normal subgroups. We find a structure theorem for cyclic groups. The contents of this chapter are indispensable for any further understanding of group theory.

4.1 FUNDAMENTALS

a. Preliminary remarks

We begin by reminding the reader of our previous results. A group is a semigroup in which every element has an inverse. Consequently we have the following.

(1) The identity is unique. (Theorem 2.1, page 31.)

(2) The inverse of an element is unique (Theorem 2.2, page 33), and if G is any group and $g, h \in G$, then $(gh)^{-1} = h^{-1}g^{-1}$.

(3) The product of n elements $a_1, \ldots, a_n$, in that order, is independent of the bracketing (Theorem 2.5, page 39).

(4) Homomorphisms, monomorphisms and isomorphisms for groups are defined as they are for groupoids. (Section 2.5, page 40.)

(5) If G is a group, and θ any homomorphism of G into a groupoid, then

$$G\theta = \{x \mid x = g\theta,\ g \in G\}$$

is a group. For by Theorem 2.6, page 44, $G\theta$ has an identity, is associative, and each element has an inverse. Note that θ maps the identity of G to the identity of $G\theta$ and that $(g^{-1})\theta = (g\theta)^{-1}$ for each g in G.

(6) If $\theta : G \to K$ is a homomorphism from the group G to the group K, and if H is a subgroup of G, then $H\theta$ is a subgroup of K. For $\theta_{|H}$ is a homomorphism of H into K and, by (5), $H\theta$ is a group.

(7) Isomorphic groups are roughly the same except for the names of their elements. (See Section 2.5d, page 45.)

The following theorem is useful.

Theorem 4.1: If a and b are two elements of a group G, then there exist unique elements x and y such that $ax = b$ and $ya = b$.

Proof: We consider first the solution of the equation $ax = b$. If we put $x = a^{-1}b$, then $a(a^{-1}b) = (aa^{-1})b = b$. Hence the equation $ax = b$ has a solution.

Suppose $ax_1 = b$ and $ax_2 = b$; then $ax_1 = ax_2$. Multiplying both sides of the equation on the left by a^{-1}, we have

$$a^{-1}(ax_1) = a^{-1}(ax_2), \quad (a^{-1}a)x_1 = (a^{-1}a)x_2 \quad \text{or} \quad x_1 = x_2$$

The argument for solving $ya = b$ is similar; in fact $y = ba^{-1}$ is a solution. Also, if $y_1a = b$ and $y_2a = b$, then $y_1a = y_2a$. Multiplying both sides by a^{-1} on the right, we get

$$(y_1a)a^{-1} = y_1 = (y_2a)a^{-1} = y_2$$

Problems

4.1. Prove that the groups given by the following multiplication tables are isomorphic.

G:

	-1	1
-1	1	-1
1	-1	1

H:

	0	1
0	0	1
1	1	0

Solution:

Let $\theta : G \to H$ be defined by $1\theta = 0$, $-1\theta = 1$; then θ is a one-to-one onto mapping. If it is also a homomorphism it will be an isomorphism. We must check that $(g_1 g_2)\theta = g_1\theta g_2\theta$ for all possible choices of g_1 and g_2 in G, i.e. we must check

(i) $(1 \cdot 1)\theta = 1\theta 1\theta$ (iii) $(-1 \cdot -1)\theta = (-1\theta)(-1\theta)$

(ii) $(-1 \cdot 1)\theta = (-1\theta)(1\theta)$ (iv) $(1 \cdot -1)\theta = (1\theta) \cdot (-1\theta)$

(i) to (iv) hold. (Thus for (i): $1 \cdot 1 = 1$ by the multiplication table. $1\theta = 0$. $1\theta \cdot 1\theta = 0 \cdot 0 = 0$. Hence (i) holds.) Therefore $G \cong H$.

Of course θ had to be some mapping of G to H. How did we know which was the right mapping to choose? Examining the multiplication table for G, it is obvious that 1 is the identity for G. 0 is the identity for H. We remarked that any homomorphism must map an identity to an identity. Thus the choice for θ was quite clear.

4.2. Prove that S_2, the symmetric group of degree 2, is isomorphic to G, where G is the group of Problem 3.5, page 53, with $m = 2$. Prove $S_3 \cong D_3$ the dihedral group of degree 3, i.e. the symmetry group of the equilateral triangle. (Difficult.)

Solution:

The multiplication table for G is

	0	1
0	0	1
1	1	0

while the multiplication table for S_2 is (Problem 3.20, page 58):

	ι	β
ι	ι	β
β	β	ι

Let $\theta : S_2 \to G$ be defined by $\iota\theta = 0$, $\beta\theta = 1$. Then it can be checked that θ is a homomorphism. As it is one-to-one and onto, $S_2 \cong G$.

The multiplication table for S_3 is on page 57, that of D_3 is in Problem 3.40, page 76. As we have used the same Greek symbols for S_3 and D_3, we face the risk of not knowing whether σ, for example, refers to an element of S_3 or to an element of D_3. To avoid such ambiguities, we will rename the elements of D_3, replacing a σ by an s and a τ by a t. The multiplication table then becomes

	s_1	s_2	s_3	t	ts_2	ts_3
s_1	s_1	s_2	s_3	t	ts_2	ts_3
s_2	s_2	s_3	s_1	ts_3	t	ts_2
s_3	s_3	s_1	s_2	ts_2	ts_3	t
t	t	ts_2	ts_3	s_1	s_2	s_3
ts_2	ts_2	ts_3	t	s_3	s_1	s_2
ts_3	ts_3	t	ts_2	s_2	s_3	s_1

Note that an element τ_j, $j = 1, 2, 3$, satisfies $\tau_j \tau_j = \iota$. If θ is an isomorphism from S_3 to D_3, then $\tau_j\theta\, \tau_j\theta = (\tau_j\tau_j)\theta = \iota\theta = s_1$. So θ can only map the τ_j, $j = 1, 2, 3$, among the elements t, ts_2, ts_3 since these are the only non-identity elements of D_3 which have the property that their squares are s_1. As θ must map ι to s_1, it maps σ_1, σ_2 onto the elements s_2, s_3.

If we know the effect of θ on σ_1, since $\sigma_1\sigma_1 = \sigma_2$, we know the effect of θ on σ_2. Also if we know the effect of θ on τ_1, then, because $\tau_2 = \tau_1\sigma_2$ and $\tau_3 = \tau_1\sigma_1$, we know the effect of θ on all the elements of S_3.

So we have to experiment. A suitable mapping $\theta : S_3 \to D_3$ must satisfy $\sigma_1\theta = s_2$ or s_3 while $\tau_1\theta = t$, ts_2 or ts_3.

We try the following definition. Let $\iota\theta = s_1$, $\sigma_1\theta = s_2$ and $\tau_1\theta = t$. Then we must have $\sigma_2\theta = s_3$, $\tau_2\theta = ts_3$, $\tau_3\theta = ts_2$ if θ is to be an isomorphism.

To check whether this mapping is an isomorphism, we must check whether this mapping is a homomorphism. As a mechanical procedure of doing this we use the following table.

	ι	σ_1	σ_2	τ_1	τ_2	τ_3
ι	s_1 / s_1	s_2 / s_2	s_3 / s_3	t / t	ts_3 / ts_3	ts_2 / ts_2
σ_1	s_2 / s_2	s_3 / s_3	s_1 / s_1	ts_3 / ts_3	ts_2 / ts_2	t / t
σ_2	s_3 / s_3	s_1 / s_1	s_2 / s_2	ts_2 / ts_2	t / t	ts_3 / ts_3
τ_1	t / t	ts_2 / ts_2	ts_3 / ts_3	s_1 / s_1	s_3 / s_3	s_2 / s_2
τ_2	ts_3 / ts_3	t / t	ts_2 / ts_2	s_2 / s_2	s_1 / s_1	s_3 / s_3
τ_3	ts_2 / ts_2	ts_3 / ts_3	t / t	s_3 / s_3	s_2 / s_2	s_1 / s_1

The entry in the second row and third column, for example, is calculated as follows: In the bottom corner we place $\sigma_1\theta \cdot \sigma_2\theta$. In the top corner we place $(\sigma_1\sigma_2)\theta$. If θ is a homomorphism, $(\sigma_1\theta)(\sigma_2\theta) = (\sigma_1\sigma_2)\theta$. Hence these two entries should be the same in each square of the table. Checking through this table, we see that the entries in each square are equal. Hence θ is a homomorphism and as it is one-to-one onto, θ is an isomorphism.

4.3. Prove that if $\theta : F \to G$ and $\phi : G \to F$ are two homomorphisms such that $\theta\phi$ = identity mapping on F and $\phi\theta$ = identity mapping on G, then θ and ϕ are isomorphisms of F onto G and of G onto F respectively.

Solution:

θ is one-to-one, for if $x\theta = y\theta$, then $x\theta\phi = y\theta\phi$. But $\theta\phi$ is the identity on F. Hence $x = y$. Similarly ϕ is one-to-one.

Next let $g \in G$; then $g\phi \in F$. $g\phi\theta = g$; hence g is the image of an element of F under θ and so θ is an onto mapping. Thus θ is an isomorphism. Similarly ϕ is an isomorphism.

4.4. Prove that if g_1, g_2, g_3 are elements of a group G, then the equation $g_1xg_2 = g_3$ has a unique solution.

Solution:

If we put $x = g_1^{-1}g_3g_2^{-1}$, we find $g_1xg_2 = g_3$. If $g_1x_1g_2 = g_1x_2g_2 = g_3$, then on multiplying by g_1^{-1} on the left and g_2^{-1} on the right we have $g_1^{-1}(g_1x_1g_2)g_2^{-1} = g_1^{-1}(g_1x_2g_2)g_2^{-1}$ or $x_1 = x_2$.

4.5. Prove that if G is a finite group and H is an infinite group, then G and H are not isomorphic.

Solution:

If $G \cong H$, there is a one-to-one mapping from G onto H. But this is not possible since G is finite and H is infinite.

4.6. Prove that $S_n \cong S_m$ if and only if $n = m$.

Solution:

S_n has order $n!$ and S_m has order $m!$ Now if $S_n \cong S_m$, then there is a one-to-one mapping of S_n onto S_m. So S_n and S_m have the same order, i.e. $n! = m!$, and this implies $n = m$. On the other hand every group G is isomorphic to itself. In fact the identity mapping of G onto G is an isomorphism. Hence $n = m$ implies $S_n \cong S_m$.

4.7. Prove that if $G \cong H$, then $H \cong G$.

Solution:

Let θ be an isomorphism from G onto H. Then θ^{-1} is an isomorphism from H to G, and so $H \cong G$. (See Problem 2.38, page 42.)

4.8. Prove that if $G \cong H$ and $H \cong K$, then $G \cong K$.

Solution:

Suppose θ is an isomorphism from G to H and ϕ an isomorphism from H to K. Then $\theta\phi$ is an isomorphism from G to K, i.e. $G \cong K$. (See Problem 2.38, page 42.)

4.9. Prove that there are infinitely many groups, no two of which are isomorphic.

Solution:

Consider the symmetric groups $S_1, S_2, \ldots$. Then by Problem 4.6, no pair of these groups is isomorphic.

4.10. Prove that if G is a finite group and H is a subgroup of G, $H \neq G$, then G and H are not isomorphic.

Solution:

We observed in the solution of Problem 4.6 that if two finite groups are isomorphic, they have the same order. Since the order of H is less than that of G, it follows that G and H are not isomorphic.

b. More about subgroups

Let $G = S_4$ and let $X = \{\sigma_8, \tau_7\}$ where

$$\sigma_8 = \begin{pmatrix} 1 & 2 & 3 & 4 \\ 2 & 1 & 4 & 3 \end{pmatrix} \quad \text{and} \quad \tau_7 = \begin{pmatrix} 1 & 2 & 3 & 4 \\ 2 & 3 & 1 & 4 \end{pmatrix}$$

(See Problem 3.21, page 59.) Suppose we wish to refer to a product such as $\sigma_8\tau_7$ or $\tau_7\sigma_8\sigma_8$ or $\sigma_8\tau_7\sigma_8^{-1}\tau_7^{-1}\tau_7^{-1}\sigma_8^{-1}$. It will be convenient to have some general notation. We will write

$$x_1^{\epsilon_1} \cdots x_n^{\epsilon_n}, \quad \text{where } \epsilon_i = \pm 1,\ x_i \in X$$

to represent the product of n elements chosen from X or the set of inverses of the elements of X, where x_i^1 will mean x_i, and x_i^{-1} will mean the inverse of x_i. For example, if $\epsilon_1 = \epsilon_2 = 1$, $\epsilon_3 = \epsilon_4 = \epsilon_5 = \epsilon_6 = -1$, and $x_1 = x_3 = x_6 = \sigma_8$, $x_2 = x_4 = x_5 = \tau_7$, then $x_1^{\epsilon_1}x_2^{\epsilon_2}x_3^{\epsilon_3}x_4^{\epsilon_4}x_5^{\epsilon_5}x_6^{\epsilon_6}$ stands for $\sigma_8\tau_7\sigma_8^{-1}\tau_7^{-1}\tau_7^{-1}\sigma_8^{-1}$.

Example 1: If $g = x_1^{\epsilon_1} \cdots x_n^{\epsilon_n}$, then $g^{-1} = h$ where $h = x_n^{-\epsilon_n} \cdots x_1^{-\epsilon_1}$.

Proof:

$$gh = x_1^{\epsilon_1} \cdots x_n^{\epsilon_n} x_n^{-\epsilon_n} \cdots x_1^{-\epsilon_1} = x_1^{\epsilon_1} \cdots x_{n-1}^{\epsilon_{n-1}} \cdot 1 \cdot x_{n-1}^{-\epsilon_{n-1}} \cdots x_1^{-\epsilon_1}$$
$$= \cdots = x_1^{\epsilon_1} x_1^{-\epsilon_1} = 1$$

Similarly $hg = 1$.

We proved in Lemma 3.1, page 55, that if H is a subgroup of G and $x_1, x_2 \in H$, then $x_1x_2^{-1} \in H$. We now generalize this and prove

Lemma 4.2: If H is a subgroup of G and $X \subseteq H$, then

$$H \supseteq \{x_1^{\epsilon_1} \cdots x_n^{\epsilon_n} \mid x_i \in X,\ \epsilon_i = \pm 1,\ n \text{ a positive integer}\}$$

Proof: Recall that as H is a group, $\iota \in H$ and $x, y \in H$ implies $y^{-1} \in H$, $xy^{-1} \in H$, and $xy \in H$. We prove the lemma by induction on n. Let $n = 1$. Then $x_1^{-1} \in H$ since $x_1 \in H$. Hence $x_1^{\epsilon_1} \in H$. Assume, by induction, that $x = x_1^{\epsilon_1} \cdots x_n^{\epsilon_n} \in H$ for $n = k$. Let $x_{k+1} \in X$. Since $x, x_{k+1}^{\epsilon_{k+1}} \in H$ where $\epsilon_{k+1} = \pm 1$,

$$xx_{k+1}^{\epsilon_{k+1}} = x_1^{\epsilon_1} \cdots x_{k+1}^{\epsilon_{k+1}} \in H$$

Hence $x_1^{\epsilon_1} \cdots x_n^{\epsilon_n} \in H$ for all n. This proves the lemma.

Now if X is "large enough", e.g. $X = H$, we may have

$$H = \{x_1^{\epsilon_1} \cdots x_n^{\epsilon_n} \mid x_i \in X,\ \epsilon_i = \pm 1,\ n \text{ a positive integer}\}$$

We ask what happens if X is not "large enough", i.e. if X is a subset of H and

$$S = \{x_1^{\epsilon_1} \cdots x_n^{\epsilon_n} \mid x_i \in X,\ \epsilon_i = \pm 1,\ n \text{ a positive integer}\}$$

is S a subgroup of G?

Lemma 4.3: Let G be a group and let X be a non-empty subset of G. Let

$$S = \{x_1^{\epsilon_1} \cdots x_n^{\epsilon_n} \mid x_i \in X,\ \epsilon_i = \pm 1,\ n \text{ a positive integer}\}$$

Then S is a subgroup of G. If H is any subgroup containing X, $H \supseteq S$.

Proof: We must prove that:

(i) $S \neq \emptyset$; this is true because there exists $x_1 \in X$ as X is non-empty.

(ii) If $f, g \in S$, then $fg^{-1} \in S$ (Lemma 3.1, page 55).

$f, g \in S$ means

$$f = x_1^{\epsilon_1} \cdots x_n^{\epsilon_n}\ (\epsilon_i = \pm 1) \quad \text{and} \quad g = y_1^{\eta_1} \cdots y_m^{\eta_m}\ (\eta_i = \pm 1)$$

where x_i and y_i are elements of X. Hence $g^{-1} = y_m^{-\eta_m} \cdots y_1^{-\eta_1}$ and

$$fg^{-1} = x_1^{\epsilon_1} \cdots x_n^{\epsilon_n} y_m^{-\eta_m} \cdots y_1^{-\eta_1} = x_1^{\epsilon_1} \cdots x_n^{\epsilon_n} x_{n+1}^{\epsilon_{n+1}} \cdots x_{n+m}^{\epsilon_{n+m}}$$

where $x_{n+1} = y_m, \ldots, x_{n+m} = y_1$ and $\epsilon_{n+1} = -\eta_m, \ldots, \epsilon_{n+m} = -\eta_1$. Therefore $fg^{-1} \in S$ and S is a subgroup of G. If $H \supseteq X$, we use the previous lemma to conclude $H \supseteq S$.

We denote S by $gp(X)$ and call S *the subgroup generated by X*.

If a group can be generated by a finite set, we call it a *finitely generated* group.

Example 2: What is $gp(\{1\})$ in the group of Problem 3.5, page 53, where $m = 3$? Recall that the multiplication table is

	0	1	2
0	0	1	2
1	1	2	0
2	2	0	1

$$S = gp(\{1\}) = \{x_1^{\epsilon_1} \cdots x_n^{\epsilon_n} \mid x_i \in \{1\},\ \epsilon_i = \pm 1,\ n \text{ a positive integer}\}$$

Now $1 \in S$ and $2 \in S$. Also, $1 \cdot 2 = 0 \in S$. Hence $gp(\{1\})$ is the whole group.

We remind the reader that, for example, in the multiplicative group of nonzero rationals the inverse of a, which we have denoted in this section by a^{-1}, is $1/a$, i.e. in this case a^{-1} has the meaning usually associated with it when a is a number. But in the additive group of rationals a^{-1} is $-a$.

Problems

4.11. Let $G = Z$, the additive group of integers. What is $gp(\{1\})$?

Solution:

$gp(\{1\}) \supseteq \underbrace{1+1+1+\cdots+1}_{n \text{ times}}$ $(n \geqq 1)$. Hence $gp(\{1\})$ contains all positive integers.

$gp(\{1\})$ contains $\underbrace{1^{-1}+1^{-1}+\cdots+1^{-1}}_{n \text{ times}} = \underbrace{-1+(-1)+\cdots+(-1)}_{n \text{ times}}$ $(n \geqq 1)$. Hence $gp(\{1\})$ contains all negative integers.

Also, $gp(\{1\})$ contains $1 \cdot 1^{-1} = 1+(-1) = 0$.

Thus $gp(\{1\}) = Z$.

4.12. Let $G = Q$ the additive group of rationals. Find $gp(\{1\})$.

Solution:

Exactly as in the last problem, $gp(\{1\}) = Z$. Since no other elements can arise as sums or differences of 1, $gp(\{1\}) \neq Q$.

4.13. Find the subgroup of the multiplicative group of rationals generated by $\{2\}$.

Solution:

$2^{-1} = \frac{1}{2}$. The elements of $gp(\{2\})$ are either of the form 2^n or 2^{-n}, n a positive integer.

4.14. Determine the subgroup H of S_3 generated by σ_1 and τ_1 of Section 3.3a, page 57.

Solution:

We use the multiplication table for S_3 shown in page 57. $\sigma_1^{-1} = \sigma_2$; hence H contains σ_2. As H contains τ_1, it contains $\tau_3 = \tau_1\sigma_1$ and $\tau_2 = \tau_1\sigma_2$. Thus H contains all the elements of S_3, and so $H = S_3$.

4.15. Determine the subgroup of the symmetry group of a square generated by those isometries that leave two vertices fixed. (Hard.) (*Hint:* To see what is happening, cut out a square from a piece of cardboard and label the four vertices. Perform the isometries on the figure.)

Solution:

We refer to Problem 3.39(ii), page 73. s_1 leaves G and I fixed; s_2 leaves H and J fixed; s_5 leaves all vertices fixed. Hence we require $S = gp(\{s_5, s_2, s_1\})$, and this must contain s_7, since $s_1s_2 = s_7$. It is easy to prove $s_1s_2 = s_7$, for the effect of s_7 and s_1s_2 is the same on three points not on a single straight line and this is sufficient by Lemma 3.7, page 71. Note that the inverses of s_2, s_1, s_7 are s_2, s_1, s_7 respectively.

Let $T = \{s_5, s_7, s_2, s_1\}$. We assert that T is a subgroup. All we must check is that $t_1, t_2 \in T$ implies $t_1t_2^{-1} \in T$. Since $t \in T$ implies $t^{-1} = t$, all we must check is that $t_1t_2 \in T$ for $t_1, t_2 \in T$. As s_5 is the identity, $t_1t_2 \in T$ if s_5 is either t_1 or t_2. If $t_1 = t_2$, then $t_1t_2 = s_5 \in T$. Therefore we need only consider the following cases: $s_1s_2 = s_2s_1 = s_7 \in T$; $s_1s_7 = s_7s_1 = s_2 \in T$. Finally, $s_2s_7 = s_7s_2 = s_1 \in T$. Then T is a subgroup of the symmetry group of the square. But $S \supseteq T$, and $T \supseteq \{s_5, s_2, s_1\}$. Hence $T \supseteq gp(s_5, s_2, s_1) = S$, by Lemma 4.3. Thus $T = S$.

4.16. Find the subgroup of M, the group of Möbius transformations (see Section 3.5a, page 78), generated by

$$\eta: z \to -z \quad (z \neq \infty), \qquad \eta: \infty \to \infty$$

$$\tau: z \to z+1 \quad (z \neq \infty), \qquad \tau: \infty \to \infty$$

(In the notation of Section 3.5a, $\eta = \sigma(-1,0,0,1)$ and $\tau = \sigma(1,1,0,1)$. This is a difficult problem.)

Solution:

Let $\sigma_{(n,\epsilon)}$ be the mapping defined by $z \to \epsilon z + n$ for $z \neq \infty$, and $\infty \to \infty$, where $\epsilon = \pm 1$ and n is any integer. In other words, $\sigma_{(n,\epsilon)} = \sigma(\epsilon, n, 0, 1)$.

We will show that the subgroup generated by η and τ consists of all $\sigma_{(n,\epsilon)}$, n any integer, $\epsilon = \pm 1$. Let $\Sigma = \{\sigma_{(n,\epsilon)} \mid n \text{ any integer}, \ \epsilon = \pm 1\}$. We claim $gp(\eta, \tau) = \Sigma$. Any element of Σ is of the form $\sigma_{(n,\epsilon)}$. Now $z\tau^{-1} = z - 1$. Also, $z\underbrace{\tau \cdots \tau}_{n} = z + n$ and $z\underbrace{\tau^{-1} \cdots \tau^{-1}}_{n} = z - n$ for any positive integer n. Hence for any arbitrary integer n, $\sigma_{(n,1)} \in gp(\eta, \tau)$. (We must check what happens to ∞, but this presents no difficulty.) Also, $\sigma_{(n,1)}\eta = \sigma_{(-n,-1)} \in gp(\eta, \tau)$ and so $\sigma_{(n,\epsilon)}$, $\epsilon = \pm 1$, n any integer, belongs to $gp(\eta, \tau)$. Thus $gp(\eta, \tau) \supseteq \Sigma$.

Note that $\eta = \sigma_{(0,-1)}$ and $\tau = \sigma_{(1,1)}$ belongs to Σ. So we need only show that Σ is a subgroup of the group of Möbius transformations to conclude that $\Sigma \supseteq gp(\eta, \tau)$. To show Σ is a subgroup, we need only show that $\sigma_{(n,\epsilon)}\sigma_{(m,\delta)}^{-1} \in \Sigma$, $\delta = \pm 1$. But $\sigma_{(m,\delta)}^{-1} = \sigma_{(-\delta m,\delta)}$, since $z\sigma_{(m,\delta)}\sigma_{(-\delta m,\delta)} = (\delta z + m)\delta - \delta m = \delta^2 z + \delta m - \delta m = \delta^2 z = z$. Then $z\sigma_{(n,\epsilon)}\sigma_{(m,\delta)}^{-1} = (\epsilon z + n)\sigma_{(-\delta m,\delta)} = \epsilon\delta z + n\delta - \delta m$, and we conclude that $\sigma_{(n,\epsilon)}\sigma_{(m,\delta)}^{-1} = \sigma_{(n\delta - \delta m, \epsilon\delta)} \in \Sigma$.

c. Exponents

We have seen in the previous section that we often are forced to consider the product of m a's $(m > 0)$, e.g. $\underbrace{a \cdot \cdots \cdot a}_{m}$. (Note that as we are dealing with groups, it is not necessary to indicate in which order the multiplication is performed. See Section 4.1a.) It is convenient to introduce the notation a^m for the product of m a's, $m > 0$. Then $a^m \cdot a^n$ is the product of m a's followed by n a's, i.e. $a^m \cdot a^n = a^{m+n}$ (Section 2.4d, page 39). Our idea is to extend the exponent notation in a sensible way to zero and negative exponents. We would naturally like the law

$$a^m \cdot a^n = a^{m+n} \tag{4.1}$$

to be true when m and n are arbitrary integers. Now if it were true that $a^0 a^m = a^m$, then multiplication by a^0 leaves a^m unchanged. Hence we have only one choice in extending the exponent notation and retaining the law (*4.1*), namely putting $a^0 = 1$, the identity. Now if $m = -n$ where $n > 0$, $m + n = 0$. Because we want (*4.1*) to be satisfied, we must have $a^{m+n} = a^0 = 1$, i.e. we must put $a^m = (a^n)^{-1}$. Note that $(a^n)^{-1} = (a^{-1})^n = a^{-n}$. Thus we have defined a^m to be

(i) the product of m a's if $m > 0$,

(ii) 1 if $m = 0$,

(iii) the product of $-m$ a^{-1}'s if $m < 0$,

hoping thus to satisfy (*4.1*) for all m and n.

(*4.1*) is true if m, n are both positive. If both are nonnegative, again by running through the possible cases (*4.1*) holds. If both m and n are negative, then

$$a^m a^n = (a^{-1})^{(-m)} \cdot (a^{-1})^{(-n)} = (a^{-1})^{(-m+-n)} = (a^{-1})^{-(m+n)} = a^{m+n}$$

If m and n are nonpositive, again the result is easily verified. If $m > 0$ and $n < 0$, then by checking the various possibilities $m > -n$, $m = -n$ and $m < -n$, we find $a^m a^n = a^m (a^{-1})^{(-n)} = a^{m+n}$.

Another result which holds for exponents is

$$(a^m)^n = a^{mn} \tag{4.2}$$

We already know that (*4.2*) holds when $n = -1$. If m, n are positive, $(a^m)^n$ is the product of n elements, each of which is the product of m a's. Hence $(a^m)^n$ is the product of mn a's. If m is negative, n positive,

$$\begin{aligned} (a^m)^n &= ((a^{-1})^{-m})^n = (a^{-1})^{-mn} \quad \text{as } -m > 0, \ n > 0 \\ &= a^{mn} \quad \text{as } mn < 0 \end{aligned}$$

If now n is negative,

$$\begin{aligned} (a^m)^n &= (a^m)^{(-1)(-n)} = ((a^m)^{-1})^{-n} = (a^{-m})^{(-n)} \\ &= a^{(-m)\cdot(-n)} \quad \text{(by our previous remarks)} \\ &= a^{mn} \end{aligned}$$

Hence (*4.2*) is proved.

In the study of groups there are two main notations for the binary operation. One is the multiplicative notation we have employed up until now. The other is the additive notation. We denote the binary composition by + in this case. The identity is denoted by zero, 0, and the inverse of a by $-a$. The result of performing $n > 0$ compositions of the same element, i.e. of taking $\underbrace{a + a + \cdots + a}_{n}$, we denote by na. The law (*4.1*) becomes

$$na + ma = (n+m)a$$

while the law (*4.2*) becomes

$$n(ma) = (nm)a$$

In other words, translation takes place according to the following dictionary:

Multiplicative notation	**Additive notation**
ab	$a + b$
1	0
a^{-1}	$-a$
a^n	na

It is immaterial which notation one uses. But additive notation is most often used for a group in which the order of the composition of two elements is irrelevant, i.e. in which $a + b = b + a$ for all a, b in the group. Such a group is called *abelian*, after the Norwegian mathematician Niels Henrik Abel, or commutative (Section 2.2, page 29).

Problems

4.17. Find 1^3, 1^{-4} where $1 \in Q$, the additive group of rationals.

Solution:

In the additive group of rationals the binary operation is the usual addition. Then 1^3 means $1 \circ 1 \circ 1$ where $\circ$ is the binary operation in Q. Hence $1^3 = 1+1+1 = 3$. Also 1^{-4} means $(1^{-1})^4$, i.e. $1^{-1} \circ 1^{-1} \circ 1^{-1} \circ 1^{-1}$ where $\circ$ is the binary operation under discussion. Now $1^{-1} = -1$ in $(Q, +)$. Thus $1^{-4} = (-1) + (-1) + (-1) + (-1) = -4$.

4.18. Find 2^2, 2^{-3} where $2 \in Q^*$, the multiplicative group of nonzero rationals.

Solution:

$2^2 = 2 \cdot 2 = 4$ and $2^{-3} = (2^{-1})^3 = (\frac{1}{2})^3 = \frac{1}{8}$.

4.19. Find σ^3, $\sigma = \begin{pmatrix} 1 & 2 & 3 & 4 \\ 2 & 1 & 4 & 3 \end{pmatrix}$, an element of S_4.

Solution:

$\sigma^2 = \iota$ and so $\sigma^3 = \iota\sigma = \sigma$.

4.20. Find τ^n, where τ is as defined in Problem 4.16.

Solution:

$\tau^n = \sigma_{(n,1)}$. See Problem 4.16.

4.2 CYCLIC GROUPS

a. Fundamentals of cyclic groups

If $gp(X) = H$, we say H is generated by X. To get an understanding of groups, a good plan is to investigate the simpler groups first. So we begin by considering groups which can be generated by a single element. We call such groups *cyclic*. Thus a group H is cyclic if we can find an element $x \in H$ such that $H = gp(\{x\})$. We will usually write $gp(x)$ instead of $gp(\{x\})$.

Lemma 4.4: $gp(x) = \{t \mid t = x^r,\ r \text{ an integer}\}$. Cyclic groups are abelian.

Proof:

$$\begin{aligned} gp(x) &= \{x_1^{\epsilon_1} \cdots x_n^{\epsilon_n} \mid x_i \in \{x\},\ \epsilon_i = \pm 1,\ n > 0\} \\ &= \{x^{\epsilon_1} \cdots x^{\epsilon_n} \mid \epsilon_i = \pm 1,\ n > 0\} \\ &= \left\{x^{\left(\sum_{i=1}^{n} \epsilon_i\right)} \,\middle|\, \epsilon_i = \pm 1,\ n > 0\right\} \\ &= \{x^r \mid r \text{ any integer}\} \end{aligned}$$

If $a, b \in gp(x)$, then $a = x^r$, $b = x^s$, $ab = x^r x^s = x^{r+s}$, $ba = x^s x^r = x^{r+s}$. Hence $ab = ba$ for any two elements of a cyclic group. Thus we have shown that cyclic groups are abelian.

Suppose now that $H = gp(x)$ and $|H| = m$ $(m < \infty)$. Then we know that the elements of H are of the form x^r for various integers r. The x^i cannot be distinct for all integers i. Consider $x^0 = 1, x, \ldots, x^{l-1}$ and suppose these are distinct but that $x^l = x^k$ for some $k < l$, $k \geqq 0$; then $x^l(x^k)^{-1} = x^{l-k} = 1$. If $k \neq 0$, $m = l - k < l$ and x^m is equal to x^0. But we assumed this was not so. Hence $k = 0$ and $x^l = x^0 = 1$.

We will show that $S = \{1, x, x^2, \ldots, x^{l-1}\}$ is actually H. This is easy. First notice that as $x^l = 1$, every positive power of x is in S. Furthermore, $x^{-1} = x^{l-1}$. Hence every negative power of x lies in S. But $H = \{x^r \mid r \text{ any integer}\}$. Therefore $H \subseteq S$ and so $S = H$ as stated.

Thus we have proved

Lemma 4.5: Let G be cyclic of order m generated by the element x. Then $G = \{x^0, x^1, \ldots, x^{m-1}\}$. Furthermore $x^m = 1$, and x^m is the least positive power of x that is 1.

We ask a simple question: do cyclic groups of order m exist for all finite integers $m > 0$? Yes! Let us consider in the symmetric group S_m of degree m the element

$$\sigma_m = \begin{pmatrix} 1 & 2 & \ldots & m-1 & m \\ 2 & 3 & \ldots & m & 1 \end{pmatrix}$$

Then

$$\sigma_m^2 = \begin{pmatrix} 1 & 2 & \ldots & m-2 & m-1 & m \\ 3 & 4 & \ldots & m & 1 & 2 \end{pmatrix}$$

$$\cdots\cdots\cdots\cdots\cdots\cdots\cdots\cdots$$

$$\sigma_m^m = \begin{pmatrix} 1 & 2 & \ldots & m \\ 1 & 2 & \ldots & m \end{pmatrix} = \iota$$

and so the elements $\iota, \sigma_m, \ldots, \sigma_m^{m-1}$ are distinct and $H = gp(\sigma_m)$ is cyclic of order m. Hence there exist cyclic groups of order m for each $m > 0$.

And now we ask another question: are there two essentially different cyclic groups of order m? Rephrasing the question, we ask: are two cyclic groups of order m isomorphic?

Lemma 4.6: Let $G = gp(x)$, $H = gp(y)$ be each of order m. Then $G \cong H$.

Proof: $G = \{x^0, x, x^2, \ldots, x^{m-1}\}$, $H = \{y^0, y^1, \ldots, y^{m-1}\}$. Let $\theta: G \to H$ be defined by

$$x^i\theta = y^i \quad (i = 0, 1, \ldots, m-1)$$

Then θ is one-to-one onto H. To prove it is an isomorphism we must show it is a homomorphism. Consider

$$(x^i x^j)\theta = (x^{i+j})\theta$$

Now $0 \leq i, j \leq m-1$. Then $0 \leq i+j \leq 2(m-1) = 2m-2$ and so $i+j = \epsilon m + r$ where $0 \leq r \leq m-1$ and $\epsilon = 0$ or 1. Hence

$$(x^{i+j})\theta = (x^{\epsilon m + r})\theta = (x^{\epsilon m}x^r)\theta = (x^r)\theta = y^r$$

But

$$(x^i\theta)(x^j\theta) = y^iy^j = y^{i+j} = y^{\epsilon m + r} = y^{\epsilon m}y^r = y^r$$

Hence $((x^i)(x^j))\theta = (x^i\theta)(x^j\theta)$. Thus θ is a homomorphism and, as it is one-to-one onto, it is an isomorphism.

We now ask the obvious question: are there any infinite cyclic groups and are two infinite cyclic groups isomorphic?

Consider the element σ in the symmetric group S_Z on Z, the set of integers, defined by

$$z\sigma = z+1, \quad z \in Z$$

As $z\sigma^n = z+n$, $\sigma^m = \sigma^n$ implies $m = n$. Then $gp(\sigma) = H$, say, has an infinite number of elements and so H is an infinite cyclic group.

Recall that $G = gp(x) = \{x^n \mid n \text{ any integer}\}$. If there exists an integer $m > 0$ such that $x^m = 1$, then G will consist of only a finite number of elements (see the remarks preceding Lemma 4.5). Consequently if G is infinite, there exists no $m \neq 0$ for which $x^m = 1$. For we have already shown that there can exist no $m > 0$ for which $x^m = 1$; while if $x^m = 1$ for $m < 0$, then $x^{(-m)} = 1$ and $(-m) > 0$. If $x^l = x^n$, $n \neq l$, then $x^{n-l} = 1$. But this contradicts the condition that there exists no m such that $x^m = 1$. Hence the elements of G are simply the powers x^n of x, and two such powers x^m and x^n are equal if and only if $m = n$.

Now we can easily prove that two infinite cyclic groups are isomorphic. Let $G = gp(x)$, $H = gp(y)$ both be infinite cyclic groups. Then each element of G is uniquely of the form x^n, n an integer, and each element of H is uniquely of the form y^n, n an integer. Define $(x^n)\theta = y^n$. θ is a one-to-one onto mapping. Furthermore, $(x^nx^m)\theta = (x^{n+m})\theta = y^{n+m}$ and $(x^n\theta)(x^m\theta) = y^ny^m = y^{n+m}$. Hence $x^n\theta x^m\theta = (x^nx^m)\theta$. Therefore θ is an isomorphism and G and H are isomorphic groups.

Collecting our results, we have proved

Theorem 4.7: There exist cyclic groups of all orders, finite and infinite. Any two cyclic groups of the same order are isomorphic. (We therefore often talk about *the* cyclic group of order m, or *the* infinite cyclic group, or sometimes *the* infinite cycle.)

If x is an element of a group G, then we define the *order of* x as the order of $gp(x)$. Note that if x is of order $m < \infty$, then $x^m = 1$ and m is the first positive integer r for which $x^r = 1$. If x is of infinite order, then $x^m = 1$ implies $m = 0$. If x is of order m, $m < \infty$, we say x is of *finite* order.

Lemma 4.8: Let x be of order $m < \infty$. If $x^r = 1$, then m divides r.

Proof: Put $r = qm + s$ where $0 \leq s < m$. Then $1 = x^r = x^{qm}x^s = x^s$. As m is the first integer greater than 0 for which $x^m = 1$, $s = 0$. Hence m divides r.

Problems

4.21. Prove that the additive group of integers is infinite cyclic.

Solution:

$Z = gp(1)$. As Z is infinite, it is infinite cyclic.

4.22. Prove that the group of Problem 3.5, page 53, is cyclic of order m.

Solution:

The group is $gp(1)$, and its order is m.

4.23. Prove that $(Z,+)$ and the subgroup of M, the group of Möbius transformations, generated by the mapping $\eta: z \to z+1$, $\infty\eta = \infty$ are isomorphic.

Solution:

By Theorem 4.7 all we need prove is that $gp(\eta)$ is infinite, as we know from Problem 4.21 that $(Z,+)$ is infinite cyclic. But since $z\eta^n = z+n$, $\eta^n = \eta^m$ implies $n = m$. Thus $gp(\eta)$ is infinite and so $gp(\eta) \cong (Z,+)$.

4.24. Find the order of (i) $\sigma = \begin{pmatrix} 1 & 2 & 3 \\ 2 & 1 & 3 \end{pmatrix} \in S_3$, (ii) $\sigma = \begin{pmatrix} 1 & 2 & 3 & 4 \\ 2 & 3 & 4 & 1 \end{pmatrix} \in S_4$, (iii) the map η of M defined by $z\eta = -z$, $\infty\eta = \infty$.

Solution:

(i) $\sigma \neq \iota$ and $\sigma^2 = \begin{pmatrix} 1 & 2 & 3 \\ 2 & 1 & 3 \end{pmatrix}\begin{pmatrix} 1 & 2 & 3 \\ 2 & 1 & 3 \end{pmatrix} = \iota$. Hence σ is of order 2.

(ii) $\sigma, \sigma^2, \sigma^3$ are not ι, but $\sigma^4 = \iota$. Thus σ is of order 4.

(iii) $\eta^2 = \iota$ and so η is of order 2.

4.25. Let G be abelian. Let $x, y \in G$ be of orders r, s respectively. Show that xy is of order rs if r and s are co-prime, i.e. have no common prime divisors.

Solution:

Note that since G is abelian, $(xy)^n = x^n y^n$ for any integer n. Since $(xy)^{rs} = x^{rs}y^{rs} = 1$, the order of xy divides rs, by Lemma 4.8. If $(xy)^m = 1$, i.e. $x^m y^m = 1$, then $x^m = y^{-m}$ and $1 = (x^m)^r = y^{-mr}$. Therefore s, the order of y, divides $-mr$. Since s does not divide r, s must divide m. Similarly we can show r divides m. Hence rs divides m. So if m is the order of xy, m is divisible by rs and also m divides rs. Thus the order of xy is rs.

4.26. Show that if G is a cyclic group of order $m < \infty$ and s is co-prime to m, then $a^s = b^s$ $(a, b \in G)$ implies $a = b$. Find a group G and a nonzero integer n such that there are two elements $a, b \in G$ with $a^n = b^n$ but $a \neq b$.

Solution:

Since G is abelian, so $(ab^{-1})^s = a^s(b^{-1})^s = 1$. Since $G = gp(x)$ and the order of G is m, then $ab^{-1} = x^r$ for some r, and $(x^r)^s = 1$. Hence $x^{rs} = 1$ and m divides rs. But s and m are co-prime; then m divides r, say $r = qm$. Now $ab^{-1} = x^{qm} = 1$ and so $a = b$.

In S_3, let $a = \begin{pmatrix} 1 & 2 & 3 \\ 2 & 1 & 3 \end{pmatrix}$, $b = \begin{pmatrix} 1 & 2 & 3 \\ 1 & 3 & 2 \end{pmatrix}$. Then $a^2 = b^2 = \iota$ but $a \neq b$.

4.27. Show that if $G = gp(x)$ and G is of finite order r and s is co-prime to r, then $gp(x^s) = G$.

Solution:

The distinct elements of $gp(x^s)$ are $1, x^s, x^{2s}, \ldots, x^{(n-1)s}$ where $(x^s)^n = x^{ns} = 1$ and n is the least such positive integer. Since $x^{ns} = 1$ and G is of order r, r divides ns. As r and s are co-prime, r divides n. Hence there are at least r distinct elements in $gp(x^s)$. But as $G \supseteq gp(x^s)$ and G itself has only r elements, $gp(x^s) = G$.

4.28. Find a group which is not abelian. (*Hint.* Consider S_3.)

Solution:

See Section 3.3a, page 57, where we pointed out that $\sigma_1\tau_1 \neq \tau_1\sigma_1$. Hence S_3 is not abelian.

4.29. Prove that a subgroup H of S_3 is cyclic if $H \neq S_3$.

Solution:

A survey of the subgroups of S_3 shows that if H is a subgroup of S_3 and $H \neq S_3$, then H is either cyclic of order 3 or cyclic of order 2 or cyclic of order 1. To obtain all the subgroups of S_3, we refer to the multiplication table for S_3 in Section 3.3a, page 57, and list all the subsets of S_3. Then we check which subsets are subgroups. Of course since a subgroup must contain the identity, there is no need to go through the process of finding all subgroups quite so crudely. Nevertheless this method will suffice.

4.30. Prove that $(Q,+)$ is not cyclic.

Solution:

If $(Q,+)$ is cyclic, there exists $q=m/n$, m and n integers $(n \neq 0)$, such that $gp(m/n)=Q$. Of course $m \neq 0$. Each element $(\neq 0)$ of Q would then be of the form $\underbrace{q+q+\cdots+q}_{r}$ with some suitable choice of the positive integer r, or else of the form $\underbrace{-q-q-\cdots-q}_{r}$.

But $1/2n \in Q$. $1/2n = \underbrace{q+\cdots+q}_{r}$ implies $1/2n = rm/n$, i.e. $1=2rm$; then $1-2rm=0$. But r and m are integers; hence the equation $1-2rm=0$ is not true. If $1/2n = -q-q-q-\cdots-q$, r terms in all, a similar argument leads to a contradiction. Therefore $(Q,+)$ is not cyclic.

4.31. Prove that an abelian group generated by a finite number of elements of finite order is finite.

Solution:

Let $G = gp(\{x_1,\ldots,x_n\})$, $n<\infty$, and suppose G is abelian. Then every element g in G is of the form

$$g = x_{i_1}^{\epsilon_1}\cdots x_{i_r}^{\epsilon_r} \qquad (x_{i_j}\in X,\ \epsilon_j=\pm 1)$$

Since G is abelian, we can rewrite g in the form

$$g = x_1^{\gamma_1}\cdots x_n^{\gamma_n} \qquad (\gamma_1,\gamma_2,\ldots,\gamma_n \text{ integers}) \tag{4.3}$$

To see this we need only observe that if $i_s = i_t$ for $s<t$ in the first expression for g, then

$$g = x_{i_1}^{\epsilon_1}\cdots x_{i_s}^{\epsilon_s+\epsilon_t}\cdots x_{i_r}^{\epsilon_r}$$

i.e. we can always "collect" all occurrences of any x in a product. Now if $x_1,x_2,\ldots,x_n$ are all of finite order, then the number of distinct elements given by (*4.3*) is finite. For if k_i is the order of x_i, $i=1,2,\ldots,n$, the distinct powers of x_i are $1,x_i,x_i^2,\ldots,x_i^{k_i-1}$. Thus the number of distinct elements given by (*4.3*) is at most $k_1k_2\ldots k_n$, and so G is finite.

b. Subgroups of cyclic groups

Before beginning the study of a new section it is a good idea to list the natural questions. If we want to know something about the subgroups of cyclic groups, we might ask:

(i) Are subgroups of cyclic groups cyclic?

(ii) Does there exist a subgroup of any given order?

(iii) How many distinct subgroups of a cyclic group (less than or equal to the order of the group) are there?

(iv) How many subgroups of a given order are there?

We tackle each of these questions.

Theorem 4.9: (i) Let H be a subgroup of $G=gp(x)$. Then H is cyclic and either $H=gp(x^l)$ where x^l is the least positive power of x which lies in H or else $H=\{1\}$. If the order of G is $m<\infty$, then $l\mid m$ and the order of H is m/l. If the order of G is infinite, H is infinite or $H=\{1\}$.

(ii) Conversely if l is any positive integer dividing m, then $S=gp(x^l)$ is of order m/l. Consequently there is a subgroup of order q for any q that divides m.

(iii) The number of distinct subgroups of G is the same as the number of distinct divisors of $m=|G|<\infty$.

(iv) There is at most one subgroup of G of any given order for G finite.

Proof:

(i) If $H \neq \{1\}$, there exists $x^n \neq 1 \in H$. As H is a subgroup, $x^{-n} \in H$. Now one of $n, -n$ is positive. Hence we can talk meaningfully about the smallest positive power $x^l \in H$. Clearly, $H \supseteq S = gp(x^l)$. Suppose $x^r \in H$; then $r = ql + s$, $0 \leq s < l$, and

$$x^r(x^{lq})^{-1} = x^s \in H$$

But x^l is the least positive power of x that belongs to H. Thus $s = 0$ and so $r = ql$ and $(x^r) = (x^l)^q \in S$. Hence $S = H$.

If the order of G is $m < \infty$, then $m = ql + s$, $0 \leq s < l$. Now

$$1 = x^m = x^{ql+s} = x^{ql}x^s \in H$$

and so $x^s \in H$. Then $s = 0$, as x^l is the least positive power of x that lies in H. Hence l divides m, and $m = lq$. Clearly $(x^l)^q = 1$, and q is the least positive integer for which this occurs. Then, by Lemma 4.5, writing $a = x^l$, we have $H = gp(a) = \{a^0, a^1, \ldots, a^{q-1}\}$ and hence $|H| = q = m/l$. If the order of G is infinite, all the powers of x are distinct, and so $x^l, x^{2l}, \ldots$ is an infinite set of distinct elements of H. Thus H is infinite.

(ii) Let $l \mid m$, $l > 0$. Put $m/l = q$ and $x^l = a$. Then $S = gp(x^l) = \{1, a, \ldots, a^{q-1}\}$, as $a^q = x^m$ is the least positive power of a which is 1 (for if $a^{q'} = 1$, $q' < q$, then $x^{lq'} = 1$ and $lq' < m$, contradicting the fact that x^m is the least positive power of x which is 1). Thus S is a subgroup of order q. Consequently if we start out with a positive integer q which divides m and we put $l = m/q$, then S is a subgroup of G of order q.

(iii) Let $l_1, l_2, \ldots, l_n$ be the distinct divisors of m. Then put $H_1 = gp(x^{l_1}), \ldots, H_n = gp(x^{l_n})$. We know $|H_i| = m/l_i$. These are n distinct subgroups of G (because their orders are different). Are there any more subgroups? By (i) any subgroup H of G will have to be generated by x^l where l is a positive integer dividing m. Hence $l = l_i$, say. Therefore $H = H_i$. Thus the subgroups of G are simply $H_1, H_2, \ldots, H_n$, as desired.

(iv) If H and K are two subgroups of G with $|H| = |K|$, then H and K are H_i and H_j of part (iii) above, for some i and j. But $|H_i| = m/l_i$, $|H_j| = m/l_j$. Since $|H| = |K|$, $l_i = l_j$ and therefore $i = j$ and $H = H_i = H_j = K$.

The reader will perceive that our knowledge of the cyclic groups is in some ways quite comprehensive. We know in the case of finite cyclic groups what the distinct subgroups are, we know they are cyclic and we know which cyclic subgroups appear. In the case of infinite cyclic groups we can easily prove there are an infinite number of subgroups. We will distinguish between them in Theorem 4.24, page 126, using the concept of index which will be introduced in Section 4.3b.

The reader might naturally be led to consider now groups generated by two elements, hoping that similar powerful conclusions can be obtained, e.g. that every subgroup of a two generator group is a two generator group. But in going from one to two generators we lose control. It has been shown that every countable group is a subgroup of a two generator group, so we can never hope for a simple account of two generator groups.

Problems

4.37. A subgroup H of a group G is called *proper* if $H \neq G$ and $H \neq \{1\}$. Let G be a cyclic group of order a prime p. Prove that G has no proper subgroups.

Solution:

We know from Theorem 4.9 that the number of subgroups of G is the same as the number of distinct divisors of p, which are p and 1. Hence the number of distinct subgroups of G is two. As $\{1\}$ and G itself are two distinct subgroups, the number of proper subgroups is zero.

4.38. Prove that the only groups which have no proper subgroups are the cyclic groups of order p and the group consisting of the identity alone.

Solution:

Let G be a group with no proper subgroups, $G \neq \{1\}$. Let $g \in G$, $g \neq 1$. Then $S = gp(g)$ is a subgroup by Lemma 4.3. Since $g \in S$, $S = G$ as G has no proper subgroups. Hence G is cyclic. If G is cyclic of order mn, $m, n \neq 1$, then, by Theorem 4.9, G has a subgroup of order m. But this is a proper subgroup. Hence G is cyclic of prime order or else possibly infinite cyclic, say $G = gp(x) = \{\ldots, x^{-2}, x^{-1}, x^0, x^1, x^2, \ldots\}$. But $H = gp(x^2)$ is a subgroup not equal to $\{1\}$, and not equal to G since $x \notin H$. Hence G can only be cyclic of order p, a prime.

4.39. Find a group with two distinct subgroups both of the same order. [*Hint.* Consider S_3.]

Solution:

Let $\tau_3 = \begin{pmatrix} 1 & 2 & 3 \\ 2 & 1 & 3 \end{pmatrix}$, $\tau_2 = \begin{pmatrix} 1 & 2 & 3 \\ 1 & 3 & 2 \end{pmatrix}$. Then $|gp(\tau_3)| = |gp(\tau_2)| = 2$.

4.40. Find a group which is of infinite order but has a subgroup of finite order. (*Hint.* Try the group of Möbius transformations, M, of Section 3.5, page 77.)

Solution:

Let $\eta : z \to 1/z$, $\infty \to \infty$. Then $gp(\eta)$ is of order 2, but M is infinite.

4.41. Let H be a subgroup of G. Let $g \in G$. Prove that the set $S = \{g^{-1}hg \mid h \in H\}$ is a subgroup of G. Prove that $\theta : H \to S$, defined by $h\theta = g^{-1}hg$, is an isomorphism of H onto S. If K is a finite cyclic subgroup of G which contains both H and S, prove that $H = S$. (Hard.)

Solution:

Since $H \neq \emptyset$, $S \neq \emptyset$. Let $g^{-1}h_1g, g^{-1}h_2g \in S$. Then

$$(g^{-1}h_1g)(g^{-1}h_2g)^{-1} \;=\; g^{-1}h_1gg^{-1}h_2^{-1}g \;=\; g^{-1}(h_1h_2^{-1})g \;\in\; S$$

because H is a subgroup implies $h_1h_2^{-1} \in H$. Thus S is a subgroup. θ is an onto map, since $h\theta = g^{-1}hg$. If $h_1\theta = h_2\theta$, then $g^{-1}h_1g = g^{-1}h_2g$. Pre-multiply by g and post-multiply by g^{-1}: $g(g^{-1}h_1g)g^{-1} = g(g^{-1}h_2g)g^{-1}$. Hence $h_1 = h_2$ and so θ is one-to-one onto. We need only check that θ is a homomorphism to conclude the proof:

$$h_1\theta h_2\theta \;=\; g^{-1}h_1g \cdot g^{-1}h_2g \;=\; g^{-1}h_1h_2g \;=\; (h_1h_2)\theta$$

Thus H and S are isomorphic.

If K is a finite cyclic subgroup containing H and S, then H and S are both of finite order; and since they are isomorphic, $|H| = |S|$. But, by Theorem 4.9, K has only one subgroup of any given order. Hence $H = S$.

4.3 COSETS

a. Introduction to the idea of coset

In this section we propose a natural question which introduces the idea of a coset. Cosets are important for other reasons: (i) With cosets we can perform useful counting arguments for finite groups. (ii) Cosets of a subgroup sometimes enable us to construct a new group from an old. We can also see how a group G is built up from one of its subgroups H and the group constructed from the cosets of H. (iii) The fundamental idea of a homomorphism can be re-interpreted in terms of the idea of a group constructed from cosets.

What is the natural question we ask? In Section 3.4c, page 67, and Section 3.4e, page 73, we defined the group I of isometries of the plane E and the isometry group I_S of a given figure S in E. Recall that an isometry σ of E belongs to I_S if for each $t \in E$, $t\sigma \in S$ implies $t \in S$, and $s \in S$ implies $s\sigma \in S$. Suppose $\sigma \in I - I_S$. We ask: which ele-

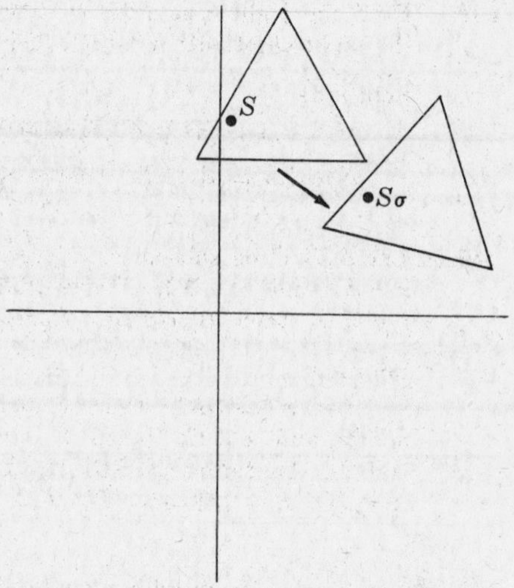

ments θ of I are such that $S\theta = S\sigma$? In the case where S is an equilateral triangle we know that $S\sigma$ is a congruent equilateral triangle by Lemma 3.6, page 71. So the question we are asking is this: which other elements θ of I send the equilateral triangle S onto the equilateral triangle $S\sigma$?

Let $S\theta = S\sigma$. How are θ and σ related? If θ were equal to σ, then $\theta\sigma^{-1} = \iota$. If θ and σ were widely different, we would expect $\theta\sigma^{-1}$ to be anything but ι. Let $\phi = \theta\sigma^{-1}$. We will show that ϕ is an element of I_S. To do so observe that ϕ is an isometry of the plane E, since it is a product of isometries of E. Now $S\theta = S\sigma$ implies that for each $s \in S$ there is a $t \in S$ such that $s\theta = t\sigma$, and conversely. Therefore for each $s \in S$,

$$s\phi = s\theta\sigma^{-1} = (t\sigma)\sigma^{-1} = t \in S$$

Suppose now that $x \in E$ and $x\phi \in S$. We must show that $x \in S$ to complete the proof that $\phi \in I_S$. Suppose $x\phi = t \in S$, i.e. $x(\theta\sigma^{-1}) = t$. Now there exists an $s \in S$ such that $t\sigma = s\theta$. Hence $x\theta = ((x\theta)\sigma^{-1})\sigma = t\sigma = s\theta$ and so $x\theta = s\theta$. As θ is one-to-one, $x = s$. Hence $x \in S$. This means $\phi \in I_S$.

We have of course $\theta = \phi\sigma$. If we write $I_S\sigma = \{\tau\sigma \mid \tau \in I_S\}$, we may put our deduction in the form $\theta \in I_S\sigma$. We have thus shown that every isometry θ of E satisfying $S\theta = S\sigma$ lies in the set of isometries $I_S\sigma$. Conversely if $\theta \in I_S\sigma$, then $\theta = \phi\sigma$ for some $\phi \in I_S$; then $S\theta = S\phi\sigma = S\sigma$. This means that $I_S\sigma$ consists of *all* the isometries θ of E for which $S\theta = S\sigma$. Such a subset $I_S\sigma$ of the group I of all isometries of E associated with the subgroup I_S of I is called a *right coset* of I_S in I. More generally we have the following

Definition: Let G be a group and H a subgroup of G. Then a *right coset* of H in G is a subset of the form $Hg = \{x \mid x = hg,\ h \in H\}$ for some g in G. We define a *left coset* of H in G to be a subset of the form $gH = \{x \mid x = gh,\ h \in H\}$.

Note that a coset is a right or left coset according as the element g is on the right or the left of H.

In the case where the group G is written additively, i.e. $+$ is used to denote the binary operation, a right coset is written $H+g$. Of course, $H+g = \{x \mid x = h+g,\ h \in H\}$.

Problems

4.42. Let G be the cyclic group of order 4 generated by $\{a\}$. Let $H = gp(a^2)$. Find all right cosets of H in G. Show that two cosets are either equal or else have no elements in common, and prove that the union of these cosets is G.

Solution:

$H \cdot 1 = H = \{1, a^2\}$ is a right coset. $Ha = \{a, a^3\}$ is a right coset. $Ha^2 = \{a^2, a^4\} = \{1, a^2\} = H$. $Ha^3 = \{a^3, a^5\} = \{a^3, a\} = Ha$. Thus the distinct cosets of H in G are H and Ha.

$H \cap Ha = \emptyset$, and $H \cup Ha = \{1, a^2, a, a^3\} = G$.

4.43. Let H be the trivial subgroup of a group G, i.e. $H = \{1\}$. Determine the distinct right cosets of H in G.

Solution:

If $g \in G$, $Hg = \{1g\} = \{g\}$. Thus the cosets are the sets consisting of single elements of G.

4.44. Find the right cosets of $H = gp(\tau_3)$ in S_3 with the notation of Section 3.3a, page 57. What are the right cosets of $K = gp(\sigma_1)$ in S_3?

Solution:

$H = (\iota, \tau_3)$, $H\iota = H$, $H\sigma_1 = \{\sigma_1, \tau_2\}$, $H\sigma_2 = \{\sigma_2, \tau_1\}$. These are all the cosets of H in S_3.

$K = \{\sigma_1, \sigma_2, \iota\}$. The cosets of K are $K\iota = K$ and $K\tau_1 = \{\tau_1, \tau_2, \tau_3\}$.

4.45. Let A, B, C be subsets of a group G. If X and Y are subsets of G, we define $XY = \{g \mid g = xy, x \in X \text{ and } y \in Y\}$. Prove that $A(BC) = (AB)C$. Hence conclude that if H is a subset of G, $f, g \in G$, then (i) $(fg)H = f(gH)$, (ii) $H(fg) = (Hf)g$, (iii) $(fH)g = f(Hg)$.

Solution:

Let $x \in A(BC)$; then $x = ad$ where $a \in A$ and $d \in BC$. But $d \in BC$ implies $d = bc$, where $b \in B$ and $c \in C$; hence $x = a(bc) = (ab)c \in (AB)C$ and so $A(BC) \subseteq (AB)C$. Similarly $(AB)C \subseteq A(BC)$. Therefore $A(BC) = (AB)C$. (i), (ii) and (iii) follow immediately, e.g. (i) is the case where $A = \{f\}$, $B = \{g\}$ and $C = H$.

4.46. Let G be a group with a subset H. Show that $fH = Hf$ implies $f^{-1}H = Hf^{-1}$.

Solution:

$fH = Hf$. Hence

$$f^{-1}(fH) = f^{-1}(Hf), \quad (f^{-1}f)H = (f^{-1}H)f \quad \text{and} \quad H = (f^{-1}H)f$$

Thus $Hf^{-1} = ((f^{-1}H)f)f^{-1} = f^{-1}H$.

Note that we have used the "associative law" proved in Problem 4.45.

b. Cosets form a partition. Lagrange's Theorem

In the problems above it is clear that any two right (or left) cosets are either disjoint or exactly the same and that the union of all the right (or left) cosets of H in G is G. We recall that a family of subsets of a set G is a *partition* of G if they are disjoint and their union is G. The examples above point to the following:

Theorem 4.10: Let H be a subgroup of a group G. Then the right (left) cosets of H in G form a partition of H in G, i.e. the union of all the right (left) cosets of H in G is G itself and any pair of distinct cosets has empty intersection.

Proof: It is easy to show that each element of G occurs in at least one right coset. (The proof for left cosets is similar and is not included here.) For if $g \in G$, then $g \in Hg$ since $1 \in H$ and $1 \cdot g = g$.

Suppose now that Ha and Hb are two cosets of H in G and that $Ha \cap Hb \neq \emptyset$, i.e. there exists $g \in Ha \cap Hb$. Then $g = h'a = h''b$, $h', h'' \in H$. Hence $a = h'^{-1}h''b = h'''b$, $h''' \in H$, since the product of two elements of H belongs to H. Therefore

$$Ha = \{ha \mid h \in H\} = \{h(h'''b) \mid h \in H\} \subseteq Hb \quad \text{as } hh'''b = h_1 b,\ h_1 \in H$$

Similarly $Hb \subseteq Ha$. Thus $Ha = Hb$ and any two cosets are either disjoint or identical.

In Section 4.3a we mentioned in (i) that cosets are useful for counting arguments; this follows from Theorem 4.10. If G is of finite order and H a subgroup of G, then, since the cosets of H in G are disjoint, the order of G is the sum of the number of elements in each coset. We use this fact in proving

Theorem 4.11 (Lagrange's Theorem): The order of a subgroup H of a finite group G divides the order of G.

Proof: Let the distinct cosets of H in G be $Hg_1, Hg_2, \ldots, Hg_n$. Since these form a partition of G,

$$|G| = |Hg_1| + |Hg_2| + \cdots + |Hg_n| \tag{4.4}$$

What is $|Hg|$? We will show that the mapping $\theta_g: H \to Hg$ defined by $h\theta = hg$ is a one-to-one onto mapping and hence $|H| = |Hg|$. Clearly θ_g is onto by the definition of right coset. If $h_1 g = h_2 g$, then multiplying by g^{-1} on the right we conclude that $h_1 = h_2$ and so $h_1\theta_g = h_2\theta_g$ implies $h_1 = h_2$. Therefore θ_g is one-to-one and onto and $|H| = |Hg|$. Thus for each i, $|H| = |Hg_i|$. Hence $|G| = n|H|$ by *(4.4)*.

Corollary 4.12: Let G be a finite group and g an element of G of order m. Then m divides $|G|$.

Proof: The order of g is the order of $gp(g)$ which is a subgroup of G. Then by Theorem 4.11, $|gp(g)|$ divides $|G|$. But $m = |gp(g)|$. Hence m divides $|G|$.

Corollary 4.13: If G is of finite order n, and $g \in G$, then $g^n = 1$.

Proof: Every element of G must be of finite order. Let $g \in G$ be of order m. Then, by the preceding corollary, m divides $|G|$ and so $|G| = qm$. Hence $g^{|G|} = g^{qm} = (g^m)^q = 1$ and the result follows.

Definition: Let the number of right cosets of H in G be called the *index* of H in G. Denote it by $[G:H]$.

Note that $[G:H]$ is read as "the index of H in G", i.e. in the opposite order to which G and H appear in $[G:H]$.

Corollary 4.14: If G is a finite group, $|G| = |H| \cdot [G:H]$.

Proof: In the proof of Theorem 4.11 we conclude with "Hence $|G| = n|H| \ldots$". Since n is the number of cosets of H in G, i.e. $n = [G:H]$, we have $|G| = |H|\,[G:H]$.

Problems

4.47. Show that: (i) S_7 has no subgroup of order 11; (ii) D_4 has no subgroup of order 3; (iii) if $g \in A_5$ and $g^7 = \iota$, then $g = \iota$.

Solution:

(i) $|S_7| = 7! = 7 \cdot 6 \cdot 5 \cdot 4 \cdot 3 \cdot 2 = 7 \cdot 5 \cdot 3^2 \cdot 2^4$. If H were a subgroup of S_7 of order 11, then, by Theorem 4.11, 11 divides $|S_7|$. But in the prime decomposition of $|S_7|$ there is no 11. Hence there is no subgroup of order 11.

(ii) $|D_4| = 8$. Since 3 does not divide 8, Theorem 4.11 tells us there is no subgroup of order 3.

(iii) If $g^7 = \iota$, then either $g = \iota$ or g is of order 7, since $g^m = 1$ implies that the order of g divides m. As $m = 7$ and is a prime, the only possibility if $g \neq \iota$ is that the order of g is 7. Now by Corollary 4.12 it would follow that 7 divides $|A_5|$, which is not true. Hence $g = \iota$.

4.48. Prove that if G is a group of prime order, then G is cyclic.

Solution:

If $G = \{1\}$, there is nothing to prove. If $1 \neq g \in G$, $gp(g) = H$ is a subgroup of G. Hence its order, by Theorem 4.11, divides the prime $|G|$. As $|H| \neq 1$, $|H| = |G|$ since the only divisors of $|G|$ are 1 and $|G|$. Thus $H = G$, as $H \subseteq G$ and H and G have the same number of elements.

4.49. Give the right and left cosets of $H = gp(\{\eta\})$ where $\eta = \sigma(0,1,1,0)$ is an element of M, the group of Möbius transformations, Section 3.5a, page 77.

Solution:

If $\sigma(a,b,c,d) \in M$, then

$$H\sigma(a,b,c,d) = \{\iota \cdot \sigma(a,b,c,d),\ \eta\sigma(a,b,c,d)\} = \{\sigma(a,b,c,d),\ \sigma(b,a,d,c)\}$$

by the rule for multiplication which is obtained in Problem 3.46, page 79.

Now $\sigma(a,b,c,d)H = \{\sigma(a,b,c,d),\ \sigma(c,d,a,b)\}$, as is easily checked. Thus we know what the right and left cosets are in terms of a, b, c, d.

c. Normal subgroups

We discussed left and right cosets of a subgroup H in G. Each gives rise to a partition of G. How do these partitions compare? In particular, are they the same? Sometimes yes. In Problem 4.43, the right cosets are just the elements of G; the left cosets can similarly be shown to be just the elements of G. Sometimes no. In Problem 4.49 the right coset containing $\sigma(a,b,c,d)$ also contains $\sigma(b,a,d,c)$. But the left coset containing $\sigma(a,b,c,d)$ contains $\sigma(c,d,a,b)$. With a suitable choice of a,b,c,d we can ensure that $\sigma(b,a,d,c) \neq \sigma(c,d,a,b)$, e.g. if $a=b=c=1$, $d=0$, then $\sigma(1,1,0,1) \neq \sigma(1,0,1,1)$. Thus the left and right cosets of H in G do not coincide.

We ask: when do the right and left cosets of a subgroup H in a group G coincide? Suppose every left coset of H is also a right coset of H in G. Let $a \in G$. aH contains a, as does Ha. Since the right cosets form a partition, the only right coset containing a is Ha. But we have assumed that there is some right coset which is the same as aH. Hence it must be Ha, and so $aH = Ha$. In other words if every left coset of H in G is a right coset, then for every $a \in H$ we must have $aH = Ha$.

Proposition 4.15: A necessary and sufficient condition for the left cosets of H in G to provide the same partition as the right is that for each $a \in G$, $aH = Ha$.

Proof: We have proved above that if every left coset is a right coset, $Ha = aH$. Let $Ha = aH$ for all $a \in G$. $\{Ha \mid a \in G\}$ is the set of all the right cosets of H in G. Therefore every right coset of H in G is a left coset of H in G. Similarly every left coset of H in G is a right coset of H in G. This completes the proof.

If $Ha = aH$ for all $a \in G$, then for each $h \in H$, $ha = ah_1$ for some $h_1 \in H$. Hence $a^{-1}ha \in H$ for each $h \in H$.

Conversely if $a^{-1}ha = h_1$, for some h_1 belonging to H, $ha = ah_1$. Hence $Ha \subseteq aH$. If $ah \in aH$ and assuming $x^{-1}hx \in H$ for all $x \in G$ and all $h \in H$, then $ah = ah(a^{-1}a) = ((a^{-1})^{-1}ha^{-1})a \in Ha$. Accordingly $Ha \supseteq aH$ and $aH = Ha$. Thus we have

Proposition 4.16: $aH = Ha$ for all $a \in G$ if and only if $a^{-1}ha \in H$ for all $h \in H$ and all $a \in G$.

Definition: A subgroup H of a group G is *normal* (also called *invariant*) in G if $g^{-1}hg \in H$ for all $g \in G$ and all $h \in H$. We write $H \lhd G$ and read it as: "H is a normal subgroup of G".

By Proposition 4.16, H is normal in G if and only if $Hg = gH$ for all $g \in G$ (equivalently, $g^{-1}Hg = H$).

In Section 4.3a we gave in (ii) and (iii) reasons for the importance of some cosets. The cosets we had in mind are those arising from normal subgroups.

Problems

4.50. Prove that every subgroup of an abelian group is a normal subgroup.

Solution:

Let G be abelian and H any subgroup of G. Then if $g \in G$ and $h \in H$, $g^{-1}hg = h$; for since G is abelian, $gh = hg$ and hence multiplying by g^{-1} on the left, $h = g^{-1}hg$. Then if $h \in H$, $g^{-1}hg \in H$. Thus H is a normal subgroup of G.

4.51. Prove that $A_n \lhd S_n$ for each positive integer n.

Solution:

Let $\sigma \in A_n$, $\tau \in S_n$. Is $\tau^{-1}\sigma\tau \in A_n$? Now τ is either odd or even. If τ is even, then τ and $\tau^{-1} \in A_n$ and so $\tau^{-1}\sigma\tau \in A_n$. If τ is odd, then τ^{-1} is also odd, and hence $\tau^{-1}\sigma$ is odd, for $\sigma \in A_n$. Since $\tau^{-1}\sigma$ and τ are odd, their product $\tau^{-1}\sigma\tau$ is even. Hence $\tau^{-1}\sigma\tau \in A_n$. Note that we used Lemma 3.2, page 62.

4.52. Let H be a finite cyclic subgroup of G, and let $H \triangleleft G$. Let K be a proper subgroup of H. Prove that $K \triangleleft G$. (Hard.)

Solution:

Let $K = gp(y)$. Note that if y is of order m, and $g \in G$, then $g^{-1}yg$ is also of order m, since $(g^{-1}yg)^m = g^{-1}y^m g = 1$ and $(g^{-1}yg)^r = 1$ implies $g^{-1}y^r g = 1$ and therefore $y^r = 1$. Hence m divides r, and the order of $g^{-1}yg$ is m. Since $H \triangleleft G$, $g^{-1}yg \in H$. Therefore $gp(g^{-1}yg)$ is a subgroup of H of order m. Then by Theorem 4.9(iv), $gp(g^{-1}yg) = K$. In particular, $g^{-1}kg \in K$ for any $k \in G$. Hence $K \triangleleft G$.

d. Commutator subgroups, centralizers, normalizers

We will now introduce some subgroups which are normal.

1. If G is a group, we define the center of G, denoted by $Z(G)$, to be

$$\{z \mid z \in G \text{ and for all } g \in G,\ gz = zg\}$$

$Z(G)$ turns out to be a normal subgroup of G (see problems below).

2. If G is a group and $x, y \in G$, then $x^{-1}y^{-1}xy$ is called the *commutator* of x and y or, more briefly, a commutator. We often write $[x, y]$ for the commutator $x^{-1}y^{-1}xy$. The subgroup of G generated by all commutators is called the *commutator subgroup* (also called the derived group) of G and is denoted by G'. Again G' turns out to be normal in G.

Proceeding along somewhat different lines, let A be a subset of a group G.

1. The centralizer $C(A)$ of A (in G) is defined by

$$C(A) = \{c \mid c \in G \text{ and for all } a \in A,\ ca = ac\}$$

$C(A)$ is a subgroup of G (see problems below). If A is an abelian subgroup, A is normal in $C(A)$ (see problems below).

2. The normalizer $N(A)$ of A in G is defined by

$$N(A) = \{n \mid n \in G \text{ and } An = nA\}$$

$N(A)$ is a subgroup of G and, if A is a subgroup of G, A is normal in $N(A)$. Furthermore, if A is a subgroup of G, A is normal in G if and only if $N(A) = G$. These facts will be proved in the problems below.

The details concerning the groups $Z(G)$, G', $C(A)$, $N(A)$ appear in the problems below. In Chapter 5 we will use the concepts we have just introduced.

Problems

4.53. Prove that the inverse of a commutator is a commutator.

Solution:

$[x, y] = x^{-1}y^{-1}xy = z$, say. So $z^{-1} = y^{-1}x^{-1}yx = [y, x]$.

4.54. Prove that G' is normal in G.

Solution:

We must show that if $g \in G$ and $h \in G'$, then $g^{-1}hg \in G'$. If h is a commutator, say $h = x^{-1}y^{-1}xy$, then

$$g^{-1}hg = g^{-1}x^{-1}gg^{-1}y^{-1}gg^{-1}xgg^{-1}yg = x_1^{-1}y_1^{-1}x_1y_1 = [x_1, y_1]$$

where $x_1 = g^{-1}xg$ (consequently $x_1^{-1} = g^{-1}x^{-1}g$) and $y_1 = g^{-1}yg$.

Now any element h of G' is a product of commutators and their inverses; and as an inverse of a commutator is a commutator, every element h of G' is a product $c_1 \cdots c_k$ of commutators. Therefore

$$g^{-1}hg = g^{-1}(c_1 \cdots c_k)g = g^{-1}c_1gg^{-1}c_2g \cdots g^{-1}c_kg = d_1d_2 \cdots d_k$$

where $d_i = g^{-1}c_ig$. But we have just shown that d_i is a commutator. Hence if $h \in G'$, $g^{-1}hg \in G'$.

4.55. Show that G is abelian if and only if $G' = \{1\}$.

Solution:

Suppose G is abelian and that $x, y \in G$. As x and y commute (i.e. $xy = yx$), $[x,y] = x^{-1}y^{-1}xy = x^{-1}x = 1$. Then G' is the subgroup of G generated by 1, and $G' = 1$. Now if $G' = \{1\}$, then in particular any commutator $[x,y] = x^{-1}y^{-1}xy = 1$. Hence $x(x^{-1}y^{-1}xy) = x$ and $y(y^{-1}xy) = yx$, i.e. $xy = xy$. Thus G is abelian.

4.56. Show that every element in A_3 is a commutator of elements in S_3. Hence show that $S_3' = A_3$.

Solution:

We use the table of Section 3.3a, page 57. $\tau_1 = \tau_1^{-1}$. $\tau_1^{-1}\sigma_1^{-1}\tau_1\sigma_1 = \tau_1\sigma_2\tau_1\sigma_1 = \sigma_1\sigma_1 = \sigma_2$. $\tau_1^{-1}\sigma_2^{-1}\tau_1\sigma_2 = \tau_1\sigma_1\tau_1\sigma_2 = \sigma_2\sigma_2 = \sigma_1$. $\tau_1^{-1}\iota^{-1}\tau_1\iota = \iota$. Thus every element of A_3 is a commutator of elements in S_3 (A_3 is listed in Problem 3.23, page 63). If we can show that all commutators belong to A_3, then $A_3 = S_3'$. This is a matter of trying all possibilities, e.g. $\tau_1^{-1}\tau_2^{-1}\tau_1\tau_2 = \tau_1\tau_2\tau_1\tau_2 = \sigma_1$. Hence the result.

4.57. Show that the commutator subgroup of M of Section 3.5a, page 77, is infinite. [*Hint.* Find an infinite cyclic subgroup generated by a commutator.]

Solution:

$[\sigma(2,0,0,1), \sigma(1,1,0,1)] = \sigma(1,-1,0,1) = \sigma$, say. Now $gp(\sigma)$ is infinite cyclic, as $\sigma^n = (1,-n,0,1)$ for each n. Since the commutator subgroup of M contains $gp(\sigma)$, it is infinite.

4.58. Prove that if G is any group, $Z(G)$ is a normal subgroup of G.

Solution:

$1 \in Z(G)$, since $1g = g1$ for all $g \in G$. Consequently $Z(G) \neq \emptyset$. If $g_1, g_2 \in Z(G)$ and $g \in G$, then $g(g_1g_2^{-1}) = (gg_1)g_2^{-1} = g_1(gg_2^{-1}) = g_1g_2^{-1}g$ since $gg_2 = g_2g$ implies $g_2^{-1}g = gg_2^{-1}$. It follows that $Z(G)$ is a subgroup of G. If $g \in G$ and $h \in Z(G)$, then $gh = hg$ and so $h = g^{-1}hg$. Hence $g^{-1}hg \in Z(G)$ for all $g \in G$ and all $h \in Z(G)$. Thus $Z(G)$ is normal in G.

4.59. Show that $C(A)$ is a subgroup of G and, if A is an abelian subgroup of G, $A \lhd C(A)$.

Solution:

$1 \in C(A)$ and so $C(A) \neq \emptyset$. If $g_1, g_2 \in C(A)$, and $a \in A$, then $g_2a = ag_2$ and hence $ag_2^{-1} = g_2^{-1}a$, i.e. $g_2^{-1} \in C(A)$. Now $ag_1g_2^{-1} = g_1ag_2^{-1} = g_1g_2^{-1}a$ and so $g_1g_2^{-1} \in C(A)$ if $g_1, g_2 \in C(A)$. Therefore $C(A)$ is a subgroup of G. If A is an abelian subgroup, then each $a \in A$ belongs to $C(A)$. Now if $g \in C(A)$, then for each $a \in A$, $ga = ag$, i.e. $g^{-1}ag = a \in A$. Accordingly $A \lhd C(A)$.

4.60. Show that if A is a subset of G, then $N(A)$ is a subgroup of G. Show that if A is a subgroup of G, then $A \lhd G$ if and only if $N(A) = G$.

Solution:

$N(A) \neq \emptyset$, since $1 \in N(A)$. Let $f, g \in N(A)$. Using the results of Problems 4.45 and 4.46, $gA = Ag$ and $fA = Af$ implies $(fg^{-1})A = f(g^{-1}A) = f(Ag^{-1}) = (fA)g^{-1} = (Af)g^{-1} = A(fg^{-1})$. Hence $f, g \in N(A)$ implies $fg^{-1} \in N(A)$. Therefore $N(A)$ is a subgroup of G. Clearly, if A is a subgroup, $A \subseteq N(A)$ and $A \lhd N(A)$.

If A is a subgroup and $A \lhd G$, then for each $g \in G$, $gA = Ag$. Hence $g \in N(A)$ and so $G \subseteq N(A)$. Therefore $G = N(A)$. If A is a subgroup and $N(A) = G$, then since $A \lhd N(A)$, $A \lhd G$.

4.61. Find all normal subgroups of S_3.

Solution:

We use the notation and multiplication table in page 57.

Clearly $\{\iota\}$ and S_3 are both normal subgroups of S_3. There are no normal subgroups of S_3 containing elements of order 2 except S_3. The elements of order 2 in S_3 are, as we readily check by using the multiplication table for S_3, τ_1, τ_2 and τ_3. Suppose for example that a normal subgroup N of S_3 contains τ_1; then $\sigma_1^{-1}\tau_1\sigma_1 = \sigma_2\tau_1\sigma_1 = \tau_2 \in N$. Similarly $\tau_3 \in N$. Hence $\tau_1\tau_2 = \sigma_2 \in N$ and $\sigma_2^2 = \sigma_1 \in N$, and so it follows in this way that $N = S_3$.

We have shown that if N is a normal subgroup of S_3, then if N contains elements of order 2, $N = S_3$. Therefore if $N \neq \{\iota\}$, then N must contain elements of order 3 (there are only elements of order 1, 2, or 3 in S_3). Now σ_1 and $\sigma_2 = \sigma_1^2$ are the only elements of S_3 of order 3. In fact $\{\iota, \sigma_1, \sigma_2\}$ is a normal subgroup of S_3. For example, $\tau_1^{-1}\sigma_1\tau_1 = \tau_1\sigma_1\tau_1 = \sigma_2 \in \{\iota, \sigma_1, \sigma_2\}$. Accordingly S_3 has precisely three distinct normal subgroups.

4.62. Show that if A is a subgroup of G and $B \lhd G$, then AB is a subgroup of G, where

$$AB = \{x \mid x = ab,\ a \in A,\ b \in B\}$$

Solution:

$AB \neq \emptyset$, as $1 = 1 \cdot 1 \in AB$. If $g_1, g_2 \in AB$, then $g_1 = a_1b_1$, $g_2 = a_2b_2$ where $a_i \in A$ and $b_i \in B$. Now

$$\begin{aligned} g_1g_2^{-1} &= a_1b_1b_2^{-1}a_2^{-1} = a_1b_3a_2^{-1} \text{ (where } b_3 \in B) \\ &= a_1a_2^{-1}a_2b_3a_2^{-1} = a_1a_2^{-1}(a_2^{-1})^{-1}b_3(a_2^{-1}) \\ &= ab, \text{ say,} \end{aligned}$$

where $a = a_1a_2^{-1} \in A$, and $b = a_2b_3a_2^{-1} \in B$ as $B \lhd G$. Thus $g_1g_2^{-1} \in AB$ and AB is a subgroup of G.

4.63. Show that the intersection of two normal subgroups is a normal subgroup.

Solution:

We refer the reader to Problem 3.15, page 55, in which we proved that the intersection of two subgroups is a subgroup. If H, K are normal subgroups of G, then $H \cap K$ is a subgroup of G. If $c \in H \cap K$ and $g \in G$, then $g^{-1}cg \in H$ as $c \in H$ and $H \lhd G$, and $g^{-1}cg \in K$ as $c \in K$ and $K \lhd G$. Thus $g^{-1}cg \in H \cap K$, and so $H \cap K$ is normal in G.

e. Factor groups

In Section 4.3a we mentioned that the concept of a coset sometimes gives rise to a new group. This occurs when, and only when, the group is normal.

Let G be a group and $N \lhd G$. Let us denote by G/N (read as "G over N", or "G factor N", or "the factor group of G by N") the set of right cosets of N in G. We turn G/N into a groupoid by defining a binary operation as follows.

Define a product of two cosets Na and Nb to be the coset Nab. This definition of multiplication depends on a and b. But it is conceivable that if $Na = Na_1$ and $Nb = Nb_1$, that $Na_1b_1 \neq Nab$. In such case, what would we take for the "product" of the two cosets, Na_1b_1 or Nab? What we must show is that the product of two cosets is uniquely defined by the formula $NaNb = Nab$ when $N \lhd G$. If $Na_1 = Na$ and $Nb_1 = Nb$, then $a_1 = na$ for some $n \in N$ and $b_1 = mb$ for some $m \in N$. Accordingly,

$$a_1b_1 = namb = n(ama^{-1})ab = lab$$

where $l = n(ama^{-1})$. Since $N \lhd G$, $ama^{-1} \in N$ and hence $l \in N$. Thus $a_1b_1 \in Nab$. Since the cosets form a partition, it follows that $Na_1b_1 = Nab$. But this is just what we wanted to prove.

Thus G/N with this binary operation is a groupoid. Is it an associative groupoid?

$$\begin{aligned} ((Na)(Nb))(Nc) &= (Nab)Nc = N(ab)c = Na(bc) \text{ (as } G \text{ is associative)} \\ &= (Na)(Nbc) = (Na)((Nb)(Nc)) \end{aligned}$$

and so G/N is an associative groupoid.

Theorem 4.17: *G/N is a group. The mapping $\nu: G \to G/N$ defined by $g\nu = Ng$ is a homomorphism of G onto G/N.*

Proof: First, G/N is an associative groupoid. $N1 = N$ is an identity, for $Na \cdot N1 = N(a \cdot 1) = Na$ and $N \cdot Na = N(1 \cdot a) = Na$. Next, each element Na has an inverse, for $NaNa^{-1} = N(aa^{-1}) = N$ while $Na^{-1}Na = N(a^{-1}a) = N$. Thus G/N is a group.

Clearly ν is a mapping of G onto G/N. Since $(g_1g_2)\nu = N(g_1g_2)$ and $(g_1\nu)(g_2\nu) = Ng_1Ng_2 = Ng_1g_2$, $(g_1g_2)\nu = (g_1\nu)(g_2\nu)$ and hence ν is a homomorphism.

ν is called the *natural* homomorphism of G onto its factor group G/N.

Note that in the case where G is a group whose binary operation is $+$, as we remarked at the end of Section 4.3a, the elements of G/N are of the form $N+g$. Instead of using the multiplicative notation for G/N, we use additive notation. Our definition of the product of two cosets is written as the sum: $(N+g_1)+(N+g_2) = N+(g_1+g_2)$.

Problems

4.64. Prove that $Z/E \cong C_2$, where E is the set of even integers and C_2 is the cyclic group of order 2.

Solution:

As Z is abelian, E is a normal subgroup of Z and Z/E makes sense. A coset of E is of the form $E+z$, $z \in Z$. $E+0$ consists of all the even integers, $E+1$ of all the odd. Since an integer is either odd or even, these are all the cosets. $gp(\{(E+1)\})$ contains $E+1$ and $(E+1)+(E+1)=E$, and so $gp(\{(E+1)\}) = Z/E$. Thus Z/E is cyclic of order 2. Hence the result.

4.65. Is $Q/Z \cong Q$, where Q is the additive group of rationals? (*Hint.* Examine the order of the elements of Q/Z.)

Solution:

A coset of Z in Q is of the form $Z+q$, $q \in Q$. As $q = m/n$ for some integers m,n, it follows that $nq \in Z$. Then

$$\underbrace{(Z+q)+(Z+q)+\cdots+(Z+q)}_{n \text{ terms in all}} = Z$$

and therefore $Z+q$ is of finite order in Q/Z. Thus every element of Q/Z is of finite order. On the other hand no element other than 0 of Q is of finite order, for $nq \neq 0$ if $q \neq 0$ and $n \neq 0$. It is this fact that we will utilize to prove that Q/Z is not isomorphic to Q. Suppose it were and that $\theta: Q/Z \to Q$ was such an isomorphism. Choose $q \in Q$ $(q \neq 0)$; then there exists an element $Z+r$ of Q/Z such that $(Z+r)\theta = q$. Now $Z+r$ is of finite order n, say. Then $(n(Z+r))\theta = nq$ and now $n(Z+r) = Z$. Since Z is the identity of Q/Z and θ is an isomorphism, θ takes Z to the identity of Q, namely 0; hence $nq = 0$. But $q \neq 0$, so $nq \neq 0$. This contradiction proves that there exists no isomorphism of Q/Z onto Q.

4.66. Find S_3/S_3'.

Solution:

From Problem 4.56, $S_3' = A_3$. Hence the cosets of S_3/S_3' are $A_3 = A_3\iota = \{\iota, \sigma_1, \sigma_2\}$ and $A_3\tau_1 = \{\tau_1, \tau_2, \tau_3\}$. Thus S_3/S_3' consists of these two cosets. The multiplication table is

	A_3	$A_3\tau_1$
A_3	A_3	$A_3\tau_1$
$A_3\tau_1$	$A_3\tau_1$	A_3

4.67. Prove that if $G \cong H$, then $G' \cong H'$ and $Z(G) \cong Z(H)$.

Solution:

Let θ be an isomorphism from G to H. Then $\Psi = \theta_{|G'}$ is a monomorphism into H, since θ is both one-to-one and a homomorphism.

What is $G'\Psi$? The elements of G' are products of commutators and their inverses. As the inverse of a commutator is a commutator, each element of G' is of the form $c_1 \cdots c_k$, where each c_i is a commutator. Then the images of G' under Ψ are of the form $c_1\Psi \cdots c_k\Psi$. If

$$c = [a,b] = a^{-1}b^{-1}ab, \quad \text{then} \quad c\Psi = (a\theta)^{-1}(b\theta)^{-1}a\theta b\theta = [a\theta, b\theta]$$

Thus the $c_i\Psi$ are commutators and so $G'\Psi \subseteq H'$.

To show $G'\Psi = H'$, we need only show that all possible commutators $[h_1, h_2]$ are images under Ψ. As θ is onto, there exists g_1, g_2 such that $g_1\theta = h_1$, $g_2\theta = h_2$. Hence $[g_1, g_2]\Psi = [h_1, h_2]$. Therefore $G'\Psi = H'$ and consequently $G' \cong H'$.

Let $\phi = \theta_{|Z(G)}$. Then ϕ is a monomorphism of $Z(G)$ into H, and we need only show that it maps onto $Z(H)$ to prove $Z(G) \cong Z(H)$. Let $z \in Z(G)$. For each $h \in H$ there exists $g \in G$ with $g\theta = h$. Hence $z\theta h = z\theta g\theta = (zg)\theta = (gz)\theta = g\theta z\theta = hz\theta$ and so $Z(G)\phi \subseteq Z(H)$. Let $x \in Z(H)$. As θ is onto, there exists $y \in G$ with $y\theta = x$. Let $g \in G$. Then $(gy)\theta = g\theta y\theta = y\theta g\theta = (yg)\theta$. As θ is one-to-one, $gy = yg$. Hence $y \in Z(G)$, and every element of $Z(H)$ is an image of an element of $Z(G)$. The result follows.

4.68. (i) If A is abelian show that $A/A' \cong A$.
(ii) Show that G/G' is abelian.
(iii) Show that G/N abelian implies $N \supseteq G'$.

Solution:

(i) A' is the subgroup generated by the commutators $x^{-1}y^{-1}xy$, $x, y \in A$. As A is abelian, $x^{-1}y^{-1}xy = y^{-1}x^{-1}xy = y^{-1}1y = 1$ and A' is the subgroup generated by 1. Since any product of 1 and its inverse is again 1, $A' = \{1\}$. Let ν be the natural homomorphism of A to A/A'. To show ν is an isomorphism we need only show that it is one-to-one. Suppose $a_1\nu = a_2\nu$, i.e. $\{1\}a_1 = \{1\}a_2$, i.e. $\{a_1\} = \{a_2\}$. Then of course $a_1 = a_2$, so ν is one-to-one. Thus ν is an isomorphism.

(ii) Let $G'x$ and $G'y$ be two elements of G/G'. Then $G'xG'y = G'xy$ while $G'yG'x = G'yx$. Now $(x^{-1})^{-1}(y^{-1})^{-1}x^{-1}y^{-1} = xyx^{-1}y^{-1} \in G'$, and so $G'yx$ contains the element $xyx^{-1}y^{-1} \cdot yx = xy$. But $G'xy$ contains xy; hence $G'xy = G'yx$. Therefore $(G'x)(G'y) = (G'y)(G'x)$ and G/G' is abelian.

(iii) We need only show that N contains every commutator. For then N contains the subgroup generated by the commutators, which is G' by definition.

If G/N is abelian and x, y are any elements of G, $(xy)N = xNyN = yNxN = (yx)N$. Hence $xy = yxn$ where $n \in N$. Multiplying on the left by y^{-1} and then by x^{-1}, we obtain $x^{-1}y^{-1}xy = n$, i.e. $[x, y] \in N$.

4.69. Show that if H is a subgroup of G containing G', then $H \lhd G$. Show that if H is of index 2 in G, then $H \lhd G$ and G/H is cyclic of order 2.

Solution:

Let $h \in H$ and consider $g^{-1}hg$. Now $g^{-1}hgh^{-1}$ is a commutator, and hence belongs to G' and thus to H. Therefore $g^{-1}hgh^{-1} \cdot h = g^{-1}hg \in H$ and $H \lhd G$.

If now H is of index 2, $G = H \cup Hg$ where $H \cap Hg = \emptyset$. Let $h \in H$. If $k \in G$, then $k = h_1$ or $k = h_1g$ $(h_1 \in H)$. Hence

$$k^{-1}hk = h_1^{-1}hh_1 \in H \qquad \text{or} \qquad k^{-1}hk = g^{-1}h_1^{-1}hh_1g = g^{-1}h_2g$$

where $h_2 \in H$. If $g^{-1}h_2g \in H$, we are through. Otherwise $g^{-1}h_2g \in Hg$, so that $g^{-1}h_2g = h'g$ $(h' \in H)$ and thus $g^{-1}h_2 = h'$. Hence $g = h_2h^{-1} \in H$. But this contradicts the assumption $H \cap Hg = \emptyset$. Thus we are forced to conclude that $k^{-1}hk \in H$ for every $h \in H$ and every $k \in G$, i.e. that $H \lhd G$. The two cosets of H in G are H and Hg for some $g \notin H$. Accordingly $gp(Hg) = G/H$ and so G/H is cyclic of order 2.

4.70. Show by considering a suitable non-normal subgroup H of S_3 that the product of two cosets cannot be defined as on page 114 without ambiguity.

Solution:

We use the notation of Section 3.3a, page 57, in dealing with the symmetric group S_3 of degree 3. Let us take $H = gp(\tau_1) = \{1, \tau_1\}$. $H\sigma_1 = \{\sigma_1, \tau_3\} = H\tau_3$ while $H\sigma_2 = \{\sigma_2, \tau_2\} = H\tau_2$. In page 114 we defined the product of the coset $H\sigma_1$ and $H\sigma_2$ as $H\sigma_1\sigma_2 = H$. However, the product of $H\tau_3$ and $H\tau_2$ would be $H\tau_3\tau_2 = H\sigma_1 \neq H$. This means that the product of cosets is not uniquely defined.

4.4 HOMOMORPHISM THEOREMS

a. Homomorphisms and factor groups: The homomorphism theorem

We now consider the connection between homomorphisms and factor groups. We have already established in Theorem 4.17 that corresponding to every factor group G/N there is a homomorphism $\nu: G \to G/N$ such that $G\nu = G/N$. What about the converse? Suppose now that θ is a homomorphism of G into a group H.

We ask: is there a normal subgroup N of G such that $G/N \cong G\theta$? Let us define the *kernel* of θ (denoted $\text{Ker}\,\theta$) by $\text{Ker}\,\theta = \{g \mid g \in G,\ g\theta = 1\}$. We will show that $\text{Ker}\,\theta$ will do the trick.

Theorem 4.18 (Homomorphism Theorem, also called the First Isomorphism Theorem): If $\theta: G \to H$ is a homomorphism of a group G into a group H, then $N = \text{Ker}\,\theta$ is a normal subgroup of G, and $\eta: g\theta \to Ng$ defines an isomorphism of $G\theta$ onto G/N.

Proof: First we will show that N is a subgroup. If $g_1, g_2 \in N$, then

$$(g_1g_2^{-1})\theta = (g_1\theta)(g_2^{-1}\theta) = (g_1\theta)(g_2\theta)^{-1} = 1 \cdot 1 = 1$$

and so $g_1g_2^{-1} \in N$. Also $1 \in N$, so $N \neq \emptyset$, the empty set. Thus N is a subgroup of G. To prove that it is a normal subgroup of G, let $n \in N$ and $g \in G$. We must show that $g^{-1}ng \in N$; this will hold if $(g^{-1}ng)\theta = 1$. But

$$(g^{-1}ng)\theta = (g\theta)^{-1}(n\theta)(g\theta) = (g\theta)^{-1}1(g\theta) = (g\theta)^{-1}(g\theta) = 1$$

Hence $g^{-1}ng \in N$ and N is a normal subgroup of G.

Next we must show that $\eta: g\theta \to Ng$ defines a mapping. It is conceivable that there exist $g_1 \neq g_2$ with $g_1\theta = g_2\theta$. We ask: is $Ng_1 = Ng_2$? For if not, η is not a mapping as it is not uniquely defined. Now $g_1g_2^{-1} \in N$, for $(g_1g_2^{-1})\theta = (g_1\theta)(g_2\theta)^{-1} = 1$ as $g_1\theta = g_2\theta$. Hence $g_1g_2^{-1} \in N$ and $g_1 = ng_2$ where $n \in N$. Thus Ng_1 and Ng_2 have g_1 in common and, as the right cosets form a partition, $Ng_1 = Ng_2$. Consequently η is a well defined mapping.

Is it a homomorphism?

$$(g_1\theta g_2\theta)\eta = ((g_1g_2)\theta)\eta = N(g_1g_2) = Ng_1Ng_2 = (g_1\theta)\eta(g_2\theta)\eta$$

and so η is a homomorphism.

Finally, is η one-to-one? If $(g_1\theta)\eta = (g_2\theta)\eta$, then $Ng_1 = Ng_2$. Then $ng_1 = g_2$ for some $n \in N$, and $g_2\theta = (ng_1)\theta = n\theta g_1\theta = 1 \cdot g_1\theta = g_1\theta$. Thus η is one-to-one and hence is an isomorphism.

Problems

4.71. Let θ be the homomorphism of Z into the multiplicative group of nonzero rational numbers defined by $x\theta = 1$ if x is even, and $x\theta = -1$ if x is odd. Find the kernel of θ and examine the claim $G/(\text{Ker}\,\theta) \cong G\theta$.

Solution:

$\text{Ker}\,\theta = \{x \mid x\theta = 1\} = \{x \mid x \text{ is even}\}$ and $G/(\text{Ker}\,\theta) = \{\text{Ker}\,\theta, \text{Ker}\,\theta + 1\}$, so $\text{Ker}\,\theta + 1$ generates $G/(\text{Ker}\,\theta)$. We have $(\text{Ker}\,\theta + 1) + (\text{Ker}\,\theta + 1) = \text{Ker}\,\theta$; hence $\text{Ker}\,\theta + 1$ is of order 2 and $G/(\text{Ker}\,\theta)$ is cyclic of order 2. Now $G\theta = \{1, -1\}$. $G\theta = gp(\{-1\})$ and, since $(-1)\cdot(-1) = 1$, $G\theta$ is also cyclic of order 2. Therefore $G/(\text{Ker}\,\theta)$ and $G\theta$ are isomorphic by Theorem 4.7, page 103.

4.72. Check that the kernel of the natural homomorphism of the additive group of integers $(Z, +)$ onto $Z/2Z$ is $2Z$, where $2Z = \{x \mid x = 2z,\ z \in Z\}$, i.e. $2Z$ is the set of even integers.

Solution:

The natural homomorphism ν is defined by $z\nu = 2Z + z$. The identity of $Z/2Z$ is the coset $2Z$. Consequently $z \in \text{Ker}\,\nu$ if and only if $2Z + z = 2Z$ and hence if and only if $z \in 2Z$. Thus $\text{Ker}\,\nu = 2Z$.

4.73. Verify that if G is any group, the subgroup generated by the squares of the elements of G, i.e. elements of the form $gg = g^2$, is a normal subgroup of G.

Solution:

Let S denote the subgroup generated by the squares of the elements of G. Let $x \in S$. Then x is a product $s_1 s_2 \cdots s_k$ of elements of G each of which is a square or the inverse of a square. Since the inverse of a square is also a square, we may assume each $s_1, \ldots, s_k$ is a square. If $g \in G$, then

$$g^{-1}xg = g^{-1}s_1 g g^{-1} s_2 g \cdots g^{-1} s_k g = t_1 t_2 \cdots t_k$$

where $t_i = g^{-1}s_i g$. We assert that each t_i is a square. Since $s_i = r_i^2$ for some r_i,

$$t_i = g^{-1}s_i g = g^{-1}r_i g g^{-1} r_i g = (g^{-1}r_i g)^2$$

and hence the t_i are squares as asserted and $g^{-1}xg \in S$. Thus $S \triangleleft G$.

4.74. Let σ be the homomorphism of Q^*, the multiplicative group of nonzero rationals into Q^* defined by $x\sigma = |x|$. Find the kernel and image of σ. Verify the homomorphism theorem directly in this case.

Solution:

σ is a homomorphism, since $(xy)\sigma = |xy| = |x|\,|y| = (x\sigma)(y\sigma)$.

$\text{Ker}\,\sigma = \{x \mid x\sigma = 1\} = \{x \mid |x| = 1\} = \{1, -1\}$.

$Q^*\sigma = \{x \mid x = y\sigma,\ y \in Q^*\} = \{x \mid x \in Q^*,\ x \text{ is positive}\}$

Let ν be the mapping of $Q^*/(\text{Ker}\,\sigma)$ into $Q^*\sigma$ that takes

$$(\text{Ker}\,\sigma)q \rightarrow q\sigma$$

Now $(\text{Ker}\,\sigma)q = \{q, -q\}$, and so the only other "representation" for the coset $(\text{Ker}\,\sigma)q$ is $(\text{Ker}\,\sigma)(-q)$. Hence the other possibility for ν as far as $(\text{Ker}\,\sigma)q$ is concerned is that $((\text{Ker}\,\sigma)q)\nu = (-q)\sigma$. But $(-q)\sigma = |q| = q\sigma$. Thus ν is a well defined mapping of $G/\text{Ker}\,\sigma$ into $Q^*\sigma$. Since

$$\begin{aligned}[(\text{Ker}\,\sigma)q_1 \cdot (\text{Ker}\,\sigma)q_2]\nu &= ((\text{Ker}\,\sigma)q_1q_2)\nu = |q_1q_2| = |q_1|\,|q_2| \\ &= ((\text{Ker}\,\sigma)q_1)\nu \cdot ((\text{Ker}\,\sigma)q_2)\nu\end{aligned}$$

ν is a homomorphism. Is ν one-to-one? If $(\text{Ker}\,\sigma)q_1\nu = (\text{Ker}\,\sigma)q_2\nu$, then we have that $q_1\sigma = q_2\sigma$, i.e. $|q_1| = |q_2|$. Thus if $q_1 \neq q_2$, $q_1 = -q_2$. Therefore $(\text{Ker}\,\sigma)q_1 = (\text{Ker}\,\sigma)q_2$. Hence ν is one-to-one and it follows that $Q^*/(\text{Ker}\,\sigma) \cong Q^*\sigma$.

4.75. Let G be the group of mappings of the real line R onto itself of the form $\alpha_{a,b}\colon x \rightarrow ax + b$, $a \neq 0$, a, b real numbers, $x \in R$. Prove that the map $\theta : \alpha_{a,b} \rightarrow \alpha_{a,0}$ is a homomorphism of G into G. Find the kernel and the image of this homomorphism and exhibit the isomorphism described in the homomorphism theorem.

Solution:

Note that $\alpha_{a,b}\alpha_{c,d} = \alpha_{ac,\,bc+d}$, since $x\alpha_{a,b} = ax + b$ and $(x\alpha_{a,b})\alpha_{c,d} = c(ax+b) + d$. Then $(\alpha_{a,b}\alpha_{c,d})\theta = (\alpha_{ac,\,bc+d})\theta = \alpha_{ac,0}$, $(\alpha_{a,b}\theta)(\alpha_{c,d}\theta) = \alpha_{a,0}\alpha_{c,0} = \alpha_{ac,0}$, and so θ is a homomorphism. If $\alpha_{a,b} \in \text{Ker}\,\theta$, we have $\alpha_{a,b}\theta = \alpha_{1,0}$ as $\alpha_{1,0} = \iota$, the identity mapping. Hence $\{\alpha_{1,b} \mid b \text{ any real number}\} = \text{Ker}\,\theta$. The image of $\theta = \{\alpha_{a,0} \mid a \text{ any nonzero real number}\}$.

A typical coset is

$$\begin{aligned}(\text{Ker}\,\theta)\alpha_{c,d} &= \{\alpha_{1,b} \mid b \text{ any real number}\}\alpha_{c,d} \\ &= \{\alpha_{c,\,bc+d} \mid b \text{ any real number}\} \\ &= \{\alpha_{c,e} \mid e \text{ any real number}\}\end{aligned}$$

The isomorphism η between $G/(\text{Ker}\,\theta)$ and $G\theta$ as given by the homomorphism theorem is the one that takes the coset $(\text{Ker}\,\theta)\alpha_{c,d}$ to $\alpha_{c,0}$.

4.76. Prove that if G is cyclic of order n and p divides n, then there is a homomorphism of G onto a cyclic group of order p. What is the kernel of this homomorphism?

Solution:

Let $G = gp(x)$ and let $n = pm$. Let H be cyclic of order p, $H = gp(y)$. We define θ to be the mapping

$$\theta : x^i \rightarrow y^i, \quad 0 \leq i \leq n-1$$

θ is well defined as we established in Lemma 4.5 that the elements x^i, $0 \leq i \leq n-1$, are all the distinct elements of G. Now let i and j be less than n. We have $(x^i x^j)\theta = (x^{i+j-\epsilon n})\theta$ where $\epsilon = 0$ if $i+j \leq n-1$ while $\epsilon = 1$ if $i+j \geq n$. Then

$$(x^i x^j)\theta = y^{i+j-\epsilon n} = y^i y^j y^{-\epsilon n}$$

Since the order of y divides n, $y^{-\epsilon n} = 1$. Hence $(x^i x^j)\theta = y^i y^j = (x^i\theta)(x^j\theta)$ and so θ is a homomorphism. The kernel of θ is the set of all x^i such that $x^i\theta = 1$, $0 \leq i \leq n-1$. Since $x^i\theta = y^i$ and $y^i = 1$ if and only if p divides i,

$$\text{Ker}\,\theta = \{x^i \mid p \text{ divides } i\} = \{x^p, x^{2p}, \ldots, x^{(m-1)p}\} = gp(x^p)$$

4.77. Let $(R^+, \cdot)$ be the multiplicative group of positive real numbers. Prove that the mapping of R^+ into $(R, +)$, the additive group of real numbers defined by $\theta: x \to \log_{10} x$, is a homomorphism. What is the kernel of θ? What is the image? Using the homomorphism theorem, prove that $(R^+, \cdot) \cong (R, +)$.

Solution:

θ is certainly a mapping of R^+ into R. Furthermore,

$$(xy)\theta = \log_{10}(xy) = \log_{10} x + \log_{10} y = x\theta + y\theta$$

and so θ is a homomorphism.

$$\text{Ker}\,\theta = \{x \mid \log_{10} x = 0,\ x \in R^+\} = \{1\}$$

We assert that $R^+\theta = R$. To see this observe that if x is any real number, then $10^x \in R^+$. Moreover, $\log 10^x = x$. Then if y is any element of R, $10^y\theta = \log 10^y = y$ and hence $R^+\theta = R$.

The homomorphism theorem states that $R^+/(\text{Ker}\,\theta) \cong R^+\theta = R$. As $\text{Ker}\,\theta = \{1\}$, all we must do is show that $R^+/\{1\} \cong R^+$. We can indeed show this in general: if G is any group, $G/\{1\} \cong G$. To do this we exhibit the isomorphism. Let ν be the natural homomorphism of G onto $G/\{1\}$. Then we need only show that ν is one-to-one. Suppose $g_1\nu = g_2\nu$, i.e. $\{1\}g_1 = \{1\}g_2$. Then $\{g_1\} = \{g_2\}$ and consequently $g_1 = g_2$. Hence ν is one-to-one and thus an isomorphism. Accordingly $R^+ \cong R^+/\{1\} \cong R$ and so $R^+ \cong R$.

4.78. Prove that the mapping $\theta: x \to e^x$ defines an isomorphism of $(R, +)$ onto $(R^+, \cdot)$, the multiplicative group of positive real numbers.

Solution:

$(x+y)\theta = e^{x+y} = e^x e^y = (x\theta)(y\theta)$ and so θ is a homomorphism. If $x\theta = y\theta$, then $e^x = e^y$ and $e^{x-y} = 1$, from which $x - y = 0$ and $x = y$, so θ is one-to-one. Is θ onto? Yes, for if y is any positive real number, the equation $e^x = y$ has a solution $x \in R$. Thus θ is an isomorphism between $(R, +)$ and $(R^+, \cdot)$.

4.79. Prove that $[fg, a] = g^{-1}[f, a]\,g[g, a]$ for any f, g and a in a given group. Suppose $G' \subseteq Z(G)$, the center of G. Let a be a fixed element of G. Prove that the mapping $\theta: g \to [g, a]$ is a homomorphism of G into G. What is $\text{Ker}\,\theta$? (Difficult.)

Solution:

Observe that $[fg, a] = g^{-1}f^{-1}a^{-1}fga$. On the other hand,

$$\begin{aligned} g^{-1}[f,a]\,g[g,a] &= g^{-1}(f^{-1}a^{-1}fa)g(g^{-1}a^{-1}ga) \\ &= g^{-1}f^{-1}a^{-1}f(agg^{-1}a^{-1})ga \\ &= g^{-1}f^{-1}a^{-1}fga = [fg, a] \end{aligned}$$

If $G' \subseteq Z(G)$, then $[f, a] \in Z(G)$ and $g^{-1}[f, a]g = [f, a]g^{-1}g = [f, a]$. Therefore $[fg, a] = [f, a][g, a]$. Hence $(fg)\theta = f\theta g\theta$ and θ is a homomorphism of G into G. $\text{Ker}\,\theta = \{g \mid [g, a] = 1, g \in G\}$. Thus $\text{Ker}\,\theta = C(gp(a))$, the centralizer in G of $gp(a)$.

4.80. Prove that if $\theta: G \to K$ is a homomorphism and $|G| < \infty$, then $|G\theta|$ divides $|G|$.

Solution:

By the homomorphism theorem, $G/(\text{Ker}\,\theta) \cong G\theta$. By Lagrange's theorem, $|\text{Ker}\,\theta|$ divides the order of G. Therefore $|G\theta| = \dfrac{|G|}{|\text{Ker}\,\theta|}$ divides $|G|$.

4.81. Show that the group M of Möbius transformations is a homomorphic image of the group

$$\mathcal{U} = \left\{ \begin{pmatrix} a & c \\ b & d \end{pmatrix} \middle| \; a, b, c, d \text{ complex numbers}, \; ad - bc = 1 \right\}$$

Find the kernel of the homomorphism.

Solution:

In Problem 3.48, page 80, we showed

$$M = \{\sigma(a, b, c, d) \mid a, b, c, d \text{ complex numbers}, \; ad - bc = 1\}$$

We define the mapping $\mu: \mathcal{U} \to M$ by $\begin{pmatrix} a & c \\ b & d \end{pmatrix} \mu = \sigma(a, b, c, d)$. Clearly μ is a mapping. Furthermore, using the results of Problem 3.46, page 79,

$$\begin{aligned} \left[\begin{pmatrix} a_1 & c_1 \\ b_1 & d_1 \end{pmatrix} \begin{pmatrix} a_2 & c_2 \\ b_2 & d_2 \end{pmatrix}\right] \mu &= \begin{pmatrix} a_1a_2 + c_1b_2 & a_1c_2 + c_1d_2 \\ b_1a_2 + d_1b_2 & b_1c_2 + d_1d_2 \end{pmatrix} \mu \\ &= \sigma(a_1a_2 + c_1b_2, b_1a_2 + d_1b_2, a_1c_2 + c_1d_2, b_1c_2 + d_1d_2) \\ &= \sigma(a_1, b_1, c_1, d_1)\sigma(a_2, b_2, c_2, d_2) = \begin{pmatrix} a_1 & c_1 \\ b_1 & d_1 \end{pmatrix} \mu \begin{pmatrix} a_2 & c_2 \\ b_2 & d_2 \end{pmatrix} \mu \end{aligned}$$

Thus μ is a homomorphism. Kernel $\mu = \left\{ \begin{pmatrix} 1 & 0 \\ 0 & 1 \end{pmatrix}, \begin{pmatrix} -1 & 0 \\ 0 & -1 \end{pmatrix} \right\}$, because $\begin{pmatrix} a & c \\ b & d \end{pmatrix} \mu = \sigma(a, b, c, d)$ and $\sigma(a, b, c, d) = \sigma(1, 0, 0, 1)$ (the identity of M) if and only if $a = d = 1$ and $b = c = 0$ or $a = d = -1$ and $b = c = 0$, by Problem 3.49, page 80. Therefore

$$M \cong \mathcal{U} \Big/ \left\{ \begin{pmatrix} 1 & 0 \\ 0 & 1 \end{pmatrix}, \begin{pmatrix} -1 & 0 \\ 0 & -1 \end{pmatrix} \right\}$$

b. Correspondence Theorem. Factor of a factor theorem

Let $\theta: G \to K$ be a homomorphism of G onto K. If H is a subgroup of G, then $H\theta$ is a subgroup of K. If H is normal in G, $H\theta$ is normal in K. (See Problem 4.82 below.)

What about the reverse of this procedure? Would the preimage of a subgroup S of K, i.e. the set $\{g \mid g \in G, \; g\theta \in S\}$, be a subgroup of G? We know that if $S = \{1\}$, then the preimage of S is $\operatorname{Ker}\theta$, and this is indeed a subgroup of G. We generalize this result in the following theorem.

Theorem 4.19 (Correspondence Theorem): Let $\theta: G \to K$ be a homomorphism of G onto K. The preimage H of any subgroup S of K is a subgroup of G containing $\operatorname{Ker}\theta$. If $S \lhd K$, then $H \lhd G$. Furthermore if H_1 is any other subgroup of G containing $\operatorname{Ker}\theta$ such that $H_1\theta = S$, then $H_1 = H$.

Proof: $H = \{g \mid g\theta \in S\}$. Since 1θ is the identity of K, and S contains the identity of K, then $1 \in H$ and so $H \neq \emptyset$, the empty set. Also, if $g, h \in H$, $(gh^{-1})\theta = g\theta(h\theta)^{-1}$. As $g\theta \in S$ and $h\theta \in S$, it follows that $(gh^{-1})\theta \in S$. Hence $gh^{-1} \in H$ and H is a subgroup of G. Since $\operatorname{Ker}\theta = \{x \mid x\theta = 1\}$ and $1 \in S$, $\operatorname{Ker}\theta \subseteq H$.

If $S \lhd K$, we must show that $H \lhd G$. Let $h \in H$, $g \in G$; then $(g^{-1}hg)\theta = (g\theta)^{-1}(h\theta)(g\theta)$. Now $h\theta \in S$, $g\theta \in K$, and $S \lhd K$ implies $(g^{-1}hg)\theta \in S$. Then $g^{-1}hg \in H$ and so $H \lhd G$.

Let H_1 be a subgroup of G containing $\operatorname{Ker}\theta$ and suppose $H_1\theta = S$. We will show that $H_1 = H$. Let $h_1 \in H_1$; then $h_1\theta \in S$. Now $H = \{g \mid g\theta \in S\}$. Therefore $h_1 \in H$ and hence $H_1 \subseteq H$. On the other hand if $h \in H$, then $h\theta = s \in S$. Choose $h_1 \in H_1$ such that $h_1\theta = s$. Then $hh_1^{-1} \in \operatorname{Ker}\theta \subseteq H_1$ and so $h \in H_1$ and $H \subseteq H_1$. Thus $H = H_1$.

Corollary 4.20: Let $\theta: G \to K$ be an onto homomorphism. Let S be a subgroup of K of index $n < \infty$. Let H be the preimage of S. Then H is of index n in G.

Proof: Let $Sk_1, Sk_2, \ldots, Sk_n$, where $k_i \in K$, be the distinct cosets of S in K. As θ is an onto homomorphism, there are elements $g_1, \ldots, g_n$ of G such that $g_i\theta = k_i$. We claim that

$$Hg_1, \ldots, Hg_n \tag{4.5}$$

are the distinct cosets of H in G.

Suppose that $Hg_i = Hg_j$; then $g_ig_j^{-1} \in H$. Hence $g_i\theta(g_j\theta)^{-1} \in S$, i.e. $k_ik_j^{-1} \in S$, from which $Sk_i = Sk_j$ and $i = j$. Thus we have shown that all the cosets in (*4.5*) are distinct.

Let $g \in G$. Then $g\theta \in K$ and so $g\theta \in Sk_i$ for some integer i. Hence $g\theta = sk_i$ with $s \in S$. Consider $x = gg_i^{-1}$. $x\theta = g\theta(g_i\theta)^{-1} = sk_ik_i^{-1} = s$. Consequently gg_i^{-1} is in the preimage of S, so that $gg_i^{-1} \in H$. This means that $g \in Hg_i$. We have thus shown that every element of G is a member of one of the cosets in (*4.5*). It follows that (*4.5*) consists of all the cosets of H in G.

Of course we can always reformulate results about homomorphisms with the aid of the homomorphism theorem as results about factor groups. Thus we have

Corollary 4.21: Let $N \triangleleft G$. Let L be a subgroup of G/N. Then we can write $L = H/N$ where H is a subgroup of G containing N. If $L \triangleleft G/N$, then $H \triangleleft G$. If $H_1/N = H/N$ where H_1 and H are subgroups of G containing N, then $H_1 = H$.

Proof: Let ν be the natural homomorphism of $G \to G/N$. Let $H = \{g \mid g\nu \in L\}$. Then by the correspondence theorem H is a subgroup of G; and if $L \triangleleft G/N$, $H \triangleleft G$. Also, $H\nu = L$. Since $\nu: g \to Ng$, $H\nu$ consists of all cosets Nh, $h \in H$. Because $H \supseteq N = \text{Ker}\,\nu$ by the correspondence theorem above, H/N makes sense and consists of all the cosets Nh, $h \in H$. Hence $H/N = H\nu = L$.

Now if $H_1 \supseteq N$ and $H_1/N = H/N$, then $H_1\nu = L$. It then follows by the correspondence theorem that $H_1 = H$.

(Problems 4.82-4.89 below may be studied before reading Theorem 4.22.)

The reader may very well wonder what happens when we take a factor group of a factor group. For example, if $N \triangleleft G$ and G/N is a group containing a normal subgroup M/N, then what is $(G/N)/(M/N)$? The next theorem tells us that this is isomorphic to a single factor group, i.e. a factor G by one of its normal subgroups.

Theorem 4.22 (Factor of a Factor Theorem, also called the Third Isomorphism Theorem): If in the factor group G/N there is a normal subgroup M/N, $M \supseteq N$, then $M \triangleleft G$ and

$$G/M \cong (G/N)/(M/N)$$

Proof: Let $\nu: G \to G/N$ be the natural homomorphism of $G \to G/N$. Let $\rho: G/N \to (G/N)/(M/N)$ be the natural homomorphism of $G/N \to (G/N)/(M/N)$. Put $\theta = \nu\rho$. Then θ is a homomorphism of $G \to (G/N)/(M/N)$; and since ν is onto G/N and ρ is onto $(G/N)/(M/N)$, $\nu\rho$ is onto $(G/N)/(M/N)$. Therefore $G/\text{Ker}\,(\nu\rho) \cong (G/N)/(M/N)$, by the homomorphism theorem. If $g \in G$, $g\nu = Ng$ and $(Ng)\rho = (M/N)(Ng)$; note that here $(M/N)(Ng)$ is a coset of the normal subgroup M/N in (G/N), i.e. an element in the group $(G/N)/(M/N)$. Now the elements of M/N are all the cosets Nm, $m \in M$. The identity of $(G/N)/(M/N)$ is (M/N). We ask, what is the kernel of $\nu\rho$? It will be all $g \in G$ such that $g\nu\rho = (M/N)Ng = M/N$. But in that case $Ng \in M/N$, i.e. $Ng = Nm$ for some $m \in M$. Hence $g = nm$ where $n \in N$. But $M \supseteq N$. Therefore $g \in M$ and so $\text{Ker}\,\nu\rho \subseteq M$. Note that if $m \in M$, $m(\nu\rho) = (M/N)Nm = M/N$. Then $\text{Ker}\,\nu\rho = M$. Thus M as a kernel of a homomorphism is normal in G and $G/M \cong (G/N)/(M/N)$, which is the required result.

Problems

4.82. Prove that if $\theta: G \to K$ is a homomorphism of G onto K, and H is a subgroup of G, then $H\theta$ is a subgroup of K. If $H \lhd G$, prove that $H\theta \lhd K$.

Solution:

Since $1\theta \in H\theta$, $H\theta \neq \emptyset$. If $x_1, x_2 \in H\theta$, $x_1 = h_1\theta$, $x_2 = h_2\theta$ for some $h_1, h_2 \in H$. Then

$$x_1x_2^{-1} = h_1\theta(h_2\theta)^{-1} = h_1\theta h_2^{-1}\theta = (h_1h_2^{-1})\theta = h\theta$$

where $h = h_1h_2^{-1} \in H$. Hence $H\theta$ is a subgroup of K. If now $H \lhd G$, then $g^{-1}hg \in H$ for all $g \in G$ and all $h \in H$. Now any element of K, say k, is of the form $g\theta$ for some $g \in G$; and any element of $H\theta$ is of the form $h\theta$. Is $(g\theta)^{-1}h\theta g\theta \in H\theta$? Yes, because $(g\theta)^{-1}(h\theta)g\theta = (g^{-1}hg)\theta$ and as $g^{-1}hg \in H$, $(g\theta)^{-1}(h\theta)g\theta \in H\theta$. Thus $H\theta \lhd H$.

4.83. Let $G = D_4 = \{\sigma_1, \sigma_2, \sigma_3, \sigma_4, \tau, \tau\sigma_2, \tau\sigma_3, \tau\sigma_4\}$. Let $K = gp(b)$ be cyclic of order 2. Then $\theta: G \to K$ defined by $\sigma_i\theta = 1$, $(\tau\sigma_i)\theta = b$, $i = 1, 2, 3, 4$, is a homomorphism of G onto K. (Take this as a fact.) Find all the subgroups of K and all their preimages. Check that the assertions of Theorem 4.19 and Corollary 4.20 hold. (See Problem 3.42, page 77, for the multiplication table.)

Solution:

The subgroups of K are $K_1 = K$ and $K_2 = \{1\}$. $G_1 =$ the preimage of $K_1 = \{g \mid g\theta \in K\} = G$. $G_2 =$ the preimage of $K_2 = \{g \mid g\theta = 1\} = \{\sigma_1, \sigma_2, \sigma_3, \sigma_4\}$.

(*a*) G_1, G_2 are subgroups of G containing $\text{Ker}\,\theta = G_2$.

(*b*) K_1 and K_2 are normal in K, and G_1 and G_2 are normal in G.

(*c*) The subgroups of G containing $\text{Ker}\,\theta$ are G_1 and G_2. We note then that $G_i\theta = G_j\theta$ implies $i = j$ where $1 \leq i, j \leq 2$.

(*d*) K_1 is of index 1 in K. G_1 is of index 1 in G. K_2 is of index 2 in K. G_2 is of index 2 in G, its cosets being $G_1 = \{\sigma_1, \sigma_2, \sigma_3, \sigma_4\}$ and $\tau G_2 = \{\tau, \tau\sigma_2, \tau\sigma_3, \tau\sigma_4\}$.

Thus (*a*), (*b*), (*c*) and (*d*) agree with Theorem 4.19 and Corollary 4.20.

4.84. Let $G = gp(a)$ be cyclic of order 12 and let $K = gp(b)$ be cyclic of order 4. Let $\theta: G \to K$ be defined by $a^i\theta = b^i$, $i = 0, 1, 2, \ldots, 11$. Then θ is a homomorphism of G onto K. (Take this as fact.) Find all the subgroups of K and all their preimages. Check that the assertions of Theorem 4.19 and Corollary 4.20 hold.

Solution:

The subgroups of K are $K_1 = K$, $K_2 = \{1, b^2\}$, $K_3 = \{1\}$. $G_1 =$ the preimage of $K_1 = \{x \mid x\theta \in K\} = G$. $G_2 =$ the preimage of $K_2 = \{x \mid x\theta = 1$ or $x\theta = b^2\}$. Clearly $1 \in G_2$. $a\theta = b \neq b^2$, hence $a \notin G_2$. $a^2\theta = b^2$, so $a^2 \in G_2$. Continuing in this fashion we conclude that $G_2 = \{1, a^2, a^4, a^6, a^8, a^{10}\}$. Finally, $G_3 =$ the preimage of $K_3 = \{x \mid x\theta = 1\} = \{1, a^4, a^8\}$.

(*a*) G_1, G_2 and G_3 are all subgroups of G containing $\text{Ker}\,\theta = G_3$.

(*b*) K_1, K_2 and K_3 are normal in K, and G_1, G_2 and G_3 are normal in G (trivially, as G is abelian).

(*c*) The subgroups of G containing the kernel of θ are those containing a^4. Hence the subgroups of G containing $\text{Ker}\,\theta$ are G_1, G_2 and G_3. Note $G_i\theta = G_j\theta$ implies $i = j$ $(1 \leq i, j \leq 3)$.

(*d*) K_1 is of index 1 in K. G_1 is of index 1 in G. K_2 is of index 2 in K, its cosets being $K_2 = \{1, b^2\}$ and $K_2b = \{b, b^3\}$. G_2 is of index 2 in G, its cosets being $G_2 = \{1, a^2, a^4, a^6, a^8, a^{10}\}$, and $G_2a = \{a, a^3, a^5, a^7, a^9, a^{11}\}$. K_3 is of index 4 in K. G_3 is of index 4 in G, the distinct cosets being $G_3 = \{1, a^4, a^8\}$, $G_3a = \{a, a^5, a^9\}$, $G_3a^2 = \{a^2, a^6, a^{10}\}$, $G_3a^3 = \{a^3, a^7, a^{11}\}$.

Thus (*a*), (*b*), (*c*) and (*d*) agree with Theorem 4.19 and Corollary 4.20.

4.85. Let $\theta: G \to K$ be an onto homomorphism. Let K be cyclic of order 10. Prove that G has normal subgroups of index 2 and 5 and 10.

Solution:

Let $K = gp(k)$. Then the subgroups $K_1 = \{1\}$, $K_2 = gp(k^5)$, $K_3 = gp(k^2)$ are normal subgroups of K of index 10, 5 and 2 respectively. Consequently their preimages, by Theorem 4.19 and Corollary 4.20, are normal of index 10, 5 and 2 respectively. Hence the result.

4.86. Let $N \lhd G$ and suppose G/N is cyclic of order 6. Let $G/N = gp(Nx)$. Find all the subgroups of G/N and express them in the form of Corollary 4.21. (Hard.)

Solution:

Let $G/N = K$. Let $K_1 = \{N\}$, $K_2 = \{N, Nx^3\}$, $K_3 = \{N, Nx^2, Nx^4\}$ and $K_4 = K$. These are all the subgroups of K. To find the corresponding subgroups of Corollary 4.21 we let $\nu: G \to G/N$ be the natural homomorphism, i.e. $g\nu = Ng$. Then let G_i be the preimage of K_i, $i = 1, 2, 3, 4$.

$$G_1 = \{g \mid g\nu = N\} = \{g \mid Ng = N\} = \{g \mid g \in N\} = N$$

$$G_2 = \{g \mid g\nu = N \text{ or } g = Nx^3\} = \{g \mid Ng = N \text{ or } Ng = Nx^3\} = N \cup Nx^3$$

$$G_3 = \{g \mid g\nu = N \text{ or } g\nu = Nx^2 \text{ or } g\nu = Nx^4\} = N \cup Nx^2 \cup Nx^4$$

$$G_4 = \{g \mid g\nu \in K\} = G. \qquad \text{Then } G_i/N = K_i \text{ for } i = 1, 2, 3, 4.$$

4.87. Use the correspondence theorem to prove that if H is a subgroup of G containing G' (the derived group of G), then $H \lhd G$ (i.e. prove Problem 4.69 by another method).

Solution:

Let $\nu: G \to G/G'$ be the natural homomorphism; then ν is onto. G/G' is abelian by Problem 4.68. Any subgroup of G/G' is therefore normal, and thus $H\nu = S$, say, is normal in G/G'. By the correspondence theorem, H is the preimage of S. Hence using the correspondence theorem once more, H is normal in G.

4.88. Let H be a subgroup of index n in G. Let $\theta: G \to K$ be a homomorphism onto K. Prove that $H\theta$ is of index n in K if $H \supseteq \operatorname{Ker} \theta$.

Solution:

It is only necessary to prove that $S = H\theta$ is of finite index in K, for then the result follows from Corollary 4.20. If $Hg_1, \ldots, Hg_n$ are the cosets of H in G, then we claim that $\{S(g_1\theta), \ldots, S(g_n\theta)\}$ is the set of all the cosets of S in K. We need only show that if $k \in K$, then $k \in S(g_i\theta)$ for some $i = 1, \ldots, n$. As θ is onto, there is a $g \in G$ such that $g\theta = k$. Let $g = hg_i$. Then $k = g\theta = h\theta(g_i\theta) \in S(g_i\theta)$. The result now follows from Corollary 4.20.

Alternatively we can show that $S(g_1\theta), \ldots, S(g_n\theta)$ are all distinct. Suppose $S(g_i\theta) = S(g_j\theta)$. Then $(g_i\theta)(g_j\theta)^{-1} = (g_ig_j^{-1})\theta \in S$. Hence $g_ig_j^{-1} \in H$ as H, by the correspondence theorem, is the preimage of S. Accordingly $i = j$ and the index of S in K is n.

4.89. Let G be a group and let N be a normal subgroup of G. Suppose further that L and M are subgroups of G/N. Then show that we can write L in the form H/N, and M in the form K/N, where H and K are subgroups of G containing N. Show also that if $L \subseteq M$, $H \subseteq K$; and if $L \lhd M$, $H \lhd K$. Show that if $L \subseteq M$ and $[M:L] = n < \infty$, then $[K:H] = n$.

Solution:

This is just an application of Corollary 4.21 to Corollary 4.20 and Theorem 4.19. That $L = H/N$ and $M = K/N$ follows from Corollary 4.21. If ν is the natural homomorphism, we recall that $H = \{g \mid g\nu \in L\}$ and $K = \{g \mid g\nu \in M\}$. Hence if $L \subseteq M$, $H \subseteq K$ follows immediately. Now if $L \lhd M$, we consider the homomorphism $\theta: K \to K/N$ defined by $k\theta = k\nu$ for all $k \in K$, i.e. $\theta = \nu_{|K}$. Clearly $K\theta = K/N$ and the preimage of L is H, the preimage of M is K. We can then conclude from the correspondence theorem that $H \lhd K$.

4.90. Let $G = D_4$. Let $M = \{\sigma_1, \sigma_3, \tau\sigma_3, \tau\}$, $N = \{\sigma_1, \sigma_3\}$. Then accept $N \lhd G$ and $M \lhd G$. Consider G/N and M/N. Find $(G/N)/(M/N)$ explicitly and check that it is isomorphic to G/M. In other words, check agreement with the factor of a factor theorem. (Use table on page 77.)

Solution:

G/N consists of the cosets $A_1 = N = \{\sigma_1, \sigma_3\}$, $A_2 = \sigma_2 N = \{\sigma_2, \sigma_4\}$, $A_3 = \tau N = \{\tau, \tau\sigma_3\}$, $A_4 = (\tau\sigma_2)N = \{\tau\sigma_2, \tau\sigma_4\}$. Now $M = \{\sigma_1, \sigma_3, \tau\sigma_3, \tau\}$, hence M/N consists of the elements A_1, A_3.

$M/N \lhd G/N$. Therefore we can talk of $(G/N)/(M/N)$. The elements of this group are the cosets of (M/N) in (G/N). These cosets are $B_1 = (M/N)A_1 = \{A_1, A_3\}$ and $B_2 = (M/N)A_2 = \{A_1A_2, A_3A_2\} = \{A_2, A_4\}$. ($A_3A_2$, for example, is calculated as follows: $A_3A_2 = (\tau N)(\sigma_2 N) = (\tau\sigma_2)N = A_4$.) Multiplication in the group $(G/N)/(M/N)$ is calculated in the usual way for cosets, e.g.,

$$B_2B_2 = (M/N)A_2(M/N)A_2 = (M/N)(A_2A_2) = (M/N)A_1 = B_1$$

as $A_2A_2 = (\sigma_2 N)(\sigma_2 N) = \sigma_3 N = N = A_1$. It is clear then that $(G/M)/(M/N) = \{B_1, B_2\}$ is the cyclic group of order 2 generated by B_2.

Now let us decide what G/M is. The elements of G/M are the cosets $C_1 = M = \{\sigma_1, \sigma_3, \tau\sigma_3, \tau\}$ and $C_2 = M\sigma_2 = \{\sigma_2, \sigma_4, \tau\sigma_4, \tau\sigma_2\}$. As $C_2C_2 = M\sigma_2M\sigma_2 = M\sigma_3 = M$, $G/M = \{C_1, C_2\}$ is the cyclic group of order 2. Therefore $G/M \cong (G/N)/(M/N)$.

4.91. Let G be the cyclic group of order 12, say $G = gp(a)$. Let $M = gp(a^2)$, $N = gp(a^6)$. Consider G/N and M/N. Find $(G/N)/(M/N)$ explicitly and check that it is isomorphic to G/M. In other words, check agreement with the factor of a factor theorem. (Difficult.)

Solution:

G/N consists of the cosets $A_1 = N = \{1, a^6\}$, $A_2 = Na = \{a, a^7\}$, $A_3 = Na^2 = \{a^2, a^8\}$, $A_4 = Na^3 = \{a^3, a^9\}$, $A_5 = Na^4 = \{a^4, a^{10}\}$, $A_6 = Na^5 = \{a^5, a^{11}\}$.

Now $M = \{1, a^2, a^4, a^6, a^8, a^{10}\}$. Hence M/N consists of the elements A_1, A_3 and A_5. $M/N \lhd G/N$, and we can talk of $(G/N)/(M/N)$. The elements of this group are the cosets of M/N in G/N. These cosets are

$$B_1 = (M/N)A_1 = \{A_1, A_3, A_5\} \quad \text{and} \quad B_2 = (M/N)A_2 = \{A_2, A_4, A_6\}$$

Multiplication is calculated in the usual way,

$$B_2B_2 = (M/N)A_2(M/N)A_2 = (M/N)(A_2A_2)$$

Now the product of A_2 and A_2 is also the product of cosets. As $A_2 = Na$, $A_2A_2 = Na^2 = A_3$. Hence $B_2B_2 = (M/N)A_3 = B_1$. It is clear then that $(G/N)/(M/N) = \{B_1, B_2\} = gp(B_2)$ is a cyclic group of order 2.

Now let us decide what G/M is. The elements of G/M are the cosets

$$C_1 = M = \{1, a^2, a^4, a^6, a^8, a^{10}\} \quad \text{and} \quad C_2 = Ma = \{a, a^3, a^5, a^7, a^9, a^{11}\}$$

Clearly $G/M = gp(C_2)$ is cyclic of order 2. Hence $G/M \cong (G/M)/(M/N)$.

4.92. Let G be any group. Let $F_1(G)$ be all possible nonisomorphic factor groups of G. Let $F_{i+1}(G) = \{$All possible factor groups of the groups in $F_i(G)\}$. In the particular case that G is cyclic of order 2^5, find all nonisomorphic groups in $\bigcup_{i=1}^{\infty} F_i(G)$, i.e. $F_1(G) \cup F_2(G) \cup \cdots$. (Hard.)

Solution:

It is sufficient to consider $F_1(G)$. For if $L \in F_2(G)$, $L = M/N$ where $M \in F_1(G)$. But then $M = G/K$, by definition of $F_1(G)$. Thus L is a factor of a factor group. Hence by the factor of a factor theorem it is isomorphic to a factor group of G.

We must therefore find the number of factor groups of G. All subgroups of G are known. G has unique subgroups of orders $1, 2, 2^2, 2^3, 2^4$ and 2^5 by Theorem 4.9, page 105. The factor groups will therefore be of orders $2^5, 2^4, 2^3, 2^2, 2$ and 1. In each case the factor groups will be cyclic. For if $G = gp(a)$ and N is a subgroup of G, $gp(Na) = G/N$. Hence the result.

c. The subgroup isomorphism theorem

In the homomorphism theorem we were able to say that the image of a homomorphism $\theta: G \to K$ was essentially a factor group of G. What can we say about the effect of θ on subgroups? Let H be a subgroup of G. Let $\theta_1 = \theta_{|H}$, i.e. θ_1 is the mapping of H to K defined by $h\theta_1 = h\theta$, $h \in H$. Then θ_1 is a homomorphism of $H \to K$, and so $H\theta_1 = H\theta \cong H/(\text{Ker}\,\theta_1)$. Now if $\text{Ker}\,\theta = N = \{x \mid x \in G,\ x\theta = 1\}$, then $\text{Ker}\,\theta_1 = \{x \mid x \in H$ and $x\theta_1 = x\theta = 1\} = H \cap N$. So $H\theta = H\theta_1 \cong H/(H \cap N)$. On the other hand, we know $H\theta$ is a subset of $G\theta$ and $G\theta \cong G/N$.

Our question is: what has $H/(H \cap N)$ got to do with G/N? It must be isomorphic to some subgroup of G/N. But which? This is what the subgroup isomorphism theorem is about.

Theorem 4.23 (Subgroup Isomorphism Theorem, also called the Second Isomorphism Theorem): Let $N \lhd G$ and let H be a subgroup of G. Then $H \cap N \lhd H$, HN is a subgroup of G, and

$$H/(H \cap N) \cong HN/N$$

$(HN = \{hn \mid h \in H,\ n \in N\})$

Proof: If $n \in H \cap N$ and $h \in H$, then $h^{-1}nh \in N$ as $N \lhd G$, and $h^{-1}nh \in H$ as $n \in H$. Therefore $h^{-1}nh \in H \cap N$ and $H \cap N \lhd H$.

HN is a subgroup, for it is not empty; and if $x_1, x_2 \in HN$, then $x_1 = h_1n_1$, $x_2 = h_2n_2$ and

$$x_1x_2^{-1} = h_1n_1n_2^{-1}h_2^{-1} = h_1n_3h_2^{-1} = h_1h_2^{-1}h_2n_3h_2^{-1} = hn_4$$

where $n_3 = n_1n_2^{-1} \in N$, $h_1h_2^{-1} = h \in H$ and $h_2n_3h_2^{-1} = n_4 \in N$ as $N \lhd G$. Hence $x_1x_2^{-1} \in HN$ and HN is a subgroup of G.

Let $\phi: H \to HN/N$ be defined by $h\phi = Nh$. Then ϕ is clearly onto, i.e. $H\phi = HN/N$. Also ϕ is a homomorphism: $(h_1h_2)\phi = N(h_1h_2) = Nh_1Nh_2 = h_1\phi h_2\phi$. By the homomorphism theorem, $H\phi \cong H/(\text{Ker}\,\phi)$. $\text{Ker}\,\phi = \{x \mid x \in H,\ x\phi = 1\} = \{x \mid x \in H,\ Nx = N\}$. If $Nx = N$, then $1x = x \in N$; and if $x \in N$, $Nx = N$. Therefore $\text{Ker}\,\phi = \{x \mid x \in H,\ x \in N\} = H \cap N$. Hence $HN/N \cong H/(H \cap N)$.

Problems

4.93. Let Q^* be the multiplicative group of rationals. Let $N = \{1, -1\}$. Let H be the subgroup generated by $\{\frac{1}{2}\}$. Find HN, HN/N and thereby verify the assertion of the subgroup isomorphism theorem that $HN/N \cong H/H \cap N$.

Solution:

The elements of H are all of the form $(\frac{1}{2})^r$, r various integers.

$$\begin{aligned} HN &= \{x \mid x = hn,\ h \in H,\ n \in N\} = \{x \mid x = h \text{ or } x = -h,\ h \in H\} \\ &= \{x \mid x = \pm(\tfrac{1}{2})^r \text{ for all integers } r\} \end{aligned}$$

A coset of HN/N is of the form $Nx = \{1, -1\}x = \{x, -x\}$ where $x \in HN$. Now if $x \in HN$, $x = \pm(\frac{1}{2})^r$. Hence each coset is of the form $\{(\frac{1}{2})^r, -(\frac{1}{2})^r\}$. Since $\underbrace{N(\tfrac{1}{2}) \cdot N(\tfrac{1}{2}) \cdot \cdots \cdot N(\tfrac{1}{2})}_{r} = N(\frac{1}{2})^r$ and $(\frac{1}{2})^r \in N(\frac{1}{2})^r$, each coset of HN/N is a power of $N(\frac{1}{2})$. Thus $gp(\{N(\frac{1}{2})\}) = HN/N$; and since $(\frac{1}{2})^r \notin N$ for $r \neq 0$, HN/N is the infinite cyclic group.

Now $H \cap N = \{x \mid x = (\frac{1}{2})^r$ for some r and $x = \pm 1\} = \{1\}$. Hence $H/(H \cap N) \cong H$. But H is infinite cyclic. Thus we have verified that $H/(H \cap N) \cong HN/N$.

4.94. Let $\sigma = \begin{pmatrix} 1 & 2 & 3 & \ldots & n \\ 2 & 1 & 3 & \ldots & n \end{pmatrix}$ and $H = gp(\{\sigma\})$. Prove that HA_n/A_n is cyclic of order 2.

Solution:

$HA_n/A_n \cong H/(H \cap A_n)$. Now σ is an odd permutation, hence $\sigma \notin A_n$. Also $H = \{\sigma, \iota\}$, so $H \cap A_n = \{\iota\}$. Therefore $HA_n/A_n \cong H/\{\iota\} \cong H$. But H is cyclic of order 2. Thus HA_n/A_n is cyclic of order 2.

4.95. Let a group G contain two normal subgroups M and N. Let H be a subgroup of G. Prove that $HM/M \cong HN/N$ if $H \cap M = H \cap N$.

Solution:

By the subgroup isomorphism theorem, $HM/M \cong H/H \cap M = H/H \cap N \cong HN/N$. Hence $HN/N \cong HM/M$ by Problem 4.8, page 97.

4.96. If in the preceding problem we know that G/M has every element of order a power of 2, show that $H/(H \cap N)$ has every element a power of 2.

Solution:

$H/(H \cap N) = H/(H \cap M) \cong HM/M \subseteq G/M$. Hence the result.

4.97. Let H and K be subgroups of G, $N \lhd G$ and $HN = KN$. Prove that $H/(H \cap N) \cong K/(K \cap N)$.

Solution:

$H/(H \cap N) \cong HN/N = KN/N \cong K/(K \cap N)$. Then by Problem 4.8, page 97, $H/(H \cap N) \cong K/(K \cap N)$.

4.98. Let $G \supseteq G_1 \supseteq \{1\}$ and $G_1 \lhd G$. Suppose G/G_1 and G_1 are abelian and H is any subgroup of G. Prove that there exists a subgroup H_1 of H such that $H_1 \lhd H$, and H/H_1 and H_1 are abelian. (Hard.)

Solution:

Let $H_1 = H \cap G_1$. Then by the subgroup isomorphism theorem, $H_1 \lhd H$, and $H/H_1 \cong HG_1/G_1$. But $HG_1/G_1 \subseteq G/G_1$ and G/G_1 is abelian. Therefore H/H_1 is abelian. As $H_1 \subseteq G_1$ and G_1 is abelian, we conclude that H_1 is abelian.

4.99. Let $G \supseteq G_1 \supseteq G_2 \supseteq \{1\}$. Let $G_1 \lhd G$, $G_2 \lhd G_1$, and suppose G/G_1, G_1/G_2 and G_2 are abelian. Prove that if H is any subgroup, then it has subgroups H_1 and H_2 such that $H_1 \lhd H$, $H_2 \lhd H_1$ and H/H_1, H_1/H_2 and H_2 are abelian. (Hard.)

Solution:

Let $H_1 = H \cap G_1$. Then, as in Problem 4.95, $H_1 \lhd H$ and H/H_1 is abelian.

Now consider H_1 as a subgroup of G_1. As $G_2 \lhd G_1$, by the subgroup isomorphism theorem $H_1 \cap G_2 \lhd H_1$ and $H_1/(H_1 \cap G_2) \cong H_1G_2/G_2 \subseteq G_1/G_2$. Since G_1/G_2 is abelian, so is $H_1/(H_1 \cap G_2)$. Consequently we put $H_2 = H_1 \cap G_2$.

Finally as $H_2 \subseteq G_2$ and G_2 is abelian, so is H_2 and the result follows.

d. Homomorphisms of cyclic groups

We return now to study cyclic groups. In Section 4.2b we could state that in a finite cyclic group there was at most one subgroup of any given order. An analogy for the infinite cyclic group would have been awkward to formulate without the concept of index which we have had at our disposal since Section 4.3a.

Recall that a subgroup H of a group G is of index n in G if there are exactly n distinct right cosets of H in G. In the case of finite cyclic groups we have proved that there is one and only one subgroup H of any order m dividing $|G|$. Since $[G:H] = |G|/m$, there is one and only one subgroup of any given index dividing the order of G. This gives the clue to

Theorem 4.24: There is one and only one subgroup of any given finite index $n > 0$ in the infinite cyclic group.

Proof: Let $G = gp(x)$ where G is infinite cyclic. Let $H_n = gp(x^n)$. Then $H_n = \{x^{nr} \mid \text{all integers } r\}$. Hence the cosets $H_n, H_nx, \ldots, H_nx^{n-1}$ are all distinct and are all the cosets of H_n in G. Hence H_n is of index n.

Next if H is a subgroup of index n, we already know that H is generated by the smallest positive power $x^r \in H$ (Theorem 4.9, page 105), and this means $H = H_r$. But H_r is of index r. Hence $r = n$ and $H_r = H_n$. This concludes the proof.

With our knowledge of cyclic groups it is easy to apply the homomorphism theorem to find all homomorphisms of cyclic groups.

Theorem 4.25: Let θ be a homomorphism of a cyclic group G. Then $G\theta$ is cyclic; and if $|G| < \infty$, $|G\theta|$ divides $|G|$.

Furthermore if H is any cyclic group such that $|H|$ divides $|G|$, there is a homomorphism of G onto H.

If G is infinite cyclic, there is a homomorphism of G onto any cyclic group.

Proof: If $G = gp(x)$, then $G = \{x^r \mid \text{all integers } r\}$, and

$$G\theta = \{x^r\theta \mid \text{all } r\} = \{(x\theta)^r \mid \text{all } r\} = gp(x\theta)$$

Thus $G\theta$ is cyclic. If $|G| < \infty$, then by the homomorphism theorem $G\theta \cong G/N$ for some normal subgroup N of G. Hence $|G\theta| = |G|/|N|$ (see Lagrange's theorem) and $|G\theta|$ divides $|G|$.

Suppose H is cyclic, $H = gp(y)$, and the order m of H divides the order of $G = gp(x)$. Say, $|G| = rm$. Let $N = gp(x^m)$. Then $N = \{1, x^m, \ldots, (x^m)^{r-1}\}$, $|N| = r$, and N is of index m. Because G is abelian, G/N makes sense and is of order m. But G is cyclic and consequently so is G/N. Hence $G/N \cong H$, and H is a homomorphic image of G. If G is infinite cyclic, then, as we saw in the proof of Theorem 4.24, $[G : G_n] = n$ if $G_n = gp(x^n)$. As $G_n \lhd G$, G/G_n is of order n. But G/G_n is the homomorphic image of a cyclic group; hence it is cyclic. Thus G has as homomorphic image any cyclic group of order n, $n > 0$. Obviously it also has the infinite cyclic group as a homomorphic image.

A look back at Chapter 4

In this chapter we have thoroughly investigated the simplest class of groups, the cyclic groups. We know that there are cyclic groups of all orders, we know their subgroups, we know that they have as homomorphic images only cyclic groups, and we know whether any cyclic group G has as homomorphic image a given cyclic group. Furthermore the subgroups of cyclic groups are again cyclic.

We have also introduced the concept of coset. The cosets form a partition of the group. Using this fact we obtained Lagrange's theorem which states that the order of a subgroup divides the order of a finite group. This enables us to eliminate certain groups as possible subgroups of a given group. We will see later on that it also enables us to find more quickly the groups of a given order.

Next we have introduced the idea of a normal subgroup. This gives rise to a new way of looking at homomorphisms, namely as factor groups (see the homomorphism theorem). The subgroup isomorphism theorem tells us that the subgroup corresponding to a given subgroup H of G in a factor group G/N is isomorphic to a factor group of H itself, namely $H/(H \cap N)$.

The factor of a factor theorem tells us that a factor group of a factor group G/N is just a factor group of G of the form G/M. Finally, the correspondence theorem associates with each subgroup of the image of a homomorphism $\theta : G \to K$ a unique subgroup of G itself.

Supplementary Problems

FUNDAMENTALS

4.100. Prove that if a and b are elements of a group G and if $a^{-1}b^2a = ba$, then $b = a$.

4.101. Suppose a and b are elements of a group G. If $a^2 = 1$ and $a^{-1}b^2a = b^3$, prove $b^5 = 1$. (Hard.)

4.102. Suppose a and b are elements of a group G. If $a^{-1}b^2a = b^3$ and $b^{-1}a^2b = a^3$, prove $a = 1 = b$. (Very hard.)

4.103. Suppose G and H are groups. Suppose that G cannot be generated by two elements but that H can. Prove G and H are not isomorphic.

4.104. Let X be a non-empty set and let $Y = \{y\}$ be disjoint from X. Prove $S_X \cong S_{X \cup Y}$ if and only if X is infinite.

CYCLIC GROUPS

4.105. Let G be a cyclic group. Prove that if N is a subgroup of G such that $G/N \cong G$, then $N = \{1\}$.

4.106. Let $G = Z \times Z$ where Z is the set of integers. Define a binary operation $\circ$ in G by

$$(k, l) \circ (m, n) = (k+m, l+n)$$

where $(k, l), (m, n) \in G$. G is a group with respect to this composition. Prove that G is not cyclic.

4.107. Let $G_1, G_2, \ldots$ be subgroups of a group G. If $G_i \subseteq G_{i+1}$, $G_i \neq G_{i+1}$ for $i = 1, 2, \ldots$, prove that $G_1 \cup G_2 \cup \cdots$ is not a cyclic group. (Hard.)

4.108. Let G be a group and let $\theta: G \to F$ be a homomorphism. Let C be a cyclic subgroup of G. Let $c\theta \in C$ for all $c \in C$. If H is any subgroup of C, prove that $h\theta \in H$ for all $h \in H$.

4.109. Let Q be the additive group of rationals with respect to addition. Prove that every two-generator subgroup ($\neq 0$) of Q is infinite cyclic.

COSETS

4.110. Let H be a subgroup of G. Prove that $\theta: Hg \to g^{-1}H$ is a matching of the right cosets of H in G with the left cosets of H in G.

4.111. Let H and K be subgroups of a group G. Show that a coset of H intersection a coset of K is a coset of $H \cap K$.

4.112. Let D be the group of Problem 3.72, page 91. Let $N = \{(0, z) \mid z \in Z\}$. Prove that $N \triangleleft D$ and D/N is infinite cyclic.

4.113. Let G be the group of Problem 3.74, page 91. Let $N = \{(0, q) \mid q \in Q\}$. Prove that $N \triangleleft G$ and G/N is infinite cyclic. Show that N is isomorphic with the additive group of rationals.

4.114. Let W be the group of Problem 3.77, page 91. Let $M = \{(0, b) \mid (0, b) \in W\}$. Show that $M \triangleleft G$ and that G/M is infinite cyclic.

4.115. Let G be a group, let H be a subgroup of G, and let g be an element of G. Prove that if $N(H)$ is the normalizer of H and $N(g^{-1}Hg)$ the normalizer of $g^{-1}Hg$, then $g^{-1}N(H)g = N(g^{-1}Hg)$. $g^{-1}Hg = \{g^{-1}hg \mid h \in H\}$.

HOMOMORPHISM THEOREMS

4.116. Let $\mathcal{G} = \left\{\begin{pmatrix} a & b \\ c & d \end{pmatrix} \middle| \ a, b, c, d \in Z\right\}$. Prove that $\mathcal{G}$ forms a group with respect to the operation $+$ defined by

$$\begin{pmatrix} a & b \\ c & d \end{pmatrix} + \begin{pmatrix} a_1 & b_1 \\ c_1 & d_1 \end{pmatrix} = \begin{pmatrix} a+a_1 & b+b_1 \\ c+c_1 & d+d_1 \end{pmatrix}$$

Let $\theta: \mathcal{G} \to Z$ be defined by $\begin{pmatrix} a & b \\ c & d \end{pmatrix}\theta = a + d$. Prove that θ is a homomorphism of $\mathcal{G}$ onto the additive group of integers and find its kernel. Consider $\mathcal{G}/(\text{Ker } \theta)$ and prove that in accordance with the homomorphism theorem it is isomorphic with the additive group of integers.

4.117. Let $\mathcal{G} = \left\{\begin{pmatrix} a & b \\ c & d \end{pmatrix} \middle| \ ad - bc \neq 0, \ a, b, c, d \text{ real numbers}\right\}$ with operation matrix multiplication. Let $\theta: \mathcal{G} \to R^*$, the nonzero real numbers, be defined by $\begin{pmatrix} a & b \\ c & d \end{pmatrix}\theta = ad - bc$. Prove that θ is a homomorphism from $\mathcal{G}$ onto the multiplicative group of nonzero real numbers and find its kernel. Prove that $\mathcal{G}/(\text{Ker } \theta) \cong R^*$ in accordance with the homomorphism theorem.

4.118. Let G be any subgroup of S_n, the symmetric group of degree n. Let $\theta: G \to \{1, -1\}$ be the mapping defined by $x\theta = 1$ if x is an even permutation and $x\theta = -1$ if x is an odd permutation. Prove that θ is a homomorphism of G into the group $\{1, -1\}$ with operation multiplication of integers. Using the homomorphism theorem, prove that the even permutations of G form a normal subgroup of G.

4.119. Let G be a group and N a normal subgroup of G. Suppose that $H = G/N$ has a sequence of subgroups $H = H_1 \supseteq H_2 \supseteq \cdots \supseteq H_n$ where $[H_i : H_{i+1}] = i+1$, for $i = 1, 2, \ldots, n-1$. Prove that G has a sequence of subgroups $G = G_1 \supseteq G_2 \supseteq \cdots \supseteq G_n$ such that $[G_i : G_{i+1}] = i+1$, $i = 1, \ldots, n-1$.

4.120. Let M and N be normal subgroups of G with $M \supseteq N$. Prove that G/N is finite if G/M and M/N are finite.

4.121. Let G be a group and N a normal subgroup of G. Suppose G/N has a factor group which is infinite cyclic. Prove that G has a normal subgroup of index n for each positive integer n.

4.122. Let G be a finite group with normal subgroups M and N. Let H be a subgroup of G. Suppose that the orders of M and H and those of N and H are co-prime. Prove that $HM/M \cong HN/N$.

4.123. Let N, M be normal subgroups of G, $N \supseteq M$. Suppose G/N is cyclic and $|N/M| = 2$. Prove that G/M is abelian.

4.124. Find a group G with normal subgroups N and M, $N \supseteq M$, G/N cyclic, N/M cyclic but G/M not abelian.

Chapter 5

Finite Groups

Preview of Chapter 5

The most important result of this chapter is a theorem of Sylow which guarantees the existence of subgroups of prime power order. We prove two other theorems of Sylow concerning subgroups of prime power order and then examine groups of prime power order. One result is that groups of prime power order always have non-trivial centers.

In order to construct a new group from any two groups G and H, we define a binary operation on the cartesian product of G and H. The resultant group is called the direct product of G and H. A simple condition enables us to conclude that a group is a direct product.

The concept of direct product together with general theorems about subgroups, e.g. the Sylow theorems, help us to classify finite groups. In this chapter we find all groups up to order 15.

We study a class of groups called solvable groups. Solvable groups are used in Galois theory to determine whether an equation is solvable in terms of nth roots.

An ambitious plan for studying finite groups is to find all simple groups, i.e. groups without proper normal subgroups, and then see how groups are built from simple groups. The Jordan-Hölder theorem shows that in a sense a group is built from simple groups in only one way. As yet the task of finding all simple groups is far from complete. We conclude the chapter by exhibiting a class of simple groups, namely A_n, for $n \geqq 5$.

5.1 THE SYLOW THEOREMS

a. Statements of the Sylow Theorems

Lagrange's theorem (Theorem 4.11, page 109) tells us that the order of a subgroup divides the order of a finite group. Conversely one might ask: if G is a finite group and $n \mid |G|$, is there always a subgroup of order n?

The answer to this question is no: A_4 is of order 12 but has no subgroup of order 6 (see Problem 5.1 below). The following important theorem, however, ensures the existence of subgroups of prime power order. In the following p will denote a prime.

Theorem 5.1 (First Sylow Theorem):

Let G be a finite group, p a prime, and p^r the highest power of p dividing the order of G. Then there is a subgroup of G of order p^r.

Suppose H is a subgroup of G of order a power of a prime p, and $|H|$ is the highest power of p that divides $|G|$. Then H is called a *Sylow p-subgroup of G*. By Theorem 5.1 every finite group has a Sylow p-subgroup.

In general a group of order a power of the prime p is called a *p-group*. A Sylow p-subgroup H of a group G is a maximal p-group in G, i.e. if $H \subseteq F \subseteq G$ where F is a p-group, then $F = H$ (see Problem 5.4).

As an illustration we find the Sylow p-subgroups of the symmetric group S_3 on $\{1,2,3\}$, for $p = 2$ and 3. The elements of S_3 given in Section 3.3a, page 57, are

$$\iota = \begin{pmatrix} 1 & 2 & 3 \\ 1 & 2 & 3 \end{pmatrix} \qquad \sigma_2 = \begin{pmatrix} 1 & 2 & 3 \\ 3 & 1 & 2 \end{pmatrix} \qquad \tau_2 = \begin{pmatrix} 1 & 2 & 3 \\ 3 & 2 & 1 \end{pmatrix}$$

$$\sigma_1 = \begin{pmatrix} 1 & 2 & 3 \\ 2 & 3 & 1 \end{pmatrix} \qquad \tau_1 = \begin{pmatrix} 1 & 2 & 3 \\ 1 & 3 & 2 \end{pmatrix} \qquad \tau_3 = \begin{pmatrix} 1 & 2 & 3 \\ 2 & 1 & 3 \end{pmatrix}$$

The order of any Sylow 2-subgroup is 2 and the order of any Sylow 3-subgroup is 3, since $|S_3| = 6 = 2 \cdot 3$. Now $\tau_1^2 = \tau_2^2 = \tau_3^2 = \iota$, so the sets $\{\iota, \tau_1\}$, $\{\iota, \tau_2\}$ and $\{\iota, \tau_3\}$ are all subgroups of order 2 and therefore, by definition, Sylow 2-subgroups of S_3. There are no other Sylow 2-subgroups of S_3 because $\sigma_1^2 = \sigma_2$, $\sigma_2^2 = \sigma_1$ implies S_3 has no other elements of order 2. $\{\iota, \sigma_1, \sigma_2\}$ is the only subgroup of order three in S_3, so it is the only Sylow 3-subgroup of S_3.

Theorem 5.2 (Second Sylow Theorem):

If H is a subgroup of a finite group G and H is a p-group, then H is contained in a Sylow p-subgroup of G.

Two subgroups S and T of a group G are called *conjugate* if there is a $g \in G$ such that $g^{-1}Sg = T$. Recall $g^{-1}Sg = \{g^{-1}sg \mid s \in S\}$.

Theorem 5.3 (Third Sylow Theorem):

Any two Sylow p-subgroups of a finite group G are conjugate. The number s_p of distinct Sylow p-subgroups of G is congruent to 1 modulo p and s_p divides $|G|$. (s_p is congruent to 1 modulo p if $s_p = 1 + kp$ for some integer k.)

Before proving the Sylow theorems, we will use them to show that, up to isomorphism, there is one and only one group of order 15. If $|G| = 15$ then, by Theorem 5.1, G has at least one subgroup of order 3 and at least one of order 5. Now Theorem 5.3 implies that there are $s_3 = 1 + 3k$ subgroups of order 3 and $s_3 \mid |G|$. But $(1+3k) \mid 15$ implies $k = 0$. Therefore G has one and only one subgroup of order 3. Similarly G has one and only one subgroup of order 5. These subgroups must be cyclic (Problem 4.48, page 110). Let $H_1 = \{1, a, a^2\}$ be the subgroup of order 3 and $H_2 = \{1, b, b^2, b^3, b^4\}$ the subgroup of order 5. $H_1 \cap H_2 = \{1\}$, because an element $\neq 1$ cannot have order 3 and 5 simultaneously. We look at the order of ab in G which must be either 1, 3, 5 or 15. If the order of ab is 1, then $ab = 1$ and $a = b^{-1}$ which is impossible, for $H_1 \cap H_2 = \{1\}$. If the order of ab is 3, then $gp(ab) = H_1$, since H_1 is unique. In this case $ab = a^i$ ($i = 0, 1$ or 2) and $b = a^{i-1}$ which is impossible. If the order of ab is 5, $gp(ab) = H_2$. Hence $ab = b^i$ ($i = 0, 1, 2, 3$ or 4) and $a = b^{i-1}$ which is impossible. Therefore the order of ab is 15 and G is the cyclic group of order 15 generated by ab.

Further applications of the Sylow theorems are given in the problems below and in Section 5.3.

Problems

5.1. Show that the alternating group A_4 has no subgroup of order 6.

Solution:

The elements of A_4 were given in Section 3.3c, page 62. We repeat them here for convenience:

$$\iota = \begin{pmatrix} 1 & 2 & 3 & 4 \\ 1 & 2 & 3 & 4 \end{pmatrix} \quad \tau_3 = \begin{pmatrix} 1 & 2 & 3 & 4 \\ 3 & 2 & 4 & 1 \end{pmatrix} \quad \tau_6 = \begin{pmatrix} 1 & 2 & 3 & 4 \\ 4 & 1 & 3 & 2 \end{pmatrix} \quad \sigma_2 = \begin{pmatrix} 1 & 2 & 3 & 4 \\ 3 & 4 & 1 & 2 \end{pmatrix}$$

$$\tau_1 = \begin{pmatrix} 1 & 2 & 3 & 4 \\ 1 & 3 & 4 & 2 \end{pmatrix} \quad \tau_4 = \begin{pmatrix} 1 & 2 & 3 & 4 \\ 4 & 2 & 1 & 3 \end{pmatrix} \quad \tau_7 = \begin{pmatrix} 1 & 2 & 3 & 4 \\ 2 & 3 & 1 & 4 \end{pmatrix} \quad \sigma_5 = \begin{pmatrix} 1 & 2 & 3 & 4 \\ 4 & 3 & 2 & 1 \end{pmatrix}$$

$$\tau_2 = \begin{pmatrix} 1 & 2 & 3 & 4 \\ 1 & 4 & 2 & 3 \end{pmatrix} \quad \tau_5 = \begin{pmatrix} 1 & 2 & 3 & 4 \\ 2 & 4 & 3 & 1 \end{pmatrix} \quad \tau_8 = \begin{pmatrix} 1 & 2 & 3 & 4 \\ 3 & 1 & 2 & 4 \end{pmatrix} \quad \sigma_8 = \begin{pmatrix} 1 & 2 & 3 & 4 \\ 2 & 1 & 4 & 3 \end{pmatrix}$$

Suppose that A_4 has a subgroup H of order 6. $\sigma_2^2 = \sigma_5^2 = \sigma_8^2 = \iota$, so σ_2, σ_5 and σ_8 are of order 2. Also τ_j is of order 3 for $j = 1, 2, \ldots, 8$. Hence the elements of order 2 are σ_2, σ_5 and σ_8, and the elements of order 3 are $\tau_1, \tau_2, \ldots, \tau_8$. Now H is of order 6, so it contains a subgroup of order 3, by Theorem 5.1. Therefore $\tau_i \in H$ for some i, say $\tau_1 \in H$; then $\iota, \tau_1, \tau_1^2 = \tau_2 \in H$. H must also contain an element of order 2, by Theorem 5.1. Hence H contains a σ_i, say H contains σ_2. Because $\sigma_2\tau_1 = \tau_4$ and $\tau_1\sigma_2 = \tau_8$, if H contains σ_2 it also contains $\tau_4, \tau_4^2 = \tau_3, \tau_8$ and $\tau_8^2 = \tau_7$. This would mean H has at least 8 distinct elements, which contradicts the assumption that $|H| = 6$. Thus $\sigma_2 \notin H$. A similar argument shows σ_5 and $\sigma_8 \notin H$. This means that H does not contain subgroups of order 2, contradicting Theorem 5.1. Therefore our initial assumption is invalid and A_4 does not contain a subgroup of order 6.

5.2. Find all Sylow p-subgroups of A_4 for $p = 2$ and 3.

Solution:

The elements of A_4 are given in Problem 5.1. The order of a Sylow 2-subgroup is 4, since 2^2 is the highest power of two dividing 12, the order of A_4. Consequently by Lagrange's theorem none of the τ's can be elements of a Sylow 2-subgroup because they are all of order 3 (see Problem 5.1) and 3 does not divide 4. Now $\sigma_i\sigma_j = \sigma_k$ where $i, j, k \in \{2, 5, 8\}$ and $\sigma_i^2 = \iota$ for $i = 2, 5, 8$. Hence $P = \{\iota, \sigma_2, \sigma_5, \sigma_8\}$ is a subgroup of A_4 of order 4. P is the only possible Sylow 2-subgroup as there are only four elements having order dividing 4, viz. $\iota, \sigma_2, \sigma_5, \sigma_8$, and these elements are in P. The order of a Sylow 3-subgroup is 3. The sets $\{\iota, \tau_1, \tau_1^2\}$, $\{\iota, \tau_3, \tau_3^2\}$, $\{\iota, \tau_5, \tau_5^2\}$ and $\{\iota, \tau_7, \tau_7^2\}$ are all subgroups of order 3. These are all the possible Sylow 3-subgroups, as they include all the elements of order 3.

Alternately we may use Theorem 5.3: $s_3 = 1 + 3k$ must divide 12. Clearly $k \neq 0$ (we already have four subgroups); and if $k > 1$, s_3 does not divide 12. Hence $k = 1$ and there are exactly four Sylow 3-subgroups.

5.3. If H is a subset of a group G and $g \in G$, then $|g^{-1}Hg| = |H|$, where $g^{-1}Hg = \{g^{-1}hg \mid h \in H\}$.

Solution:

We define a matching $\alpha: H \to g^{-1}Hg$ by $\alpha: h \to g^{-1}hg$ for $h \in H$. α is clearly an onto mapping. To show α is also one-to-one, we must prove $h_1 = h_2$ $(h_1, h_2 \in H)$ if and only if $g^{-1}h_1g = g^{-1}h_2g$. Let $h_1 = h_2$. Then by multiplying on the left by g^{-1} and on the right by g we get $g^{-1}h_1g = g^{-1}h_2g$. Similarly $g^{-1}h_1g = g^{-1}h_2g$ implies $h_1 = h_2$. Hence α is a matching and $|g^{-1}Hg| = |H|$.

5.4. Let $|G| = p^rm$ $(r \geq 1$ and $p \nmid m)$ and let P be a Sylow p-subgroup of G. Prove that if H is a p-group such that $P \subseteq H \subseteq G$, then $H = P$.

Solution:

Suppose $|H| = p^t$, $t \geq 0$. By Lagrange's theorem, $p^t \mid p^rm$. Since $p \nmid m$, $t \leq r$. But $P \subseteq H$ and $|P| = p^r$. Hence $t = r$ and $|P| = |H|$, and so $P = H$.

5.5. If H is a Sylow p-subgroup of G, then $g^{-1}Hg$ is also a Sylow p-subgroup of G.

Solution:

Suppose $|G| = p^rm$ $(r \geq 0$ and $p \nmid m)$; then $|H| = p^r$. But $|g^{-1}Hg| = |H|$ by Problem 5.3. Hence $g^{-1}Hg$ is a Sylow p-subgroup of G if it is a subgroup. To prove $g^{-1}Hg$ is a subgroup, observe that $(g^{-1}h_1g)(g^{-1}h_2g)^{-1} = g^{-1}h_1h_2^{-1}g \in g^{-1}Hg$. From Lemma 3.1, page 55, $g^{-1}Hg$ is therefore a subgroup.

5.6. Prove that a finite group G is a p-group if and only if every element of G has order a power of p.

Solution:

If $|G| = p^r$ then, as every element of G must have order dividing the order of the group, every element has order a power of p. To prove the converse let every element of G have order a power of p and assume the order of G is not a power of p. Then there is some prime q, $q \neq p$, such that $q \mid |G|$. But by the first Sylow theorem, G has a subgroup H of order a nonzero power of q. So H contains an element $g \neq 1$. By Lagrange's theorem, the order of g is a nonzero power of q and hence the order of g is not a power of p. This contradicts the assumption that all elements have order a power of p. Hence $|G| = p^r$ for some $r \geq 0$.

5.7. If G has only one Sylow p-subgroup H, then $H \lhd G$.

Solution:

If $g \in G$, $g^{-1}Hg$ is a Sylow p-subgroup by Problem 5.5. But G has only one Sylow p-subgroup. Thus $g^{-1}Hg = H$ for all $g \in G$ and $H \lhd G$.

5.8. If $|G| = pq$, where p and q primes and $p < q$, then G has one and only one subgroup of order q. Furthermore if $q \neq 1 + kp$ for any integer k, then G is the cyclic group of order pq.

Solution:

By Theorem 5.3 G has $s_q = 1 + kq$ Sylow q-subgroups of order q, with $k \geqq 0$. Also $1 + kq$ divides pq. There can only be four possibilities for $1 + kq$ as the expression of pq as a product of primes is unique: $1 + kq = q$ or $1 + kq = p$ or $1 + kq = pq$ or $1 + kq = 1$. As q does not divide $1 + kq$, we are left with the possibilities that $1 + kq = p$ or $1 + kq = 1$. Since $q > p$, $1 + kq \neq p$ and hence $k = 0$. Thus there is only one subgroup of order q, say H.

There are $s_p = 1 + kp$ subgroups of order p. Again we have the possibilities $1 + kp = 1$, $1 + kp = p$, $1 + kp = q$, or $1 + kp = pq$ as s_p divides $|G|$. Clearly p does not divide $1 + kp$, so $1 + kp = 1$ or $1 + kp = q$. The last is not true by assumption, so again there is only one subgroup K of order p.

It follows from Problem 5.7 that $H \lhd G, K \lhd G$. Also $H \cap K = \{1\}$ as the nonunit elements of H are of order q and those of K are of order p. If $h \in H$ and $k \in K$ $(h \neq 1,\ k \neq 1)$, then

$$\begin{aligned} h^{-1}k^{-1}hk &= h^{-1}(k^{-1}hk) \in H \quad \text{as } H \lhd G \\ &= (h^{-1}k^{-1}h)k \in K \quad \text{as } K \lhd G \end{aligned}$$

Hence $h^{-1}k^{-1}hk = 1$ and h and k commute. By Lagrange's theorem, the order of hk is p, q or pq. But $(hk)^p = h^pk^p$ as h and k commute, so $(hk)^p = h^p \neq 1$. Similarly $(hk)^q = k^q \neq 1$. Therefore hk is of order pq, and so G is cyclic.

Instead of the argument of the last paragraph, we note that $H = \{1, h, h^2, \ldots, h^{q-1}\}$, $K = \{1, k, k^2, \ldots, k^{p-1}\}$. Now hk is an element of order $1, p, q$ or pq. If hk is of order p, since there is only one subgroup of order p, $gp(hk) = K$, i.e. $hk = k^i$ for some i, $0 \leqq i \leqq p-1$. But then $h \in K$, which contradicts $H \cap K = \{1\}$. Similarly $gp(hk)$ is not of order q, nor of order 1. Thus $gp(hk)$ is of order pq, and so G is cyclic.

5.9. Show that if $q = 1 + kp$ in Problem 5.8, G is not necessarily cyclic.

Solution:

Consider S_3, the symmetric group on $\{1, 2, 3\}$. $|S_3| = 6 = 3 \cdot 2$, $3 = 1 + 1 \cdot 2$ and S_3 is not a cyclic group.

5.10. If $|G| = 2p$, p an odd prime, then G has one and only one subgroup of order p and either G has exactly p subgroups of order 2 or it has exactly one subgroup of order 2.

Solution:

From Problem 5.8 we know G has one and only one Sylow p-subgroup. Because p is itself the highest power of p dividing $|G|$, the Sylow p-subgroup of G is of order p. Thus there is precisely one subgroup of G of order p. The number of Sylow 2-subgroups of G is $s_2 = 1 + k2$ for some integer k. Again $1 + 2k = 1, 2, p$ or $2p$. As 2 does not divide $1 + 2k$, then $1 + 2k = 1$ or $1 + 2k = p$ and the number of Sylow 2-subgroups is either 1 or p.

b. Two lemmas used in the proof of the Sylow theorems

Lemmas 5.4 and 5.6 will provide the tools for proving the Sylow theorems (see Section 5.1c).

Throughout this section G will be a fixed group and H a subgroup of G. As usual we denote subsets of G by A, B, C, etc.

A generalization of the concept of normalizer as defined in Section 4.3d, page 112, will be essential. We point out once more that if A is a non-empty subset of a group G and $g \in G$, then $g^{-1}Ag = \{g^{-1}ag \mid a \in A\}$.

Definition: Let A be a non-empty subset of a group G. The set $\{h \mid h^{-1}Ah = A,\ h \in H\}$ is called the normalizer of A in H and is written $N_H(A)$.

It is easy to prove that $N_H(A)$ is a subgroup of H (Problem 5.11). When $H = G$, $N_G(A)$ is the normalizer of A as defined in Chapter 4.

Definition: Let A and B be non-empty subsets of G. B is said to be an *H-conjugate of* A if $h^{-1}Ah = B$ for some $h \in H$. (Note that if $H = G$, then A and B are conjugate as defined in Section 5.1a.)

The next lemma gives us a formula for calculating the number of distinct subsets of G which are H-conjugates of A.

Lemma 5.4: If G is a finite group with subgroup H and non-empty subset A, the number of distinct H-conjugates of A is the index of $N_H(A)$ in H, i.e. $[H : N_H(A)]$.

Proof: Since $[H : N_H(A)]$ is the number of distinct right cosets of $N_H(A)$ in H, we need only define a one-to-one mapping, α, of the right cosets of $N_H(A)$ in H onto the distinct H-conjugates of A. Let α be defined by

$$\alpha : N_H(A)h \to h^{-1}Ah \quad (h \in H)$$

To show that α is a one-to-one mapping, we must prove that for $h_1, h_2 \in H$,

$$N_H(A)h_1 = N_H(A)h_2 \quad \text{if and only if} \quad h_1^{-1}Ah_1 = h_2^{-1}Ah_2$$

(i) Let $h_1^{-1}Ah_1 = h_2^{-1}Ah_2$. Then $A = h_1h_2^{-1}Ah_2h_1^{-1} = (h_2h_1^{-1})^{-1}A(h_2h_1^{-1})$. Hence $h_2h_1^{-1} \in N_H(A)$ and so $h_2 \in N_H(A)h_1$. Since two right cosets are equal or disjoint, we conclude $N_H(A)h_1 = N_H(A)h_2$. Thus $h_1^{-1}Ah_1 = h_2^{-1}Ah_2$ implies $N_H(A)h_1 = N_H(A)h_2$.

(ii) If $N_H(A)h_1 = N_H(A)h_2$, then $h_1 \in N_H(A)h_2$, i.e. $h_1 = nh_2$ for some $n \in N_H(A)$. Therefore

$$h_1^{-1}Ah_1 = (nh_2)^{-1}Anh_2 = h_2^{-1}n^{-1}Anh_2 = h_2^{-1}Ah_2$$

because $n^{-1}An = A$ by definition of $N_H(A)$. Hence $N_H(A)h_1 = N_H(A)h_2$ implies $h_1^{-1}Ah_1 = h_2^{-1}Ah_2$. α is clearly onto, so the proof is complete.

Most of our arguments are concerned with sets whose elements are subsets of G. We denote such sets by script letters $\mathcal{A}$, $\mathcal{B}$, etc.

For example, let G be the cyclic group of order 6, $G = \{1, a, \ldots, a^5\}$. Subsets of G are, for example, $A = \{1, a\}$, $B = \{a^2, a^3, a^6\}$, $C = \{a\}$. An example of a set whose elements are subsets of G is the set whose elements are A and B. We write $\mathcal{A} = \{A, B\}$. Another such set would be $\mathcal{B} = \{A, B, C\}$.

Proposition 5.5: Let $\mathcal{A}$ be a set of subsets of G. We define for $A, B \in \mathcal{A}$, $A \sim B$ if B is an H-conjugate of A (i.e. if there exists an element $h \in H$ such that $h^{-1}Ah = B$). Then $\sim$ is an equivalence relation on $\mathcal{A}$ (see Problem 5.16 for the proof).

We will make a few observations about $\sim$, which follow because it is an equivalence relation on $\mathcal{A}$. Recall that if $A \in \mathcal{A}$, $A\sim = \{X \mid X \in A \text{ and } X \sim A\}$, i.e. $A\sim$ is the equivalence class containing A (see Section 1.2c, page 9). Recall that the distinct equivalence classes are disjoint and that their union is $\mathcal{A}$ (Theorem 1.2, page 10).

By a *set of representatives of the equivalence classes* we mean a set $\mathcal{R}$ which contains one and only one element from each of the distinct equivalence classes. It follows that $\mathcal{A}$ is the disjoint union of the sets $R\sim$, $R \in \mathcal{R}$. Hence $|\mathcal{A}| = \sum_{R \in \mathcal{R}} |R\sim|$. We are now in a position to prove our main lemma.

Lemma 5.6: Let $\mathcal{A}$ $(\neq \emptyset)$ be a set of subsets of G. Suppose that for each $A \in \mathcal{A}$ and each $h \in H$, $h^{-1}Ah \in \mathcal{A}$. Let $\sim$ denote the equivalence relation defined by $A \sim B$ if B is an H-conjugate of A. Let $\mathcal{R}$ be a set of representatives of the equivalence classes. Then

$$|\mathcal{A}| = \sum_{R \in \mathcal{R}} [H : N_H(R)]$$

Proof: We know from the remarks above that

$$|\mathcal{A}| = \sum_{R \in \mathcal{R}} |R\sim|$$

But $R\sim = \{X \mid X = h^{-1}Rh$ for some $h \in H\}$ since $h^{-1}Rh \in \mathcal{A}$ for every $h \in H$. So $R\sim$ is the set of H-conjugates of R. The number of such H-conjugates is, by Lemma 5.4, $[H : N_H(R)]$. Hence $|\mathcal{A}| = \sum_{R \in \mathcal{R}} [H : N_H(R)]$, as claimed.

Corollary 5.7: Let P $(\neq \emptyset)$ be a subset of G. Let $\mathcal{A} = \{g^{-1}Pg \mid g \in G\}$. Let $\mathcal{R}$, H and $\sim$ be as in Lemma 5.6. Then

$$|\mathcal{A}| = \sum_{R \in \mathcal{R}} [H : N_H(R)] = [G : N_G(P)]$$

Proof: Clearly $|\mathcal{A}|$ is the number of G-conjugates of P, and the result follows from Lemma 5.4.

Corollary 5.8: Let $\mathcal{A} = \{A \mid A$ is a subset of G and A has precisely one element$\}$. Let $\sim$ be the equivalence relation in $\mathcal{A}$ when $H = G$, and let $\mathcal{R}$ be a set of representatives of the equivalence classes. Let $\mathcal{R}^* = \{R \mid R \cap Z(G) = \emptyset, R \in \mathcal{R}\}$. Then

$$|G| = |Z(G)| + \sum_{R \in \mathcal{R}^*} [G : N_G(R)]$$

(We remind the reader that $Z(G) = \{x \mid xg = gx$ for all $g \in G\}$)

Proof: Clearly $|\mathcal{A}| = |G|$; hence

$$|G| = \sum_{R \in \mathcal{R}} [G : N_G(R)] \qquad (5.1)$$

If $z \in Z(G)$, then $\{z\} \in \mathcal{A}$ and the number of G-conjugates of $\{z\}$ is one, namely $\{z\}$ itself. Consequently $\{z\} \in \mathcal{R}$ for each $z \in Z(G)$. Note that $N_G(\{z\}) = G$ if $z \in Z(G)$. Hence adding first the contribution made by all $R \in \mathcal{R}$ with $R \cap Z(G) \neq \emptyset$ in (*5.1*), we obtain $|Z(G)|$ and the result follows.

Note that as $R = \{r\}$,

$$\begin{aligned} N_G(R) &= \{g \mid g \in G \text{ and } g^{-1}rg \in R\} \\ &= \{g \mid g \in G \text{ and } g^{-1}rg = r\} \\ &= C(R) \end{aligned}$$

(For the definition of $C(R)$, the centralizer of R in G, see Section 4.3d, page 112.)

Hence Corollary 5.8 takes the form

$$|G| = |Z(G)| + \sum_{R \in \mathcal{R}^*} [G : C(R)] \qquad (5.2)$$

(*5.2*) is called the *class equation of G*.

Problems

5.11. If A is a non-empty subset and H a subgroup of a group G, then $N_H(A)$ is a subgroup of G.

Solution:

$N_H(A)$ is clearly a subset of G. $N_H(A) \neq \emptyset$, since $1 \in H$ and $1^{-1}A1 = A$ implies $1 \in N_H(A)$. Let $n \in N_H(A)$. $n^{-1}An = A$ implies $nAn^{-1} = A$, or $(n^{-1})^{-1}An^{-1} = A$. Furthermore $n^{-1} \in H$, since H is a subgroup; hence $n^{-1} \in N_H(A)$. If $m, n \in N_H(A)$, $(mn)^{-1}A(mn) = n^{-1}(m^{-1}Am)n = n^{-1}An = A$; hence $mn \in N_H(A)$. Accordingly $N_H(A)$ is a subgroup of G.

5.12. Check Lemma 5.4 by direct computation when $G = S_3$ and (using the notation of Section 3.3a, page 57) $A = \{\tau_1\}$, $H = \{\iota, \tau_2\}$.

Solution:

The H-conjugates of A are $\iota^{-1}\{\tau_1\}\iota = \{\tau_1\}$, $\tau_2^{-1}\{\tau_1\}\tau_2 = \{\tau_3\}$. Thus the number of H-conjugates of A is 2. Lemma 5.4 requires that $2 = [H : N_H(A)]$. But $N_H(A) = \{x \mid x \in H$ and $x^{-1}Ax = A\} = \{\iota\}$. Hence $[H : N_H(A)] = 2$, as required.

5.13. Check Corollary 5.7 when $G = D_8$, the dihedral group of degree 8 $(G = \{b^i, ab^i, 0 \leq i < 8\}$, where $a^2 = b^8 = 1$ and $a^{-1}ba = b^{-1}$; see page 75, with $a = \tau$, $b = \sigma_2)$, given $P = \{a\}$ and $H = \{1, b^2, b^4, b^6\}$.

Solution:

$\mathcal{A}$ of Corollary 5.7 is given by $\mathcal{A} = \{g^{-1}Pg \mid g \in G\} = \{P_1, P_2, P_3, P_4\}$ where $P_1 = \{a\}$, $P_2 = \{ab^2\}$, $P_3 = \{ab^4\}$, $P_4 = \{ab^6\}$; thus $|\mathcal{A}| = 4$. Using the equivalence relation $\sim$ of the corollary,

$$P_1\sim \;=\; \{h^{-1}P_1h \mid h \in H\} \;=\; \{P_1, P_3\}$$

$$P_2\sim \;=\; \{h^{-1}P_2h \mid h \in H\} \;=\; \{P_2, P_4\}$$

For $\mathcal{R}$ choose one representative from each of the equivalence classes, e.g. choose $\mathcal{R} = \{P_3, P_2\}$. Then Lemma 5.6 claims that $|\mathcal{A}| = 4 = [H : N_H(P_3)] + [H : N_H(P_2)]$. Now $N_H(P_3) = \{1, b^4\} = N_H(P_2)$. Hence $[H : N_H(P_3)] = [H : N_H(P_2)] = 2$ and the required equation of Corollary 5.7 holds. We must also show that $[G : N_G(P)] = 4$. As

$$\begin{aligned} N_G(P) &= \{x \mid x \in G \text{ and } x^{-1}Px = P\} \\ &= \{x \mid x \in G \text{ and } x^{-1}ax = a\} \\ &= \{1, a, b^4, ab^4\} \end{aligned}$$

the index of $N_G(P)$ in G is 4, the required number.

5.14. Check Corollary 5.8 when $G = S_3$. Use the notation of page 57.

Solution:

$\mathcal{A}$ (of Corollary 5.8) $= \{P_1, P_2, P_3, P_4, P_5, P_6\}$ where $P_1 = \{\iota\}$, $P_2 = \{\sigma_1\}$, $P_3 = \{\sigma_2\}$, $P_4 = \{\tau_1\}$, $P_5 = \{\tau_2\}$, $P_6 = \{\tau_3\}$. Let $\sim$ be as in Corollary 5.8. Consequently $P_1\sim = \{P_1\}$, $P_2\sim = \{P_2, P_3\}$, $P_4\sim = \{P_4, P_5, P_6\}$. To define $\mathcal{R}$, we choose one element from each of these equivalence classes. Let us take $\mathcal{R} = \{P_1, P_2, P_4\}$. As $Z(G) = \{\iota\}$, $P_j \cap Z(G) = \emptyset$ except for $j = 1$. Therefore $\mathcal{R}^* = \{P_2, P_4\}$. Now $N_G(P_2) = \{\iota, \sigma_1, \sigma_2\}$ and $N_G(P_4) = \{\iota, \tau_1\}$. Hence, as required,

$$|Z(G)| + [G : N_G(P_2)] + [G : N_G(P_4)] \;=\; 1 + 2 + 3 \;=\; 6 \;=\; |S_3|$$

5.15. Show that $N_H(A) = N_G(A) \cap H$ for any non-empty subset A and subgroup H of a group G.

Solution:

Let $n \in N_H(A)$; then $n \in H$ and $n^{-1}An = A$. But $H \subseteq G$, so that $n \in G$ and by definition $n \in N_G(A)$. Consequently $N_H(A) \subseteq N_G(A) \cap H$. If $n \in N_G(A) \cap H$, then $n^{-1}An = A$ and $n \in H$. Thus $N_G(A) \cap H \subseteq N_H(A)$ and the equality follows.

5.16. Prove Proposition 5.5, page 134.

Solution:

As H is a subgroup of G, H contains the identity, so $A = 1^{-1}A1$ and thus $A \sim A$. If $A \sim B$, then there is an element $h \in H$ such that $h^{-1}Ah = B$. Consequently $(h^{-1})^{-1}B(h^{-1}) = A$ and $B \sim A$. Finally, if $A \sim B$ and $B \sim C$, then there exist $h, g \in H$ such that $h^{-1}Ah = B$ and $g^{-1}Bg = C$. It follows that $hg \in H$ and $(hg)^{-1}A(hg) = g^{-1}(h^{-1}Ah)g = g^{-1}Bg = C$, and so $A \sim C$. Hence $\sim$ is an equivalence relation on $\mathcal{A}$.

5.17. Let A, B be subsets of G. Suppose B is an H-conjugate of A, where H is a subgroup of G. Prove that $[H : N_H(A)] = [H : N_H(B)]$.

Solution:

Let $\mathcal{A} = \{X \mid X = g^{-1}Ag$ or $X = g^{-1}Bg$ for some $g \in G\}$. We use Proposition 5.5. $A\sim = B\sim$, as B is an H-conjugate of A. $|A\sim|$ is therefore the number of H-conjugates of A, and also the number of H-conjugates of B. By Lemma 5.4, $[H : N_H(A)] = [H : N_H(B)]$.

c. Proofs of the Sylow theorems

First we prove a weak form of the first Sylow theorem.

Proposition 5.9: If G is a finite abelian group and p is a prime dividing the order of G, then G has an element of order p.

Proof: We will prove the proposition by induction on the order of G. If $|G| = 1$, there is nothing to prove. Assume the proposition is true for all groups of order less than n, the order of G, where $n > 1$. Recall from Section 4.2b, page 105, that if G is cyclic there is a subgroup of order any integer that divides $|G|$. Thus if G is cyclic the theorem holds, and we may therefore assume G is not cyclic. If n is a prime, G is cyclic; hence n is not a prime.

Suppose $h\,(\neq 1) \in G$, h of order m. Clearly $m < n$. Let H be the cyclic group generated by h. H is a proper subgroup of G. Now if $p \mid m$, by the induction assumption, H has an element of order p. If $p \nmid m$, form the factor group G/H (every subgroup of an abelian group is a normal subgroup so $H \lhd G$). Since $|H| > 1$, $|G/H| < |G|$. As $|G/H| = |G|/|H|$, $p \mid |G|/|H|$. Therefore by the induction assumption, G/H has an element $\bar{g}$ of order p.

Let $\nu: G \to G/H$ be the natural homomorphism of a group onto its factor group (see Theorem 4.17, page 114) and g be a preimage of $\bar{g}$ under ν. Now $(g^p)\nu = \bar{g}^p =$ the identity of G/H, so $g^p \in H$. As H is of order m, $(g^m)^p = (g^p)^m = 1$. Therefore g^m has order p or $g^m = 1$. If $g^m = 1$, then $g^m\nu = \bar{g}^m = 1$. Since $\bar{g}$ has order p this implies p divides m, contrary to our assumption. Therefore g^m is an element of G of order p.

We are now in a position to prove the Sylow theorems. For convenience we repeat the statement of each theorem.

The First Sylow Theorem: Let G be a finite group, p a prime and p^r the highest power of p dividing the order of G. Then there is a subgroup of G of order p^r.

Proof: We will prove the theorem by induction on the order n of G. For $|G| = 1$ the theorem is trivial. Assume $n > 1$ and that the theorem is true for groups of order $< n$. Suppose $|Z(G)| = c$. We have two possibilities: (i) $p \mid c$ or (ii) $p \nmid c$.

(i) Suppose $p \mid c$. $Z(G)$ is an abelian group. By Proposition 5.9, $Z(G)$ has an element of order p. Let N be a cyclic subgroup of $Z(G)$ generated by an element of order p. $N \lhd G$, since any subgroup of $Z(G)$ is normal in G. Consider G/N. Then $|G/N| = n/p$ by Corollary 4.14, page 110. Hence by our induction assumption, G/N has a subgroup $\bar{H}$ of order p^{r-1}.

By Corollary 4.21, page 121, there exists a subgroup H of G such that $H/N = \bar{H}$. As $p^{r-1} = |\bar{H}| = |H|/|N| = |H|/p$, we conclude that $|H| = p^r$. Thus in this case, G has a subgroup of order p^r.

(ii) Suppose $p \nmid c$. The class equation for G is (see Equation (*5.2*) of Corollary 5.8, page 135)

$$|G| \;=\; |Z(G)| + \sum_{R \in \mathcal{R}^*} [G : C(R)]$$

Since $p \mid |G|$ and $p \nmid c$, we have $p \nmid \sum_{R \in \mathcal{R}^*} [G : C(R)]$. Therefore for at least one $R \in \mathcal{R}^*$, $p \nmid [G : C(R)]$. But $|G| = [G : C(R)]\,|C(R)|$ by Corollary 4.14 to Lagrange's theorem, page 110. Hence $p^r \mid |C(R)|$, since $p^r \mid |G|$. Now $|C(R)| \neq |G|$; for if $|C(R)| = |G|$, then $C(R) = G$ and $R \cap Z(G) = R$, contrary to the assumption that $R \cap Z(G) = \emptyset$. Thus by the induction assumption, $C(R)$ has a subgroup H of order p^r. Consequently so does G.

In either case we have found a subgroup H of order p^r. The proof is complete.

The following gives a simple formula for the normalizer of a Sylow p-subgroup P in a subgroup H of G, where $|H|$ is a power of p. It will be used in the proof of the second Sylow theorem.

Lemma 5.10: If G is a finite group, P a Sylow p-subgroup of G, and H is a subgroup of G of order a power of p, then

$$N_H(P) = H \cap P$$

Proof: $P \cap H \subseteq N_H(P)$, as conjugation by an element of P sends P to itself. We show $N_H(P) \subseteq P \cap H$. $N_H(P) \subseteq N_G(P)$ and $P \lhd N_G(P)$ (see Problem 5.15 and Problem 4.60, page 113), so that by the subgroup isomorphism theorem (Theorem 4.23, page 125) we have: $N_H(P)P$ is a subgroup of G and

$$N_H(P)P/P \cong N_H(P)/N_H(P) \cap P$$

Consequently $[N_H(P)P : P] = [N_H(P) : N_H(P) \cap P]$. But $N_H(P)$ is a p-group, i.e. a group of order a power of p, since it is a subgroup of the p-group H. Thus $[N_H(P) : N_H(P) \cap P]$ is a power of p. $[N_H(P)P : P]$ is therefore also a power of p and, as P is a p-group, $|N_H(P)P|$ is a power of p. Accordingly, $N_H(P)P$ is a p-group. But $P \subseteq N_H(P)P$ and P is a Sylow p-subgroup. Hence $P = N_H(P)P$, for P cannot be a proper subgroup of any other p-subgroup of G (see Problem 5.4, page 132). $N_H(P)$ is therefore a subgroup of P. As $N_H(P) \subseteq H$, we conclude $N_H(P) \subseteq H \cap P$.

The Second Sylow Theorem: Let H be a subgroup of a finite group G, and let P be a Sylow p-subgroup of G. If H is a p-group, then H is contained in a G-conjugate of P.

Proof: We apply Corollary 5.7, page 135, to $\mathcal{A} = \{g^{-1}Pg \mid g \in G\}$ to conclude

$$|\mathcal{A}| = \sum_{R \in \mathcal{R}} [H : N_H(R)] = [G : N_G(P)]$$

By Lemma 5.10, $N_H(R) = H \cap R$ for each $R \in \mathcal{R}$. Hence

$$[G : N_G(P)] = \sum_{R \in \mathcal{R}} [H : H \cap R] \tag{5.3}$$

If $H \cap R \neq H$ for all $R \in \mathcal{R}$, as H is a p-group, the right-hand side of equation (*5.3*) is divisible by p. Hence $[G : N_G(P)]$ is divisible by p. But $P \subseteq N_G(P)$, so that p does not divide $[G : N_G(P)]$. This contradiction implies that $H \cap R = H$ for at least one $R \in \mathcal{R}$. But as $R \in \mathcal{A}$, R is a G-conjugate of P. The result follows.

The Third Sylow Theorem: (i) Any two Sylow p-subgroups of a finite group G are G-conjugate. (ii) The number s_p of distinct Sylow p-subgroups of G is congruent to 1 modulo p. (iii) $s_p \mid |G|$.

Proof:

(i) Let P and P' be two Sylow p-subgroups of G. By the second Sylow theorem, P', as a p-group, is contained in some G-conjugate R of P. But $|P'| = |R|$, by Problem 5.3, page 132. Hence $P' = R$ and P' is conjugate to P under G.

(ii) Let P be any Sylow p-subgroup of G. Since any other Sylow p-subgroup is conjugate to P and any conjugate of a Sylow p-subgroup is a Sylow p-subgroup (Problem 5.5, page 132), we conclude by Lemma 5.4 that

$$s_p = [G : N_G(P)]$$

But on putting $P = H$ in Equation (*5.3*), we have

$$s_p = \sum_{R \in \mathcal{R}} [P : P \cap R]$$

Now for exactly one $R \in \mathcal{R}$, $R = P$; for the only P-conjugate of P is P itself and so P is the only possible representative of its equivalence class. In all other cases, $P \cap R \neq P$. Therefore $[P : P \cap R]$ is a power of p for all $R \in \mathcal{R}$ except one, and for this one $[P : P \cap R] = 1$. Hence

$$s_p = 1 + kp$$

(iii) By Corollary 4.14 to Lagrange's theorem, $|G| = [G : N_G(P)]\,|N_G(P)|$. Since $s_p = [G : N_G(P)]$, $s_p \mid |G|$.

5.2 THEORY OF p-GROUPS

a. The importance of p-groups in finite groups

Suppose that G is a finite group. In Section 5.1a we saw that G has a Sylow p-subgroup for any prime p. (p will be a prime throughout this section.) One reason why the study of p-groups (groups of order a power of p) is so important is that the structure of the Sylow p-subgroups of G partly determines the structure of G. One instance is the following theorem: If G is a finite group whose Sylow p-subgroups are all cyclic, then G has a normal subgroup N such that G/N and N are both cyclic. (M. Hall, Jr., *The Theory of Groups*, Macmillan, 1959, Theorem 9.4.3, page 146.)

In this section we shall determine some of the elementary properties of p-groups.

b. The center of a p-group

A very important property of finite p-groups is given by

Theorem 5.11: If $G \neq \{1\}$ and G is a finite p-group, then $Z(G)$, the center of G, is not of order 1.

Proof: We make use of the class equation (equation (*5.2*), page 135)

$$|G| = |Z(G)| + \sum_{R \in \mathcal{R}^*} [G : C(R)] \tag{5.2}$$

It follows immediately from the definition of $C(R)$ and $Z(G)$ that $C(R) = G$ if and only if $R \subseteq Z(G)$. Because the sum on the right side of (*5.2*) is taken over all R such that $R \cap Z(G) = \emptyset$ and because $|G| = p^r$, $p \mid [G : C(R)]$ for all $R \in \mathcal{R}^*$. Hence $p \mid \sum_{R \in \mathcal{R}^*} [G : C(R)]$. Since $p \mid |G|$, we can conclude that $p \mid |Z(G)|$, which means $Z(G) \neq \{1\}$.

Corollary 5.12: If G is a group of order p^r, $r \geqq 1$, then G has a normal subgroup of order p^{r-1}.

Proof: The proof is by induction on r. The statement is clearly true for $r = 1$. Suppose the corollary is true for all $k < r$ where $r > 1$. By Theorem 5.11, $Z(G) \neq \{1\}$. Because $p \mid |Z(G)|$, Proposition 5.9 implies $Z(G)$ has an element g of order p. Let $N = gp(g)$. $N \lhd G$, since any subgroup of $Z(G)$ is normal in G. Consider G/N. $|G/N| = p^{r-1}$. Therefore by the induction assumption, G/N has a normal subgroup $\bar{H}$ of order p^{r-2}. By Corollary 4.21, page 121, there exists a subgroup H of G which contains N and such that $H/N = \bar{H}$. Then $|H| = p^{r-1}$. Furthermore, again by Corollary 4.21, $H \lhd G$. Thus G has a normal subgroup of order p^{r-1} and the proof is complete.

Clearly we could repeat this argument until we obtain a sequence of subgroups of G

$$\{1\} = H_0 \subseteq H_1 \subseteq \cdots \subseteq H_{r-1} \subseteq H_r = G \tag{5.4}$$

where $H_i \lhd H_{i+1}$ $(i = 0, 1, \ldots, r-1)$ and $|H_i| = p^i$ $(i = 0, 1, 2, \ldots, r)$.

Problems

5.18. Suppose G is a group with S a subgroup of the center $Z(G)$. Prove G is abelian if G/S is cyclic.

Solution:

Suppose G/S is cyclic. Then we can find $a \in G$ such that every element of G/S is a power of aS. Then if $g, h \in G$,

$$g = a^i z, \qquad h = a^j z'$$

for a suitable choice of the integers i and j, with z and z' in S. Then

$$gh = a^i z a^j z' = a^i a^j z z' = a^j a^i z' z = a^j z' a^i z = hg$$

Thus every pair of elements of G commute. Hence G is abelian.

5.19. Prove that a group of order p^2 is abelian (p a prime).

Solution:

Let G be of order p^2 and let Z be the center of G. By Theorem 5.11, $Z \neq \{1\}$. If $Z = G$, G is abelian. Suppose $Z \neq G$; then $|G/Z| = p$, so G/Z is cyclic. By Problem 5.18, it follows that G is abelian.

5.20. Let $A = \{0, 1, \ldots, p-1\}$, where p is a prime. Then under addition modulo p, A is an abelian group. Let

$$G = \{(a, b, c) \mid a, b, c \in A\}$$

be the set of all triples (a, b, c) of elements of A. Define

$$(a, b, c) \cdot (a', b', c') = (a + a',\ b + b',\ c + c' - ba')$$

Prove that with respect to this binary operation, G is a non-abelian group of order p^3.

Solution:

It is clear that $|G| = p^3$. To prove that G is a group, we check first the associative law:

$$\begin{aligned}((a, b, c)(a', b', c')) \cdot (a'', b'', c'') &= (a + a', b + b', c + c' - ba')(a'', b'', c'') \\ &= (a + a' + a'', b + b' + b'', c + c' + c'' - ba' - (b + b')a'')\end{aligned}$$

On the other hand,

$$\begin{aligned}(a, b, c)((a', b', c')(a'', b'', c'')) &= (a, b, c)(a' + a'', b' + b'', c' + c'' - b'a'') \\ &= (a + a' + a'', b + b' + b'', c + c' + c'' - b'a'' - b(a' + a''))\end{aligned}$$

We check that

$$c + c' + c'' - ba' - (b + b')a'' = c + c' + c'' - b'a'' - b(a' + a'')$$

which is true. Thus G is a semigroup. Now

$$(a, b, c) \cdot (0, 0, 0) = (a, b, c) = (0, 0, 0) \cdot (a, b, c)$$

and so $(0, 0, 0)$ is the unit element of G. Finally,

$$(a, b, c) \cdot (-a, -b, -c - ba) = (0, 0, 0) = (-a, -b, -c - ba)(a, b, c)$$

and hence every element of G has an inverse.

Our last task is to prove that G is non-abelian. Now

$$(1, 0, 0)(0, 1, 0) = (1, 1, 0), \qquad (0, 1, 0)(1, 0, 0) = (1, 1, -1)$$

and thus $(1, 0, 0)(0, 1, 0) \neq (0, 1, 0)(1, 0, 0)$.

5.21. Let A be the additive group of integers modulo p, $A = \{0, 1, \ldots, p-1\}$; and let B be the additive group of integers modulo p^2, $B = \{0, 1, \ldots, p^2 - 1\}$. Let G be the set of all pairs (i, j), $i \in A$, $j \in B$. Prove that under the binary operation

$$(i, j) \cdot (i', j') = (i + i', j + j' + ji'p)$$

G is a non-abelian group of order p^3.

Solution:

Clearly G is of order p^3. We check that G is a semigroup.

$$((i,j)(i',j'))(i'',j'') = (i+i', j+j'+ji'p)(i'',j'')$$
$$= (i+i+i'', j+j'+j''+ji'p+(j+j'+ji'p)i''p)$$

$$(i,j)((i',j')(i'',j'')) = (i,j)(i'+i'', j'+j''+j'i''p)$$
$$= (i+i'+i'', j+j'+j''+j'i''p+j(i'+i'')p)$$

To prove that the binary operation in G is associative, we need check only that

$$ji'p + (j+j'+ji'p)i''p = j'i''p + j(i'+i'')p$$

Since $p^2 = 0$ in B, this equality is readily verified.

The identity element of G is $(0,0)$. The inverse of (i,j) is $(-i, -j+jip)$. Thus G is a group.

Finally,

$$(1,0)(0,1) = (1,1), \qquad (0,1)(1,0) = (1,1+p)$$

and therefore G is non-abelian.

5.22. Prove that the group G in Problem 5.20 has the property that for all $g \in G$, $g^p = 1$ (i.e. $(0,0,0)$) if p is odd. Is this true if $p = 2$?

Solution:

Let $(a,b,c) \in G$. Then

$$(a,b,c)^2 = (a,b,c)(a,b,c) = (2a, 2b, 2c - ba)$$

Continuing, we find

$$(a,b,c)^3 = (a,b,c)^2(a,b,c) = (2a, 2b, 2c-ba)(a,b,c)$$
$$= (3a, 3b, 3c - ba - 2ba)$$

By induction it follows that

$$(a,b,c)^p = (pa, pb, pc - ba - 2ba - \cdots - (p-1)ba)$$

But $pa = 0$, $pb = 0$, $pc = 0$. Finally,

$$ba + 2ba + \cdots + (p-1)ba = \frac{p(p-1)}{2}ba$$

since $1+2+\cdots+p-1 = \frac{1}{2}p(p-1)$. If p is odd, $p-1$ is even. Therefore $\frac{1}{2}p(p-1)$ is an integer divisible by p. Hence $\frac{1}{2}p(p-1)ba = 0$. Thus we have $(a,b,c)^p = 1$.

If $p = 2$, then

$$(a,b,c)^2 = (2a, 2b, 2c - ba) = (0, 0, -ba)$$

In particular if $a = 1$, $b = 1$ and $c = 0$, we have $(1,1,0)^2 = (0,0,-1)$. Thus not every element of G is of order 2. This result could have been observed by noting that a group G satisfying $g^2 = 1$ for all g in G is abelian. To see this let $g, h \in G$. Then as $(gh)^2 = 1$,

$$gh = (gh)^{-1} = h^{-1}g^{-1} = h^2h^{-1}g^2g^{-1} = hg$$

and so G is abelian. But as G is not abelian, not every element is of order 2.

5.23. If p is odd, does the group G of Problem 5.21 satisfy $g^p = 1$ (i.e. $(0,0)$) for all $g \in G$?

Solution:

No, since $(0,1)^2 = (0,1)(0,1) = (0,2)$. Inductively, $(0,1)^p = (0,p) \neq (0,0)$.

5.24. Prove that if G is a group such that $g^p = 1$ for all $g \in G$, then every homomorphic image H has the same property, i.e. $h^p = 1$ for all $h \in H$.

Solution:

Let θ be a homomorphism of G onto H. Then if $h \in H$, we can find $g \in G$ such that $g\theta = h$. Therefore $h^p = (g\theta)^p = (g^p)\theta = 1\theta = 1$.

5.25. Prove that if p is an odd prime, then the groups in Problems 5.20 and 5.21 are not isomorphic.

Solution:

Let G be the group defined in Problem 5.20 and let H be the group defined in Problem 5.21, for p an odd prime. Then by Problem 5.22, if $g \in G$, $g^p = 1$. But if $G \cong H$, it follows from Problem 5.24 that $h^p = 1$ for all $h \in H$. But by Problem 5.23, $(0,1)^p \neq 1$. Therefore G is not isomorphic to H.

5.26. Prove that a non-abelian group G of order p^3 has a center of order p (p a prime).

Solution:

Let Z be the center of G. By Theorem 5.11, $Z \neq \{1\}$. Also, $Z \neq G$ since G is non-abelian. Now if $|Z| = p^2$, then $|G/Z| = p$. Therefore G/Z would be cyclic and hence, by Problem 5.18, G would be abelian, a contradiction. Thus $|Z| = p$.

c. The upper central series

Suppose G is a group. We shall define a series

$$\{1\} = Z_0 \subseteq Z_1 \subseteq \cdots$$

of subgroups $Z_0, Z_1, \ldots$ of G, called the *upper central series of G*. We begin by defining $Z_0 = \{1\}$, and Z_1 to be the center of G. Next we define Z_2. We look at G/Z_1. Since every subgroup of G/Z_1 is uniquely of the form H/Z_1 where H is a subgroup of G containing Z_1, the center of G/Z_1 is of the form Z_2/Z_1 (we are using Corollary 4.21, page 121). Notice that as the center of a group is a normal subgroup, Z_2/Z_1 is a normal subgroup of G/Z_1. Therefore by Corollary 4.21, Z_2 is a normal subgroup of G.

In general, once Z_i has been defined and proved to be a normal subgroup of G, we define Z_{i+1}/Z_i to be the center of G/Z_i. By Corollary 4.21 it follows that $Z_{i+1} \triangleleft G$.

We shall call a group G *nilpotent* if its upper central series ascends to G in a finite number of steps.

Our objective in this section is to prove

Theorem 5.13: A finite p-group G is nilpotent.

Proof: If $G = \{1\}$, there is nothing to prove. If $G \neq \{1\}$, then $Z_1 \neq \{1\}$ by Theorem 5.11. If G/Z_1 is not the identity, the center of $G/Z_1 = Z_2/Z_1 \neq Z_1/Z_1$, again by Theorem 5.11. Notice that if $Z_1 \neq G$, then $Z_2 \neq Z_1$. Similarly if $G \neq Z_2$, $Z_3 \neq Z_2$. By induction we can show that if $Z_i \neq G$, $Z_{i+1} \neq Z_i$ and thus

$$1 = Z_0 \subset Z_1 \subset \cdots \subset Z_i \subset Z_{i+1}$$

Since G is finite, $Z_k = G$ for some k. Therefore G is nilpotent.

Problems

5.27. Prove that if a non-abelian group G is of order p^3, then $Z_2 = G$.

Solution:

$Z_1 \neq \{1\}$ by Theorem 5.11. So if $Z_1 \neq G$, then G/Z_1 is of order p or p^2. Since G/Z_1 is cyclic only if $Z_1 = G$ (Problem 5.18), we find G/Z_1 is of order p^2 and hence abelian (Problem 5.19). Therefore $Z_2/Z_1 = G/Z_1$, i.e. $Z_2 = G$.

5.28. Let D_n be the dihedral group of order $2n$. Prove that D_n is nilpotent if n is a power of 2.

Solution:

D_n has the property that it contains two elements a and b such that $a^2 = 1$, $b^n = 1$, $a^{-1}ba = b^{-1}$ and every element of D_n is uniquely expressible in the form a^ib^j where $i = 0, 1$ and $j = 0, 1, \ldots, n-1$. (See Section 3.4f, page 75, where $a = \tau$ and $b = \sigma_2$.)

Method 1. Suppose $n = 2^m$. Then $b^{2^{m-1}}$ is of order 2. Hence

$$a^{-1}b^{2^{m-1}}a = (a^{-1}ba)^{2^{m-1}} = (b^{-1})^{2^{m-1}} = b^{2^{m-1}}$$

So $b^{2^{m-1}}$ commutes with a; clearly $b^{2^{m-1}}$ commutes with b. Therefore $b^{2^{m-1}}$ commutes with every element of G, and so $b^{2^{m-1}} \in Z_1$. If $m = 1$, then $Z_1 = D_n$; otherwise $a \notin Z_1$.

What other elements can be in the center? If $a^i b^j \in Z_1$, then clearly $a^{-1}(a^i b^j)a = a^i b^j$. On the other hand, $a^{-1}(a^i b^j)a = a^i a^{-1} b^j a = a^i b^{-j}$. Hence $b^j = b^{-j}$ and $(b^j)^2 = 1$, i.e. $b^j = b^{2^{m-1}}$. Thus the only possible element in the center other than $b^{2^{m-1}}$ is $ab^{2^{m-1}}$. But as $b^{2^{m-1}} \in Z_1$, this implies $a \in Z_1$, which is not so. Consequently $Z_1 = \{1, b^{2^{m-1}}\}$.

Similarly Z_2 consists of the powers of $b^{2^{m-2}}, \ldots, Z_i$ consists of the powers of $b^{2^{m-i}}$. Therefore Z_{m-1} consists of the powers of b^2. Now this means $|G/Z_{m-1}| = 4$, and so G/Z_{m-1} is abelian. Thus $Z_m = G$ and G is nilpotent.

Method 2. $|D_n| =$ some power of 2. Hence we can apply Theorem 5.13.

5.29. Prove that A_4 is not nilpotent.

Solution:

$Z_1 = \{1\}$ in A_4, as a direct check shows. Hence $Z_1 = Z_2 = \cdots$, and thus $Z_n \neq A_4$ for every n.

5.3 DIRECT PRODUCTS AND GROUPS OF LOW ORDER

a. Direct products of groups

In Chapter 1 we defined the cartesian product $H \times K$ of two sets H and K as the set of all ordered pairs (h, k), $h \in H$ and $k \in K$. If H and K are groups, we can define a multiplication of elements of $H \times K$ as follows. Let $(h_1, k_1), (h_2, k_2) \in H \times K$ and define

$$(h_1, k_1) \cdot (h_2, k_2) = (h_1h_2, k_1k_2) \tag{5.5}$$

where h_1h_2 and k_1k_2 are the products in the groups H and K respectively. The multiplication defined in (*5.5*) is clearly a binary operation. The set $H \times K$ with binary operation (*5.5*) is a group. To see that $H \times K$ is a group, let $(h_1, k_1), (h_2, k_2), (h_3, k_3) \in H \times K$. Then

$$\begin{aligned}
[(h_1, k_1) \cdot (h_2, k_2)] \cdot (h_3, k_3) &= (h_1h_2, k_1k_2) \cdot (h_3, k_3) \\
&= ((h_1h_2)h_3, (k_1k_2)k_3) \\
&= (h_1(h_2h_3), k_1(k_2k_3)) \\
&= (h_1, k_1) \cdot (h_2h_3, k_2k_3) \\
&= (h_1, k_1) \cdot [(h_2, k_2) \cdot (h_3, k_3)]
\end{aligned}$$

Multiplication is therefore an associative binary operation on $H \times K$. If 1 stands simultaneously for the identity element of H and of K and $(h, k) \in H \times K$,

$$(1,1)(h,k) = (h,k) = (h,k)(1,1)$$

so that $(1,1)$ is an identity of $H \times K$. It is clear that if $(h, k) \in H \times K$, then (h^{-1}, k^{-1}) is its inverse, for $(h,k)(h^{-1}, k^{-1}) = (hh^{-1}, kk^{-1}) = (1,1)$. The group $H \times K$ with binary operation (*5.5*) is called the *external direct product of the groups H and K*. We often refer to $H \times K$ as just the direct product. We define the internal direct product after Proposition 5.19. If H and K are finite groups, then it is clear that

$$|H \times K| = |H|\,|K|$$

If $H \neq \{1\}$ and $K \neq \{1\}$ are finite groups, then $H \times K$ is neither isomorphic to H nor to K, because $|H \times K| \neq |H|$ and $|H \times K| \neq |K|$. Therefore the direct product gives us a simple way of constructing new finite groups. For example, let C_2 be the cyclic group of order 2 generated by g.

$$C_2 \times C_2 = \{(g,1), (g,g), (1,g), (1,1)\}$$

Now $|C_2 \times C_2| = 4$, so we have (as we shall soon see) two non-isomorphic groups of order 4, namely: the cyclic group of order 4, $C_4 = \{1, b, b^2, b^3\}$ where $b^4 = 1$, and the group $C_2 \times C_2$.

The multiplication table for $C_2 \times C_2$ is

	$(1,1)$	$(1,g)$	$(g,1)$	(g,g)
$(1,1)$	$(1,1)$	$(1,g)$	$(g,1)$	(g,g)
$(1,g)$	$(1,g)$	$(1,1)$	(g,g)	$(g,1)$
$(g,1)$	$(g,1)$	(g,g)	$(1,1)$	$(1,g)$
(g,g)	(g,g)	$(g,1)$	$(1,g)$	$(1,1)$

Note that all the elements of $C_2 \times C_2$ are of order 2. Hence C_4 is not isomorphic to $C_2 \times C_2$. $C_2 \times C_2$ is called the *Klein four group*, or simply the *four group.*

Theorem 5.14: If $G = H \times K$ is the direct product of the groups H and K, then the sets

$$\hat{H} = \{(h,1) \mid h \in H,\ 1 \text{ the identity of } H\}$$
$$\hat{K} = \{(1,k) \mid k \in K,\ 1 \text{ the identity of } K\}$$

are subgroups of G. Furthermore, $H \cong \hat{H}$, $K \cong \hat{K}$; and if $\hat{a} \in \hat{H}$ and $\hat{b} \in \hat{K}$, $\hat{a}\hat{b} = \hat{b}\hat{a}$. Finally, $G = \hat{H}\hat{K}$ and $\hat{H} \cap \hat{K} = \{(1,1)\}$, the identity subgroup of G.

Proof: If $(h_1,1), (h_2,1) \in \hat{H}$, then $(h_1,1)(h_2,1)^{-1} = (h_1,1)(h_2^{-1},1) = (h_1h_2^{-1},1) \in \hat{H}$, since $h_1h_2^{-1} \in H$. $\hat{H}$ is clearly non-empty. Therefore $\hat{H}$ is a subgroup of G. Similarly $\hat{K}$ is a subgroup of G. The mapping α of H onto $\hat{H}$ defined by

$$\alpha: h \to (h,1),\ h \in H$$

is clearly an isomorphism. Similarly K and $\hat{K}$ are isomorphic. Now let $\hat{a} = (h,1) \in H$ and $\hat{b} = (1,k) \in \hat{K}$; then

$$\hat{a}\hat{b} = (h,1)(1,k) = (h,k) = (1,k)(h,1) = \hat{b}\hat{a}$$

Now $\hat{H} \cap \hat{K} = \{(1,1)\}$. Any element (h,k) of G can be written as $(h,1)(1,k)$, so $G \subseteq \hat{H}\hat{K}$. Clearly $\hat{H}\hat{K} \subseteq G$. Hence $G = \hat{H}\hat{K}$ and Theorem 5.14 follows.

Corollary 5.15: Let $G = H \times K$ and $\hat{H}, \hat{K}$ be as in Theorem 5.14. Then every $g \in G$ can be written uniquely as a product $\hat{h}\hat{k}$ where $\hat{h} \in \hat{H}$, $\hat{k} \in \hat{K}$.

Proof: If $\hat{g} = (h,k)$, then $\hat{g} = (h,1)(1,k)$ is an expression for $\hat{g}$ as the product of an element in $\hat{H}$ by an element in $\hat{K}$. If we also have $\hat{g} = (h_1,1)(1,k_1)$, then clearly $h_1 = h$ and $k_1 = k$. Thus the expression is unique.

As a converse we have

Theorem 5.16: Let G be a group with subgroups H and K such that $H \cap K = \{1\}$, the elements of H commute with those of K, and $HK = G$. Then $G \cong H \times K$.

Proof: We first show that any element $g \in G$ can be written uniquely in the form $g = hk$ where $h \in H$ and $k \in K$. Since $G = HK$, $g = hk$ for some $h \in H$ and $k \in K$. Suppose $g = h_1k_1$ and $g = h_2k_2$ where $h_1, h_2 \in H$ and $k_1, k_2 \in K$. $h_1k_1 = h_2k_2$ implies $h_2^{-1}h_1 = k_2k_1^{-1}$. But $H \cap K = \{1\}$, and so $h_2^{-1}h_1 = 1$ and $k_2k_1^{-1} = 1$. Hence $h_1 = h_2$ and $k_1 = k_2$.

We define the mapping $\alpha: G \to H \times K$ by $g\alpha = (h,k)$ where $g = hk \in G$. α is a one-to-one mapping, for we have shown that there is one and only one way of writing g in the form $g = hk$, and the elements of $H \times K$ are of the unique form (h,k). To prove α is a homomorphism we must demonstrate that if $g_1 = h_1k_1$ and $g_2 = h_2k_2$ are any two elements in G, then

$$(g_1g_2)\alpha = g_1\alpha g_2\alpha \quad \text{or} \quad (h_1k_1h_2k_2)\alpha = (h_1k_1)\alpha(h_2k_2)\alpha$$

Now $\quad (h_1k_1h_2k_2)\alpha = (h_1h_2k_1k_2)\alpha = (h_1h_2, k_1k_2) = (h_1, k_1)(h_2, k_2) = (h_1k_1)\alpha(h_2k_2)\alpha$

Hence α is a homomorphism and the result follows.

Note that if H, K are normal subgroups of a group G with $H \cap K = \{1\}$, then H and K commute elementwise. For if $h \in H$, $k \in K$,

$$h^{-1}k^{-1}hk = (h^{-1}k^{-1}h)k \in K = h^{-1}(k^{-1}hk) \in H$$

Therefore $h^{-1}k^{-1}hk \in H \cap K = \{1\}$, and so H and K commute elementwise. Clearly $G = HK$ and H and K commute elementwise implies H and K are normal in G.

Consequently Theorem 5.16 can be stated as follows:

Corollary 5.17: Let G be a group with normal subgroups H and K, and suppose $H \cap K = \{1\}$, and $HK = G$. Then $G \cong H \times K$.

The hypothesis of Theorem 5.16 asserts G must equal HK. But if G is a finite group and $|HK| = |G|$, we can conclude, since $HK \subseteq G$, that $HK = G$. It is useful to be able to count the number of elements in HK. We therefore prove the following proposition.

Proposition 5.18: If G is a finite group with subgroups H and K, then

$$|HK| = \frac{|H| \cdot |K|}{|H \cap K|}$$

Proof: Let $I = H \cap K$. I is a subgroup of G and, since $I \subseteq K$, I is a subgroup of K. Let $Ik_1, Ik_2, \ldots, Ik_n$ be the n distinct cosets of I in K. Thus

$$K = Ik_1 \cup Ik_2 \cup \cdots \cup Ik_n$$

and, by Corollary 4.14, page 110, $n = |K|/|I| = |K|/|H \cap K|$.

We claim now that

$$HK = Hk_1 \cup Hk_2 \cup \cdots \cup Hk_n$$

For if $hk \in HK$, then $k = lk_j$ for some $l \in I$, j an integer between 1 and n. Hence $hk = (hl)k_j = h'k_j$ where $h' \in H$, as both h, l belong to H. Thus $HK = Hk_1 \cup Hk_2 \cup \cdots \cup Hk_n$.

Now suppose $Hk_i \cap Hk_j \neq \emptyset$ for some integers i and j. Then $hk_i = h'k_j$ for some $h, h' \in H$. Consequently $h'^{-1}h = k_jk_i^{-1}$, so $k_jk_i^{-1} \in I = H \cap K$. But $k_jk_i^{-1} \in I$ implies that $k_j \in Ik_i$. Since two cosets are either equal or disjoint, $Ik_j = Ik_i$. Hence $k_i = k_j$. Thus $Hk_i \cap Hk_j = \emptyset$ for $i \neq j$ and

$$|HK| = |Hk_1| + |Hk_2| + \cdots + |Hk_n|$$

Now $|Hk_i| = |H|$, because $h_1k_i = h_2k_i$ if and only if $h_1 = h_2$. Therefore

$$|HK| = n\,|H| = \frac{|H|\,|K|}{|H \cap K|}$$

since $n = |K|/|H \cap K|$.

To illustrate the use of Proposition 5.18, let G be a group of order 28 and H_1 and H_2 subgroups of G of orders 7 and 4 respectively. $H_1 \cap H_2 = \{1\}$, because an element in H_1 and also in H_2 must have order dividing 7 and 4. Accordingly,

$$|H_1H_2| = \frac{|H_1|\,|H_2|}{|H_1 \cap H_2|} = 28 = |G|$$

and $G = H_1H_2$.

Using Proposition 5.18, we can replace Theorem 5.16 in the case of finite groups by

Theorem 5.16′: Let G be a finite group with normal subgroups H and K where $|H|\,|K| = |G|$. If either (i) $H \cap K = \{1\}$ or (ii) $HK = G$, then $G \cong H \times K$.

Proof:

(i) $H \cap K = \{1\}$ and $|H|\,|K| = |G|$ implies, by Proposition 5.18, $|HK| = |H|\,|K|/|H \cap K| = |G|$. Since $HK \subseteq G$, we can conclude $HK = G$. But then the hypotheses of Corollary 5.17 are fulfilled and $G \cong H \times K$.

(ii) If $HK = G$, then $|HK| = |G|$. Therefore

$$|G| \;=\; |HK| \;=\; \frac{|H|\,|K|}{|H \cap K|} \qquad \text{or} \qquad |H \cap K|\,|G| \;=\; |H|\,|K|$$

But $|H|\,|K| = |G|$ by hypothesis. Hence $H \cap K = \{1\}$ and, by Corollary 5.17, $G \cong H \times K$.

The concept of direct product can easily be generalized to the direct product of a finite number of groups, $G_1, G_2, \ldots, G_n$ ($n \geqq 2$). Let $G = G_1 \times G_2 \times \cdots \times G_n$ be the cartesian product of n groups. Define a multiplication in G by $(g_1, g_2, \ldots, g_n)(g_1', g_2', \ldots, g_n') = (g_1g_1', g_2g_2', \ldots, g_ng_n')$ for $(g_1, g_2, \ldots, g_n), (g_1', g_2', \ldots, g_n') \in G$. G is then a group (see Problem 5.30 below) called the (external) direct product of the groups $G_1, G_2, \ldots, G_n$. We denote G by $\prod_{i=1}^{n} G_i$.

In Chapter 6 we will define the direct product of an infinite number of groups differently. Proposition 5.19 below and Corollary 5.15 will provide a link between the two definitions.

Proposition 5.19: Let H and K be subgroups of a group G. If

(i) $hk = kh$ for all $h \in H$ and $k \in K$

and

(ii) every element $g \in G$ is a unique product of an element in H and an element in K, (i.e. $g = hk$, $h \in H$, $k \in K$; and if $g = h_1k_1$, $h_1 \in H$, $k_1 \in K$, then $h = h_1$ and $k = k_1$),

then $G \cong H \times K$.

Proof: We need only prove $H \cap K = \{1\}$ to fulfill the hypotheses of Theorem 5.16. Suppose $g \in H \cap K$. Then $g = h \cdot 1 = 1 \cdot k$ for some $h \in H$ and $k \in K$. But condition (ii) implies $h = 1$ and $k = 1$. Therefore $g = 1$ and $H \cap K = \{1\}$.

If G is a group with subgroups H and K satisfying conditions (i) and (ii), G is said to be the *internal direct product* of H and K, and we write $G = H \otimes K$. By Proposition 5.19, $H \otimes K \cong H \times K$.

Problems

5.30. Let $G = G_1 \times G_2 \times \cdots \times G_n$ be the cartesian product of n groups. Define a multiplication in G by

$$(g_1, g_2, \ldots, g_n)(g_1', g_2', \ldots, g_n') \;=\; (g_1g_1', g_2g_2', \ldots, g_ng_n')$$

for $(g_1, g_2, \ldots, g_n), (g_1', g_2', \ldots, g_n') \in G$. Show that G is a group.

Solution:

The multiplication is clearly an associative binary operation in G, since multiplication is associative in each G_i. If 1 stands simultaneously for the identity of G_i, $i = 1, 2, \ldots, n$, then $(1, 1, \ldots, 1)$ is clearly the identity of G. If $g = (g_1, g_2, \ldots, g_n) \in G$, then $(g_1^{-1}, g_2^{-1}, \ldots, g_n^{-1})$ is the inverse of g in G.

5.31. If $H \cong \bar{H}$ and $K \cong \bar{K}$, where $H, \bar{H}, K$ and $\bar{K}$ are groups, then $H \times K \cong \bar{H} \times \bar{K}$.

Solution:

If $\alpha: H \to \bar{H}$ and $\beta: K \to \bar{K}$ are isomorphisms, we define $\gamma: H \times K \to \bar{H} \times \bar{K}$ by $\gamma: (h, k) \to (h\alpha, k\beta)$, $h \in H$, $k \in K$. γ is a one-to-one mapping, for $(h\alpha, k\beta) = (h'\alpha, k'\beta)$ if and only if $h\alpha = h'\alpha$

and $k\beta = k'\beta$. Since α and β are one-to-one mappings, $h\alpha = h'\alpha$ and $k\beta = k'\beta$ if and only if $h = h'$ and $k = k'$. To show γ is a homomorphism, let $(h,k),(h',k') \in H \times K$. Then

$$[(h,k)\cdot(h',k')]\gamma = (hh',kk')\gamma = ((hh')\alpha,(kk')\beta) = (h\alpha h'\alpha, k\beta k'\beta)$$
$$= (h\alpha,k\beta)\cdot(h'\alpha,k'\beta) = (h,k)\gamma(h',k')\gamma$$

Finally, it is clear that γ is an onto mapping.

5.32. Show that $G = H \times K$ is an abelian group if and only if H and K are both abelian groups.

Solution:

Suppose H and K are abelian groups. Letting $(h,k),(h',k') \in G$,

$$(h,k)\cdot(h',k') = (hh',kk') = (h'h,k'k) = (h',k')\cdot(h,k)$$

and so G is abelian.

Conversely, suppose G is abelian. Let $h,h' \in H$. Then if 1 is the identity of K, $(h,1)(h',1) = (h',1)(h,1)$ or $(hh',1) = (h'h,1)$. But this implies $hh' = h'h$. Hence H is abelian. Similarly we can show K is abelian.

5.33. Consider the groups $C_2 \times K_4$, $C_4 \times C_2$, C_8 where C_n is the cyclic group of order n and K_4 the four group, i.e. the non-cyclic group of order 4, described above. Are any two of these groups isomorphic? Is any one non-abelian?

Solution:

Let $C_2 = gp(a)$, $C_4 = gp(b)$, and $C_8 = gp(g)$. We look at the set of elements of order 2 in each group. Since every isomorphism maps elements of order 2 onto elements of order 2, if there are more elements of order 2 in one group than in another, these groups cannot be isomorphic. Every element ($\neq 1$) of $C_2 \times K_4$ is of order 2, for $(c,k)^2 = (c^2,k^2) = (1,1)$ and $(1,k)^2 = (1,k^2) = (1,1)$ for any $k \in K_4$. C_8 on the other hand has only one element of order 2, namely g^4, because $(g^i)^2 \neq 1$ if $i \neq 4$, $0 \leq i \leq 7$. Now $C_4 \times C_2$ has at least one element of order 4, $(b,1)$, and at least two elements of order 2, (b^2,a) and $(1,a)$. Therefore no two of the groups are isomorphic. As C_2, C_4, K_4 and C_8 are abelian, Problem 5.32 implies $C_2 \times K_4$, $C_4 \times C_2$ and C_8 are also abelian. Thus we have exhibited three non-isomorphic abelian groups of order 8.

5.34. If C_n and C_m are the cyclic groups of order n and m respectively and $(n,m) = 1$, then $C_n \times C_m \cong C_{nm}$, the cyclic group of order nm.

Solution:

Say $C_n = gp(g)$ and $C_m = gp(h)$. Consider the order of the element (g,h) in $C_n \times C_m$. We claim that the order of (g,h) is nm. If $(g,h)^k = (1,1)$ for some k, then $(g^k,h^k) = (1,1)$ and so $g^k = 1$ and $h^k = 1$. Since the order of g is n and the order of h is m, $m \mid k$ and $n \mid k$. Hence k is divisible by nm. On the other hand, $(g,h)^{nm} = (g^{nm},h^{nm}) = 1$ and so the order is nm. Accordingly, $C_n \times C_m = gp((g,h))$. Therefore $C_n \times C_m \cong C_{nm}$, since all cyclic groups of the same order are isomorphic (Theorem 4.7, page 103).

5.35. Show that C_{s^2} is not isomorphic to $C_s \times C_s$ (where C_n is the cyclic group of order n), for any integer $s > 1$.

Solution:

Since C_{s^2} is a cyclic group it has, by Theorem 4.9, page 105, one and only one subgroup of order s. But $C_s \times C_s$ has two subgroups of order s, namely $gp((1,g))$ and $gp((g,1))$, where g is the generator of C_s. Since subgroups of a given order are mapped onto subgroups of the same order by any isomorphism, C_{s^2} cannot be isomorphic to $C_s \times C_s$.

5.36. Show that for any prime p there are exactly two non-isomorphic groups of order p^2.

Solution:

By Problem 5.19, page 140, we know that any group of order p^2 is abelian. C_{p^2}, the cyclic group of order p^2, and $C_p \times C_p$, where C_p is the cyclic group of order p, are two non-isomorphic groups of order p^2 (Problem 5.35). To see that these are the only possible groups of order p^2, consider a subgroup H of order p in G, a group of order p^2. Such a subgroup exists by Corollary 5.12. H is cyclic, since p is a prime. Let $a \notin H$. The order of a is either p or p^2. If the order of a is p^2, G is a cyclic group generated by a. If $|gp(a)| = p$, then $gp(a) \cap H = \{1\}$, for $b\ (\neq 1) \in gp(a) \cap H$ implies $H = gp(b)$ and $gp(a) = gp(b)$, since a group of prime order has no proper subgroups. Also $|gp(a)||H| = p^2$ and, as G is abelian (Problem 5.19), $gp(a) \triangleleft G$ and $H \triangleleft G$. Therefore, using Theorem 5.16′, we conclude $G \cong gp(a) \times H$. But $gp(a) \times H \cong C_p \times C_p$ by Problem 5.31. Hence $G \cong C_p \times C_p$.

5.37. Show $(H_1 \times H_2) \times H_3 \cong H_1 \times H_2 \times H_3$.

Solution:

Define $\Psi: (H_1 \times H_2) \times H_3 \to H_1 \times H_2 \times H_3$ by $\Psi: ((h_1, h_2), h_3) \to (h_1, h_2, h_3)$ for $h_1 \in H$, $h_2 \in H_2$ and $h_3 \in H_3$. Clearly Ψ is an onto mapping. If $((h_1, h_2), h_3)\Psi = ((\bar{h}_1, \bar{h}_2), \bar{h}_3)\Psi$, then $(h_1, h_2, h_3) = (\bar{h}_1, \bar{h}_2, \bar{h}_3)$ and consequently $h_1 = \bar{h}_1$, $h_2 = \bar{h}_2$ and $h_3 = \bar{h}_3$. Therefore Ψ is one-to-one. To show Ψ is a homomorphism, let $((h_1, h_2), h_3)$ and $((\bar{h}_1, \bar{h}_2), \bar{h}_3) \in (H_1 \times H_2) \times H_3$. Then

$$\begin{aligned}
[((h_1, h_2), h_3)((\bar{h}_1, \bar{h}_2), \bar{h}_3)]\Psi &= ((h_1, h_2)(\bar{h}_1, \bar{h}_2), h_3\bar{h}_3)\Psi = ((h_1\bar{h}_1, h_2\bar{h}_2), h_3\bar{h}_3)\Psi \\
&= (h_1\bar{h}_1, h_2\bar{h}_2, h_3\bar{h}_3) = (h_1, h_2, h_3)(\bar{h}_1, \bar{h}_2, \bar{h}_3) \\
&= ((h_1, h_2), h_3)\Psi((\bar{h}_1, \bar{h}_2), \bar{h}_3)\Psi
\end{aligned}$$

and so Ψ is an isomorphism.

b. Groups of small order: orders p and $2p$

As an application of the Sylow theorems and the theorems of Section 5.3a we will find, up to isomorphism, all groups of order less than 16. We will use C_n to denote the cyclic group of order n, and K_4 to denote the four group. Recall that K_4 is defined to be $C_2 \times C_2$. We refer to the notation of Section 5.3a. Let us put $1 = (1, 1)$, $x = (1, g)$, $y = (g, 1)$ and $z = (g, g)$. The multiplication table for K_4 is

	1	x	y	z
1	1	x	y	z
x	x	1	z	y
y	y	z	1	x
z	z	y	x	1

We note that $xy = z$, $xz = y$ and $yz = x$. Notice that the multiplication table is symmetric in x, y and z. If we put $x = a$ and $y = b$, then $z = ab$ and we can write the multiplication table in the form

	1	a	b	ab
1	1	a	b	ab
a	a	1	ab	b
b	b	ab	1	a
ab	ab	b	a	1

There is, up to isomorphism, clearly only one group of order 1.

If p is a prime, any group of order p is cyclic (Problem 4.48, page 110). Up to isomorphism, there is one and only one cyclic group of order p (see Theorem 4.7, page 103). *Thus there is one and only one group of order p, p a prime.* In particular, the only groups of order 2, 3, 5, 7, 11 and 13 are cyclic.

There are precisely two non-isomorphic groups of order 4, namely C_4 and K_4 (Section 5.3a and Problem 5.36).

Next we show there are precisely two groups in each case of order 6, 10 or 14. Note that $6 = 2 \cdot 3$, $10 = 2 \cdot 5$ and $14 = 2 \cdot 7$, so these groups are of order $2p$ for some prime $p \neq 2$. Let G be a group of order $2p$, p an odd prime. By Problem 5.10, G has exactly one subgroup K of order p, and either

(i) exactly one subgroup H of order 2

or

(ii) precisely p subgroups of order 2.

(i) In this case the group G is a cyclic group of order $2p$. To see this notice that H is a unique Sylow 2-group and K is a unique Sylow p-subgroup; so, by Problem 5.7, page 133, $H \lhd G$ and $K \lhd G$. Furthermore $H \cap K = \{1\}$, for any element common to H and K must have order dividing 2 and p and hence is the identity. Clearly $|H|\,|K| = 2p = |G|$. Therefore by Theorem 5.16′, $G \cong H \times K$. But H and K are cyclic groups of order 2 and p respectively. Thus by Problem 5.34, G is cyclic of order $2p$.

(ii) Let $K = gp(a)$ where $a^p = 1$. Since K is the only subgroup of order p, $b \notin K$ implies $b^2 = 1$. Clearly, $G = K \cup bK$. Hence G consists of the distinct elements

$$1, a, a^2, \ldots, a^{p-1}, \quad b, ba, ba^2, \ldots, ba^{p-1} \tag{5.6}$$

Now if $i = 0, 1, \ldots, p-1$, then

$$(ba^i)^2 = 1 \quad \text{and} \quad ba^i = a^{p-i}b \tag{5.7}$$

since $ba^i \notin K$ and each element of G outside K is of order 2. Also

$$(ba^i)^2 = (ba^i)(ba^i) = 1 \quad \text{implies} \quad ba^i = (ba^i)^{-1} = (a^i)^{-1}b^{-1} = a^{p-i}b$$

Now if $\bar{G}$ is any group of order $2p$, then it is either of type (i) or (ii). If $\bar{G}$ is of type (i), then by our analysis it must be a cyclic group of order $2p$. By Theorem 4.7, page 103, cyclic groups of the same order are isomorphic. Hence all groups of order $2p$ having property (i) are isomorphic.

Suppose $\bar{G}$ is of type (ii). Then, arguing as above, $\bar{G}$ has a subgroup $\bar{K} = gp(\bar{a})$ of order p and an element $\bar{b}$ of order 2 such that

$$\bar{G} = \{1, \bar{a}, \ldots, \bar{a}^{p-1}, \bar{b}, \bar{b}\bar{a}, \ldots, \bar{b}\bar{a}^{p-1}\}$$

where for $i = 0, 1, \ldots, p-1$,

$$(\bar{b}\bar{a}^i)^2 = 1 \quad \text{and} \quad \bar{b}\bar{a}^i = \bar{a}^{p-i}\bar{b} \tag{5.8}$$

The mapping $\alpha: G \to \bar{G}$ defined by

$$\alpha: a^i \to \bar{a}^i, \quad \alpha: b \to \bar{b}, \quad \alpha: ba^i \to \bar{b}\bar{a}^i \quad (i \text{ any integer})$$

is an isomorphism. First, α is a mapping; for if $a^i = a^j$, p divides $i - j$ and hence $\bar{a}^i = \bar{a}^j$. Consequently α is well defined on a^i. If $ba^i = ba^j$, then $a^i = a^j$ and so p divides $i - j$ and $\bar{a}^i = \bar{a}^j$. Hence α is well defined on the ba^i. As $\alpha: a^i \to \bar{a}^i, ba^i \to \bar{b}\bar{a}^i$ $(i = 0, 1, \ldots, p-1)$, α is one-to-one and onto. α is also a homomorphism, for $g_1, g_2 \in G$ implies $g_1 = b^ja^i$ and $g_2 = b^sa^t$ for some choice of $j, s \in \{0, 1\}$ and $i, t \in \{0, 1, 2, \ldots, p-1\}$. Using equations (5.7) and (5.8), we obtain, when $s = 0$,

$$(g_1g_2)\alpha = [(b^ja^i)(a^t)]\alpha = (b^ja^{i+t})\alpha = \bar{b}^j\bar{a}^{i+t} = \bar{b}^j\bar{a}^i\bar{a}^t = (b^ja^i)\alpha(a^t)\alpha = g_1\alpha g_2\alpha$$

and when $s = 1$,

$$\begin{aligned}(g_1g_2)\alpha &= [(b^ja^i)(ba^t)]\alpha = (b^jba^{p-i}a^t)\alpha = (b^{j+1}a^{p-i+t})\alpha = \bar{b}^{j+1}\bar{a}^{p-i+t} = \bar{b}^j\bar{b}\bar{a}^{p-i}\bar{a}^t \\ &= \bar{b}^j\bar{a}^i\bar{b}\bar{a}^t = (b^ja^i)\alpha(ba^t)\alpha = g_1\alpha g_2\alpha\end{aligned}$$

Therefore any two groups of order $2p$ of type (ii) are isomorphic.

So far we have shown that up to isomorphism there are at most two possible groups of order $2p$, p a prime. This does not mean that there exist two non-isomorphic groups of order $2p$, for each prime p. But from Theorem 4.7, page 103, we know that for each positive integer n there exists a cyclic group of order n, and from Section 3.4f, page 75, we know that for each n the order of the dihedral group D_n is $2n$ and D_n is not cyclic for $n > 2$ (it is not even abelian). *There are therefore, up to isomorphism, exactly two groups of order* $2p$ *for each prime* $p \neq 2$: *one a cyclic group and the other* D_p. In particular there are exactly two non-isomorphic groups of order i, $i = 6, 10, 14$.

It is worthwhile summarizing our method of finding all groups of order $2p$. We first showed that if a group had order $2p$ it had to be isomorphic to one of two possible groups. The isomorphism in case (ii) was shown by using the fact that the elements of such a group had to satisfy equations (*5.7*) and (*5.8*). (As those equations determine the group up to isomorphism, they are usually called defining relations for the groups; they will be discussed in detail in Chapter 8.) After demonstrating the isomorphism, we proved that each of the possible groups exists by exhibiting a group of each type.

c. Groups of small order: orders 8 and 9

Let G be a group of order 8. There are at least three non-isomorphic abelian groups of order 8: $C_8, C_2 \times C_4$ and $C_2 \times K_4$ (see Problem 5.33). We show that if G is abelian it is isomorphic to one of these three groups.

If G has an element of order 8, G is cyclic. If G has no element of order 8 but has an element a of order 4, let $H = gp(a)$. Let $b \in G - H$. If b is of order 4, b^2 is an element of order 2 and lies in H (since the coset decomposition of G is $H \cup bH$). As a^2 is the only element of H of order 2, $b^2 = a^2$. Hence $(ab)^2 = a^2b^2 = a^2a^2 = 1$. Since $ab \notin H$, we may assume that there exists an element $x \in G - H$ of order 2 ($x = b$ if b is of order 2 or else $x = ab$). Let $X = gp(x)$. $X \cap H = \{1\}$. Therefore by Theorem 5.16′, $G \cong X \times H$. Since $X \cong C_2$ and $H \cong C_4$, we conclude $G \cong C_2 \times C_4$.

If G has no elements of order 4 or 8, all its non-identity elements are of order 2. Let a, b be distinct elements of order 2 in G. Let $A = gp(a)$, $B = gp(b)$. Then AB is a group by Problem 4.62, page 114. Now $A \cap B = \{1\}$ and $|A|\,|B| = |AB|$. So, by Theorem 5.16′, $AB \cong A \times B$ and consequently $AB \cong C_2 \times C_2$. Let $c \in G - AB$ and $C = gp(c)$; then $C \cap AB = \{1\}$. Thus $G \cong (C_2 \times C_2) \times C_2 = K_4 \times C_2$.

We conclude that there are, up to isomorphism, exactly three abelian groups of order 8.

Assume G is a non-abelian group of order 8. G has an element of order 4 and no elements of order 8; for if $g \in G$ is of order 8, $G \cong C_8$. On the other hand if all elements of G are of order 2, then $(ab)^2 = 1$ for any $a, b \in G$ and consequently

$$ba = a^2bab^2 = a(abab)b = ab$$

contrary to our assumption that G is non-abelian. Let $a \in G$ be an element of order 4, and put $H = gp(a)$. Then $G = H \cup Hb$ for some $b \in G$. Also $H \lhd G$, as it is of index 2 (Problem 4.69, page 116). $b^2 \in H$; for if not, the cosets H, Hb, Hb^2 would be distinct, and this would contradict $[G:H] = 2$. We have four possibilities for b^2: (i) $b^2 = a$, (ii) $b^2 = a^2$, (iii) $b^2 = a^3$, or (iv) $b^2 = 1$.

If (i) or (iii) occurs, clearly $G = gp(b)$, contrary to our assumption. Thus (ii) and (iv) are the only possibilities.

(ii) $b^2 = a^2$. Since $H \lhd G$, $b^{-1}ab \in H$. As a is of order 4, so is $b^{-1}ab$. Thus $b^{-1}ab = a$ or a^3. If $b^{-1}ab = a$, then $ab = ba$. But every element of G can be written as a^ib or a^i for some integer i, since $G = H \cup Hb$. Hence $ab = ba$ implies G is abelian, contrary to our assumption. Thus $b^{-1}ab = a^3$ or

$$ab = ba^3 \tag{5.9}$$

Since $G = H \cup Hb$, the elements of G can be expressed as $1, a, a^2, a^3, b, ab, a^2b, a^3b$. If a group of this type actually exists, then equation (*5.9*), $b^2 = a^2$ and $a^4 = 1$ provide us with enough information to construct its multiplication table.

	1	a	a^2	a^3	b	ab	a^2b	a^3b
1	1	a	a^2	a^3	b	ab	a^2b	a^3b
a	a	a^2	a^3	1	ab	a^2b	a^3b	b
a^2	a^2	a^3	1	a	a^2b	a^3b	b	ab
a^3	a^3	1	a	a^2	a^3b	b	ab	a^2b
b	b	a^3b	a^2b	ab	a^2	a	1	a^3
ab	ab	b	a^3b	a^2b	a^3	a^2	a	1
a^2b	a^2b	ab	b	a^3b	1	a^3	a^2	a
a^3b	a^3b	a^2b	ab	b	a	1	a^3	a^2

Table 5.1

To calculate the products in the table, we used the fact that $ab = ba^3$ and $b^2 = a^2$ imply $ba = a^3b$ since $a^3b = a^2(ba^3) = b^3a^3 = b(a^2a^3) = ba$.

If $\bar{G}$ is another non-abelian group of order 8 with an element $\bar{a}$ of order 4 and an element $\bar{b} \notin gp(\bar{a})$ such that $\bar{b}^2 = \bar{a}^2$, then as in our argument above,

$$\bar{G} = \{\bar{1}, \bar{a}, \bar{a}^2, \bar{a}^3, \bar{b}, \bar{a}^2\bar{b}, \bar{a}\bar{b}, \bar{a}\bar{b}^3\}$$

with the elements satisfying the equations

$$\bar{a}^4 = 1, \quad \bar{a}^2 = \bar{b}^2, \quad \bar{a}\bar{b} = \bar{b}\bar{a}^3$$

from which we find a multiplication table for $\bar{G}$ which is identical to Table 5.1 except that $\bar{a}$ is substituted for a and $\bar{b}$ for b. The mapping $\alpha: G \to \bar{G}$ defined by $\alpha: a \to \bar{a}$ and $\alpha: a^ib \to \bar{a}^i\bar{b}$, $i = 0, 1, 2, 3$, is clearly an isomorphism.

Table 5.1 also shows that a group of order 8 of this type actually does exist, for the table *defines* a group. To see this, notice that the product of any two elements is again an element, i.e. the table defines a binary operation, 1 is an identity element, and every element has an inverse. The only difficulty is checking that the binary operation is associative. This involves much calculation (the reader should check some of the calculations himself). We shall give another description of this group in Problem 5.40. This group is called the Quaternion group and has the interesting property that all its subgroups are normal and yet it itself is not abelian (Problem 5.43).

We now move on to a discussion of (iv).

(iv) $b^2 = 1$. Let $K = gp(b)$; then $H \cap K = \{1\}$ and $G = HK$. Now $H \lhd G$ so that $b^{-1}ab \in H$ and, since a is of order 4, we have $b^{-1}ab = a$ or a^3. As in (ii), $b^{-1}ab = a$ implies G is abelian. Hence $b^{-1}ab = a^3$, which leads to

$$ba = a^3b \tag{5.10}$$

The elements $1, a, a^2, a^3, b, ab, a^2b, a^3b$ are the distinct elements of G. Equation (*5.10*), $b^2 = 1$ and $a^4 = 1$ enable us to construct the following multiplication table.

	1	a	a^2	a^3	b	ab	a^2b	a^3b
1	1	a	a^2	a^3	b	ab	a^2b	a^3b
a	a	a^2	a^3	1	ab	a^2b	a^3b	b
a^2	a^2	a^3	1	a	a^2b	a^3b	b	ab
a^3	a^3	1	a	a^2	a^3b	b	ab	a^2b
b	b	a^3b	a^2b	ab	1	a^3	a^2	a
ab	ab	b	a^3b	a^2b	a	1	a^3	a^2
a^2b	a^2b	ab	b	a^3b	a^2	a	1	a^3
a^3b	a^3b	a^2b	ab	b	a^3	a^2	a	1

Table 5.2

As in part (ii), any non-abelian group of order 8 having property (iv) is isomorphic to G. Such a group exists, for Table 5.2 also defines a group. This group is isomorphic to the dihedral group D_4, the group of symmetries of a square (see Problem 5.38).

The two groups given in Tables 5.1 and 5.2 are not isomorphic because the group of Table 5.1 has exactly one element a^2 of order 2, whereas the group of Table 5.2 has five elements of order 2: a^2, b, ab, a^2b and a^3b. Thus we have shown that there are exactly two non-isomorphic non-abelian groups of order 8.

To summarize, *there are five non-isomorphic groups of order 8, three abelian and two non-abelian.*

Since $9 = 3^2$, we know by Problem 5.36 that *there are two and only two non-isomorphic groups of order 9, namely C_9 and $C_3 \times C_3$.*

d. Groups of small order: orders 12 and 15

To complete our list of all groups up to order 15, we must find all possible groups of order 12 and 15. Because $12 = 3 \cdot 2^2$, we know that a group G of order 12 has at least one Sylow 2-subgroup of order 2^2 and at least one Sylow 3-subgroup of order 3. The third Sylow theorem tells us that the number of Sylow 2-subgroups is congruent to one modulo 2 (i.e. $s_2 = 1 + 2k$ for some integer k) and s_2 divides $|G|$. When $k = 0$, $s_2 = 1$; and when $k = 1$, $s_2 = 3$. If $k > 1$ it is clear that $1 + 2k$ does not divide 12. We therefore have two possibilities: G has exactly one Sylow 2-subgroup or G has exactly three Sylow 2-subgroups. A similar argument shows that G has exactly one Sylow 3-subgroup or G has exactly four Sylow 3-subgroups. Therefore we have four possibilities:

(i) $s_2 = 1$ and $s_3 = 1$

(ii) $s_2 = 1$ and $s_3 = 4$

(iii) $s_2 = 3$ and $s_3 = 1$

(iv) $s_2 = 3$ and $s_3 = 4$

Notice that because the Sylow 2-subgroup has order 4 it must be isomorphic to C_4 or K_4, and the Sylow 3-subgroup must be isomorphic to C_3 (Section 5.3b). We treat each case separately.

(i) Let F be the Sylow 2-subgroup and T the Sylow 3-subgroup of G. Then $F \lhd G$ and $T \lhd G$, since a Sylow p-subgroup is a normal subgroup if it is unique (Problem 5.7, page 133). Furthermore, $F \cap T = \{1\}$ since any element in the intersection must have order dividing 3 and 4 and so must be the identity. Moreover, $|F|\,|T| = 12$. Hence by Theorem 5.16′, $G \cong F \times T$. We have two possibilities for F: (a) $F \cong C_4$ or (b) $F \cong K_4$. Thus $G \cong C_4 \times C_3 \cong C_{12}$ (by Problem 5.34) or $G \cong K_4 \times C_3$; both these groups are abelian.

These are the only abelian groups of order 12, for if G is abelian any two conjugate subgroups are equal. Hence by Theorem 5.3, page 131, $s_2 = s_3 = 1$.

In cases (ii) through (iv) we assume G is a non-abelian group. Furthermore, let F be any Sylow 2-subgroup of G and T any Sylow 3-subgroup of G. Then $F \cap T = \{1\}$ and, by Proposition 5.18, page 145, $|FT| = |F|\,|T|/|F \cap T| = |F|\,|T| = |G|$. Thus $G = FT$. If $ft = tf$ for all $f \in F$ and $t \in T$, then G is abelian since $g_1, g_2 \in G$ implies $g_1 = f_1t_1$ and $g_2 = f_2t_2$ for some $f_1, f_2 \in F$ and $t_1, t_2 \in T$. Now as F and T are both abelian groups,

$$g_1g_2 \;=\; f_1t_1f_2t_2 \;=\; f_2t_2f_1t_1 \;=\; g_2g_1$$

Hence we also assume if F is any Sylow 2-subgroup of G and T any Sylow 3-subgroup of G, that it is not the case that $ft = tf$ for all $f \in F$ and $t \in T$.

(ii) $s_2 = 1$ and $s_3 = 4$. Let F be the (unique) Sylow 2-subgroup and T be a Sylow 3-subgroup. There are two possibilities: (a) $F \cong C_4$ and (b) $F \cong K_4$. We treat each case separately.

(a) Let $F = \{1, a, a^2, a^3\}$ where $a^4 = 1$, and $T = \{1, b, b^2\}$ where $b^3 = 1$. $s_2 = 1$ implies $F \lhd G$. Thus $b^{-1}ab \in F$. If $b^{-1}ab = a$, then $ab = ba$. But this implies every element of F commutes with every element of T, contrary to our assumption. Therefore since a has order 4, $b^{-1}ab \neq a^2$ and $b^{-1}ab \neq 1$ so that $b^{-1}ab = a^3$ or $ab = ba^3$. We show that under these assumptions $gp(ba) = G$. $(ba)^2 = b(ab)a = b^2a^4 = b^2$. So $(ba)^3 = (ba)^2ba = b^2ba = a$. Hence $gp(ba)$ contains a and b and thus coincides with G. Then G is cyclic, contrary to assumption. There is therefore no non-abelian group of order 12 with $s_2 = 1$, $s_3 = 4$ and Sylow 2-subgroup isomorphic to C_4.

(b) Let $F = \{1, x, y, z\}$ be the four group as given in Section 5.3b, and $T = \{1, c, c^2\}$ where $c^3 = 1$. As in part (a), $F \lhd G$ so that $c^{-1}fc \in F$ for all $f \in F$. Now by assumption $c^{-1}fc \neq f$ for at least one $f \in F$. Suppose $c^{-1}xc \neq x$ (the other cases are similar). We may assume $c^{-1}xc = y$. (The other case, $c^{-1}xc = z$, is argued similarly.) Let us, as in Section 5.3b, put $x = a$, $y = b$ and $z = ab$. Then $ac = cb$, which implies $a = cbc^{-1}$. Now $c^{-1}bc \neq a$, for $c^{-1}bc = a$ implies $c^{-1}bc = cbc^{-1}$ or $b = c^2bc^{-2}$. Then, as $c^2 = c^{-1}$ and $c^{-2} = c$, $b = c^{-1}bc$. Hence $a = b$, a contradiction. Similarly $c^{-1}bc \neq b$ and $c^{-1}bc \neq 1$. Then $c^{-1}bc = ab$ and $c^{-1}(ab)c = (c^{-1}ac)(c^{-1}bc) = bab = b^2a = a$. Consequently the equations

$$ac = cb, \quad bc = cab, \quad abc = ca, \quad a^2 = b^2 = 1, \quad c^3 = 1, \quad ab = ba \tag{5.11}$$

hold in G. Now $1, c, c^2$ determine distinct cosets of F in G. Therefore the elements of G are

$$1,\ c,\ c^2,\ a,\ b,\ ab,\ ca,\ cb,\ cab,\ c^2a,\ c^2b,\ c^2ab$$

Using equations *(5.11)*, we can write down the multiplication table for G, as shown in Table 5.3 below.

	1	c	c^2	a	b	ab	ca	cb	cab	c^2a	c^2b	c^2ab
1	1	c	c^2	a	b	ab	ca	cb	cab	c^2a	c^2b	c^2ab
c	c	c^2	1	ca	cb	cab	c^2a	c^2b	c^2ab	a	b	ab
c^2	c^2	1	c	c^2a	c^2b	c^2ab	a	b	ab	ca	cb	cab
a	a	cb	c^2ab	1	ab	b	cab	c	ca	c^2b	c^2a	c^2
b	b	cab	c^2a	ab	1	a	cb	ca	c	c^2	c^2ab	c^2b
ab	ab	ca	c^2b	b	a	1	c	cab	cb	c^2ab	c^2	c^2a
ca	ca	c^2b	ab	c	cab	cb	c^2ab	c^2	c^2a	b	a	1
cb	cb	c^2ab	a	cab	c	ca	c^2b	c^2a	c^2	1	ab	b
cab	cab	c^2a	b	cb	ca	c	c^2	c^2ab	c^2b	ab	1	a
c^2a	c^2a	b	cab	c^2	c^2ab	c^2b	ab	1	a	cb	ca	c
c^2b	c^2b	ab	ca	c^2ab	c^2	c^2a	b	a	1	c	cab	cb
c^2ab	c^2ab	a	cb	c^2b	c^2a	c^2	1	ab	b	cab	c	ca

Table 5.3

By a similar argument to that used in the discussion of the non-abelian groups of order 8, any group of order 12 with $s_2 = 1$ and $s_3 = 3$ is isomorphic to G. Moreover, Table 5.3 defines a group: the table defines a binary operation, the identity is 1, and every element clearly has an inverse. The associativity of the operation must also be checked, an even more tedious task than in the case of a group of order 8. The alternating group A_4 is a group of this type (Problem 5.38).

(iii) $s_2 = 3$ and $s_3 = 1$. Let F be a Sylow 2-subgroup and $T = \{1, c, c^2\}$ $(c^3 = 1)$ be the Sylow 3-subgroup. Again we have two possibilities: (*a*) $F = \{1, a, a^2, a^3\} \cong C_4$ and (*b*) $F = \{1, x, y, z\} \cong K_4$.

(*a*) T, being a unique Sylow 3-subgroup, is normal in G and so $a^{-1}ca \in T$. We may assume $a^{-1}ca \neq c$, otherwise the group is abelian. Hence $a^{-1}ca = c^2$ and $ca = ac^2$. Also, $c^2a = cac^2 = ac^4 = ac$. The equations which determine a multiplication table for this group are

$$ca = ac^2, \quad c^2a = ac, \quad c^3 = 1, \quad a^4 = 1 \tag{5.12}$$

The distinct elements of G are then

$$1,\ a,\ a^2,\ a^3,\ c,\ c^2,\ ac,\ a^2c,\ a^3c,\ ac^2,\ a^2c^2,\ a^3c^2$$

and we obtain Table 5.4 below.

We conclude, by an argument similar to that used in the discussion of the non-abelian groups of order 8, that any group of order 12 with $s_2 = 3$, $s_3 = 1$ and in which the Sylow 2-subgroups are cyclic of order 4, is isomorphic to the group G defined in the table. Again it can be checked that the table defines a group, so that a group of this type exists. We have as yet not encountered an example of this type of group, but in Problem 5.41 we show that there is a group of 2 by 2 matrices which is isomorphic to G.

	1	a	a^2	a^3	c	c^2	ac	a^2c	a^3c	ac^2	a^2c^2	a^3c^2
1	1	a	a^2	a^3	c	c^2	ac	a^2c	a^3c	ac^2	a^2c^2	a^3c^2
a	a	a^2	a^3	1	ac	ac^2	a^2c	a^3c	c	a^2c^2	a^3c^2	c^2
a^2	a^2	a^3	1	a	a^2c	a^2c^2	a^3c	c	ac	a^3c^2	c^2	ac^2
a^3	a^3	1	a	a^2	a^3c	a^3c^2	c	ac	a^2c	c^2	ac^2	a^2c^2
c	c	ac^2	a^2c	a^3c^2	c^2	1	a	a^2c^2	a^3	ac	a^2	a^3c
c^2	c^2	ac	a^2c^2	a^3c	1	c	ac^2	a^2	a^3c^2	a	a^2c	a^3
ac	ac	a^2c^2	a^3c	c^2	ac^2	a	a^2	a^3c^2	1	a^2c	a^3	c
a^2c	a^2c	a^3c^2	c	ac^2	a^2c^2	a^2	a^3	c^2	a	a^3c	1	ac
a^3c	a^3c	c^2	ac	a^2c^2	a^3c^2	a^3	1	ac^2	a^2	c	a	a^2c
ac^2	ac^2	a^2c	a^3c^2	c	a	ac	a^2c^2	a^3	c^2	a^2	a^3c	1
a^2c^2	a^2c^2	a^3c	c^2	ac	a^2	a^2c	a^3c^2	1	ac^2	a^3	c	a
a^3c^2	a^3c^2	c	ac^2	a^2c	a^3	a^3c	c^2	a	a^2c^2	1	ac	a^2

Table 5.4

(b) $F = \{1, x, y, z\}$ and $T = \{1, c, c^2\}$. Since $T \lhd G$, we have $f^{-1}cf \in T$ for all $f \in F$. By assumption, for at least one $f \in F$, $f^{-1}cf \neq c$. We may therefore assume, without loss of generality, that $x^{-1}cx = c^2$. Again, as in Section 5.3b, put $x = a$, $y = b$ and $z = ab$. Then $ca = ac^2$. Note that $c^2a = c(ca) = c(ac^2) = (ca)c^2 = ac^2c^2 = ac$, i.e. $c^2a = ac$.

We claim that $S = \{1, c, c^2, a, ca, c^2a\}$ is a subgroup. We leave to the reader the task of checking that the product of any two elements in S is again in S ($ca = ac^2$ and $c^2a = ac$ make this task easy). The identity 1 is in S, and on inspection we find every element in S has an inverse in S: $c^{-1} = c^2$, $(c^2)^{-1} = c$, $a^{-1} = a$, $(ca)^{-1} = ca$, $(c^2a)^{-1} = c^2a$. Hence S is a subgroup of G. $|S| = 6$ for $c^ia = c^j$ ($i = 1$ or 2; and $j = 0, 1$ or 2) implies $a \in T$, a contradiction; $c^ia = a$ ($i = 1$ or 2) implies $c^i = 1$, a contradiction; and $ca = c^2a$ implies $c = 1$, a contradiction. S is non-abelian ($ac = c^2a$ is not equal to ca) and hence is isomorphic to D_3 since there is only one non-abelian group of order 6 (up to isomorphism). Now $[G:S] = 2$ implies $S \lhd G$ (by Problem 4.69, page 116). Hence $b^{-1}cb \in S$. As $b^{-1}cb$ is an element of order 3, it is either c or c^2, as all other elements of S are of order 2. We shall now choose an element $h \in F$, $h \notin S$, such that $h^{-1}ch = c$. If $b^{-1}cb = c$, let $h = b$. If on the other hand $b^{-1}cb = c^2$, let $h = ab$. Recall that $a^{-1}ca = c^2$. Then $(ab)^{-1}c(ab) = b^{-1}(a^{-1}ca)b = b^{-1}c^2b = b^{-1}cb \cdot b^{-1}cb = c^2 \cdot c^2 = c$. Hence there exists an element $h \notin S$ in F (i.e. b or ab) such that $h^{-1}ch = c$. Consider $H = gp(h)$. Clearly $S \cap H = \{1\}$, S and H commute elementwise, and $|S|\,|H| = |G|$, and so $G \cong S \times H$ by Theorem 5.16. But $S \cong D_3$ and $H \cong C_2$. We therefore conclude that any group G with $s_2 = 3$, $s_3 = 1$ and Sylow 2-subgroup isomorphic to K_4, is isomorphic to $D_3 \times C_2$. The dihedral group D_6 is a group of this type (Problem 5.38).

(iv) $s_2 = 3$, $s_3 = 4$. Since distinct cyclic groups of order 3 intersect in the identity element, the four Sylow 3-subgroups have together 9 distinct elements. A Sylow 2-subgroup is of order 4. Since the intersection of a group of order 4 and a group of order 3 can only be the identity, it follows that the number of distinct elements in the 4 3-Sylow subgroups and a single 2-Sylow subgroup is 12.

But $|G| = 12$, so there cannot be another distinct Sylow 2-subgroup. Thus there is no group of type (iv).

To summarize, we have shown that there are up to isomorphism exactly three non-abelian groups of order 12. They are the groups

(ii) (*b*) with $s_2 = 1$. $s_3 = 4$ and the Sylow 2-subgroup $\cong K_4$; see Table 5.3. (Such a group is isomorphic to A_4.)

(iii) (*a*) with $s_2 = 3$, $s_3 = 1$ and the Sylow 2-subgroup $\cong C_4$; see Table 5.4. (Such a group is isomorphic to a group of 2 by 2 matrices given in Problem 5.41.)

(*b*) with $s_2 = 3$, $s_3 = 1$ and the Sylow 2-subgroup $\cong K_4$. (Such a group is isomorphic to $D_3 \times C_2$ which is isomorphic to D_6. See Problem 5.38.)

Clearly no two of these groups are isomorphic. The abelian groups of order 12 are $K_4 \times C_3$ and C_{12}. *Thus including the abelian groups, there are five non-isomorphic groups of order 12.*

If G is of order 15, we have seen that G is cyclic (Section 5.1a). The following table gives the number of non-isomorphic groups of order 1 through 15.

Order of group	1	2	3	4	5	6	7	8	9	10	11	12	13	14	15
No. of groups	1	1	1	2	1	2	1	5	2	2	1	5	1	2	1

The reader will agree that finding all groups of a given order is difficult. Indeed there is not even a general method of determining how many non-isomorphic groups of a given order there can be.

Problems

5.38. Show that

(i) The dihedral group D_4 is a group of order 8 isomorphic to the group given in Table 5.2, page 152.

(ii) The alternating group A_4 is isomorphic to the group given in Table 5.3.

(iii) The dihedral group D_6 is isomorphic to $D_3 \times C_2$.

Solution:

(i) D_4 is a non-abelian group of order 8 and as such is isomorphic to one of the groups given in Section 5.3c, Tables 5.1 and 5.2. A check of Table 5.1 shows that there is only one element of order 2, namely a^2. But D_4 is the symmetry group of the square (Section 3.4f, page 75). A reflection τ is of order 2 and if σ is a rotation of 90°, $\tau\sigma$ is of order 2 as can be seen in the discussion of these groups in Chapter 3. Therefore D_4 must be isomorphic to the group given in Table 5.2.

(ii) A_4 is a non-abelian group of order 12. Hence it is either isomorphic to the group of Table 5.3 or 5.4 or to $D_3 \times C_2$ (see page 155). As can be seen from the multiplication table for A_4 given in Chapter 3, page 63, A_4 has exactly three elements of order 2, namely σ_2, σ_5 and σ_8. Now the group of Table 5.4 has only one element a^2 of order 2, so that it could not be isomorphic to A_4. We have shown in Problem 5.1, page 131, that A_4 has no subgroup of order 6 and D_3 is isomorphic to a subgroup of $D_3 \times C_2$ by Theorem 5.14. Hence A_4 is not isomorphic to $D_3 \times C_2$. Thus it must be isomorphic to the group given in Table 5.3.

(iii) D_6 is of order 12, and is not abelian. D_6 is the symmetry group of the hexagon and therefore has a subgroup of order 6, namely the rotation of 60° about the center. So it is not the case that $D_6 \cong A_4$, and hence D_6 cannot be isomorphic to the group of Table 5.3. Also, a reflection followed by a rotation is an element of order 2. Since there are six such elements in D_6, it cannot be isomorphic to the group of Table 5.4 as this group has only one element of order 2. The only other possibility is that $D_6 \cong D_3 \times C_2$.

5.39. Find a cyclic subgroup of order 6 in $D_3 \times C_2$.

Solution:

Let $a \in D_3$ be of order 3 and $b\,(\neq 1) \in C_2$. Consider the element $(a, b) \in D_3 \times C_2$. $(a, b)^2 = (a^2, 1)$, $(a, b)^3 = (1, b)$, $(a, b)^4 = (a, 1)$, $(a, b)^5 = (a^2, b)$, and $(a, b)^6 = (1, 1)$. Hence $gp((a, b))$ is a cyclic subgroup of $D_3 \times C_2$ of order 6.

5.40. Consider the matrices $A = \begin{pmatrix} 0 & i \\ i & 0 \end{pmatrix}$ and $B = \begin{pmatrix} 0 & 1 \\ -1 & 0 \end{pmatrix}$ where $i = \sqrt{-1}$. A and B have nonzero determinants and thus are elements in the group of all 2×2 matrices with nonzero determinants. Show that $gp(A, B)$ is a group of order 8 which is isomorphic to the quaternion group (Table 5.1, page 151).

Solution:

By direct calculation we find

$$A^2 = \begin{pmatrix} -1 & 0 \\ 0 & -1 \end{pmatrix} \qquad A^3 = \begin{pmatrix} 0 & -i \\ -i & 0 \end{pmatrix} \qquad A^4 = \begin{pmatrix} 1 & 0 \\ 0 & 1 \end{pmatrix} = I, \text{ (identity matrix)}$$

$$B^2 = \begin{pmatrix} -1 & 0 \\ 0 & -1 \end{pmatrix} \qquad AB = \begin{pmatrix} -i & 0 \\ 0 & i \end{pmatrix} \qquad A^2B = \begin{pmatrix} 0 & -1 \\ 1 & 0 \end{pmatrix} \qquad A^3B = \begin{pmatrix} i & 0 \\ 0 & -i \end{pmatrix}$$

$$B^2A = A^3, \quad B^3A = AB \quad \text{and} \quad A^3B = BA$$

Let $G = \{I, A^2, A^3, B, AB, A^2B, A^3B\}$. We claim that $G = \{I, A, A^2, A^3, B, AB, A^2B, A^3B\} = gp(A, B)$. Clearly, $G \subseteq gp(A, B)$. To show $G = gp(A, B)$, we need only show G is a group, as $A, B \in G$. Note G is a subset of the group of 2×2 matrices with nonzero determinant. Hence the elements of G satisfy the associative law. By direct calculation we can show that G is closed under matrix multiplication; the equation $BA = A^3B$ simplifies the calculations, e.g.,

$$(A^2B)(A^3B) = A^2(BA)A^2B = A^2A^3BA^2B = AA^3BAB = A^3B^2 = A$$

Furthermore every element of G has an inverse in G, e.g. using $B^3A = AB$ we have

$$(A^3B)^{-1} = B^{-1}A^{-3} = B^3A = AB$$

Checking all these details enables us to conclude that G is a group of order 8 and $G = gp(A, B)$. G is non-abelian, since $AB \neq BA$, so G is either isomorphic to the group of Table 5.1 or Table 5.2. Because G has only one element of order 2, it cannot be isomorphic to the group of Table 5.2. Thus G is isomorphic to the quaternion group of Table 5.1.

5.41. Consider the matrices $A = \begin{pmatrix} 0 & i \\ i & 0 \end{pmatrix}$ and $B = \begin{pmatrix} \epsilon & 0 \\ 0 & \epsilon^2 \end{pmatrix}$, where $i = \sqrt{-1}$ and ϵ is a nonreal complex cube root of 3 (so, in particular, $\epsilon^3 = 1$ and $\epsilon \neq 1$). A and B have nonzero determinants and thus are elements in the group of all 2×2 matrices with nonzero determinants (see Section 3.5b, page 81). Show that $gp(A, B)$ is a group of order 12 which is isomorphic to the group given in Table 5.4, page 155.

Solution:

We find by direct calculation that

$$A^2 = \begin{pmatrix} -1 & 0 \\ 0 & -1 \end{pmatrix} \qquad B^2 = \begin{pmatrix} \epsilon^2 & 0 \\ 0 & \epsilon \end{pmatrix} \qquad A^2B = \begin{pmatrix} -\epsilon & 0 \\ 0 & -\epsilon^2 \end{pmatrix} \qquad A^2B^2 = \begin{pmatrix} -\epsilon^2 & 0 \\ 0 & -\epsilon \end{pmatrix}$$

$$A^3 = \begin{pmatrix} 0 & -i \\ -i & 0 \end{pmatrix} \qquad B^3 = \begin{pmatrix} 1 & 0 \\ 0 & 1 \end{pmatrix} \qquad A^3B = \begin{pmatrix} 0 & -i\epsilon^2 \\ -i\epsilon & 0 \end{pmatrix} \qquad A^3B^2 = \begin{pmatrix} 0 & -i\epsilon \\ -i\epsilon^2 & 0 \end{pmatrix}$$

$$A^4 = \begin{pmatrix} 1 & 0 \\ 0 & 1 \end{pmatrix} \qquad AB = \begin{pmatrix} 0 & i\epsilon^2 \\ i\epsilon & 0 \end{pmatrix} \qquad AB^2 = \begin{pmatrix} 0 & i\epsilon \\ i\epsilon^2 & 0 \end{pmatrix}$$

Let $H = \{A, A^2, A^3, A^4, B, B^2, AB, A^2B, A^3B, AB^2, A^2B^2, A^3B^2\}$. We claim $H = gp(A, B)$. Clearly $H \subseteq gp(A, B)$. To prove $H = gp(A, B)$ we need only show H is a group, as $A, B \in H$. Note that H is a subset of the 2×2 matrices with nonzero determinant. Hence the elements of H satisfy the associative law. To check that H is closed under matrix multiplication, first note that $A^{-1}BA = B^2$. Then, for example,

$$(A^2B)(A^3B) = A^2A \cdot (A^{-1}BA)A^2B = A^3B^2A^2B = A^4(A^{-1}BA)(A^{-1}BA)AB$$

$$= B^2B^2AB = BAB = AA^{-1}BAB = AB^2B = A$$

Also the inverse of, for example, (A^3B) is given by

$$(A^3B)^{-1} = B^2A = AA^{-1}B^2A = A(A^{-1}BA)(A^{-1}BA) = AB^2B^2 = AB \in H$$

Checking all these details enables us to conclude that H is a group of order 12 and $H = gp(A, B)$. The mapping $\alpha: a \to A$, and $c \to B$ is an isomorphism of the group of Table 5.4 and H. This obtains because H satisfies the equations

$$BA = AB^2, \quad B^2A = AB, \quad B^3 = I, \quad A^4 = I$$

(I the identity matrix). These are the exact counterparts of equations (*5.12*), page 154. Consequently the multiplication table for H is obtained from that for the group of Table 5.4 by renaming via α.

5.42. Show that a group G of order 48 has a normal subgroup $\neq \{1\}$ or G. (Very difficult.)

Solution:

By the first Sylow theorem, G has a Sylow 2-subgroup of order 16. By the third Sylow theorem, $s_2 = 1 + k2$ for some integer k and $s_2 \mid 48$. The only odd divisor of 48 is 3, hence $s_2 = 1$ or 3. If $s_2 = 1$, then the Sylow 2-subgroup is unique and therefore normal (Problem 5.7, page 133). Suppose $s_2 = 3$. Let H and K be two of the Sylow 2-subgroups. As $H \cap K$ is a proper subgroup of H, $|H \cap K| \mid 16$. Then $|H \cap K| = 8$; for if $|H \cap K| \leqq 4$, then, by Proposition 5.18, page 145, $|HK| = |H|\,|K|/|H \cap K| \geqq 16 \times 16/4 = 64$, which contradicts our assumption that $|G| = 48$. Since both H and K are of order 16, $H \cap K$, as a subgroup of index 2, is normal in both H and K (Problem 4.69, page 116). Hence $H \subseteq N_G(H \cap K)$ and $K \subseteq N_G(H \cap K)$. Letting $N = N_G(H \cap K)$, we have $HK \subseteq N$. Thus $|N| \geqq |HK| = |H|\,|K|/|H \cap K| = 32$. As $|N|$ divides 48 and $|N| \geqq 32$, $|N| = 48$ and so $N = G$. Because a group is normal in its normalizer, we have $H \cap K \lhd G$.

5.43. Show that all subgroups of the quaternion group G are normal subgroups. (G is given in Table 5.1, page 151.)

Solution:

It is sufficient to check that the cyclic groups are normal in G. For if S is any subgroup, $s \in S$, and $x \in G$, then $x^{-1}sx \in gp(s)$ implies $x^{-1}sx \in S$.

Using the multiplication table we can check that for $x = a$ or b and any $s \in S$, $x^{-1}sx \in gp(s)$. (We leave this check to the reader.)

This is sufficient to prove the result, for every element of G is of the form a^ib or a^i for $i = 0, 1, 2, 3$.

5.4 SOLVABLE GROUPS

a. Definition of solvable groups

To introduce our concepts we will begin with an example. If P is a group of order p^r where p is a prime, then we showed that P has a series of subgroups P_i with

$$\{1\} = P_0 \subseteq P_1 \subseteq \cdots \subseteq P_r = P \tag{5.13}$$

where each $P_i \lhd P_{i+1}$ and $[P_{i+1}: P_i] = p$ for each integer $i = 0, \ldots, r-1$ (see equation (*5.4*), page 139).

Let G be a group and suppose it has a series of subgroups

$$\{1\} = G_0 \subseteq G_1 \subseteq \cdots \subseteq G_r = G \tag{5.14}$$

If each $G_i \lhd G_{i+1}$ for $i = 1, \ldots, r-1$, then (*5.14*) is called a *subnormal series of* (*for*) G. With this definition, (*5.13*) is a subnormal series of P.

If (*5.14*) is a subnormal series for G and $[G_{i+1}: G_i]$ is some prime (dependent on i), for $i = 0, \ldots, r-1$, G is called a *solvable group* and (*5.14*) is called a *solvable series* for G. Accordingly we conclude that P is solvable and that (*5.13*) is a solvable series for P.

If (*5.14*) is a subnormal series for G and G_{i+1}/G_i is *simple*, i.e. G_{i+1}/G_i has no normal subgroups other than G_{i+1}/G_i and the identity, for $i = 0, \ldots, r-1$, then (*5.14*) is said to be a *composition series* for G. To see that (*5.13*) is a composition series for P, note that P_{i+1}/P_i is a cyclic group of order p and hence is simple. We call the factor groups G_i/G_{i+1} of the subnormal series (*5.14*) the *factors* of (*5.14*).

We shall discuss composition series in greater detail in Section 5.5. We remark that not all groups have a composition series but finite groups do (Section 5.5a). Our main concern in this section is the concept of solvable group.

Historically, solvable groups arose in the attempt to find a formula for the roots of an nth degree polynomial

$$f(x) = a_n x^n + a_{n-1}x^{n-1} + \cdots + a_1 x + a_0 \qquad (5.15)$$

in terms of the coefficients a_i. The formula sought was one which involved the coefficients $a_0, \ldots, a_n$ of the polynomial, integers, and the operations addition, subtraction, multiplication and division, and a finite number of extraction of roots. For example, if $n = 2$, then $x = \dfrac{-a_1 \pm \sqrt{a_1^2 - 4a_2a_0}}{2a_2}$ is a formula giving the roots of (*5.15*). If the roots of $f(x)$ can be obtained by such a formula, we say $f(x) = 0$ is *solvable by radicals*.

From the Fundamental Theorem of Algebra we know that an nth degree polynomial with complex coefficients has n complex roots. Let F be the "smallest" field (see Section 3.6b, page 86, for a definition of field) of complex numbers containing the coefficients a_i of $f(x)$. By saying F is the *smallest field* we mean that if H is a field containing the coefficients a_i, then $F \subseteq H$. Let E be the smallest field containing F and the roots of $f(x)$. Now the set of automorphisms of E forms a group under the composition of mappings (see Theorem 3.15, page 87). The automorphisms of E which map every element $f \in F$ onto itself is a subgroup $\mathcal{G}$ of the group of all automorphisms of E. The group $\mathcal{G}$ is called the Galois group of the polynomial $f(x)$. In the beginning of the 19th century, the French mathematician E. Galois proved (essentially) the following extraordinary theorem: An equation $f(x) = 0$ is solvable by radicals if and only if the Galois group of $f(x)$ is solvable. It turns out that not all equations of degree ≥ 5 are solvable by radicals because the symmetric group S_n is not solvable for $n \geq 5$. (For details see Birkhoff and MacLane, *A Survey of Modern Algebra*, Macmillan, 1953.) In Section 5.5e we will prove that S_n is not solvable.

Problems

5.44. Show that the symmetric group S_n is solvable for $n = 1, 2, 3$.

Solution:

$S_1 = \{\iota\}$; then S_1 has the solvable series $\{\iota\} \subseteq S_1$ and is therefore solvable.

$$S_2 = \left\{\begin{pmatrix}1 & 2\\1 & 2\end{pmatrix}, \begin{pmatrix}1 & 2\\2 & 1\end{pmatrix}\right\}$$

then $\{\iota\} \subseteq S_2$ is a solvable series for S_2 and so S_2 is solvable.

Let $S_3 = \{\iota, \sigma_1, \sigma_2, \tau_1, \tau_2, \tau_3\}$ where

$$\iota = \begin{pmatrix}1 & 2 & 3\\1 & 2 & 3\end{pmatrix} \qquad \sigma_2 = \begin{pmatrix}1 & 2 & 3\\3 & 1 & 2\end{pmatrix} \qquad \tau_2 = \begin{pmatrix}1 & 2 & 3\\3 & 2 & 1\end{pmatrix}$$

$$\sigma_1 = \begin{pmatrix}1 & 2 & 3\\2 & 3 & 1\end{pmatrix} \qquad \tau_1 = \begin{pmatrix}1 & 2 & 3\\1 & 3 & 2\end{pmatrix} \qquad \tau_3 = \begin{pmatrix}1 & 2 & 3\\2 & 1 & 3\end{pmatrix}$$

Now $H = \{\iota, \sigma_1, \sigma_2\}$ is a cyclic subgroup of S_3. Also, $[S_3 : H] = 2$. Hence by Problem 4.69, page 116, $H \triangleleft S_3$. Thus $\{\iota\} \subseteq H \subseteq S_3$ is a solvable series for S_3, since $[H : \{\iota\}] = 3$ and $[S_3 : H] = 2$, and so S_3 is solvable.

5.45. Show that S_4 is solvable.

Solution:

The alternating group A_4 is a subgroup of order 12 in S_4. Then $[S_4 : A_4] = 2$ and $A_4 \triangleleft S_4$ by Problem 4.69, page 116. We have seen in Problem 5.1, page 131, that A_4 has no subgroup of order 6. Now A_4 is a group with a unique Sylow 2-subgroup F of order 4 and $F \cong K_4$, the four group (see Problem 5.38(ii), page 156, and Section 5.3d, page 153). Since F is a unique Sylow 2-subgroup, $F \triangleleft A_4$ and $[A_4 : F] = 3$. F, being a four group, has a normal subgroup K of order 2. Accordingly,

$$\{1\} \subseteq K \subseteq F \subseteq A_4 \subseteq S_4$$

is a subnormal series for S_4. As $[K:\{1\}] = 2$, $[F:K] = 2$, $[A_4:F] = 3$ and $[S_4:A_4] = 2$, S_4 is a solvable group.

b. Properties of, and alternative definition for, solvable groups

An important property of solvable groups is given in

Theorem 5.20: If G is a finite group and $N \lhd G$ is such that N and G/N are solvable groups, then G is also solvable.

Proof: Let $\{N\} \subseteq \bar{H}_1 \subseteq \bar{H}_2 \subseteq \cdots \subseteq \bar{H}_k = G/N$ be a subnormal series for G/N with $[\bar{H}_{i+1}:\bar{H}_i] = p_i$, p_i a prime. By the correspondence theorem (Theorem 4.19, page 120, and Corollaries 4.20 and 4.21), there are subgroups H_i in G such that $H_i \lhd H_{i+1}$, $H_i/N = \bar{H}_i$ and $[H_{i+1}:H_i] = [\bar{H}_{i+1}:\bar{H}_i] = p_i$ $(i = 0, 1, \ldots, k-1)$. Therefore

$$N = H_0 \subseteq H_1 \subseteq \cdots \subseteq H_k = G \qquad (5.16)$$

is a series of subgroups of G with $H_i \lhd H_{i+1}$ and $[H_{i+1}:H_i] = p_i$. Now N is also a solvable group. Hence N has a series

$$\{1\} = K_0 \subseteq K_1 \subseteq K_2 \subseteq \cdots \subseteq K_l = N \qquad (5.17)$$

where $[K_{i+1}:K_i]$ is a prime number $(i = 1, 2, \ldots, l-1)$. Putting the series (*5.16*) and (*5.17*) together, we obtain

$$\{1\} = K_0 \subseteq K_1 \subseteq \cdots \subseteq K_l \subseteq H_1 \subseteq H_2 \subseteq \cdots \subseteq H_k = G$$

which is a solvable series for G. The proof is complete.

Note. In contrast to Theorem 5.20, it is not always true that a group G has a property if both a normal subgroup N and G/N have the property; for example, the four group K_4 has a normal cyclic subgroup N of order 2 which is cyclic and K_4/N is cyclic, but K_4 itself is not cyclic.

Corollary 5.21: If G is a finite abelian group, G is solvable.

Proof: We use induction on the order n of G. If $|G| = 2$, the result holds trivially. Assume that any abelian group of order less then n is solvable. Suppose $p \mid n$ for some prime p. Then by Proposition 5.9, page 137, G has an element of order p. Let a be such an element. If $p \neq n$, then $|gp(a)| < |G|$ so that $gp(a)$ is solvable by the induction assumption. Furthermore, since G is abelian, $gp(a)$ is a normal subgroup of G and $|G/gp(a)| < |G|$. Hence $G/gp(a)$ is solvable by our induction assumption, and so, by Theorem 5.20, G is solvable. If $p = n$, then $\{1\} \subseteq G$ is a solvable series and G is therefore solvable in this case too.

The following theorem leads to an alternative definition of solvability.

Theorem 5.22: G is a solvable group if and only if G is finite and has a subnormal series

$$\{1\} = K_0 \subseteq K_1 \subseteq \cdots \subseteq K_n = G \qquad (5.18)$$

where K_{i+1}/K_i is abelian $(i = 0, 1, \ldots, n-1)$.

Proof: Let G be a solvable group. Then G has a subnormal series with factors of prime order and hence cyclic. Since a cyclic group is abelian, the solvable series of G is a series of type (*5.18*). Conversely, assume G has a subnormal series (*5.18*) with K_{i+1}/K_i abelian. We prove that G is solvable by induction on n, the length of the subnormal series (*5.18*). If $n = 1$, then G is abelian since $G \cong K_1/K_0$ which is abelian by assumption. Hence by Corollary 5.21, G is solvable. Assume that any finite group which has a subnormal series of length less than n in which the factor groups of consecutive terms of the series are abelian, is solvable. Let G have subnormal series (*5.18*) of length n. Then K_{n-1} has a subnormal series of length $n-1$, namely

$$K_0 \subseteq K_1 \subseteq \cdots \subseteq K_{n-2} \subseteq K_{n-1}$$

with K_{i+1}/K_i abelian for $i = 0, 1, \ldots, n-2$. Hence by the induction assumption, K_{n-1} is solvable. But G/K_{n-1} is abelian and hence solvable. Using Theorem 5.20, we conclude G is solvable.

In the theory of infinite groups one usually defines a group G to be solvable if it has a subnormal series (*5.18*) with K_{i+1}/K_i an abelian group $(i = 0, 1, \ldots, n-1)$. By Theorem 5.22 this is equivalent to our original definition for finite groups. Since this formulation of solvability is more general, we shall henceforth use it as our definition of solvability. Note that the infinite cyclic group is an example of a group that does not fit the old definition but does fit the new.

Using this new definition we prove

Theorem 5.23: Let G be a solvable group. Then (i) any subgroup of G is solvable and (ii) if $N \lhd G$ then G/N is solvable.

Proof:

(i) Let $\{1\} = H_0 \subseteq H_1 \subseteq \cdots \subseteq H_n = G$ be a subnormal series of G with H_{i+1}/H_i abelian for $i = 0, 1, \ldots, n-1$. We show that if K is a subgroup of G,

$$\{1\} = (K \cap H_0) \subseteq (K \cap H_1) \subseteq \cdots \subseteq (K \cap H_n) = K \qquad (5.19)$$

is a subnormal series with $K \cap H_{i+1}/K \cap H_i$ abelian. First we notice that $K \cap H_i = (K \cap H_{i+1}) \cap H_i$ $(i = 0, 1, \ldots, n-1)$ and that $K \cap H_n = K$. Now $H_i \lhd H_{i+1}$, and $K \cap H_{i+1}$ is a subgroup of H_{i+1}. Applying the subgroup isomorphism theorem (Theorem 4.23, page 125) inside the group H_{i+1} with subgroup $K \cap H_{i+1}$ and normal subgroup H_i, we obtain

$$(K \cap H_{i+1}) \cap H_i \lhd K \cap H_{i+1}$$

and

$$(K \cap H_{i+1})/((K \cap H_{i+1}) \cap H_i) \cong ((K \cap H_{i+1})H_i)/H_i$$

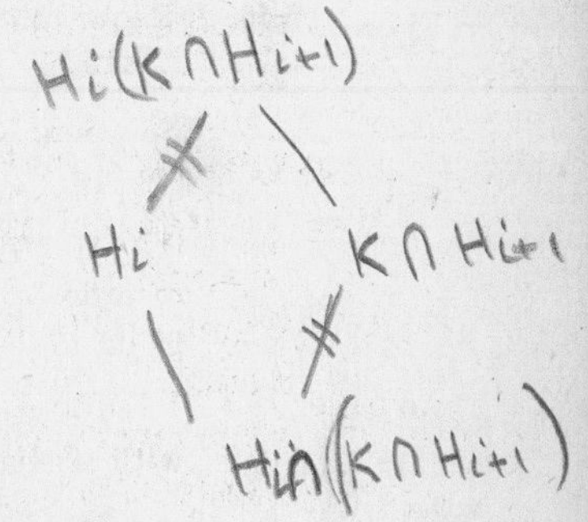

Since $(K \cap H_{i+1}) \cap H_i = K \cap H_i$, it follows that $K \cap H_i \lhd K \cap H_{i+1}$ and

$$(K \cap H_{i+1})/(K \cap H_i) \cong ((K \cap H_{i+1})H_i)/H_i$$

But $(K \cap H_{i+1})H_i \subseteq H_{i+1}$, so that we have

$$\frac{(K \cap H_{i+1})H_i}{H_i} \subseteq \frac{H_{i+1}}{H_i}$$

Now H_{i+1}/H_i is abelian by assumption and hence so is $(K \cap H_{i+1})H_i/H_i$. Therefore (*5.19*) is a subnormal series for K with abelian factors $\dfrac{K \cap H_{i+1}}{K \cap H_i}$ and consequently K is solvable.

(ii) Let G have subnormal series

$$\{1\} = H_0 \subseteq H_1 \subseteq H_2 \subseteq \cdots \subseteq H_n = G$$

where H_{i+1}/H_i is abelian. Now $N \lhd G$. Consider the natural homomorphism

$$\nu : G \to G/N$$

Any subgroup of G is mapped by ν onto a subgroup of G/N. In particular let $\bar{H}_i = H_i\nu$. We assert $\bar{H}_i \lhd \bar{H}_{i+1}$. But this follows from Problem 4.82, page 122 (with $\theta = \nu_{|H_{i+1}}$ and $H_{i+1} = G$).

Next we assert that $\bar{H}_{i+1}/\bar{H}_i$ is abelian. Let $\bar{x} = x\nu$, $\bar{y} = y\nu$ $(x, y \in H_{i+1})$ be two elements of $\bar{H}_{i+1}$. Then since H_{i+1}/H_i is abelian, $xy = yxd$ where $d \in H_i$. Thus

$$(xy)\nu = (x\nu)(y\nu) = (y\nu)(x\nu)(d\nu)$$

But $d_v \in \bar{H}_i$. Consequently

$$\bar{x}\bar{H}_i\bar{y}\bar{H}_i = \bar{x}\bar{y}\bar{H}_i = (xy)_v\bar{H}_i = (y_v)(x_v)d_v\bar{H}_i = (y_v)(x_v)H_i = \bar{y}\bar{H}_i\bar{x}\bar{H}_i$$

Therefore $\bar{H}_{i+1}/\bar{H}_i$ is abelian. Thus

$$\{N\} = \bar{H}_0 \lhd \bar{H}_1 \lhd \cdots \lhd \bar{H}_n = G/N$$

is a subnormal series of G/N with abelian factors, and so G/N is solvable.

Now that we have a definition of solvable group that applies to infinite groups, we will extend Theorem 5.20 to infinite groups.

Theorem 5.24: If $N \lhd G$ and G/N and N are solvable groups, then so is G.

Proof: Let $\{N\} = \bar{H}_0 \subseteq \bar{H}_1 \subseteq \bar{H}_2 \subseteq \cdots \subseteq \bar{H}_k = G/N$ be a subnormal series for G/N in which $\bar{H}_{i+1}/\bar{H}_i$ is abelian, $i = 0, \ldots, k-1$.

By Corollary 4.21, page 121, there are subgroups H_i in G such that $H_i \lhd H_{i+1}$ and $H_i/N = \bar{H}_i$ $(i = 0, \ldots, k)$. By the factor of a factor theorem (Theorem 4.22, page 121), $\bar{H}_{i+1}/\bar{H}_i = (H_{i+1}/N)/(H_i/N) \cong H_{i+1}/H_i$. Hence the factors H_{i+1}/H_i are abelian. Also, N has a series

$$\{1\} = K_0 \subseteq K_1 \subseteq K_2 \subseteq \cdots \subseteq K_l = N$$

with K_{i+1}/K_i abelian for $i = 0, 1, \ldots, l-1$. Therefore

$$\{1\} = K_0 \subseteq K_1 \subseteq \cdots \subseteq K_l = N = H_0 \subseteq H_1 \subseteq \cdots \subseteq H_k = G$$

is a subnormal series whose factors are abelian. Thus G is solvable.

Problems

5.46. Show that all groups G of order p^2, pq or p^2q, where p and q are distinct primes, are solvable. (Hard.)

Solution:

If $|G| = p^2$, then G is abelian (Problem 5.19, page 140) and G is solvable.

If $|G| = pq$ then from Problem 5.8, page 133, if $p < q$, G has one and only one subgroup H of order q. By Problem 5.7, page 133, $H \lhd G$. Now $|G/H| = p$, hence G/H is abelian. As H is of order q, it is abelian. Therefore we have the subnormal series

$$\{1\} \subseteq H \subseteq G$$

with abelian factors and so G is solvable.

If $|G| = p^2q$ then $s_p = 1 + kp$ divides p^2q, and so the prime factors of $1 + kp$ must be p or q. Clearly p does not divide $1 + kp$. Therefore $1 + kp = 1$ or q. If $1 + kp = 1$, then the Sylow p-subgroup H is normal in G (Problem 5.7). As H is of order p^2, H is abelian (Problem 5.19). Thus we have a subnormal series

$$\{1\} \subseteq H \subseteq G$$

with G/H abelian ($|G/H| = q$) and $H/\{1\}$ abelian. Hence G is solvable.

Suppose, however, that $1 + kp = q$; then $q > p$. Let K be a Sylow q-subgroup of G. The number of such Sylow q-subgroups is $1 + lq$. Again $1 + lq$ is not divisible by q, and so $1 + lq = 1, p$ or p^2. But as $q > p$, the only possibilities are $1 + lq = 1$ or p^2.

Case (i): $1 + lq = 1$. In this case there is only one Sylow q-subgroup K and (by Problem 5.7) $K \lhd G$. $|K| = q$ and $|G/K| = p^2$. Hence K is abelian and G/K is abelian (Problem 5.19), and it follows that G is solvable.

Case (ii): $1 + lq = p^2$. We will show that this case does not arise by showing that G would contain too many elements. We have assumed that G has q Sylow p-subgroups (of order p^2) and p^2 Sylow q-subgroups (of order q). Any two distinct subgroups of order q intersect in the identity, so there are $p^2(q-1) = p^2q - p^2$ distinct elements of order q in G. Also, G has at least 2 Sylow p-subgroups and hence there are at least p^2 distinct elements in G of order p or p^2. In the above calculations we have not counted the identity, so in all G has at least $p^2q - p^2 + p^2 + 1 = p^2q + 1$ elements, which is absurd, and we conclude that case (ii) does not arise.

5.47. Show that every nilpotent group is solvable.

Solution:

Consider the upper central series

$$\{1\} = Z_0 \subseteq Z_1 \subseteq Z_2 \subseteq \cdots \subseteq Z_n = G$$

of G, a given nilpotent group. Z_{i+1} is defined by the fact that Z_{i+1}/Z_i is the center of G/Z_i and of course this implies Z_{i+1}/Z_i is abelian. Hence G is solvable.

5.48. Prove that the converse of Problem 5.47 is false, i.e. not all solvable groups are nilpotent.

Solution:

The symmetric group S_3 is solvable (Problem 5.44). A check of the multiplication table for S_3 on page 57 shows that the center of S_3 is just the identity $\{\iota\}$. But this implies that the upper central series for G never ascends to G. Thus S_3 is not nilpotent.

5.49. Let G be any group and let $G^{(i)}$ be defined for all positive integers i by $G^{(1)} = G'$, the derived group of G, and $G^{(i+1)} = (G^{(i)})'$. Prove that G is solvable if and only if $G^{(n)} = \{1\}$ for some integer n.

Solution:

Let $G^{(n)} = \{1\}$. Then

$$\{1\} = G^{(n)} \subseteq \cdots \subseteq G^{(1)} \subseteq G$$

is a subnormal series for G and $G^{(i)}/G^{(i+1)}$ is abelian. Hence G is solvable.

Now let G be solvable. Then there exists a subnormal series

$$\{1\} = H_r \subseteq \cdots \subseteq H_0 = G$$

with H_i/H_{i+1} abelian.

By Problem 4.68, page 116, H_i/H_{i+1} abelian implies $H_{i+1} \supseteq H_i'$. We prove, by induction on i, that $H_i \supseteq G^{(i)}$, $i = 1, 2, \ldots, r$. For $i = 1$ this is true since $H_1 \supseteq H_0' = G' = G^{(1)}$. Suppose our assertion is true for $i = n$, i.e. $H_n \supseteq G^{(n)}$. Then $H_n' \supseteq (G^{(n)})' = G^{(n+1)}$. But $H_{n+1} \supseteq H_n'$, so $H_{n+1} \supseteq G^{(n+1)}$. Therefore $H_i \supseteq G^{(i)}$ for all i. In particular then, $\{1\} = H_r \supseteq G^{(r)}$. Accordingly, $G^{(r)} = \{1\}$ and the result follows.

5.5 COMPOSITION SERIES AND SIMPLE GROUPS

a. The Jordan-Hölder Theorem

In Section 5.4a we introduced the idea of a composition series for a finite group G. We recall that a series of subgroups of G

$$\{1\} = G_0 \subseteq G_1 \subseteq \cdots \subseteq G_k = G \tag{5.20}$$

is a composition series for G if $G_i \triangleleft G_{i+1}$ for $i = 0, 1, \ldots, k-1$ and G_{i+1}/G_i is simple, i.e. has precisely two different normal subgroups. This latter statement carries with it the implication that $G_i \neq G_{i+1}$ for $i = 0, \ldots, k-1$.

We observe first that every finite group G has a composition series. The easiest way to see this is by induction on the order, $|G|$, of G. If $|G| = 1$, then G has precisely one composition series:

$$\{1\} = G_0 = G$$

Suppose then that $|G| \neq 1$ and that every group of order less than $|G|$ has a composition series. Now if G is simple, then

$$\{1\} = G_0 \subseteq G_1 = G$$

is the only composition series for G. If G is not simple, let N be a normal subgroup of G, $N \neq \{1\}$, $N \neq G$. We may suppose that N is the largest normal subgroup of G, that is, if $M \triangleleft G$ and $M \neq G$, then $|M| \leq |N|$. By induction, N has a composition series

$$\{1\} = N_0 \subseteq N_1 \subseteq \cdots \subseteq N_l = N$$

We claim that

$$\{1\} = N_0 \subseteq N_1 \subseteq \cdots \subseteq N_l \subseteq G$$

is a composition series for G and note that to prove this we need only show that G/N_l is simple. But if G/N_l is not simple, it has a non-trivial normal subgroup. By the corollary to the correspondence theorem this subgroup is of the form K/N_l where K is a normal subgroup of G. But as $K \supset N_l$, this means that $|K| > |N_l|$ which contradicts the choice of N_l. Therefore every finite group has a composition series.

This proof does not suggest that if a group has two composition series then they are related. Surprisingly they are. In order to explain this relationship we associate with the composition series (*5.20*) two notions. First we term k the *length* of the series. Second we call the factor groups G_{i+1}/G_i the *composition factors* of the series (*5.20*). The relationship between composition series is given by

Theorem 5.25 (Jordan-Hölder): Every finite group G has at least one composition series. The lengths of all composition series for G are equal. Finally if

$$\{1\} = G_0 \subseteq \cdots \subseteq G_k = G$$

and

$$\{1\} = H_0 \subseteq \cdots \subseteq H_k = G$$

are a pair of composition series for G, then their respective composition factors can be paired off in such a way that paired factors are isomorphic.

We have already proved the first statement of Theorem 5.25. Before illustrating Theorem 5.25 we restate its last assertion as follows: There is a permutation π of $\{1, \ldots, k\}$ such that

$$G_{i+1}/G_i \cong H_{(i+1)\pi}/H_{(i+1)\pi-1}$$

for $i = 0, \ldots, k-1$.

Example 1: Suppose that S_3 is the symmetric group on $\{1, 2, 3\}$. Then the series

$$\{1\} \subseteq \left\{\begin{pmatrix}1 & 2 & 3\\1 & 2 & 3\end{pmatrix}, \begin{pmatrix}1 & 2 & 3\\2 & 3 & 1\end{pmatrix}, \begin{pmatrix}1 & 2 & 3\\3 & 1 & 2\end{pmatrix}\right\} \subseteq S_3$$

is a composition series for S_3. Notice that the composition factors are of orders 3 and 2. This is actually the only composition series for S_3, since

$$\left\{\begin{pmatrix}1 & 2 & 3\\1 & 2 & 3\end{pmatrix}, \begin{pmatrix}1 & 2 & 3\\2 & 3 & 1\end{pmatrix}, \begin{pmatrix}1 & 2 & 3\\3 & 1 & 2\end{pmatrix}\right\}$$

is the only normal subgroup of S_3 which is neither S_3 nor the identity subgroup.

Problem

5.50. Let n be a positive integer. What relevance does the Jordan-Hölder theorem have to the factorization of n into a product of primes? (*Hint*: Let G be the additive group of integers modulo n $(n > 1)$.)

Solution:

Let

$$\{1\} = G_0 \subseteq G_1 \subseteq \cdots \subseteq G_l = G$$

be a composition series for G. Each composition factor G_{i+1}/G_i $(i = 0, \ldots, l-1)$ is simple. As G_l is abelian, each factor G_{i+1}/G_i is abelian. Hence if G_{i+1}/G_i has any proper subgroup, it would not be simple. So G_{i+1}/G_i has no proper subgroups. In particular it has no cyclic proper subgroups, so it must be cyclic of order a prime. The number l is therefore the total number of primes (allowing for repetitions) dividing n. By Theorem 5.25, l is uniquely determined. So, as we well know, the total number of prime divisors of n is a constant. Moreover, the uniqueness of the composition factors (asserted in Theorem 5.25) simply means that these prime divisors themselves are unique. Putting these two facts together gives the well-known fact that every integer $n > 1$ can be written uniquely as a product of primes, if the order in which it is written is disregarded.

Example 2: Let S_n be the symmetric group on n letters, $n \geq 5$. Then

$$\{1\} \subseteq A_n \subseteq S_n$$

is a composition series for S_n. For we shall show that A_n is simple (in Section 5.5e, Theorem 5.34). But $A_n \lhd S_n$ and S_n/A_n is cyclic of order two. Hence this series is indeed a composition series.

b. Proof of Jordan-Hölder theorem

Suppose G is a finite group and suppose

$$1 = G_0 \subseteq G_1 \subseteq \cdots \subseteq G_k = G \tag{5.21}$$

and

$$1 = H_0 \subseteq H_1 \subseteq \cdots \subseteq H_l = G \tag{5.22}$$

are two composition series for G. We have to prove that $k = l$ and that the composition factors G_{i+1}/G_i of (5.21) can be paired off with the composition factors H_{i+1}/H_i of (5.22) so that paired composition factors are isomorphic.

The proof is by induction on the order $|G|$ of G. If $|G| = 1$, then both assertions are clear. Thus we assume that $|G| > 1$ and that the theorem holds for all groups of order less than $|G|$. It is useful to observe that if $k = 1$, then G is simple. Hence $l = 1$ also, and again the desired conclusion holds. So we assume, in addition to $|G| > 1$, that $k > 1$ (and hence $l > 1$).

There are two cases to consider: $G_{k-1} = H_{l-1}$ and $G_{k-1} \neq H_{l-1}$.

Case 1: $G_{k-1} = H_{l-1}$.

It is then clear that

$$1 = G_0 \subseteq \cdots \subseteq G_{k-1} \tag{5.23}$$

and

$$1 = H_0 \subseteq \cdots \subseteq H_{l-1} = G_{k-1} \tag{5.24}$$

are composition series for G_{k-1}. But $|G_{k-1}| < |G|$. Hence by our induction assumption, the composition series (5.23) and (5.24) have the same length, i.e. $k - 1 = l - 1$, and so $k = l$. Furthermore the composition factors of (5.23) can be paired with the composition factors of (5.24) so that paired factors are isomorphic. But the composition factors of (5.21) are those of (5.23) together with G/G_{k-1}. Similarly the composition factors of (5.22) are those of (5.24) together with G/G_{k-1}. Thus it is clear then that the composition factors of (5.21) can be paired off with the composition factors of (5.22) so that paired factors are isomorphic. This concludes the proof of Case 1.

Case 2: $G_{k-1} \neq H_{l-1}$.

Our method of proof is to produce a composition series, (5.26), which has isomorphic composition factors to those of (5.21) (by Case 1) and a composition series, (5.27), which has isomorphic composition factors to those of (5.22) (by Case 1). We will then show that (5.26) and (5.27) have isomorphic factors and this will be sufficient to prove the result.

First we will show that $G_{k-1}H_{l-1} = G$. Observe that both G_{k-1} and H_{l-1} are normal subgroups of G, and so $G_{k-1}H_{l-1}$ is also a normal subgroup of G. Obviously $G_{k-1}H_{l-1}$ contains G_{k-1} properly, so $G_{k-1}H_{l-1}/G_{k-1}$ is a non-trivial normal subgroup of G/G_{k-1} by the correspondence theorem. But G/G_{k-1} is simple, so $G_{k-1}H_{l-1}/G_{k-1} = G/G_{k-1}$. This means that

$$G_{k-1}H_{l-1} = G \tag{5.25}$$

We now put $F = G_{k-1} \cap H_{l-1}$ and note that $F \lhd G$. Let

$$\{1\} = F_0 \subseteq \cdots \subseteq F_m = F$$

be a composition series for F. Then we claim that

$$\{1\} = F_0 \subseteq \cdots \subseteq F_m \subseteq G_{k-1} \subseteq G \tag{5.26}$$

and

$$\{1\} = F_0 \subseteq \cdots \subseteq F_m \subseteq H_{l-1} \subseteq G \tag{5.27}$$

are both composition series for G. The only facts that need be verified are that G_{k-1}/F_m and H_{l-1}/F_m are simple. Now $F_m = G_{k-1} \cap H_{l-1}$ and by (*5.25*), $G = G_{k-1}H_{l-1}$. Therefore by the subgroup isomorphism theorem (Theorem 4.23, page 125),

$$G/G_{k-1} = G_{k-1}H_{l-1}/G_{k-1} \cong H_{l-1}/(H_{l-1} \cap G_{k-1}) = H_{l-1}/F_m \tag{5.28}$$

and similarly

$$G/H_{l-1} = G_{k-1}H_{l-1}/H_{l-1} \cong G_{k-1}/(H_{l-1} \cap G_{k-1}) = G_{k-1}/F_m \tag{5.29}$$

Since both G/G_{k-1} and G/H_{l-1} are simple, it follows that H_{l-1}/F_m and G_{k-1}/F_m are also both simple.

Let us compare the composition series (*5.21*) and (*5.26*). By Case 1 it follows that they have the same length and their composition factors can be paired off so that paired factors are isomorphic. Similarly for the composition series (*5.22*) and (*5.27*). Let us now compare the composition series (*5.26*) and (*5.27*). They obviously have the same length, $m+2$. Thus the series (*5.21*) and (*5.22*) have the same length. What are the composition factors of the series (*5.26*) and (*5.27*)? The composition factors of (*5.26*) are

$$F_1/F_0, \ldots, F_m/F_{m-1}, G_{k-1}/F_m, G/G_{k-1}$$

while those of (*5.27*) are

$$F_1/F_0, \ldots, F_m/F_{m-1}, H_{l-1}/F_m, G/H_{l-1}$$

Now by (*5.28*), $H_{l-1}/F_m \cong G/G_{k-1}$; and by (*5.29*), $G_{k-1}/F_m \cong G/H_{l-1}$. Thus the composition factors of (*5.26*) and (*5.27*) can be paired off so that paired factors are isomorphic. It follows that the composition factors of (*5.21*) and (*5.22*) can be paired off so that paired factors are isomorphic, since the factors of (*5.21*) can be so paired off with those of (*5.26*) and the factors of (*5.22*) can similarly be paired off with those of (*5.27*). This completes the proof of the Jordan-Hölder theorem.

From the Jordan-Hölder theorem we know that the length of a composition series and its factors are uniquely determined for the group. This suggests the following scheme for studying groups which have a composition series. First, find the structure of all groups with composition series of length 1. These are all the simple groups. Assuming now that we know all about groups with a composition series of length n, let G be a group with composition series of length $n+1$. Let

$$\{1\} = G_0 \subseteq G_1 \subseteq \cdots \subseteq G_n \subseteq G_{n+1} = G$$

be a composition series for G. If F is a group with a normal subgroup N and $F/N \cong H$, we say that *F is an extension of N by H.* In this language then, G is an extension of a group with a composition series of length n, viz. G_n, by a simple group. Hence what we must know is how the structure of a group which is an extension of one group by another is determined.

In the next few sections we shall prove that if $n > 5$, A_n is simple. The groups A_n are not all the simple groups and indeed there is no classification of simple groups as yet. This is one of the basic problems of finite group theory.

The question of how a group G is built from H and K if G is an extension of H by K, is considered in Chapter 7. Here too our knowledge is far from complete.

Problem

5.51. If two groups have the same composition factors, are they isomorphic?

Solution:

No. There is a cyclic group G of order 6 and a non-abelian group H of order 6. They have the same composition factors, but they are not isomorphic.

c. Cycles and products of cycles

We begin with an example of a new notation. Let us consider S_6. Let $\theta \in S_6$ be defined by

$$\begin{pmatrix} 1 & 2 & 3 & 4 & 5 & 6 \\ 1 & 2 & 5 & 6 & 4 & 3 \end{pmatrix}$$

Note that the effect of θ is to take $3 \to 5$, $5 \to 4$, $4 \to 6$ and $6 \to 3$, and to leave the elements 1, 2 unaltered. θ is called a cycle. We denote it by $(3, 5, 4, 6)$.

More generally, consider S_n. If $a_1, \ldots, a_m$ are distinct integers in $\{1, 2, \ldots, n\}$, $(a_1, a_2, \ldots, a_m)$ stands for the permutation that maps each integer in $\{1, 2, \ldots, n\} - \{a_1, \ldots, a_m\}$ to itself, and maps

$$a_1 \to a_2,\ a_2 \to a_3,\ \ldots,\ a_{m-1} \to a_m,\ a_m \to a_1$$

We call such a permutation a *cycle of length m*. As a convention take a cycle of length 1 to denote the identity element. A cycle of length 2 is called a *transposition*.

The inverse of the cycle $\alpha = (a_1, \ldots, a_m)$ is the cycle $\beta = (a_m, a_{m-1}, \ldots, a_2, a_1)$, since

$$a_i(\alpha\beta) = (a_i\alpha)\beta = a_{i+1}\beta = a_i \quad \text{if } i \neq m$$

while

$$a_m\alpha\beta = (a_m\alpha)\beta = a_1\beta = a_m$$

If $j \in \{1, 2, \ldots, n\} - \{a_1, \ldots, a_m\}$,

$$j(\alpha\beta) = (j\alpha)\beta = j\beta = j$$

Hence $\alpha\beta = \iota$. Similarly $\beta\alpha = \iota$.

It is clear that not every permutation is a cycle. Nor is the product of two cycles necessarily a cycle; for example, in S_4, $(1,2)(3,4) = \begin{pmatrix} 1 & 2 & 3 & 4 \\ 2 & 1 & 4 & 3 \end{pmatrix}$ which is not a cycle. An obvious question is: can we express every element of S_n as a product of cycles? The following theorem answers this question.

Theorem 5.26: Every element of S_n can be written as the product of disjoint cycles. (Cycles $(a_1, \ldots, a_m)$ and $(b_1, \ldots, b_k)$ are disjoint if the a_i and b_j are distinct, i.e. if $\{a_1, \ldots, a_m\} \cap \{b_1, \ldots, b_k\} = \emptyset$.)

Proof: Let $\pi \in S_n$. We say that $i \in \{1, \ldots, n\}$ is fixed by π if $i\pi = i$, and we say it is moved if $i\pi \neq i$.

We shall argue by induction on the number of integers moved by the permutation π. If π moves none of the integers $\{1, 2, \ldots, n\}$, then π is the identity permutation. But then $\pi = (1)$, as the 1-cycles are all the identity permutation.

Hence we have a basis for induction. So let $\pi \neq \iota$ and suppose that every element of S_n which moves fewer integers than π, can be written as the product of disjoint cycles. Now suppose $a_1 \in \{1, \ldots, n\}$ and that $a_1\pi \neq a_1$. Let us define $a_2, a_3, \ldots$ by $a_2 = a_1\pi$, $a_3 = a_2\pi, \ldots, a_s = a_{s-1}\pi, \ldots$. Let m be the first integer such that $a_m\pi = a_i$ for some integer i with $1 \leq i < m$. (As the images of π belong to $\{1, 2, \ldots, n\}$, the terms of the sequence $a_1, a_2, \ldots$ cannot all be different.) We shall prove, using the minimality of m, that $i = 1$. Suppose to the contrary that $i \neq 1$. Then $a_i = a_{i-1}\pi$ and $a_m\pi = a_i = a_{i-1}\pi$. As π is a one-to-one mapping, $a_{i-1} = a_m$. But this contradicts the choice of the integer m. Hence $i = 1$ and $a_m\pi = a_1$. We consider now the cycle $(a_1, \ldots, a_m)$ and note that $m > 1$. Let

$$\tau = (a_1, \ldots, a_m)^{-1}\pi$$

τ is a permutation that leaves $a_1, \ldots, a_m$ fixed, since $a_i(a_1, \ldots, a_m)^{-1}\pi = a_{i-1}\pi = a_i$ for $i \neq 1$, while $a_1(a_1, \ldots, a_m)^{-1}\pi = a_m\pi = a_1$. Also if $j \in \{1, 2, \ldots, n\}$, then $j\tau \neq j$ implies $j\pi \neq j$; for if $j \in \{a_1, \ldots, a_m\}$, $j\tau = j$. Hence $j \notin \{a_1, \ldots, a_m\}$ and so $j\tau = j(a_1, \ldots, a_m)^{-1}\pi = j\pi \neq j$. It follows that τ moves *fewer* integers than π. Therefore inductively τ can be written as the product of disjoint cycles, say

$$\tau = \tau_1 \cdots \tau_l$$

Now if τ_j involves an a_i, $a_i = a_i\tau = a_i(\tau_1\tau_2 \cdots \tau_l) = a_i\tau_j$. Hence $\tau_j = (a_i) = \iota$. Let

$$\tau_{i_1}, \tau_{i_2}, \ldots, \tau_{i_k} \quad (i_1 < i_2 < \cdots < i_k)$$

be those τ_j which do not involve any of the integers $a_1, \ldots, a_m$. Clearly

$$\tau = \tau_{i_1}\tau_{i_2} \cdots \tau_{i_k} \quad \text{or} \quad \tau = \iota$$

If $\tau = \iota$, $\pi = (a_1, \ldots, a_m)$ and we have proved that π is actually a cycle. Otherwise $\pi = (a_1, \ldots, a_m)\tau_{i_1} \cdots \tau_{i_k}$ and π has been expressed as the product of disjoint cycles. Hence the result.

Corollary 5.27: If $\pi \in S_n$ and $a_1, a_2, \ldots, a_m$ are chosen as in the proof of the theorem (i.e. $a_2 = a_1\pi$, $\ldots$, and $a_m\pi = a_1$ and $a_1, \ldots, a_m$ distinct), then

$$\pi = (a_1, \ldots, a_m)\tau$$

where $a\tau = a\pi$ if $a \notin \{a_1, \ldots, a_m\}$, while $a_j\tau = a_j$ for $j = 1, \ldots, m$.

Proof: This is precisely what we showed in the proof of Theorem 5.26.

This corollary provides us with a method of computing the decomposition of an element $\pi \in S_n$ into the product of disjoint cycles. For example, let us write

$$\pi = \begin{pmatrix} 1 & 2 & 3 & 4 & 5 & 6 & 7 & 8 & 9 & 10 & 11 & 12 \\ 3 & 4 & 2 & 1 & 8 & 7 & 9 & 11 & 12 & 10 & 5 & 6 \end{pmatrix}$$

as the product of disjoint cycles. Since $1\pi \neq 1$, we may take 1 for a_1. Then $a_1 = 1$, $a_2 = 3$, $a_3 = 2$, $a_4 = 4$, $a_5 = 1$. So $m = 4$. Hence by the corollary,

$$\pi = (1, 3, 2, 4)\begin{pmatrix} 1 & 2 & 3 & 4 & 5 & 6 & 7 & 8 & 9 & 10 & 11 & 12 \\ 1 & 2 & 3 & 4 & 8 & 7 & 9 & 11 & 12 & 10 & 5 & 6 \end{pmatrix}$$

Here the second factor τ on the right is obtained from π by letting it leave $1, 2, 3, 4$ unchanged, and letting it act as π does on the remaining integers. Applying the same technique to τ, we find $\tau = (5, 8, 11)(6, 7, 9, 12)$. Hence

$$\pi = (1, 3, 2, 4)(5, 8, 11)(6, 7, 9, 12)$$

In practice it is not necessary to use the corollary rigorously. We need only find the disjoint cycles by choosing a_1 such that $a_1\pi \neq a_1$, and then taking the cycle $(a_1, \ldots, a_m)$ where $a_{i+1} = a_i\pi$, $a_m\pi = a_1$, and the $a_1, \ldots, a_m$ are distinct. Then we choose $b_1 \in \{1, 2, \ldots, n\} - \{a_1, \ldots, a_m\}$ where $b_1\pi \neq b_1$, and find the cycle $(b_1, \ldots, b_k)$ with $b_{i+1} = b_i\pi$, $b_k\pi = b_1$ and $b_1, \ldots, b_k$ distinct, and so on.

Not only is every element of S_n a product of cycles, but each element can also be expressed as the product of transpositions, as can be seen from the following proposition.

Proposition 5.28: Every cycle is a product of transpositions.

Proof: We assert

$$(a_1, a_2, \ldots, a_k) = (a_1, a_2)(a_1, a_3) \cdots (a_1, a_k)$$

For if a is an integer not in $\{a_1, \ldots, a_k\}$, both the left-hand side and the right-hand side leave it unchanged.

If $i \neq k$, then $a_i(a_1, \ldots, a_k) = a_{i+1}$ while

$$a_i(a_1, a_2)(a_1, a_3) \cdots (a_1, a_k) = a_i(a_1, a_i)(a_1, a_{i+1}) \cdots (a_1, a_k)$$
$$= a_{i+1}(a_1, a_{i+2}) \cdots (a_1, a_k) = a_{i+1}$$

If $i = k$, then $a_k(a_1, \ldots, a_k) = a_1$ while

$$a_k(a_1, a_2) \cdots (a_1, a_k) = a_k(a_1, a_k) = a_1$$

Thus the effect of the left-hand side and the right-hand side is the same, and so the permutations are equal.

Problems

5.52. In S_6 compute $\alpha = (1,2,3,4)(2,5)$ and $\beta = (1,2,3,4)^{-1}(2,5)$.

Solution:

$1\alpha = (1(1,2,3,4))(2,5) = 2(2,5) = 5.$

$i\alpha = i+1$ if $i = 2,3$; $4\alpha = 1$; $5\alpha = 2$; $6\alpha = 6$. Hence $\alpha = \begin{pmatrix} 1 & 2 & 3 & 4 & 5 & 6 \\ 5 & 3 & 4 & 1 & 2 & 6 \end{pmatrix}$.

$(1,2,3,4)^{-1} = (4,3,2,1)$. Then $\beta = (4,3,2,1)(2,5)$.

$1\beta = 4$, $2\beta = 1$, $3\beta = 5$, $4\beta = 3$, $5\beta = 2$, $6\beta = 6$. Hence $\beta = \begin{pmatrix} 1 & 2 & 3 & 4 & 5 & 6 \\ 4 & 1 & 5 & 3 & 2 & 6 \end{pmatrix}$.

5.53. Express α and β of Problem 5.52 as products of disjoint cycles.

Solution:

We note that $1\alpha = 5$, $5\alpha = 2$, $2\alpha = 3$, $3\alpha = 4$, $4\alpha = 1$. Hence $\alpha = (1,5,2,3,4)$.

Since $1\beta = 4$, $4\beta = 3$, $3\beta = 5$ and $5\beta = 2$, $2\beta = 1$, $\beta = (1,4,3,5,2)$.

5.54. Write $\alpha = \begin{pmatrix} 1 & 2 & 3 & 4 & 5 & 6 & 7 & 8 & 9 & 10 & 11 & 12 & 13 & 14 \\ 2 & 4 & 6 & 8 & 10 & 12 & 14 & 1 & 3 & 5 & 7 & 9 & 11 & 13 \end{pmatrix}$ as the product of disjoint cycles.

Solution:

$1\alpha = 2$, $2\alpha = 4$, $4\alpha = 8$, $8\alpha = 1$. $3\alpha = 6$, $6\alpha = 12$, $12\alpha = 9$, $9\alpha = 3$. $5\alpha = 10$, $10\alpha = 5$. $7\alpha = 14$, $14\alpha = 13$, $13\alpha = 11$, $11\alpha = 7$. Hence

$$\alpha = (1,2,4,8)(3,6,12,9)(5,10)(7,14,13,11)$$

5.55. If $\alpha, \beta \in S_n$ are such that $i\alpha \neq i$ implies both $(i\alpha)\beta = i\alpha$ and $i\beta = i$, and $i\beta \neq i$ implies $(i\beta)\alpha = i\beta$, then α and β commute. Hence prove that disjoint cycles commute.

Solution:

Let $i \in \{1,2,\ldots,n\}$. If $i\alpha \neq i$, then $i(\alpha\beta) = (i\alpha)\beta = i\alpha$ while $i(\beta\alpha) = (i\beta)\alpha = i\alpha$. If $i\alpha = i$, then $i\alpha\beta = i\beta$ while

$$\begin{aligned}(i\beta)\alpha &= i\beta \quad \text{if } i\beta \neq i \\ &= i\alpha \quad \text{if } i\beta = i \\ &= i\end{aligned}$$

Hence $\alpha\beta = \beta\alpha$. Now if α and β are disjoint cycles, then $i\alpha \neq i$ implies α moves i and $i\alpha$ and hence β does not move either i or $i\alpha$, and so $i\beta = i$ and $(i\alpha)\beta = i\alpha$. Similarly if $i\beta \neq i$, β moves i and $i\beta$, and so $(i\beta)\alpha = i\beta$. Thus α and β commute.

5.56. Express α of Problem 5.54 as the product of transpositions. Is the expression of an element of S_n as a product of transpositions unique?

Solution:

$$\begin{aligned}\alpha &= (1,2,4,8)(3,6,12,9)(5,10)(7,14,13,11) \\ &= (1,2)(1,4)(1,8)(3,6)(3,12)(3,9)(5,10)(7,14)(7,13)(7,11)\end{aligned}$$

As $(1,2)(1,2) = \iota$, if $\alpha = \alpha_1\alpha_2 \cdots \alpha_m$ is the expression of α as a product of transpositions, then $\alpha = (\alpha_1\alpha_2 \cdots \alpha_n)(1,2)(1,2)$ is another expression of α as a product of transpositions. Thus the expression as a product of transpositions is not unique.

5.57. Find the order of the cycle $(a_1, \ldots, a_m)$. (Hard.)

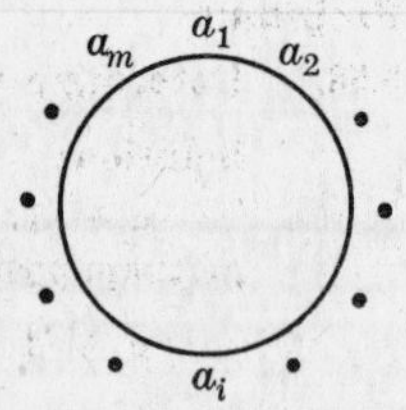

Solution:

If we write $a_1, \ldots, a_m$ in a circle, as shown on the right, it is clear that $\alpha = (a_1, \ldots, a_m)$ moves each a_i one place clockwise, α^2 moves each a_i two places further, and in general α^j moves each a_i j-places further. Hence α^m moves each a_i m-places further, i.e. α^m moves each a_i back to itself. To put this more formally, define $a_{m+1} = a_1,\ a_{m+2} = a_2,\ \ldots,\ a_{m+m} = a_m$. We will prove by induction on j that for $0 \leqq j \leqq m$, α^j maps a_i to a_{i+j}, $i = 1, 2, \ldots, m$. For $j = 0$, $\alpha^j = \iota$ and the result is true. If the result is true for $j = r$, consider $j = r + 1 \leqq m$. Then $a_i\alpha^{r+1} = (a_i\alpha^r)\alpha = a_{i+r}\alpha$. If $i + r < m$, $a_{i+r}\alpha = a_{i+r+1}$ by the action of α. If $i + r = m$, $a_m\alpha = a_1 = a_{m+1} = a_{i+r+1}$. If $i + r > m$, then $a_{i+r} = a_{i+r-m}$ and $i + r - m < m$ (as $1 \leqq i \leqq m$ and $r < m$). Hence $a_{i+r}\alpha = a_{i+r-m}\alpha = a_{i+r-m+1} = a_{i+r+1}$.

In particular, $\alpha^m = \iota$. On the other hand, $a_1\alpha^s = a_{1+s} \neq a_1$ for $1 \leqq s < m$. Therefore m is the smallest power of α that yields the identity. Accordingly m is the order of α.

d. Transpositions, and even and odd permutations

In this section we will produce another way of deciding whether a permutation is even or odd. (See the definition in Section 3.3b, page 60.) By Proposition 5.28 and Theorem 5.26 every permutation is the product of transpositions. However, this decomposition is not unique (Problem 5.56). What we shall show is that a permutation π is a product of an even number of transpositions if π is even, and the product of an odd number of transpositions if π is odd.

Our first task is to show that any transposition is odd. We have already noted that $(1, 2)$ is odd (Section 3.3d, page 64). We will use the following lemma.

Lemma 5.29: Let $\theta \in S_n$ and let $(a_1, \ldots, a_m)$ be a cycle. Then

$$\theta^{-1}(a_1, \ldots, a_m)\theta = (a_1\theta, \ldots, a_m\theta)$$

Proof: Let $x \in \{1, 2, \ldots, n\}$. If $x = a_i\theta$ for some a_i,

$$\begin{aligned} x\theta^{-1}(a_1, \ldots, a_m)\theta &= a_i\theta\theta^{-1}(a_1, \ldots, a_m)\theta \\ &= a_i(a_1, \ldots, a_m)\theta \\ &= a_{i+1}\theta \quad \text{if } i \neq m \\ &= a_1\theta \quad \text{if } i = m \end{aligned}$$

If $x \neq a_i\theta$ for some a_i, $x\theta^{-1} \notin \{a_1, \ldots, a_m\}$. Then

$$x\theta^{-1}(a_1, \ldots, a_m)\theta = (x\theta^{-1})\theta = x$$

Thus $\theta^{-1}(a_1, \ldots, a_m)\theta = (a_1\theta, \ldots, a_m\theta)$.

Theorem 5.30: All transpositions are odd. If θ is an even permutation and is written as the product of transpositions, the number of transpositions is even. If θ is odd and is written as the product of transpositions, the number of transpositions is odd.

Proof: Let (a, b) be any transposition. We know that $(1, 2)$ is odd. Let θ be any permutation such that $1\theta = a$, $2\theta = b$. Then, by Lemma 5.29,

$$(a, b) = \theta^{-1}(1, 2)\theta$$

If θ is even, θ^{-1} is even, $\theta^{-1}(1, 2)$ is odd, and so $\theta^{-1}(1, 2)\theta$ is odd (Lemma 3.2, page 62). If θ is odd, θ^{-1} is odd, $\theta^{-1}(1, 2)$ is even, and hence $\theta^{-1}(1, 2)\theta$ is odd (again by Lemma 3.2). Therefore (a, b) is odd.

Now let θ be a permutation and let $\theta = \alpha_1 \cdots \alpha_m$ be the product of transpositions. Then by Lemma 3.2, θ is even if and only if m is even, while θ is odd if and only if m is odd. This proves the theorem.

Problems

5.58. Determine whether α and β of Problem 5.52 are odd or even, using Theorem 5.30.

Solution:

$\alpha = (1,5,2,3,4) = (1,5)(1,2)(1,3)(1,4)$. Since α is expressed as the product of an even number of transpositions, α is even.

$\beta = (1,4,3,5,2) = (1,4)(1,3)(1,5)(1,2)$, and so β is even.

5.59. Determine whether α of Problem 5.54 is even by using Theorem 5.30.

Solution:

$$\begin{aligned}\alpha &= (1,2,4,8)(3,6,12,9)(5,10)(7,14,13,11)\\ &= \{(1,2)(1,4)(1,8)\}\{(3,6)(3,12)(3,9)\}\{(5,10)\}\{(7,14)(7,13)(7,11)\}\end{aligned}$$

Thus α is even.

5.60. Determine whether $(a_1, \ldots, a_m)$ is even or odd.

Solution:

$(a_1, \ldots, a_m) = (a_1, a_2)(a_1, a_3) \cdots (a_1, a_m)$, so $(a_1, \ldots, a_m)$ is the product of $m-1$ transpositions. Thus $(a_1, \ldots, a_m)$ is even or odd according as m is odd or even.

5.61. Let $(a_1, \ldots, a_m)$, $(b_1, \ldots, b_m)$ be two cycles of S_n. Prove that there exists $\theta \in S_n$ such that $\theta^{-1}(a_1, \ldots, a_m)\theta = (b_1, \ldots, b_m)$.

Solution:

Let θ be defined by:

(1) $a_i\theta = b_i$ for $i = 1, \ldots, m$;

(2) if $\{c_1, \ldots, c_{n-m}\} = \{1, 2, \ldots, n\} - \{a_1, \ldots, a_m\}$ and if $\{d_1, \ldots, d_{n-m}\} = \{1, 2, \ldots, n\} - \{b_1, \ldots, b_m\}$, put $c_i\theta = d_i$ for $i = 1, \ldots, n-m$.

Then $\theta \in S_n$ and by Lemma 5.29, $\theta^{-1}(a_1, \ldots, a_m)\theta = (b_1, \ldots, b_m)$.

5.62. If G is a group, the set of all elements conjugate to x is called the *conjugacy class containing x*. Write down the conjugacy classes of S_4 using cycle notation. (Hard.)

Solution:

The conjugacy class containing an element θ is the set of all conjugates of θ. Our first task is to express each element of S_4 as the product of disjoint cycles, and then to take all distinct conjugates of each element. We obtain the following classes:

(i) $\{(1)\}$. (As $(1) = \iota$, there are no other conjugates of (1)).

(ii) What conjugates of $(1,2)$ are there? We know that $\theta^{-1}(1,2)\theta = (1\theta, 2\theta)$. As θ runs through S_n, $1\theta, 2\theta$ run through all distinct pairs (a,b). As $(a,b) = (b,a)$, the conjugacy class containing $(1,2)$ is

$$\{(1,2), (1,3), (1,4), (2,3), (2,4), (3,4)\}$$

(iii) What conjugates of $(1,2,3)$ are there? As $\theta^{-1}(1,2,3)\theta = (1\theta, 2\theta, 3\theta)$, we will get all possible 3-cycles. Hence the conjugacy class containing $(1,2,3)$ is

$$\{(1,2,3), (1,2,4), (1,3,2), (1,3,4), (1,4,2), (1,4,3), (2,3,4), (2,4,3)\}$$

(iv) What is the conjugacy class containing $(1,2)(3,4)$?

$$\theta^{-1}(1,2)(3,4)\theta = \theta^{-1}(1,2)\theta\theta^{-1}(3,4)\theta = (1\theta, 2\theta)(3\theta, 4\theta)$$

As θ runs through S_n, we will obtain the product of all pairs of disjoint cycles. We recall that disjoint cycles commute; for example, $(1,2)(3,4) = (3,4)(1,2)$. Thus the distinct elements in the required conjugacy class are

$$\{(1,2)(3,4), (1,3)(2,4), (1,4)(2,3)\}$$

(v) What is the conjugacy class containing $(1,2,3,4)$? Again we see that we shall obtain all 4-cycles. Hence the required conjugacy class will be

$$\{(1,2,3,4), (1,2,4,3), (1,3,2,4), (1,3,4,2), (1,4,2,3), (1,4,3,2)\}$$

e. The simplicity of A_n, $n \geqq 5$.

In this section we aim to prove that A_n is simple for $n \geqq 5$. What we must do is to show that if $H \lhd A_n$ and $H \neq \{\iota\}$, then $H = A_n$. Although the proof involves much calculation, the ideas are easy.

Lemma 5.31: Every element of A_n is the product of 3-cycles, $n \geqq 5$.

Proof: Every transposition $(a_1, a_2) = (e, a_1)(e, a_2)(e, a_1)$ where e, a_1, a_2 are distinct. Now every element of A_n is the product of an even number of transpositions, and therefore the product of products of pairs of transpositions. So it is enough to prove that a product of two transpositions is a product of 3-cycles. Consider $(a_1, a_2)(b_1, b_2)$. We may assume that the four integers a_1, a_2, b_1, b_2 are distinct, for otherwise $(a_1, a_2)(b_1, b_2)$ is either the identity or is itself a 3-cycle. As $n \geqq 5$, there exists an integer e such that $1 \leqq e \leqq n$ and a_1, a_2, b_1, b_2, e are all distinct. Hence

$$\begin{aligned}(a_1, a_2)(b_1, b_2) &= (e, a_1)(e, a_2)(e, a_1)(e, b_1)(e, b_2)(e, b_1)\\ &= \{(e, a_1)(e, a_2)\}\{(e, a_1)(e, b_1)\}\{(e, b_2)(e, b_1)\} = (e, a_1, a_2)(e, a_1, b_1)(e, b_2, b_1)\end{aligned}$$

Lemma 5.32: Let $H \lhd A_n$. If H contains a 3-cycle, then $H = A_n$.

Proof: Let $(a, b, c) \in H$. Then if x, y, z are distinct elements of $\{1, 2, \ldots, n\}$ and θ is an element of S_n that sends a to x, b to y and c to z, we have $\theta^{-1}(a, b, c)\theta = (x, y, z)$ by Lemma 5.29, page 170. If θ is even, $(x, y, z) \in H$ as $H \lhd A_n$. If θ is odd, then there exists e, f distinct from a, b, c as $n \geqq 5$. Therefore $\Psi = (e, f)\theta$ is even.

$$\begin{aligned}\Psi^{-1}(a, b, c)\Psi &= \theta^{-1}(e, f)(a, b, c)(e, f)\theta\\ &= \theta^{-1}(a, b, c)\theta = (x, y, z)\end{aligned}$$

Hence by normality, $(x, y, z) \in H$. Thus H contains all 3-cycles of S_n; and by Lemma 5.31, $H = A_n$.

Lemma 5.33: Let $H \lhd A_n$. If H contains the product of two disjoint transpositions, then $H = A_n$.

Proof: Let $\alpha = (a_1, a_2)(a_3, a_4) \in H$. Since $n \geqq 5$, there exists $e \in \{1, \ldots, n\}$ distinct from a_1, a_2, a_3, a_4. Let $\theta = (a_1, a_2, e)$; then $\theta \in A_n$.

$$\begin{aligned}\theta^{-1}\alpha\theta &= (a_2, e)(a_3, a_4)\\ \alpha^{-1}\theta^{-1}\alpha\theta &= (a_1, a_2)(a_3, a_4)(a_2, e)(a_3, a_4)\\ &= (a_1, a_2)(a_2, e) = (a_1, e, a_2)\end{aligned}$$

Thus H contains a 3-cycle, and the result follows from Lemma 5.32.

Theorem 5.34: A_n is simple for $n \geqq 5$.

Proof: Let $H \lhd A_n$, $H \neq \{\iota\}$. Then there exists $\alpha \in H$, $\alpha \neq \iota$. Let $\alpha = \alpha_1 \cdots \alpha_k$ where $\alpha_1, \ldots, \alpha_k$ are disjoint cycles. We may suppose without loss of generality that $\alpha_1, \ldots, \alpha_k$ are arranged so that the length of $\alpha_i \geqq$ length of α_{i+1} for $i = 1, \ldots, k-1$, as the $\alpha_1, \ldots, \alpha_k$ commute by Problem 5.55, page 169.

Case 1:

Suppose $\alpha_1 = (a_1, \ldots, a_m)$ with $m > 3$. Let $\sigma = (a_1, a_2, a_3)$. Clearly $\sigma \in A_n$. Also σ commutes with $\alpha_2, \ldots, \alpha_k$, as they move different integers. As $\sigma^{-1} \in A_n$ and $H \lhd A_n$, $\beta = \alpha^{-1}\sigma^{-1}\alpha\sigma \in H$. Note that

$$\sigma^{-1}\alpha\sigma = (a_2, a_3, a_1, a_4, \ldots, a_m)\alpha_2 \cdots \alpha_k$$

and so

$$\begin{aligned}\beta &= \alpha_k^{-1} \cdots \alpha_1^{-1}(a_2, a_3, a_1, a_4, \ldots, a_m)\alpha_2 \cdots \alpha_k\\ &= \alpha_1^{-1}(a_2, a_3, a_1, a_4, \ldots, a_m)\\ &= (a_m, a_{m-1}, \ldots, a_1)(a_2, a_3, a_1, a_4, \ldots, a_m)\\ &= (a_1, a_2, a_4)\end{aligned}$$

Since $\beta \in H$, the result follows from Lemma 5.32.

Case 2:

Suppose $m = 3$, and α_2 is also a 3-cycle. Let $\alpha_1 = (a_1, a_2, a_3)$ and $\alpha_2 = (a_4, a_5, a_6)$. Let $\sigma = (a_2, a_3, a_4) \in A_n$. Then H contains

$$\sigma^{-1}\alpha\sigma = (\sigma^{-1}\alpha_1\sigma)(\sigma^{-1}\alpha_2\sigma)\cdots(\sigma^{-1}\alpha_k\sigma) = (\sigma^{-1}\alpha_1\sigma)(\sigma^{-1}\alpha_2\sigma)\alpha_3\alpha_4\cdots\alpha_k$$

Thus H contains

$$\begin{aligned}\alpha^{-1}\sigma^{-1}\alpha\sigma &= \alpha_1^{-1}\alpha_2^{-1}\cdots\alpha_k^{-1}(\sigma^{-1}\alpha_1\sigma)(\sigma^{-1}\alpha_2\sigma)\alpha_3\cdots\alpha_k \\ &= \alpha_1^{-1}\alpha_2^{-1}(\sigma^{-1}\alpha_1\sigma)(\sigma^{-1}\alpha_2\sigma) = \alpha_1^{-1}\alpha_2^{-1}(a_1, a_3, a_4)(a_2, a_5, a_6) \\ &= (a_3, a_2, a_1)(a_6, a_5, a_4)(a_1, a_3, a_4)(a_2, a_5, a_6) = (a_1, a_4, a_2, a_3, a_5)\end{aligned}$$

and the result follows from Case 1.

Case 3:

Suppose $m = 3$ and $\alpha_2, \ldots, \alpha_k$ are transpositions. Let $\alpha_1 = (a_1, a_2, a_3)$. Then

$$\begin{aligned}\alpha^2 &= (a_1, a_2, a_3)\alpha_1\alpha_2\cdots\alpha_k(a_1, a_2, a_3)\alpha_1\alpha_2\cdots\alpha_k \\ &= (a_1, a_2, a_3)^2\alpha_1^2\alpha_2^2\cdots\alpha_k^2 = (a_1, a_2, a_3)^2 = (a_1, a_3, a_2)\end{aligned}$$

and the result follows from Lemma 5.32.

Case 4:

Suppose all the α_i are of length 2. Then m is even, and $\alpha_1 = (a_1, a_2)$, $\alpha_2 = (a_3, a_4)$. Put $\sigma = (a_2, a_3, a_4)$. Then H contains

$$\sigma^{-1}\alpha\sigma = (a_1, a_3)(a_4, a_2)\alpha_3\alpha_4\cdots\alpha_k$$

Thus H contains $\quad \sigma^{-1}\alpha\sigma\alpha^{-1} = (a_1, a_4)(a_2, a_3)$

and the result follows by Lemma 5.33.

Corollary 5.35: The symmetric group S_n is not solvable for $n \geqq 5$.

Proof: $A_n \lhd S_n$ and $\{\iota\} \subseteq A_n \subseteq S_n$ is a composition series of S_n for $n \geqq 5$ (Example 2, page 165). By the Jordan-Hölder theorem this is, up to isomorphism, the only possible composition series. Now $|A_n| = n!/2$, which is even for $n \geqq 4$. Hence $[A_n : \{\iota\}]$ is not a prime. But if any subnormal series of S_n with factors of prime order existed, it would be a composition series. Therefore S_n is not solvable.

Problems

5.63. Prove that if $G = A_n$, $n \geqq 5$, then the derived group G' of G is G.

Solution:

We know that $G' \lhd G$. Hence $G' = G$ or else $G' = \{\iota\}$ as G is simple by Theorem 5.34. If $G' = \{\iota\}$, G is abelian. But A_n is not abelian for $n \geqq 5$; for example,

$$(1,2,3)(3,4,5) = \begin{pmatrix} 1 & 2 & 3 & 4 & 5 & 6 & \ldots & n \\ 2 & 4 & 1 & 5 & 3 & 6 & \ldots & n \end{pmatrix}$$

but

$$(3,4,5)(1,2,3) = \begin{pmatrix} 1 & 2 & 3 & 4 & 5 & 6 & \ldots & n \\ 2 & 3 & 4 & 5 & 1 & 6 & \ldots & n \end{pmatrix}$$

Therefore $G' = G$.

5.64. If $G = A_n$ and $n \geqq 5$, prove that $Z(G) = \{\iota\}$.

Solution:

As $Z(G) \lhd G$, $Z(G) = G$ or $Z(G) = \{\iota\}$. As $Z(G)$ is abelian but G is not (see Problem 5.63), $Z(G) \neq G$.

5.65. Without using Theorem 5.34, prove that $G' = G = A_n$ for $n \geq 5$, where $G = S_n$.

Solution:

We use Lemma 5.33. Let $\alpha = (2,3,4)$, $\sigma = (1,2,3)$. $\beta = \alpha^{-1}\sigma^{-1}\alpha\sigma = (1,4)(2,3) \in G'$. Hence as $G' \lhd G$, the result follows from Lemma 5.33.

5.66. Are any of A_1, A_2, A_3, A_4 simple?

Solution:

Both A_1 and A_2 are of order 1, so they are not simple. A_3 is of order 3, so A_3 is cyclic of prime order and therefore simple since groups of prime order are simple. Finally A_4 is of order 12. We have already seen that a group of order 12 is solvable. Indeed we saw in Problem 5.45 that its composition factors have orders 3, 2, 2 respectively. Hence A_4 is not simple.

5.67. Prove that S_n has no non-trivial element in its center for $n \geq 3$. (Hard.)

Solution:

Let $\alpha \in S_n$, $\alpha \neq \iota$. Let $\alpha = \alpha_1 \cdots \alpha_k$ be the decomposition of α into disjoint non-identity cycles. Let $\alpha_1 = (a_1, \ldots, a_m)$. If $m \geq 3$, let $\beta = (a_1, a_2)$. Then $\alpha^{-1}\beta\alpha = (a_1\alpha, a_2\alpha) = (a_2, a_3) \neq \beta$. Hence $\alpha \notin Z(S_n)$. If $m = 2$, let $\beta = (a_1, a_2, a_3)$. Then $\alpha^{-1}\beta\alpha = (a_1\alpha, a_2\alpha, a_3\alpha) = (a_2, a_1, b)$ where $b = a_3\alpha$ and b is an integer different from a_1 and a_2. No matter what b is, $\alpha^{-1}\beta\alpha \neq \beta$ since $a_2\beta = a_3$ but $a_2\alpha^{-1}\beta\alpha = a_1$. Hence $\alpha \notin Z(S_n)$. Thus no non-trivial element of S_n belongs to $Z(S_n)$.

5.68. Prove that A_n, S_n and $\{\iota\}$ are the only normal subgroups of S_n for $n \geq 5$. (Hard.)

Solution:

Let $H \lhd S_n$. Then $H \cap A_n \lhd A_n$. Hence $A_n \cap H = A_n$ or else $A_n \cap H = \{\iota\}$, as A_n is simple by Theorem 5.34. If $A_n \cap H = A_n$, then $A_n \subseteq H$. If $H \neq A_n$, as A_n is of index 2 in S_n, $H = S_n$.

If $A_n \cap H = \{\iota\}$, suppose $H \neq \{\iota\}$. If $\theta \in H$ and $\theta \neq \iota$, θ is odd. As θ^2 is even, $\theta^2 \in H \cap A_n$ and so $\theta^2 = \iota$. Let $\tau \in H$, $\tau \neq \iota$. Then τ is odd, and as $\theta\tau$ is even, $\tau = \theta^{-1} = \theta$. Hence H consists of only two elements, ι and θ. As S_n has no center (Problem 5.67), there exists $\mu \in S_n$ such that $\mu^{-1}\theta\mu \neq \theta$. But $\mu^{-1}\theta\mu \in H$, as $H \lhd S_n$; and since $H = \{\iota, \theta\}$, this is impossible. This contradicts the assumption that $H \neq \{\iota\}$.

5.69. Prove that $S_n' = A_n$ for $n \geq 5$.

Solution:

As S_n/A_n is abelian (cyclic of order 2 in fact), $A_n \supseteq S_n'$ (Problem 4.68(iii), page 116). As S_n is not abelian, $S_n' \neq \{\iota\}$. But $S_n' \lhd S_n$. Hence $S_n' = A_n$ by Problem 5.68. Alternatively, $S_n' \supseteq A_n' = A_n$ by Problem 5.63 or 5.65.

A look back at Chapter 5.

In this chapter we proved the three Sylow theorems. The first gives the existence of Sylow p-subgroups; the second states that a subgroup of prime power order is a subset of one of the Sylow p-subgroups; and the third states that all the Sylow p-subgroups are conjugate.

The proofs of the Sylow theorems used a standard technique of finite group theory: induction on the order of the group. It is also worth noting our counting arguments, e.g. in the proof of the class equation.

Using the class equation we proved that a group of prime power order has a non-trivial center. Then we proved as a consequence that a group of order p^r (where p is a prime) has a normal subgroup of order p^{r-1}. By repeatedly taking the center of factor groups, we defined the upper central series of a group. A group is nilpotent if a term of the upper central series is the group itself.

Next we gave a method of constructing a group from the cartesian product of two given groups H and K. This group had isomorphic copies $\hat{H}, \hat{K}$ of H and K respectively which satisfied $\hat{H}\hat{K} = G$ and $\hat{H} \cap \hat{K} = \{1\}$. Reversing this analysis we showed that if a group G had normal subgroups H and K with $HK = G$ and $H \cap K = \{1\}$, then $G \cong H \times K$. Using this result and the Sylow theorems, we classified groups of orders $1, 2, \ldots, 15$ up to isomorphism.

Then we defined solvable groups, so called because they led to a criterion for the solvability of equations. We noted that our first definition involving a subnormal series with factors of prime order did not extend to infinite groups. So we chose a criterion involving subnormal series with abelian factors for our definition of solvability. We showed that subgroups and factor groups of solvable groups were solvable, and an extension of a solvable group by a solvable group was solvable.

We next considered composition series (subnormal series with simple factors) and proved that every finite group has a composition series. In the Jordan-Hölder theorem we showed that a composition series has a unique length and unique factors up to isomorphism.

In our final section we proved that the groups A_n for $n \geqq 5$ are simple. To do this we needed to express permutations as products of disjoint cycles. This led to a method of determining whether a permutation was even or odd. As a consequence of the fact that A_n is simple, we concluded that S_n is not solvable for $n \geqq 5$.

Supplementary Problems

SYLOW THEOREMS

5.70. Prove that the Sylow 17-subgroup is normal in a group of order $255 = 3 \cdot 5 \cdot 17$.

5.71. Prove that the Sylow 13-subgroup is normal in a group of order $2 \cdot 5 \cdot 13$.

5.72. Let $A = \{0,1\}$ be the set of integers modulo 2, $B = \{0,1,\ldots,64\}$ be the set of integers modulo 65, and G be the set of all pairs (a,b), $a \in A$ and $b \in B$. Define the multiplication

$$(a_1,b_1)(a_2,b_2) = (a_1+a_2, (-1)^{a_2}b_1 + b_2)$$

for $(a_1,b_1),(a_2,b_2) \in G$. Prove that G is a group of order $2 \cdot 5 \cdot 13$ with respect to this multiplication. Is a Sylow 2-subgroup normal?

5.73. Verify the class equation (i.e. equation (*5.2*), page 135) for the group in Problem 5.72.

5.74. Prove that the Sylow p-subgroup is always normal in a group of order $4p$, where p is a prime $\geqq 5$. Is this true when $p = 3$?

5.75. Prove that the normalizer of a Sylow p-subgroup coincides with its own normalizer.

5.76. Let $|G| = n$ and $|H| = m$ where H is a subgroup of G. If $H \cap g^{-1}Hg = \{1\}$ for all $g \in G-H$, then there are precisely $n/m - 1$ elements in G which are not in any conjugate of H. (*Hint*: Examine $N_G(H)$.)

THEORY OF p-GROUPS

5.77. Show that every subgroup of index p in a finite p-group is a normal subgroup.

5.78. Let G be a finite p-group and H be a proper subgroup of G. Show that $N_G(H) \neq H$.

5.79. Prove that if G is nilpotent and G/G' is cyclic, then $G' = \{1\}$. (G' is the derived group of G).

5.80. Use Problem 5.79 to show that elements of co-prime order commute in a nilpotent group.

5.81. Find an example of a group G with a normal subgroup N such that G/N and N are nilpotent but G is not.

5.82. Let $G_{(1)} = G$ and $G_{(i+1)} = gp(\{[g, x] \mid g \in G_{(i)} \text{ and } x \in G\})$. The sequence of subgroups $G_{(1)} \supseteq G_{(2)} \supseteq \cdots \supseteq G_{(i)} \supseteq \cdots$ is called the lower central series of G. Prove that a group G is nilpotent if and only if $G_{(n)} = \{1\}$ for some positive integer n.

5.83. Find the lower central series for D_{2^n} for positive integers n. (See Problem 5.82 for the definition of lower central series.)

DIRECT PRODUCTS AND GROUPS OF LOW ORDER

5.84. Prove that a finite nilpotent group is isomorphic to a direct product of any one of its Sylow p-subgroups and some other subgroup.

5.85. Employ the results of Problem 5.70 to prove that a group of order 255 is isomorphic to a direct product of its Sylow 17-subgroup and another of its subgroups. Thereby show that a group of order 255 is cyclic.

5.86. Show that the direct product of two nilpotent groups is a nilpotent group.

5.87. Let $G = gp\left(\begin{pmatrix} 0 & 1 \\ -1 & 0 \end{pmatrix}, \begin{pmatrix} 0 & 1 \\ 1 & 0 \end{pmatrix}\right)$. Show that $|G| = 8$. Which group of order 8 is isomorphic to G?

5.88. Show that D_n is isomorphic to $G = gp\left(\begin{pmatrix} 0 & 1 \\ 1 & 0 \end{pmatrix}, \begin{pmatrix} \epsilon & 0 \\ 0 & \epsilon^{-1} \end{pmatrix}\right)$ where $\epsilon = e^{2\pi i/n}$.

5.89. Suppose G is a finite group with all its Sylow p-subgroups normal. Show that G is nilpotent.

SOLVABLE GROUPS

5.90. Prove that a group of order pqr is solvable when p, q, r are primes and $r > pq$.

5.91. Show that the direct product of two solvable groups is solvable.

5.92. Prove that a group of order $4p$, p a prime, is solvable.

5.93. In Problem 5.49, page 163, we showed that a group G is solvable if and only if $G^{(n)} = \{1\}$ for some integer n. Use this fact to give alternate proofs of Theorems 5.23, page 161, and 5.24, page 162.

5.94. Show that D_n is solvable for all positive integers n.

COMPOSITION SERIES AND SIMPLE GROUPS

5.95. Find the composition factors of a finite p-group.

5.96. Show that there is no simple group of order $p^r m$, where p is a prime and $m < p$.

5.97. Prove S_n is not solvable for $n \geqq 5$, without using the fact that A_n is simple. (*Hint.* Consider all 3-cycles of S_n.)

5.98. Find a composition series for the quaternion group.

5.99. Show that, except for groups of prime order, there are no simple groups of order < 60. (*Hint.* A difficult case occurs for order 36. See Problem 5.43, page 158.)

Chapter 6

Abelian Groups

Preview of Chapter 6

A group G with binary operation $\cdot$ is abelian if, for all $g, h \in G$, $g \cdot h = h \cdot g$.

It is customary to use "$+$" for the binary operation in abelian groups. We will begin this chapter with a preliminary section in which we restate some of our results and definitions in additive notation.

One of the concepts which we will reformulate additively and generalize is the concept of "direct product" considered in Chapter 5. In abelian groups it is customary to talk about "direct sum" instead of "direct product". We note that the direct sum of abelian groups is again abelian.

We call direct sums of infinite cyclic groups *free abelian* groups. Direct sums satisfy an important homomorphism property which gives rise to the important fact that every abelian group is a homomorphic image of a free abelian group.

We then consider classifying abelian groups according to the orders of their elements. An abelian group G which has every element of finite order can be expressed as a direct sum of p-groups, i.e. groups every element of which is of order a power of the prime p. An important p-group that we shall introduce here is the p-Prüfer group.

At this point we have three types of abelian groups:

(*a*) the cyclic groups, (*b*) the additive group Q of rationals, (*c*) the p-Prüfer groups.

The rest of the chapter is devoted to showing that many abelian groups are direct sums of these groups.

Recall that a group G is finitely generated if it contains a finite subset X with $G = gp(X)$. We show that every finitely generated abelian group is a direct sum of cyclic groups. Furthermore we associate a set of integers to each finitely generated abelian group. This set of integers, which we call the type of the abelian group, completely classifies finitely generated abelian groups; in other words, two finitely generated abelian groups are isomorphic if and only if they have the same type. This theorem is of great importance in many branches of mathematics.

The additive group of rationals Q has the property that if $g \in Q$ and n is any nonzero integer, then there exists $f \in Q$ such that $nf = g$. We express this by saying that Q is *divisible.* The p-Prüfer groups are also divisible. Note that these are not the only divisible groups, e.g. the additive group of reals is also divisible. We obtain the pleasing result that if A is any divisible abelian group, then A is a direct sum of p-Prüfer groups and groups isomorphic to the additive group of rationals.

Note. Any reader who would like a briefer account of abelian groups may refer to Sections 6.1a, 6.1c and 6.3. This will bring him quickly to the fundamental theorem of abelian groups, i.e. every finitely generated abelian group is the direct sum of cyclic groups.

6.1 PRELIMINARIES

Here we will practice expressing our ideas in additive notation.

a. Additive notation and finite direct sums

In this chapter *all* groups will be abelian, and we will use additive notation throughout. In terms of additive notation an abelian group is a non-empty set G together with a binary operation $+$ such that

(i) $(a+b)+c = a+(b+c)$ for all $a, b, c \in G$.

(ii) $a+b = b+a$

(iii) There exists an identity element, denoted by 0, such that $a+0 = a$ for all $a \in G$. The identity element 0 is often termed the *zero of G*.

(iv) Corresponding to each $a \in G$ there exists an element b such that $a+b = 0$. This b is unique and is denoted by $-a$. The element $-a$ is often termed the *negative* of a.

Standard abbreviations are as follows:

(i) $g+(-h)$ is written as $g-h$.

(ii) If n is a positive integer we write ng for $g+\cdots+g$ (n times). If $n=0$ we write ng for 0. If $n<0$ we write ng for $-g+\cdots+-g$ ($-n$ times).

If G is an abelian group and H is a subgroup, then automatically $H \lhd G$ and we may talk of the factor group G/H. (*Warning*: Some authors write $G-H$ for G/H.) Note that, in additive notation, a coset is simply a set of the form $g+H$. Instead of talking of multiplication of cosets, we talk of addition of cosets. Thus the sum of two cosets g_1+H and g_2+H is, by definition,

$$(g_1+H)+(g_2+H) = (g_1+g_2)+H$$

The following table is useful in "translating" multiplicative notation into additive notation. a and b are elements of a group G, and H and K are subgroups of G.

Multiplicative	ab	a^{-1}	1	a^n	ab^{-1}	HK	aH
Additive	$a+b$	$-a$	0	na	$a-b$	$H+K$	$a+H$

In Section 5.3a, page 146, we defined the internal direct product of two groups H and K. Now, in additive terminology, we speak of a direct sum rather than a direct product. Instead of writing $H \otimes K$, we write $H \oplus K$. If $G = H \oplus K$, H is called a *direct summand of G*. Here we are interested in extending the concept of direct sum from two subgroups to a finite number of subgroups.

Definition: An abelian group G is said to be the direct sum of its subgroups $G_1, \ldots, G_n$ if each $g \in G$ can be expressed uniquely in the form

$$g = g_1 + \cdots + g_n$$

where $g_i \in G_i$, $i = 1, \ldots, n$. In this case, we write $G = G_1 \oplus \cdots \oplus G_n$ or $G = \sum_{i=1}^{n} G_i$.

If $n=2$ we obtain the same definition of internal direct product that we gave in Chapter 5. (The condition (i) of Proposition 5.19, page 146, falls away, as all groups studied here are abelian.)

If $G = G_1 \oplus \cdots \oplus G_n$, then $G_i \cap G_j = \{0\}$ for $i \neq j$. For otherwise if $x \in G_i \cap G_j$ and $x \neq 0$, then

$$x = g_1 + \cdots + g_n$$

with $g_1 = \cdots = g_{i-1} = g_{i+1} = \cdots = g_n = 0$ and $g_i = x$, and also with $g_1 = \cdots = g_{j-1} = g_{j+1} = \cdots = g_n = 0$ and $g_j = x$. But this contradicts the definition of direct sum.

Note that if $G = H \oplus K$ and $H = L \oplus M$, then $G = L \oplus M \oplus K$ (see Problem 6.15).

The following theorem provides a simple criterion for determining when a group is the direct sum of its subgroups.

Theorem 6.1: Let $G_1, \ldots, G_n$ be subgroups of a group G and suppose each element of G can be expressed as the sum of elements from the subgroups $G_1, \ldots, G_n$. Suppose also that an equation

$$0 = g_1 + \cdots + g_n$$

with $g_i \in G_i$ for $i = 1, \ldots, n$, holds only if $g_1 = g_2 = \cdots = g_n = 0$. Then G is the direct sum of the subgroups $G_1, \ldots, G_n$.

Proof: If $g \in G$, then

$$g = g_1 + \cdots + g_n$$

with $g_i \in G_i$, $i = 1, \ldots, n$. We need only show that this expression is unique. Suppose

$$g = g_1^* + \cdots + g_n^*$$

is another such expression. Then

$$0 = (g_1 - g_1^*) + \cdots + (g_n - g_n^*)$$

By our hypothesis, $(g_1 - g_1^*) = (g_2 - g_2^*) = \cdots = (g_n - g_n^*) = 0$. Hence $g_i = g_i^*$ for $i = 1, \ldots, n$ and the two expressions for g are identical.

The question arises: if $G_1, \ldots, G_n$ are abelian groups, does there exist a group G which is the direct sum of isomorphic copies of the groups G_i? This question is answered in the following theorem.

Theorem 6.2: Let $G_1, \ldots, G_n$ be abelian groups. Then there exists a group G which is the direct sum of isomorphic copies of $G_1, \ldots, G_n$.

Proof: The proof follows closely the argument of Problem 5.30, page 146, so we will be brief. Here we will use additive notation. Let G be the cartesian product $G_1 \times G_2 \times \cdots \times G_n$. If $(g_1, \ldots, g_n)$ and $(h_1, \ldots, h_n) \in G$, define $(g_1, \ldots, g_n) + (h_1, \ldots, h_n) = (g_1 + h_1, \ldots, g_n + h_n)$. Then G is a group. Furthermore, let $\hat{G}_i = \left\{ \overbrace{(0, \ldots, g_i, 0, \ldots, 0)}^{n \text{ components}} \mid g_i \in G_i \right\}$. Then $\hat{G}_i$ is a subgroup of G, and $\hat{G}_i \cong G_i$ for $i = 1, \ldots, n$. It is clear that every element $(g_1, \ldots, g_n)$ of G is uniquely of the form $(g_1, 0, \ldots, 0) + (0, g_2, 0, \ldots, 0) + \cdots + (0, 0, \ldots, 0, g_n)$. Hence $G = \hat{G}_1 \oplus \cdots \oplus \hat{G}_n$ and the result follows.

An important result which we will prove in Section 6.1c states that if G is the direct sum of $G_1, \ldots, G_n$ and H is the direct sum of $H_1, \ldots, H_n$ and $G_i \cong H_i$ for $i = 1, \ldots, n$, then $G \cong H$.

In Section 6.1b we will define the concept of an indexed family G_i, $i \in I$. The reader who studies Section 6.1c without reading Section 6.1b may take $I = \{1, \ldots, n\}$ and G_i, $i \in I$ as shorthand for $G_1, \ldots, G_n$. Then $\sum_{i \in I} G_i$ is simply $\sum_{i=1}^{n} G_i$.

Problems

6.1. Prove that the negative of $a+b$ is $-a-b$.

Solution:

$$\begin{aligned}(a+b)+(-a-b) &= (a+b)+[(-a)+(-b)] = [(a+b)+(-a)]+(-b)\\ &= [a+(b+(-a))]+(-b) = [a+((-a)+b)]+(-b)\\ &= [(a+(-a))+b]+(-b) = (0+b)+(-b) = b+(-b) = 0\end{aligned}$$

6.2. Prove that if G is abelian and H is a subgroup of G, then G/H is abelian.

Solution:

$$(f+H)+(g+H) = (f+g)+H = (g+f)+H = (g+H)+(f+H)$$

6.3. Let H be a subset of an abelian group G. Prove that H is a subgroup of G if and only if $f,g \in H$ implies $f-g \in H$.

Solution:

This is exactly the same argument as in Lemma 3.1, page 55.

6.4. Let n be an integer and G an abelian group. Prove that if $g,h \in G$, then $n(g+h) = ng+nh$.

Solution:

If $n=0$, $n(g+h)=0$ by definition. Furthermore $ng+nh=0+0=0$. Thus $n(g+h) = ng+nh$ when $n=0$.

If $n>0$, let $n=m+1$. Then $m \geqq 0$. Inductively we may assume that $m(g+h) = mg+mh$. Keeping this in mind,

$$\begin{aligned}n(g+h) &= (m+1)(g+h) = m(g+h)+(g+h) = mg+mh+g+h\\ &= mg+g+mh+h = (m+1)g+(m+1)h = ng+nh\end{aligned}$$

Finally if $n<0$, $n=-m$ for $m>0$, and so

$$n(g+h) = m(-(g+h)) = m((-g)+(-h)) = m(-g)+m(-h) = ng+nh$$

6.5. Let n be any integer and G an abelian group. Prove that the mapping θ which sends g to ng for each g in G is a homomorphism.

Solution:

Now, in additive notation, to say that θ is a homomorphism means that $(g+h)\theta = g\theta+h\theta$ for all $g,h \in G$. But, by Problem 6.4, $(g+h)\theta = n(g+h) = ng+nh = g\theta+h\theta$.

6.6. Let G be any abelian group and let $X \subseteq G$ $(X \neq \emptyset)$. Prove that

$$gp(X) = \{y \mid y = r_1x_1+r_2x_2+\cdots+r_nx_n,\ r_i \in Z,\ x_i \in X\} \qquad (6.1)$$

In particular then $gp(x) = \{rx \mid r \in Z\}$.

Solution:

Let H be the right-hand side of (6.1). Then H is a subgroup of G, for $H \neq \emptyset$. Moreover if $h = r_1x_1+\cdots+r_nx_n$ and $k = s_1y_1+\cdots+s_py_p$ belong to H $(r_i, s_j \in Z$ and $x_i, y_j \in X)$, then clearly

$$h-k = r_1x_1+\cdots+r_nx_n+(-s_1)y_1+\cdots+(-s_p)y_p \in H$$

Thus H is a subgroup of G. It follows from the definition of H that $X \subseteq H$. Since $gp(X)$ is the smallest subgroup of G containing X, we have then $gp(X) \subseteq H$. But if $x_1,\ldots,x_n \in X$, then $r_1x_1+\cdots+r_nx_n \in gp(X)$ for every choice of $r_i \in Z$. Hence $gp(X) \supseteq H$, and so $gp(X) = H$.

When $X = \{x\}$, then $gp(X)$ is, by (6.1), the set of all multiples of x.

6.7. Let H and K be subgroups of G. Prove that $H+K$ is a subgroup of G. If $H \cap K = \{0\}$ and $G = H+K$, prove that $G = H \oplus K$. $(H+K = \{h+k \mid h \in H$ and $k \in K\}$.)

Solution:

$H+K \neq \emptyset$ since both $H \neq \emptyset$ and $K \neq \emptyset$. So we have to prove that if $u,v \in H+K$, then $u-v \in H+K$. Now $u = h+k$, $v = h'+k'$ $(h,h' \in H,\ k,k' \in K)$. Thus $u-v = (h-h')+(k-k') \in H+K$ since H and K are subgroups of G. If $H \cap K = \{0\}$ and if we consider two expressions $h_1+k_1 = h_2+k_2$ where $h_1,h_2 \in H$ and $k_1,k_2 \in K$, then $x = h_1-h_2 = k_2-k_1$ belongs to both H and K. Therefore $x=0$ and $h_1=h_2$, $k_1=k_2$. Hence the expression of an element in the form $h+k$ is unique. Since $G = H+K$, it follows that $G = H \oplus K$.

6.8. Let G have a subgroup H and suppose that G/H is infinite cyclic. Prove that H is a direct summand of G.

Solution:

Let $G/H = gp(g+H)$ where $g \in G$ and let $S = gp(g)$. Consider $x \in S \cap H$. Then $x = ng$ for some $n \in Z$. As $x \in H$, $n(g+H) = ng + H = H$. But $g + H$ is of infinite order in G/H. Thus $n = 0$ and so $S \cap H = \{0\}$. If $x \in G$, $x \in ng + H$, i.e. $x = ng + h$ for some $n \in Z$ and $h \in H$. Hence $G = S + H$. Therefore, by Problem 6.7, $G = S \oplus H$ and H is a direct summand of G.

6.9. Let $G = A \oplus B$ and let H be a subgroup containing A. Prove that $H = A \oplus (B \cap H)$.

Solution:

As every element of G is uniquely of the form $a + b$ where $a \in A$ and $b \in B$, $A + (B \cap H) = A \oplus (B \cap H)$. What we must prove then is that $A + (B \cap H) = H$. If $h \in H$, $h = a + b$ where $a \in A$ and $b \in B$. Hence $b = h - a$. But, as $A \subseteq H$, $h - a \in H$. Thus $b \in B \cap H$ and $H \subseteq A + (B \cap H)$. But $H \supseteq A$ and $H \supseteq B \cap H$. Hence $H = A \oplus (B \cap H)$.

6.10. Let G be abelian and let $a \in G$ be of order n, $b \in G$ of order m. Show that the order of $a + b$ divides the least common multiple l of m and n.

Solution:

Since both m and n divide l, let $l = qm = rn$ for some integers q and r. Then $l(a+b) = la + lb = qma + rnb = 0 + 0 = 0$. Thus the order of $a + b$ divides l.

6.11. Let $G = H \oplus K$. Prove that $G/H \cong K$.

Solution:

$G = H + K$. Then by the subgroup isomorphism theorem (Theorem 4.23, page 125), $G/H = (H+K)/H \cong K/(K \cap H)$. Now $H \cap K = \{0\}$, and so $G/H \cong K$ as required.

6.12. Prove that if G is an abelian group, then the set S of elements of G of order a power of a fixed prime p is a subgroup. Deduce that a finite abelian group has one Sylow p-subgroup for each prime p dividing $|G|$.

Solution:

If a and b are of order a power of p, then $-b$ is of order a power of p. So, by Problem 6.10, $a - b$ is of order a power of p since the least common multiple of two powers of a prime p is a power of p. Hence S is a subgroup.

If G is finite, then S is *the* Sylow p-subgroup of G. For if P is any subgroup of G of order a power of p, by the definition of S, $P \subseteq S$. So every Sylow p-subgroup of G is contained in S. S itself is of order a power of p (Problem 5.6, page 132). Since the order of a Sylow p-subgroup is the maximal power of p dividing the order of G, every Sylow p-subgroup of G must coincide with S. Thus, for each prime p, there is precisely one Sylow p-subgroup of G.

6.13. Let G be an abelian group of order 36. Prove that:

(i) If $g \in G$, then $g = g_1 + g_2$ where the order of g_1 divides 4 and the order of g_2 divides 9;

(ii) $G = A \oplus B$ where A is the Sylow 2-subgroup and B is the Sylow 3-subgroup of G.

Solution:

(i) Let g be of order $2^t 3^s$. Put $2^t = m$ and $3^s = n$. Then, since $(m, n) = 1$, there exist integers a, b such that $am + bn = 1$. Hence

$$g = (am + bn)g = (am)g + (bn)g = g_1 + g_2, \text{ say}$$

Now since $ng_1 = n(am)g = a(nm)g = 0$, and similarly $mg_2 = 0$, (i) is proved.

(ii) Clearly $A \cap B = \{0\}$, so $A + B = A \oplus B$. Now if $g \in G$, then, by (i), $g = g_1 + g_2$ where g_1 is of order dividing 4 and g_2 is of order dividing 9. By the preceding problem, the set of all elements of order a power of 2 is the Sylow 2-subgroup A, and so $g_1 \in A$. Similarly $g_2 \in B$. Hence $g \in A + B$ and we conclude that $G = A \oplus B$.

6.14. Show that the group C of complex numbers is with respect to addition the direct sum of the subgroup consisting of all the reals and the subgroup consisting of all the pure imaginary numbers.

Solution:

Let $R = \{a + i0 \mid a \text{ an arbitrary real number}\}$, and let $I = \{0 + ib \mid b \text{ an arbitrary real number}\}$ (here $i^2 = -1$). Then clearly R and I are subgroups of C, $R \cap I = \{0\}$, and, if $a + ib \in C$, $a + ib = (a + i0) + (0 + ib)$ belongs to $R + I$. Thus $C = R \oplus I$.

6.15. Let $G = H \oplus K$ and $H = L \oplus M$. Prove that $G = L \oplus M \oplus K$.

Solution:

Every element g of G can be expressed in the form $g = h + k$ where $h \in H$ and $k \in K$. But $h = l + m$ where $l \in L$ and $m \in M$. Hence $g = l + m + k$. Now if $g = l_1 + m_1 + k_1$ with $l_1 \in L$, $m_1 \in M$ and $k_1 \in K$, put $l_1 + m_1 = h_1 \in H$. Then $g = h + k = h_1 + k_1$ and consequently $h = h_1$ and $k = k_1$. As $h = l + m = l_1 + m_1$, $l = l_1$ and $m = m_1$. Hence the result.

b. Infinite direct sums (See *Note* on page 177.)

It is convenient to label the subsets of a set X with the elements of a second set. We are already familiar with such a device, e.g. in labeling a collection of sets $A_1, A_2, \ldots$ we have labeled with $I = Z$. In general, if I is an arbitrary set we shall denote by A_i, $i \in I$, such a collection of labeled sets. A collection of labeled sets A_i, $i \in I$, is called an *indexed family*. More formally, let $\theta : I \to X$ be an onto mapping. Then θ is said to be an *indexing with the elements of I of the set X*. We will denote the image of i by X_i, i.e. $X_i = i\theta$. The collection X_i, $i \in I$, is then called a *family of indexed sets*.

We generalize our definition of direct sum to apply to the direct sum of an infinite number of subgroups.

Definition: An abelian group G is said to be the direct sum of its subgroups G_i, $i \in I$, if for each $g \in G$, $g \neq 0$, there is a unique expression (but for order) for g of the form

$$g = g_1 + \cdots + g_k$$

where $g_j \in G_{j'}$, $j = 1, \ldots, k$, with $1', 2', \ldots, k'$ distinct elements of I and no g_i is zero.

Note that as G is abelian,

$$g_1 + \cdots + g_k = g_k + g_{k-1} + \cdots + g_1$$

for example; hence the uniqueness of the expression is understood to be without regard to the order of the elements $g_1, \ldots, g_k$. We write $G = \sum_{i \in I} G_i$.

If I is finite, it is easy to see that a group which is the direct sum in the sense of the above definition is also the direct sum in the sense of the definition of Section 6.1a, and conversely. Usually we will use the definition of Section 6.1a whenever I is finite.

We note that if $G = \sum_{i \in I} G_i$ and $i, j \in I$, $i \neq j$, then $G_i \cap G_j = \{0\}$. For if $x \in G_i \cap G_j$ and $x \neq 0$, then x is expressible as

$$x = g_1 \quad \text{where} \quad g_1 \in G_i$$

and

$$x = g_2 \quad \text{where} \quad g_2 \in G_j$$

But this contradicts the definition of direct sum.

The analog of Theorem 6.1 is the following:

Theorem 6.1′: Let G_i, $i \in I$ be subgroups of the abelian group G. Suppose each element of G can be expressed as the sum of elements of the subgroups G_i. Suppose also that if $1', 2', \ldots, k'$ are distinct elements of I and that the equation

$$0 = g_1 + \cdots + g_k$$

where $g_j \in G_{j'}$ holds if and only if $g_1 = g_2 = \cdots = g_k = 0$. Then $G = \sum_{i \in I} G_i$.

As the proof is similar to that of Theorem 6.1, we omit it.

Again, as in Section 6.1a, the question arises: if G_i, $i \in I$ is an indexed family of abelian groups, does there exist a group G which is the direct sum of isomorphic copies of the groups G_i? This question is answered in the following theorem:

Theorem 6.3: Let G_i, $i \in I$ be an indexed family of abelian groups. Then there exists an abelian group G which is the direct sum of groups isomorphic to G_i.

Proof: Let $G = \left\{\theta \mid \theta : I \to \bigcup_{i \in I} G_i,\ i\theta \in G_i \text{ for all } i \in I, \text{ and } i\theta \text{ is the zero element of } G_i \text{ for all but a finite number of } i \in I\right\}$. If $\theta, \phi \in G$, define $\Psi = \theta + \phi$ by $i\Psi = i\theta + i\phi$. We assert that G is a group.

First, if $\theta, \phi \in G$, then $\theta + \phi$ is clearly a mapping of I into $\bigcup_{i \in I} G_i$ and $\theta + \phi$ maps all but a finite number of the elements of I onto zero elements. Hence $\theta + \phi \in G$. Note that $\theta + \phi = \phi + \theta$ so that G is abelian.

Next, G is associative. For if $\phi_1, \phi_2, \phi_3 \in G$, and if $i \in I$,

$$\begin{aligned} i((\phi_1 + \phi_2) + \phi_3) &= i(\phi_1 + \phi_2) + i\phi_3 = i\phi_1 + i\phi_2 + i\phi_3 = i\phi_1 + (i\phi_2 + i\phi_3) \\ &= i\phi_1 + i(\phi_2 + \phi_3) = i(\phi_1 + (\phi_2 + \phi_3)) \end{aligned}$$

Hence $(\phi_1 + \phi_2) + \phi_3 = \phi_1 + (\phi_2 + \phi_3)$.

The mapping $\eta : i \to 0 \in G_i$ for all i in I is the identity of G. For if $\theta \in G$, then for all i in I, $i(\eta + \theta) = i\eta + i\theta = 0 + i\theta = i\theta$ so that $\eta + \theta = \theta$.

Finally, if $\theta \in G$, define $\phi : i \to -(i\theta)$. Then $i(\theta + \phi) = i\theta + i\phi = i\theta + (-(i\theta)) = 0 = i\eta$. Thus $\theta + \phi = \eta$ and ϕ is the inverse of G.

Now we prove that if $\hat{G}_i = \{\theta \mid \theta \in G$ and $j\theta$ is the zero of G_j for all j in I with perhaps the exception of $i\theta\}$, then $\hat{G}_i \cong G_i$. Note that $\hat{G}_i$ is a subgroup of G; for if $\theta, \phi \in G$, then on putting $\Psi = \theta - \phi$ we note that if $j \neq i$, $j \in I$, $j\Psi = j\theta - j\phi = 0 - 0 = 0$. Thus $\Psi \in \hat{G}_i$.

Next, let $\nu_i : \hat{G}_i \to G_i$ be defined by $\theta\nu_i = i\theta$, $\theta \in \hat{G}_i$. Clearly ν_i is a mapping of $\hat{G}_i$ into G_i. To see that ν_i is one-to-one, suppose $\theta, \phi \in \hat{G}_i$ and $\theta\nu_i = \phi\nu_i$. This means that $i\theta = i\phi$. But $j\theta = j\phi = 0$ for every $j \in I$, $j \neq i$ and so $\theta = \phi$. Next we prove ν_i is onto. Let $a \in G_i$ and define $\theta : I \to \bigcup_{i \in I} G_i$ by $i\theta = a$ and $j\theta = 0$ if $j \neq i$. Then $\theta \in \hat{G}_i$ and $\theta\nu_i = a$, and so ν_i is one-to-one and onto. ν_i is a homomorphism; for if $\theta, \phi \in \hat{G}_i$ and $\Psi = \theta + \phi$, then

$$(\theta + \phi)\nu_i = \Psi\nu_i = i\Psi = i\theta + i\phi = \theta\nu_i + \phi\nu_i$$

Therefore ν_i defines an isomorphism.

Finally we show that $G = \sum_{i \in I} \hat{G}_i$. We already noted that G is abelian. We need to show that if $\theta \in G$, $\theta = \theta_1 + \cdots + \theta_k$ where each θ_i belongs to one of the subgroups $\hat{G}_j$. If $\theta \in G$, then $i\theta$ is not the zero of G_i for only a finite number of elements of I, say $i_1, \ldots, i_k$. Let $\theta_1 \in \hat{G}_{i_1}$ be such that $i_1\theta_1 = i_1\theta$, $\theta_2 \in G_{i_2}$ be such that $i_2\theta_2 = i_2\theta$, $\ldots$, $\theta_k \in G_{i_k}$ be such that $i_k\theta_k = i_k\theta$. It is clear that

$$\theta = \theta_1 + \cdots + \theta_k$$

Now suppose that η, the zero of G, is of the form

$$\eta = \theta_1 + \cdots + \theta_k$$

where $\theta_j \in \hat{G}_{j'}$ and $1', 2', \ldots, k'$ are distinct elements of I. Then for $1 \leq j \leq k$,

$$j'\eta = 0 = j'(\theta_1 + \cdots + \theta_k) = j'\theta_j$$

(since $1', 2', \ldots, k'$ are distinct) from which $\theta_j = \eta$ for every $j = 1, \ldots, k$. Thus

$$G = \sum_{i \in I} \hat{G}_i$$

(Remark: Problem 6.17 shows how the mappings of this proof link up with cartesian products.)

The existence of direct sums is a very powerful result, and we will use it repeatedly.

We note the important result that: given $G_i \cong H_i$ for each $i \in I$, then if G is the direct sum of its subgroups G_i and H is the direct sum of its subgroups H_i, we can conclude $G \cong H$ (Section 6.1c).

Problems

6.16. Let G be an abelian group with subgroups G_i, $i \in I$. Suppose that $G = gp\left(\bigcup_{i \in I} G_i\right)$ and that $G_j \cap gp\left(\bigcup_{\substack{i \in I \\ i \neq j}} G_i\right) = \{0\}$ for each $j \in I$. Prove that G is the direct sum of its subgroups G_i, $i \in I$.

Solution:

We need only show that if

$$0 = g_1 + \cdots + g_k$$

where $g_i \in G_{i'}$ and $1', 2', \ldots, k'$ are distinct elements of I, then

$$g_1 = g_2 = \cdots = g_k = 0$$

Suppose, if possible, that some g_i is not zero, say $g_1 \neq 0$. Then

$$-g_1 = g_2 + \cdots + g_k$$

But $-g_1 \in G_{1'}$, say, and $g_2 + \cdots + g_k \in gp\left(\bigcup_{\substack{i \in I \\ i \neq 1'}} G_i\right)$. By hypothesis then, $-g_1 = 0$. Hence $g_1 = g_2 = \cdots = g_k = 0$ and the result follows.

6.17. Compare the construction of a direct product involving cartesian products with the construction involving mappings by showing that an ordered pair can be thought of as a mapping. Hence show that the two groups obtained are isomorphic.

Solution:

An ordered pair can be thought of as a mapping from $\{1, 2\}$. The image of 1 gives the entry in the first position, the image of 2 gives the entry in the second position. Let $G = G_1 \times G_2$ and let $\bar{G}$ be the group as constructed in Theorem 6.3 with $I = \{1, 2\}$. Let $\nu : \bar{G} \to G$ be defined by $\theta\nu = (1\theta, 2\theta)$ for any $\theta \in \bar{G}$. Then ν is clearly a one-to-one and onto mapping.

Also ν is a homomorphism. For if $\Psi = \theta + \phi$,

$$(\theta + \phi)\nu = \Psi\nu = (1\Psi, 2\Psi) = (1\theta + 1\phi, 2\theta + 2\phi)$$
$$= (1\theta, 2\theta) + (1\phi, 2\phi) = \theta\nu + \phi\nu$$

6.18. Let π be the ratio of the circumference of a circle to its diameter. Let G be the subgroup of the additive group of the real numbers generated by the numbers $\pi, \pi^2, \pi^3, \ldots$. Let $G_i = gp(\pi^i)$, $i = 1, 2, \ldots$. Prove that $G = \sum_{i \in P} G_i$ where P is the set of positive integers. Use the fact that π is not the root of any polynomial with integer coefficients.

Solution:

Clearly each element of G is of the form $g_1 + \cdots + g_n$ where each g_i belongs to one of the groups $G_1, G_2, \ldots$. Then we need only show that $0 = g_1 + \cdots + g_n$ implies that $g_1 = g_2 = \cdots =$

$g_n = 0$ where the $g_i \in G_{i'}$, with $1', \ldots, n'$ distinct elements of $\{1, 2, \ldots\}$. Now $g_i = z_i \pi^{i'}$ where $z_i \in Z$ and $i = 1, \ldots, n$. Then

$$0 = z_1\pi^{1'} + z_2\pi^{2'} + \cdots + z_n\pi^{n'}$$

If not all z_i are zero, π is the root of a polynomial with integer coefficients, contrary to the statement in the problem. Hence the result.

c. The homomorphic property of direct sums and free abelian groups

Let $G = A \oplus B$ and let H be a group which contains isomorphic copies $\bar{A}$ and $\bar{B}$ of A and B respectively. Suppose that $H = \bar{A} + \bar{B}$ (but not necessarily that $H = \bar{A} \oplus \bar{B}$). What connection, if any, is there between G and H?

It turns out that H is a homomorphic image of G. This follows from Theorem 6.4. This theorem when applied to particular cases leads also to important results: (1) Theorem 6.5 and (2) the concept of a free abelian group.

Theorem 6.4: Let $G = A \oplus B$ and let H be any group. Let θ, ϕ be homomorphisms of A into H and B into H respectively. Then there exists a homomorphism $\zeta: G \to H$ such that $\zeta_{|A} = \theta$, $\zeta_{|B} = \phi$.

Proof: If $g \in G$, then $g = a + b$ uniquely where $a \in A$, $b \in B$. Define $g\zeta = a\theta + b\phi$. ζ is uniquely defined and so is a mapping of G to H. Note that if $g_i = a_i + b_i$ where $a_i \in A$ and $b_i \in B$ $(i = 1, 2)$, then

$$\begin{aligned}(g_1 + g_2)\zeta &= ((a_1 + b_1) + (a_2 + b_2))\zeta = ((a_1 + a_2) + (b_1 + b_2))\zeta \\ &= (a_1 + a_2)\theta + (b_1 + b_2)\phi = a_1\theta + a_2\theta + b_1\phi + b_2\phi \\ &= a_1\theta + b_1\phi + a_2\theta + b_2\phi = (a_1 + b_1)\zeta + (a_2 + b_2)\zeta = g_1\zeta + g_2\zeta\end{aligned}$$

Hence ζ is the required homomorphism as $\zeta_{|A} = \theta$, $\zeta_{|B} = \phi$.

In exactly the same way we can prove that if $G = \sum_{i \in I} G_i$ and if for each $i \in I$, $\theta_i: G_i \to H$ is a homomorphism of G_i to H, then there exists a homomorphism $\theta: G \to H$ such that $\theta_{|G_i} = \theta_i$. We shall often say that θ *extends* the mappings θ_i, or that θ is an *extension* of the mappings θ_i.

We use this result to prove:

Theorem 6.5: Let $G_i \cong H_i$, $i \in I$. If G is the direct sum of its subgroups G_i and H is the direct sum of its subgroups H_i, then $G \cong H$.

Proof: Let $\theta_i: G_i \to H_i$ be an isomorphism for all $i \in I$. Then by Theorem 6.4 there exists a homomorphism $\theta: G \to H$ such that $\theta_{|G_i} = \theta_i$ for each $i \in I$. To prove θ is an isomorphism, we need only show that its kernel is trivial since θ is clearly onto. If $g \in G$, then $g = g_1 + \cdots + g_n$ where g_j belongs to the subgroup $G_{j'}$ with $1', 2', \ldots, n'$ distinct elements of I. Thus

$$g\theta = g_1\theta_{1'} + \cdots + g_n\theta_{n'} = h_1 + \cdots + h_n, \quad \text{where } h_i \in H_{i'}$$

Then $g\theta = 0$ only if each $h_j = 0$, since H is the direct sum of its subgroups H_i. Since $h_j = g_j\theta_{j'}$ and $\theta_{j'}$ is an isomorphism, we have $g_j = 0$ and $g = 0$. Thus $\operatorname{Ker}\theta = \{0\}$, and so $G \cong H$.

We will now apply Theorem 6.4 when each G_i is infinite cyclic. To begin with, let $C = gp(c)$ be infinite cyclic and H any abelian group. Note that if ϕ is a mapping of $\{c\}$ into H, then there exists a homomorphism $\theta: C \to H$ such that $\theta_{|\{c\}} = \phi$. The homomorphism θ is simply defined by putting $(rc)\theta = r(c\phi)$ for each $r \in Z$. It is readily seen that this does define a homomorphism.

Now let each G_i be infinite cyclic, $G_i = gp(c_i)$, $i \in I$. Let $X = \{c_i \mid i \in I\}$. If $\theta : X \to H$ where H is any abelian group, then there exists a homomorphism $\theta^* : \sum_{i \in I} G_i \to H$ such that $\theta^*_{|X} = \theta$. For, corresponding to each G_i, we know from the remark above that there exists a homomorphism $\theta_i : G_i \to H$ such that $c_i\theta_i = c_i\theta$. Hence it follows by Theorem 6.4 that there exists a homomorphism $\theta^* : \sum_{i \in I} G_i \to H$ which agrees with θ_i on G_i. So we have the required result.

Corollary 6.6: The direct sum $G = \sum_{i \in I} G_i$ satisfies the following condition: for every mapping $\theta : X \to H$, H any abelian group, there exists a homomorphism $\theta^* : G \to H$ such that $\theta^*_{|X} = \theta$.

A group G which contains a subset X such that

(i) $G = gp(X)$,

(ii) for every mapping $\theta : X \to H$, H any abelian group, there exists a homomorphism $\theta^* : G \to H$ such that $\theta^*_{|X} = \theta$,

is called *a free abelian group*. G is said to be *freely generated by* X and X is called a *basis* for G.

We have shown that the direct sum of infinite cyclic groups is a free abelian group. Conversely we have the following

Theorem 6.7: If G is a free abelian group freely generated by a set $X = \{x_i \mid i \in I\}$, then G is the direct sum of its subgroups $G_i = gp(x_i)$ and each G_i is infinite cyclic for all $i \in I$.

Proof: This theorem is proved by showing that G is isomorphic to a direct sum of infinite cyclic groups. To this end let H be the direct sum of its subgroups H_i,

$$H = \sum_{i \in I} H_i$$

where $H_i = gp(h_i)$ is an infinite cyclic group generated by h_i. (We know such a direct sum exists by Theorem 6.3.) Let $\theta : X \to H$ be the mapping defined by $x_i\theta = h_i$. Then θ can be extended to a homomorphism θ^* of G into H, by the definition of a free abelian group.

On the other hand H is the direct sum of the infinite cyclic groups H_i. Thus by Corollary 6.6, the mapping $\phi : \{h_i \mid i \in I\} \to X$ defined by $h_i\phi = x_i$ can be extended to a homomorphism ϕ^* of H into G. Actually ϕ^* and θ^* are inverse isomorphisms. To see this, suppose $g \in G$. Then $g = n_1x_{1'} + \cdots + n_rx_{r'}$ where $1', \ldots, r' \in I$ and $n_1, \ldots, n_r \in Z$. Accordingly,

$$\begin{aligned}(g\theta^*)\phi^* &= [n_1(x_{1'}\theta^*) + \cdots + n_{r'}(x_{r'}\theta^*)]\phi^* = (n_1h_{1'} + \cdots + n_rh_{r'})\phi^* \\ &= n_1(h_{1'}\phi) + \cdots + n_r(h_{r'}\phi) = n_1x_{1'} + \cdots + n_rx_{r'} = g\end{aligned}$$

and so $\theta^*\phi^*$ is the identity mapping on G. Similarly $\phi^*\theta^*$ is the identity mapping on H. This implies that θ^* is a one-to-one mapping of G onto H. For if $g, g' \in G$, then $g\theta^* = g'\theta^*$ implies that $(g\theta^*)\phi^* = (g'\theta^*)\phi^*$. Since $(g\theta^*)\phi^* = g$ and $(g'\theta^*)\phi^* = g'$, we have $g = g'$. Furthermore if $h \in H$, then $h = (h\phi^*)\theta^*$. Thus θ^* is one-to-one and onto.

Note that each G_i is infinite cyclic, since θ^* is an isomorphism and $G_i\theta^* = H_i$. Finally we show that G is the direct sum of its subgroups G_i. If $1', 2', \ldots, r'$ are distinct elements of I and $g_1, \ldots, g_r$ are nonzero elements of $G_{1'}, \ldots, G_{r'}$ respectively, then if

$$g_1 + \cdots + g_r = 0 \tag{6.2}$$

it follows that $g_1\theta^* + \cdots + g_r\theta^* = 0$. But then, as $g_i\theta^* \in H_{i'}$ and $g_i\theta^* \neq 0$, we have a contradiction as $H = \sum_{i \in I} H_i$. Hence (*6.2*) does not hold. Finally $gp(G_i \mid i \in I) = G$, and so $G = \sum_{i \in I} G_i$.

Corollary 6.8: Every abelian group is the homomorphic image of some free abelian group.

Proof: If G is an arbitrary group whose elements are g_i, $i \in I$, and $gp(x_i)$, $i \in I$, is infinite cyclic, then as we have seen, $F = \sum_{i \in I} gp(x_i)$ is free abelian and the mapping $\theta : x_i \to g_i$ extends to a homomorphism of F onto G.

Problems

6.19. If $|gp(a)| = m$, $|gp(b)| = n$ and $(m,n) = 1$, then $G = gp(a) \oplus gp(b)$ is cyclic of order mn.

Solution:

We show that $G = gp(a+b)$. If l is the order of $a+b$, then $l(a+b) = la + lb = 0$ implies $la = lb = 0$, by the definition of a direct sum. Consequently l is divisible by the order of a and the order of b. Since $(m,n) = 1$, $mn \mid l$ and we conclude $mn = l$, so that $G = gp(a+b)$.

6.20. Find $|G|$ where G is the direct sum of n cyclic groups of order 3.

Solution:

Let $G_1, \ldots, G_n$ be subgroups of G and suppose $G = G_1 \oplus \cdots \oplus G_n$, where each G_i is of order 3. Each expression of the form $g_1 + g_2 + \cdots + g_n$, where $g_i \in G_i$, gives rise to unique elements of G. There are 3 different possible choices for g_1, 3 for g_2, etc. Hence the total number of possible choices is $3 \cdot 3 \cdot \cdots \cdot 3 = 3^n$. Hence $|G| = 3^n$.

6.21. Prove that A is a direct summand of G if and only if there exists a homomorphism θ of G onto A such that $\theta_{|A}$ is the identity on A. (*Hint.* Use $K = \text{Ker}\,\theta$. Also consider $g - g\theta$ to prove $g \in A + K$.)

Solution:

Suppose that $G = A \oplus B$. Then the identity homomorphism on A and the trivial homomorphism on B extend to a homomorphism $\theta : G \to A$ which satisfies the required conditions (Theorem 6.4).

Conversely, if such a homomorphism exists, let $K = \text{Ker}\,\theta$ and let $x \in A \cap K$. Then $x = x\theta$, as $\theta_{|A} =$ identity on A. But $x\theta = 0$, since $x \in \text{Ker}\,\theta$. Hence $A \cap K = \{0\}$ and so $gp(A,K) = A \oplus K$. Now letting $g \in G$, $g\theta = a$ for some $a \in A$. Then $(g-a)\theta = g\theta - a\theta = a - a = 0$ and $g - a \in K$. Thus $g \in A \oplus K$ and $G = A \oplus K$.

6.22. A group G is said to be of *exponent* n if $x \in G$ implies $nx = 0$ and n is the smallest positive integer with this property. Let H be the direct sum of two cyclic groups of order n, generated by x_1 and x_2 respectively. Put $X = \{x_1, x_2\}$. Prove that if G is any group of exponent n and θ is a mapping of X into G, then there exists a homomorphism θ^* of H into G such that $\theta^*_{|X} = \theta$.

Solution:

Let $H_1 = gp(x_1)$ and $H_2 = gp(x_2)$. There is a homomorphism $\theta_1 : H_1 \to G$ satisfying $x_1\theta_1 = x_1\theta$, for $gp(x_1\theta)$ is cyclic and is of order dividing n. Similarly there is a homomorphism $\theta_2 : H_2 \to G$ satisfying $x_2\theta_2 = x_2\theta$. Thus there is a homomorphism $\theta^* : H_1 \oplus H_2 \to G$ such that $\theta^*_{|H_1} = \theta_1$ and $\theta^*_{|H_2} = \theta_2$, by Theorem 6.4. The result follows.

6.23. Let G be freely generated by a finite set X, $|X| = n$. Prove that every element of G is uniquely of the form $m_1x_1 + \cdots + m_nx_n$, where $m_i \in Z$ and $x_i \in X$.

Solution:

Let $G_i = gp(x_i)$. Now by the theorem on free abelian groups we know that $G = G_1 \oplus \cdots \oplus G_n$, and each G_i is infinite cyclic. Then each element of G is uniquely of the form $g_1 + \cdots + g_n$ where $g_i \in G_i$. But we know from the theory of infinite cyclic groups that $g_i = m_ix_i$ uniquely. The result follows.

6.24. Let G be the direct sum of cyclic groups G_i of order 2 where $i \in P$, the set of positive integers. Let E denote the set of even positive integers. Then $H = gp(G_i \mid i \in E)$ is clearly a proper subgroup of G. Prove that $H \cong G$.

Solution:

Let $\theta_i : G_i \to G_{2i}$ for all $i \in P$ be an isomorphism. (Such an isomorphism exists, since all the G_i are cyclic of order 2.) We apply Theorem 6.5 to obtain the result.

6.25. Let $G = A \oplus B$ where A is cyclic of order 3^2 and B is cyclic of order 5^2. Prove that $\text{aut}(G) \cong \text{aut}(A) \oplus \text{aut}(B)$ (hard) and hence compute $|\text{aut}(G)|$. ($\text{aut}(G)$ is the automorphism group of G (see Section 3.6a, page 83).)

Solution:

If $\alpha \in \text{aut}(A)$, then α can be used to form an element of $\text{aut}(G)$ by letting it act as the identity on B. (Theorem 6.4 states that α extends to a homomorphism. We must check that it is one-to-one and onto.) Similarly for elements β of $\text{aut}(B)$. We use the symbols α^*, β^* to represent corresponding automorphisms of G. Note that

$$(a+b)\alpha^*\beta^* \;=\; (a\alpha+b)\beta^* \;=\; a\alpha + b\beta \;=\; (a+b)\beta^*\alpha^*$$

from which $\alpha^*\beta^* = \beta^*\alpha^*$. Note that $\alpha^* = \beta^*$ implies $\alpha^* = \beta^* = \iota$, the identity automorphism.

The mapping $\alpha \to \alpha^*$ is an isomorphism of $\text{aut}(A)$ into a subgroup of $\text{aut}(G)$. The mapping $\beta \to \beta^*$ is an isomorphism of $\text{aut}(B)$ into a subgroup of $\text{aut}(G)$. As $(\text{aut}(A))^* \cap (\text{aut}(B))^* = \{\iota\}$, and as the elements of $(\text{aut}(A))^*$ commute with the elements of $(\text{aut}(B))^*$, we have $(\text{aut}(A))^* + (\text{aut}(B))^* = (\text{aut}(A))^* \oplus (\text{aut}(B))^*$, by Theorem 5.16, page 144. Now let $\theta \in \text{aut}(G)$. $\theta_{|A}$ induces an automorphism θ_A on A, for $A\theta$ must go into a subgroup of order 9 and by Sylow's theorem there is only one subgroup of order 9 (as G is abelian). Similarly $\theta_{|B}$ induces an automorphism θ_B on B. Then

$$(a+b)\theta_A^*\theta_B^* \;=\; (a\theta+b)\theta_B^* \;=\; a\theta + b\theta \;=\; (a+b)\theta$$

from which $\theta_A^*\theta_B^* = \theta$. Thus $\text{aut}(G) = (\text{aut}(A))^* \oplus (\text{aut}(B))^*$ and, by Theorem 6.5,

$$\text{aut}(G) \;\cong\; \text{aut}(A) \oplus \text{aut}(B)$$

To compute $|\text{aut}(G)|$ we must compute $|\text{aut}(A)|$ and $|\text{aut}(B)|$. Let $\alpha \in \text{aut}(A)$ and let $A = gp(a)$. α must take a onto an element of order 9; so if $a\alpha = a^r$, $(r,3) = 1$. Hence the possibilities are $r = 1,2,4,5,7,8$. Each of these gives rise to an automorphism of A, as can be checked. Thus $|\text{aut}(A)| = 6$. Similarly $|\text{aut}(B)| = 20$. Accordingly, $|\text{aut}(G)| = 6 \times 20 = 120$.

6.2 SIMPLE CLASSIFICATION OF ABELIAN GROUPS, AND STRUCTURE OF TORSION GROUPS

a. Tentative classifications: torsion, torsion-free, and mixed

Consider the following three examples of abelian groups: Q, the additive group of rationals; Q/Z the factor group of the additive group of the rationals by the integers; C, the multiplicative group of complex numbers. Each of these groups is not isomorphic to the others, but how would we prove that? One way is to examine the orders of the elements of the groups. Now every element of Q except 0 is of infinite order and every element of Q/Z is of finite order. For if $r \in Q$, $r = m/n$ where m, n are two integers. Thus $n(r+Z) = nr + Z = m + Z = Z$. Let us, to avoid confusion, continue to use the multiplicative notation for C. We assert that C has elements of infinite order and also elements of finite order. Recall that the identity of C is 1. Note that $(-1)^2 = 1$ implies that -1 is of order 2 and $3^r = 1$ if and only if $r = 0$. Hence -1 is of finite order and 3 is of infinite order. Summarizing, we have

(i) Q has every element but the identity of infinite order.
(ii) Q/Z has every element of finite order.
(iii) C has elements of finite order and elements of infinite order.

It is then easy to see that the three groups are not isomorphic.

If G is a group in which every element other than the identity is of infinite order, G is said to be *torsion-free*. If G is a group in which every element is of finite order, G is said to be a *torsion group*. If G has both an element of infinite order and an element (not equal to the identity) of finite order, G is said to be *mixed*. These three concepts provide us with a rough classification of abelian groups and, as we have seen above, distinguish between Q, Q/Z and C.

Problems

6.26. Let G be the direct sum of torsion groups. Prove that G is a torsion group.

Solution:

Let $G = \sum_{i \in I} G_i$. If $g \in G$, $g = x_1 + \cdots + x_n$ for some integer n and x_j belongs to some $G_{j'}$. If the order of x_j is p_j, then

$$(p_1 \cdot \cdots \cdot p_n)g = (p_1 \cdot \cdots \cdot p_n)x_1 + \cdots + (p_1 \cdot \cdots \cdot p_n)x_n = 0 + \cdots + 0 = 0$$

6.27. Prove that Q/Z and G, the direct sum of groups G_i $(i \in Z)$ where each G_i is cyclic of order 2, are not isomorphic.

Solution:

Following the method of Problem 6.26, it is easy to prove that every nonzero element of G is of order 2. Since $\frac{1}{3} + Z$ is of order 3, Q/Z and G are not isomorphic.

6.28. Prove that Q/Z and G, the direct sum of groups G_i $(i \in Z)$ where each G_i is cyclic of order 3^i, are not isomorphic.

Solution:

G is easily shown to have every element of order some power of 3 by following the method of Problem 6.26. Since $\frac{1}{4} + Z$ is of order 4 in Q/Z, G and Q/Z are not isomorphic.

b. The torsion subgroup

In Section 6.2a we introduced a tentative classification of abelian groups into torsion-free, torsion and mixed groups. In this section we consider the question of whether it would not be possible to split a mixed group into a torsion-free group and a torsion group. This would provide the following program for investigating abelian groups:

(1) Investigate torsion-free groups.
(2) Investigate torsion groups.
(3) Investigate how they may be put together to form mixed groups.

Such a program is found to be too difficult to accomplish completely, but it does lead to some significant results.

Let $T(G)$ be the set of all elements of G of finite order. Then $T(G)$ is a subgroup of G, as we show in the following

Theorem 6.9: $T(G)$ is a subgroup of G (termed the *torsion subgroup* of G). $G/T(G)$ is torsion-free.

Proof: Let $a, b \in G$ be of order m, n respectively. Then

$$mn(a - b) = mna - mnb = 0 - 0 = 0$$

Thus if $a, b \in T(G)$, $a - b \in T(G)$ and $T(G)$ is a subgroup of G.

Now consider $G/T(G)$. Assume $g + T(G)$ is of finite order n, i.e. $n(g + T(G)) = ng + T(G) = T(G)$. It follows that $ng \in T(G)$. As $T(G)$ consists of all the elements of G of finite order, there exists m such that $m(ng) = 0$. Then g is of finite order and $g \in T(G)$; hence $g + T(G) = T(G)$. Therefore the only element of finite order in $G/T(G)$ is the zero $T(G)$. Thus $G/T(G)$ is torsion-free.

Problems

6.29. Prove that if G is a group and H a subgroup of G such that G/H is torsion-free, then H contains the torsion subgroup of G.

Solution:

Let g in G be of finite order. Then $g + H$ is of finite order in G/H. Since G/H is torsion-free, $g + H = H$. This means $g \in H$, and so every element of finite order in G is contained in H, i.e. the torsion subgroup of G is contained in H.

6.30. Is the set consisting of 0 and of all elements of infinite order of a group automatically a subgroup?

Solution:

No. For example, let $G = H \oplus K$ where $H = gp(h)$ is infinite cyclic, and $K = gp(k)$ is of order 2. Now $h+k$ and $-h$ are both of infinite order. However $(h+k)+(-h) = k$ is of finite order.

6.31. Prove that if the set consisting of 0 and the elements of infinite order of a group G constitutes a subgroup, then G is either a torsion-free group or a torsion group.

Solution:

Suppose G is mixed and $H = \{h \mid h \in G$ and h is either 0 or of infinite order$\}$. Since G is mixed, we have $g \in G$ of infinite order and $g'(\neq 0) \in G$ of finite order. Now $g' - g$ is of infinite order so that $(g'-g) \in H$. But as H is a subgroup, $(g'-g)+g = g' \in H$. Therefore g' is 0 or an element of infinite order, contradicting the choice of g'. Hence G is not mixed.

6.32. Prove that if T is the torsion subgroup of G, then $T \cap H$ is the torsion subgroup of any given subgroup H of G.

Solution:

If $a \in H$ is of finite order, then $a \in T$. Hence $a \in T \cap H$, and so the torsion subgroup of H is contained in $T \cap H$. Conversely $T \cap H$ consists of elements of finite order, and so $T \cap H$ is contained in the torsion subgroup of H. Thus we have proved that $T \cap H$ is the torsion subgroup of H.

6.33. Find the torsion subgroup of R/Z where R is the group of real numbers under addition and Z is the subgroup of integers.

Solution:

Suppose $r+Z$ is of finite order $(r \in R)$. Then for some nonzero integer n, $n(r+Z) = Z$. But $n(r+Z) = nr+Z$, and so $nr \in Z$. This means that r is a rational number. Thus $T(R/Z) \subseteq Q/Z$, where Q is the subgroup of rational numbers. On the other hand, if $a+Z \in Q/Z$, then $a = m/n$ where $m, n \in Z$ and $n \neq 0$. So

$$n(a+Z) = n(m/n+Z) = n(m/n)+Z = m+Z = Z$$

Hence $a+Z$ is of finite order and $Q/Z \subseteq T(R/Z)$. Thus we have proved that $Q/Z = T(R/Z)$.

c. Structure of torsion groups. Prüfer groups

A group G is called a *p-group* or a *p-primary* group for some prime p if every element of G is of order a power of p. (If G is finite, it follows that the order of G is a power of p; see Problem 5.6, page 132. The definition of p-group given here thus coincides with that of Chapter 5 when the p-group is finite.) In this section we show that a torsion group is built out of p-groups. Thus the study of torsion groups becomes essentially the study of p-groups.

Theorem 6.10: Let G be any torsion group and let $G_p = \{g \mid g$ has order a power of $p\}$, p any prime. Then if Π is the set of all primes,

$$G = \sum_{p \in \Pi} G_p$$

Proof: We let the reader show that each G_p is a subgroup of G.

Suppose that $g \in G$ and is of order $p_1^{r_1} p_2^{r_2} \cdots p_n^{r_n}$, $p_1, \ldots, p_n$ distinct primes and $r_1, \ldots, r_n$ positive integers. Let $q = p_1^{r_1} \cdots p_{n-1}^{r_{n-1}}$; then $(q, p_n^{r_n}) = 1$. Thus there exist integers a and b such that $aq + bp_n^{r_n} = 1$. It follows that $g = aqg + bp_n^{r_n}g$. Now aqg is of order $p_n^{r_n}$, and so $aqg \in G_{p_n}$. $bp_n^{r_n}g$ is of order q. Since q is less then the order of g, we may assume inductively that $bp_n^{r_n}g$ is the sum of elements belonging to $G_{p_{n-1}}, G_{p_{n-2}}, \ldots, G_{p_1}$. Thus every element of G can be expressed as the sum of elements belonging to the G_p, i.e. G is generated by the subgroups G_p.

To show that $G = \sum_{p \in \Pi} G_p$, we must prove that $g_1 + \cdots + g_n = 0$ (where $g_i \in G_{p_i}$ and $p_1, \ldots, p_n$ are distinct primes) occurs only for $g_1 = g_2 = \cdots = g_n = 0$.

We proceed by induction on n. For $n = 1$ it is certainly true. If true for n, consider

$$g_1 + \cdots + g_{n+1} = 0$$

and let g_{n+1} be of order p_{n+1}^r. Then

$$p_{n+1}^r g_1 + p_{n+1}^r g_2 + \cdots + p_{n+1}^r g_{n+1} = 0$$

and

$$p_{n+1}^r g_1 + \cdots + p_{n+1}^r g_n = 0$$

By the inductive hypothesis, we have

$$p_{n+1}^r g_1 = p_{n+1}^r g_2 = \cdots = p_{n+1}^r g_n = 0$$

Now $p_{n+1}^r g_1 = 0$ implies $g_1 = 0$ as g_1 is of order some power of p_1 and $p_1 \neq p_{n+1}$. Arguing similarly for $g_2, \ldots, g_n$, we have $g_1 = g_2 = \cdots = g_n = 0$. Hence also $g_{n+1} = 0$ and we conclude that $G = \sum_{p \in \Pi} G_p$.

The subgroups G_p are called the *p-components* of G.

Example 1: Let us apply this theorem to Q/Z, i.e. the additive group of rationals modulo the integers. Q/Z is clearly a torsion group.

$$(Q/Z)_p = \{x + Z \mid x + Z \text{ of order a power of } p\} = \{x + Z \mid p^r x \in Z\}$$
$$= \{m/p^r + Z \mid \text{for various integers } r \text{ and } 0 \leq m < p^{r-1}\}$$

By the theorem, $Q/Z = \sum_{p \in \Pi} (Q/Z)_p$.

$(Q/Z)_p$ is called the *p-Prüfer group* (also called a *group of type* p^∞). In Section 6.4 the p-Prüfer groups will be fundamental. Note that the p-Prüfer group $(Q/Z)_p = \bigcup_{r=1}^{\infty} C_r$ where $C_r = gp(1/p^r + Z)$, since $(Q/Z)_p = \{m/p^r + Z \mid$ for various integers r and $0 \leq m < p^{r-1}\}$. Clearly $(Q/Z)_p \supseteq C_r$ and $(Q/Z)_p \subseteq \bigcup_{r=1}^{\infty} C_r$. The result follows.

We now have at our disposal cyclic groups of all orders, the additive group of rationals, and the p-Prüfer groups, together with all their direct sums. These, as we shall prove, constitute a large class of abelian groups.

Problems

6.34. Use Theorem 6.10 to prove that an abelian group of order pq, where p and q are different primes, is the direct sum of a cyclic group of order p and a cyclic group of order q.

Solution:

Let G be of order pq. Then by Theorem 6.10, $G = G_p \oplus G_q$; for if r is any prime other than p or q, $G_r = \{0\}$. Why does $G_r = \{0\}$? If $g \in G_r$, then, as G is of order pq, $pqg = 0$. Hence r divides pq, which is not the case. Note that $G_p \neq \{0\}$, $G_q \neq \{0\}$ by Proposition 5.9, page 137. $|G_p|$ divides pq. Since $G_q \neq \{0\}$, $|G_p| = p$ or q. As the elements of G_p are of order a power of p, it follows that $|G_p| = p$. Similarly $|G_q| = q$. Hence, as the only group of prime order is cyclic, we have the result.

6.35. Show that a direct sum of p-groups is again a p-group.

Solution:

Let $G = \sum_{i \in I} G_i$ where each G_i is a p-group. Let $g \in G$. Then $g = g_1 + \cdots + g_n$ where each $g_i \in G_{i'}$, $1', 2', \ldots, n' \in I$. Let p^r be the maximum order of the g_i. Then $p^r g = 0$. Thus the order of g is a power of p, and so G is a p-group.

6.36. Show that if G is a p-Prüfer group, for each integer $k \neq 0$ and each element $g \in G$ there exists an element $h \in G$ such that $kh = g$. (We shall call a group with this property *divisible.*)

Solution:

Let $g \in G$. Then $g = m/p^r + Z$ where $0 \leqq m < p^r$. Let $k = p^s l$ where p and l are co-prime, and let $h_1 = m/p^{r+s} + Z$. Then $p^s h_1 = g$. As l and p^{r+s} are co-prime, there exist integers a and b such that $al + bp^{r+s} = 1$. Therefore

$$h_1 = (al + bp^{r+s})h_1 = alh_1 + bp^{r+s}h_1 = alh_1$$

Put $h = ah_1$. Then

$$kh = p^s lh = p^s lah_1 = p^s h_1 = g$$

6.37. Let $G = \sum_{i=1}^{\infty} G_i$ where G_i is a cyclic group of order p_i and $p_1, p_2 \ldots$ are the primes in ascending order of magnitude. Let H be a p-Prüfer group for any prime p. Prove that G is not isomorphic to H.

Solution:

Let p_j be a prime different from p. Then the p_jth component $G_{p_j} \neq \{0\}$ but $H_{p_j} = \{0\}$. Thus G is not isomorphic to H.

d. Independence and rank

We introduce here an important concept in abelian group theory, the concept of rank.

Let G be an abelian group. A subset X of G is called *independent* if whenever $x_1, \ldots, x_n$ are distinct elements of X and $n_1, \ldots, n_r$ are integers such that

$$n_1 x_1 + \cdots + n_r x_r = 0 \qquad (6.3)$$

then $n_1 x_1 = \cdots = n_r x_r = 0$.

Note that if G is torsion-free and X is independent, then equation (*6.3*) implies $n_1 = n_2 = \cdots = n_r = 0$. Suppose $X = \{x_i \mid i \in I\}$ and $x_i \neq x_j$ for $i \neq j$. Then if X is an independent set, it follows readily that $gp\,(X) = \sum_{i \in I} gp(x_i)$.

We need one further definition. An element x in a group G is *dependent* on a subset X of G if

$$nx + n_1 x_1 + \cdots + n_r x_r = 0$$

for some choice of $x_1, \ldots, x_r \in X$, n and $n_j \in Z$ and $nx \neq 0$. In other words, x is dependent on the subset X if there is an integer n with $nx \neq 0$ and $nx \in gp(X)$. We say $Y \subseteq G$ is dependent on $X \subseteq G$ if every element of Y is dependent on X. Observe also that if G is torsion-free with subsets X, Y and W and if X is dependent on Y and Y is dependent on W, then X is dependent on W. For if $x \in X$, then for some integer $n \neq 0$, and integers $n_1, \ldots, n_r$, $nx = n_1 y_1 + \cdots + n_r y_r$ $(y_i \in Y)$. Since Y is dependent on W we can find integers $m_1 \neq 0, \ldots, m_r \neq 0$ such that $m_i y_i \in gp(W)$ for $i = 1, \ldots, r$. Put $m = m_1 \cdots m_r$. Then clearly $mn_i y_i \in gp(W)$ for $i = 1, \ldots, r$, and consequently $mnx \in gp(W)$. Furthermore, since G is torsion-free, $mnx \neq 0$. Thus every element of X depends on W.

The main result that we shall now prove is called the Steinitz Exchange Theorem.

Theorem 6.11: Let G be torsion-free and let $A = \{a_1, \ldots, a_m\}$ be an independent subset of G. Suppose $B = \{b_1, \ldots, b_n\}$ is another subset of G such that A is dependent on B. Then $n \geqq m$ and B depends on $A \cup C$ where C is a subset of B and $|C| = n - m$.

Proof: We will use induction on m. For $m = 1$ it is clear that $n \geqq m$. Now a_1 depends on B means that there exist integers $z, \ldots, z_n$ such that $z \neq 0$,

$$za_1 + z_1 b_1 + \cdots + z_n b_n = 0$$

and thus for some integer i, $z_i \neq 0$, $1 \leqq i \leqq n$. Hence B depends on $\{a_1\} \cup (B - \{b_i\})$. Thus the result holds for $m = 1$, with $C = B - \{b_i\}$.

Next we assume that the result holds for $m = r$, and consider the case $m = r+1$. Then by the inductive hypothesis, $n \geqq r$. Now $\{a_1, \ldots, a_r\}$ depends on B, and so inductively B depends on $\{a_1, \ldots, a_r\} \cup D$ where $D \subseteq B$ and $|D| = n - r$. But $\{a_{r+1}\}$ depends on B and B depends on $\{a_1, \ldots, a_r\} \cup D$. Then by our remarks above, $\{a_{r+1}\}$ depends on $\{a_1, \ldots, a_r\} \cup D$. Thus we can find integers $y \neq 0, y_1, \ldots, y_r, z_1, \ldots, z_s$ such that

$$ya_{r+1} = y_1 a_1 + \cdots + y_r a_r + z_1 d_1 + \cdots + z_s d_s, \qquad d_1, \ldots, d_s \in D$$

Suppose, if possible, that $z_1 = z_2 = \cdots = z_s = 0$. Then $ya_{r+1} = y_1 a_1 + \cdots + y_r a_r$ implies that the elements $a_1, \ldots, a_{r+1}$ are not independent. So some $z_i \neq 0$. Let $C = D - \{d_i\}$. Then it is clear that $\{a_1, \ldots, a_r\} \cup D$ depends on $\{a_1, \ldots, a_{r+1}\} \cup C$. Since B depends on $\{a_1, \ldots, a_r\} \cup D$, then B depends on $\{a_1, \ldots, a_{r+1}\} \cup C$. Finally $C \subseteq B$ and $|C| = |D| - 1$. Thus

$$|C| = (n-r) - 1 = n - (r+1) = n - m$$

as desired, and the proof of Steinitz's theorem is complete.

Let us call a subset S of a torsion-free abelian group G a *maximal independent* set

(i) if S is independent and

(ii) if $g \in G$ and $g \notin S$, then $S \cup \{g\}$ is not an independent set.

Suppose now that G, a torsion-free abelian group, has a maximal independent set S that is finite. Let T be any other finite maximal independent set. By the Steinitz exchange theorem, $|S| \leqq |T|$. Also by the same theorem, $|T| \leqq |S|$. Hence we can without ambiguity define the *rank* of a torsion-free abelian group G which has a finite maximal independent set S to be $|S|$. If G does not have a finite maximal independent set S, we shall say G is of *infinite rank.*

It is easy to see that if G and H are isomorphic groups, then they have equal ranks.

As a consequence of these remarks we obtain a result concerning free abelian groups.

If F is free abelian with a finite set of free generators X, then X is a maximal independent set of F (see Problem 6.41 below). Hence the rank of F is $|X|$. Similarly if F is also freely generated by a finite set Y, then rank of $F = |Y|$. Hence $|Y| = |X|$. Thus we have

Corollary 6.12: If F is a group freely generated by two finite sets X and Y, then $|X| = |Y|$.

We have as yet not proved that all abelian groups have maximal independent sets. To do so we need a result called Zorn's lemma. Before stating the lemma, we consider the following examples:

(*a*) Let $\mathcal{P} = \{A, B, C, D\}$, where $A = \{0,1\}$, $B = \{1,2\}$, $C = \{0,2\}$, $D = \{0,1,2,3\}$.

We inquire: does $\mathcal{P}$ have a *largest element,* i.e. one that contains all the elements of $\mathcal{P}$? Clearly it does, for $D \supseteq A$, $D \supseteq B$, $D \supseteq C$ and $D \supseteq D$. Thus D is a largest element.

(*b*) Let $\mathcal{P} = \{A, B, C, E\}$, where A, B, C are as in (*a*), and $E = \{1,2,3\}$. Now there is no largest element of $\mathcal{P}$. We search for some concept replacing that of a largest element. Note that although E is not a largest element, no element other than itself contains it. Then E is called a *maximal* element. Similarly C and A are maximal elements, whereas B is not.

(*c*) If $|\mathcal{P}|$ is finite, then it is clear that $\mathcal{P}$ has maximal elements. For we choose any element A_1 of $\mathcal{P}$. If there is an element of $\mathcal{P}$ that contains A_1 properly, we call it A_2. If there is an element of $\mathcal{P}$ that contains A_2 properly, we call it A_3. Continuing in this way we get a *chain* of elements of $\mathcal{P}$, $A_1 \subset A_2 \subset \cdots \subset A_i \subset \cdots$. As $\mathcal{P}$ has only a finite number of elements, this chain ends at A_n, say. Clearly A_n is a maximal element.

(*d*) On the other hand, not all sets $\mathcal{P}$ of sets have maximal elements. For example, let $A_n = \{0, 1, \ldots, n\}$ for $n = 1, 2, \ldots$. Let $\mathcal{P} = \{A_i \mid i \in P$, the positive integers$\}$. Then $\mathcal{P}$ has no maximal element. For if $X \in \mathcal{P}$, then $X = A_i$ for some i, and $X \subseteq A_{i+1}$, but $X \neq A_{i+1}$.

Zorn's lemma establishes a criterion for determining whether a set $\mathcal{P}$ of sets has a maximal element. What one needs is some condition for handling an ascending sequence of sets such as the A_i in (*d*). To state the criterion we need some definitions.

(1) We define A to be a *maximal* set in $\mathcal{P}$ if for each $X \in \mathcal{P}$, $X \supseteq A$ implies $X = A$.

(2) Let $\mathcal{C}$ be a subset of $\mathcal{P}$ with the property that if $X, Y \in \mathcal{C}$, then either $X \subseteq Y$ or $Y \subseteq X$. Then $\mathcal{C}$ is called a *chain* in $\mathcal{P}$ (in (*d*) above, $\mathcal{P}$ itself is a chain).

We are now in a position to state Zorn's lemma: Let $\mathcal{P}$ be a set of sets. Suppose that for every chain $\mathcal{C}$ in $\mathcal{P}$, $\bigcup_{X \in \mathcal{C}} X$ is an element of $\mathcal{P}$. Then $\mathcal{P}$ has a maximal element.

We will not prove Zorn's lemma. We take it as an axiom. We could assume a more innocent sounding axiom instead, namely the axiom of choice, which says that an element from each set may be chosen from a collection of sets. The proof of Zorn's lemma can be derived from the axiom of choice (see Problem 6.42 for a sketch of the proof).

Using Zorn's lemma we prove

Theorem 6.13: Let G be any abelian group. Then G has a maximal independent set.

Proof: Let $\mathcal{P} = \{X \mid X \subseteq G$ and X an independent set$\}$. Let $\mathcal{C}$ be any chain in $\mathcal{P}$. Let $\mathcal{C} = \{X_i \mid i \in I\}$. To apply Zorn's lemma we must show that $U = \bigcup_{i \in I} X_i \in \mathcal{P}$. Clearly $U \subseteq G$. Is U an independent set? If not, U is a dependent set. This means that it is possible to find distinct elements $u_1, \ldots, u_n \in U$ and integers $r_1, \ldots, r_n$ such that

$$r_1 u_1 + \cdots + r_n u_n = 0$$

with at least one $r_i u_i \neq 0$. As $U = \bigcup_{i \in I} X_i$, $u_1 \in X_{1'}$, $u_2 \in X_{2'}$ where $1'$, $2'$ are elements of I. Then since $\mathcal{C}$ is a chain, either $X_{1'} \subseteq X_{2'}$ or $X_{2'} \subseteq X_{1'}$. Thus u_1, u_2 both belong to some element of $\mathcal{C}$. Continuing the argument in this way we find that $u_1, \ldots, u_n$ all belong to some $X_i \in \mathcal{C}$. But this is a contradiction, since every element of $\mathcal{C}$ is independent. So U is independent and $U \in \mathcal{P}$. We conclude, using Zorn's lemma, that $\mathcal{P}$ has a maximal element and this is precisely the maximal independent set required. Hence the result follows.

From Theorem 6.13 we conclude that if G is of infinite rank, then G has an infinite maximal independent subset.

Problems

6.38. Find the rank of the additive group Q of rationals.

Solution:

If m_1/n_1 and m_2/n_2 are elements of Q, m_1, m_2, n_1, n_2 integers, then

$$(m_2 n_1)(m_1/n_1) + (-m_1 n_2)(m_2/n_2) = 0$$

Thus every set of two elements is dependent. Accordingly the rank of Q is 1.

6.39. Show that the p-Prüfer group has no independent set consisting of two elements.

Solution:

Let x, y be elements of G, a p-Prüfer group, $x \neq 0$, $y \neq 0$. Then $x, y \in C_r$, say, for some r (see Example 1, page 191, for the notation C_r). Let $C_r = gp(g)$ and let $gp(x)$ be of order p^t. Since $p^{r-t}g$ is of order p^t, $gp(p^{r-t}g) = gp(x)$. Thus $gp(x) = gp(p^i g)$ and $gp(y) = gp(p^j g)$ for some i, j. If $i \geqq j$, it follows that $gp(y) \supseteq gp(x)$. (If $i \leqq j$, we merely reverse the roles of x and y.) Consequently $x = ry$ for some integer $0 < r <$ order of y. Thus $(-r)y + 1 \cdot x = 0$ and x, y are not independent.

6.40. Prove that if H and K are torsion-free groups of finite rank m and n respectively, then $G = H \oplus K$ is of rank $m+n$. (Difficult.)

Solution:

Let $h_1, \ldots, h_m$ be a set of independent elements of H, and $k_1, \ldots, k_n$ a set of independent elements of K. Then the set $h_1, \ldots, h_m, k_1, \ldots, k_n$ is independent. For if $r_1h_1 + \cdots + r_mh_m + s_1k_1 + \cdots + s_nk_n = 0$, then

$$r_1h_1 + \cdots + r_mh_m = -s_1k_1 - \cdots - s_nk_n$$

But $H \cap K = \{0\}$, and so

$$r_1h_1 + \cdots + r_mh_m = -s_1k_1 - \cdots - s_nk_n = 0$$

Thus, by the independence of $h_1, \ldots, h_m$ and $k_1, \ldots, k_n$,

$$r_1 = r_2 = \cdots = r_m = s_1 = s_2 = \cdots = s_n = 0$$

Now suppose that $\{h_1, \ldots, h_m, k_1, \ldots, k_n\}$ is not maximal; say, there exists an element $h+k$ where $h \in H$ and $k \in K$ such that $\{h_1, \ldots, h_m, k_1, \ldots, k_n, h+k\}$ is independent. Since $\{h_1, \ldots, h_m, h\}$ is not independent, there exist integers $t_1, \ldots, t_m$, t not all zero such that

$$t_1h_1 + \cdots + t_mh_m + th = 0$$

If $t = 0$ we have a contradiction to $\{h_1, \ldots, h_m\}$ being independent, as then at least one of $t_1, \ldots, t_m$ is nonzero. Hence $t \neq 0$.

Next, $\{k, k_1, \ldots, k_n\}$ is not independent, i.e. there exist $s, s_1, \ldots, s_n$, not all zero, such that $sk + s_1k_1 + \cdots + s_nk_n = 0$. Hence arguing as above, it follows that $s \neq 0$. Thus

$$\begin{aligned} &st_1h_1 + \cdots + st_mh_m + st(h+k) + ts_1k_1 + \cdots + ts_nk_n \\ &\quad = s(th + t_1h_1 + \cdots + t_mh_m) + t(sk + s_1k_1 + \cdots + s_nk_n) = 0 \end{aligned}$$

But as $st \neq 0$, $\{h_1, h_2, \ldots, h_m, k_1, \ldots, k_n, h+k\}$ is not independent. Thus $\{h_1, h_2, \ldots, h_m, k_1, \ldots, k_n\}$ is a maximal independent set and rank $H \oplus K = m+n$.

6.41. (*a*) If F is free abelian with a finite set of free generators X, prove that the rank of F is $|X|$.

(*b*) Prove that F cannot be generated by fewer than $|X|$ elements.

Solution:

(*a*) Let $X = \{x_1, \ldots, x_n\}$. By Theorem 6.7, page 186,

$$G = gp(x_1) \oplus gp(x_2) \oplus \cdots \oplus gp(x_n)$$

We proceed by induction on n. For $n = 1$, G is infinite cyclic, and the rank of G is clearly 1. If true for $n = r$, suppose $n = r+1$. Then $gp(x_1) \oplus \cdots \oplus gp(x_r)$ is of rank r, and G is the direct sum of a torsion-free group of rank r and a group of rank 1. By Problem 6.40, G is thus of rank $r+1$, and the result follows by induction for all n.

(*b*) Let $G = gp(g_1, \ldots, g_r)$. X is an independent set. Then by the Steinitz exchange theorem (Theorem 6.11), as X is dependent on $\{g_1, \ldots, g_r\}$, $n \leq r$. Thus we obtain the result.

6.42. Let X be a set and $\mathcal{P}$ a collection of subsets of X. Suppose that if $A \in \mathcal{P}$, all subsets of A belong to $\mathcal{P}$.

(*a*) Prove that if $A \in \mathcal{P}$ is not a maximal element, there exists a set $A^* = \{A, x\} \in \mathcal{P}$ with $x \notin A$.

(*b*) Assume that $A^* = A$ if A is maximal, otherwise that A^* has been defined equal to $\{A, x\}$ with $x \notin A$ as stated in (*a*). Let $\mathcal{C}$ be a chain in $\mathcal{P}$. Suppose that if C_i, $i \in I$, is a family of elements in $\mathcal{C}$, then $\bigcup_{i \in I} C_i \in \mathcal{C}$. Suppose also that if $A \in \mathcal{C}$, $A^* \in \mathcal{C}$. Prove that $\mathcal{P}$ has a maximal element.

Solution:

(*a*) If $A \in \mathcal{P}$ is not maximal, there exists a set $B \in \mathcal{P}$ such that $B \neq A$ and $B \supseteq A$. Hence there exists $x \in B - A$. Since $\{A, x\}$ is a subset of B, $\{A, x\} \in \mathcal{P}$.

(*b*) Let $M = \bigcup_{C \in \mathcal{C}} C$. Then $M \in \mathcal{C}$, and so $M^* \in \mathcal{C}$. But $M^* \supseteq M$. However, M contains every element of $\mathcal{C}$; in particular, it contains M^*. Therefore $M = M^*$ and we conclude that M is a maximal element of $\mathcal{P}$.

Remarks. (1) In assuming that A^* can be defined we used implicitly the axiom of choice. (2) The proof of Zorn's lemma requires converting the theorem into this problem. For details see P. R. Halmos, *Naive Set Theory*, Van Nostrand, 1960.

6.43. Let G be an arbitrary non-abelian group. Prove that G has a maximal abelian subgroup (i.e. one that is not properly contained in an abelian subgroup of G).

Solution:

We use multiplicative notation for G since it is not abelian. Let $\mathcal{P}$ be the set of all abelian subgroups of G. Let $\mathcal{C}$ be a chain in $\mathcal{P}$ and let $U = \bigcup_{X \in \mathcal{C}} X$. Then U is a subgroup of G. For if $g, h \in U$ and if g belongs to $X_1 \in \mathcal{C}$ and h to $X_2 \in \mathcal{C}$, then as either $X_1 \subseteq X_2$ or $X_2 \subseteq X_1$, it follows that g, h belong to some element X of $\mathcal{C}$. Hence $gh^{-1} \in X$ as X is a subgroup, and $gh^{-1} \in U$. Also, $gh = hg$ as X is an abelian subgroup of G. Consequently U is abelian. Hence $U \in \mathcal{P}$. By Zorn's lemma, $\mathcal{P}$ has a maximal element M, say. M is the maximal abelian group sought.

6.3 FINITELY GENERATED ABELIAN GROUPS

a. Lemmas for finitely generated free abelian groups

In Section 6.3b we will show that all finitely generated abelian groups are direct sums of cyclic groups. We will do this by using a lemma (Lemma 6.15) about subgroups of free abelian groups. The relationship between Lemma 6.15 and finitely generated abelian groups is easily obtained by noting that all abelian groups are factor groups of free abelian groups.

Lemma 6.14: Let $G = gp(a_1) \oplus \cdots \oplus gp(a_n)$ be the direct sum of infinite cyclic groups. If $b_1 = a_1 + r_2a_2 + \cdots + r_na_n$, where $r_2, \ldots, r_n$ are any integers, then

$$G = gp(b_1) \oplus gp(a_2) \oplus \cdots \oplus gp(a_n)$$

Proof: As $gp(b_1, a_2, \ldots, a_n) = gp(a_1, \ldots, a_n) = G$, we must only show that if $s_1, \ldots, s_n$ are any integers, then

$$s_1b_1 + s_2a_2 + \cdots + s_na_n = 0 \tag{6.4}$$

implies all s_i are 0.

Substituting $b_1 = a_1 + r_2a_2 + \cdots + r_na_n$ into (6.4) and collecting terms, we obtain

$$s_1a_1 + (s_2 + s_1r_2)a_2 + \cdots + (s_n + s_1r_n)a_n = 0$$

As $G = \{a_1\} \oplus \cdots \oplus \{a_n\}$,

$$s_1 = s_2 + s_1r_2 = \cdots = s_n + s_1r_n = 0$$

Thus $s_1 = s_2 = \cdots = s_n = 0$ and the proof is complete.

The next lemma is a crucial one. We recall that a basis $c_1, \ldots, c_n$ for a finitely generated free abelian group G is a set of elements such that $G = gp(c_1) \oplus \cdots \oplus gp(c_n)$ (see Section 6.1c).

Lemma 6.15: Let G be free abelian, the direct sum of n cyclic groups. Let H be a subgroup of G. Then there exists a basis $c_1, \ldots, c_n$ of G and integers $u_1, \ldots, u_n$ such that $H = gp(u_1c_1, u_2c_2, \ldots, u_nc_n)$.

Proof: We use a, b, c to denote basis elements of G, h, k, l to denote elements of H, q, r, s, t, u, v to denote integers. We prove the result by induction on n. For $n = 1$, G is cyclic and the result is a consequence of Theorem 4.9, page 105. Assume the result is true for free abelian groups of rank less than n where $n > 1$. Let G be free abelian of rank n. We assume also that $H \neq \{0\}$. For if $H = \{0\}$, we may take an arbitrary basis $c_1, \ldots, c_n$ for G. Then $H = gp(u_1c_1, \ldots, u_nc_n)$ where $u_1 = \cdots = u_n = 0$.

To every basis we associate an integer, called its *size* (*with respect to H*). Let $\{a_1, \ldots, a_n\}$ be a basis for G and let q be the smallest nonnegative integer such that there exists $h \in H$ with

$$h = qa_1 + q_2a_2 + \cdots + q_na_n, \quad q_2, \ldots, q_n \text{ integers} \tag{6.5}$$

Then q is termed the *size* of the basis $\{a_1, a_2, \ldots, a_n\}$.

Assume $\{a_1, \ldots, a_n\}$ is a basis of smallest size, i.e. if $\{b_1, \ldots, b_n\}$ is a basis of G, then the size of $\{b_1, \ldots, b_n\}$ is not less than q.

Let h be as in equation (6.5). We show that q divides $q_2, \ldots, q_n$. From the division algorithm, if q_i is not divisible by q, $q_i = r_i q + s_i$ where $0 < s_i < q$. Hence

$$h = q(a_1 + r_i a_i) + \cdots + s_i a_i + \cdots + q_n a_n$$

But if we put $b_1 = a_i$, $b_2 = a_2$, ..., $b_i = a_1 + r_i a_i$, ..., $b_n = a_n$, we obtain a basis by Lemma 6.14. Furthermore this basis is of *smaller* size than the size of $\{a_1, \ldots, a_n\}$, contrary to our assumption. Thus $s_i = 0$ and q divides q_i for $i = 2, \ldots, n$. Let $q_i = r_i q$. Then

$$h = q(a_1 + r_2 a_2 + \cdots + r_n a_n)$$

Let $c_1 = a_1 + r_2 a_2 + \cdots + r_n a_n$. Then, by Lemma 6.14, $\{c_1, a_2, \ldots, a_n\}$ is a basis for G. Also

$$h = qc_1 \tag{6.6}$$

If $k = t_1 a_1 + \cdots + t_n a_n \in H$, it follows that t_1 is divisible by q. For if $t_1 = uq + v$ with $0 \leqq v < q$, then $l = k - uh \in H$ has v as its coefficient of a_1. As $v < q$, by the minimality of q, $v = 0$. Therefore

$$l = k - uh \in gp(a_2, \ldots, a_n)$$

Hence $l \in gp(a_2, \ldots, a_n) \cap H = L$, say. From this we conclude that if $k \in H$, then

$$k = uh + l \tag{6.7}$$

where $l \in L$.

By the inductive hypothesis there exist a basis $c_2, \ldots, c_n$ and integers $u_2, \ldots, u_n$ such that L is generated by $u_2 c_2, \ldots, u_n c_n$. Hence by (6.7) every element of H belongs to $gp(h, u_2 c_2, \ldots, u_n c_n)$. On the other hand, H contains $h, u_2 c_2, \ldots, u_n c_n$. Thus

$$H = gp(h, u_2 c_2, \ldots, u_n c_n)$$

Put $u_1 = q$. By (6.6),

$$H = gp(u_1 c_1, u_2 c_2, \ldots, u_n c_n)$$

Also, $c_1, \ldots, c_n$ is a basis for G. Hence the result follows.

Note that if any u_i is negative, we can replace c_i by its inverse $-c_i$. In this manner we can assume that the u_i are nonnegative.

Lemma 6.16: Suppose $G = A \oplus B$. Let A_1, B_1 be subgroups with $A_1 \subseteq A$, $B_1 \subseteq B$ and $N = A_1 + B_1$. Then $G/N \cong A/A_1 \oplus B/B_1$.

Proof: Let $K = A/A_1 \oplus B/B_1$, and let $\theta: A \to A/A_1$ and $\phi: B \to B/B_1$ be the natural homomorphisms. θ, ϕ extend to a homomorphism Ψ of G into K. Then $\operatorname{Ker}\Psi \supseteq \operatorname{Ker}\theta = A_1$ and $\operatorname{Ker}\Psi \supseteq \operatorname{Ker}\phi = B_1$. Thus $\operatorname{Ker}\Psi \supseteq A_1 + B_1$. Now let $x \in \operatorname{Ker}\Psi$. Then $x = a + b$, $a \in A$, $b \in B$. $x\Psi = (a + A_1) + (b + B_1)$ and this is the identity element only if $a \in A_1$ and $b \in B_1$. Hence $x \in A_1 + B_1$, and so $\operatorname{Ker}\Psi = A_1 + B_1$. By the homomorphism theorem (Theorem 4.18, page 117) $G/N \cong K$ and the result follows.

Corollary 6.17: Let G be free abelian with basis $c_1, \ldots, c_n$. Let $H = gp(u_1 c_1, \ldots, u_n c_n)$ where $u_1, \ldots, u_n$ are nonnegative integers. Then G/H is the direct sum of cyclic groups of orders $u_1', \ldots, u_n'$, where $u_i' = u_i$ if $u_i \neq 0$ and $u_i' = \infty$ if $u_i = 0$.

Proof: The result follows by repeated application of Lemma 6.16.

b. Fundamental theorem of abelian groups

The following theorem is called the fundamental theorem of abelian groups.

Theorem 6.18: Let G be a finitely generated abelian group. Then G is the direct sum of a finite number of cyclic groups.

Proof: $G \cong F/H$ where F is a finitely generated free abelian group (Section 6.1c). By Lemma 6.15, F has a basis $c_1, \ldots, c_n$ such that $H = gp(u_1c_1, \ldots, u_nc_n)$ for some nonnegative integers $u_1, \ldots, u_n$. We now apply Corollary 6.17 to conclude that $G \cong F/H$ is the direct sum of cyclic groups.

Corollary 6.19: If G is finitely generated, it is the direct sum of a finite number of infinite cyclic groups and cyclic groups of prime power order.

Proof: It is only necessary to show that a cyclic group of composite order is the direct sum of cyclic groups of prime power order. This we have already done in Problem 6.19, page 187.

Corollary 6.20: If G is a group with no elements of finite order and G is finitely generated, then G is free abelian.

Proof: G is the direct sum of a finite number of cyclic groups each of which must be infinite cyclic as G has no elements of finite order. Thus the result follows.

Problems

6.44. Prove that every finitely generated torsion group is finite.

Solution:

By the fundamental theorem of finitely generated abelian groups, if G is finitely generated it is the direct sum of a finite number of cyclic groups. If G is a torsion group, then it is the direct sum of a finite number of finite cyclic groups. Hence G is finite. (Compare with Problem 4.31, page 105.)

6.45. Let $G = \sum_{i=1}^{\infty} G_i$ where G_i is a cyclic group of order 2 for $i = 1, 2, \ldots$. Prove that G is not finitely generated.

Solution:

Every element in G is of finite order, for if $g(\neq 1) \in G$, $g = g_1 + g_2 + \cdots + g_n$, $g_i \in G_{i'}$ $(i' \in Z)$, and $2g = 2g_1 + \cdots + 2g_n = 0 + \cdots + 0 = 0$. Thus G is a torsion group. If G were finitely generated, G would be finite by the preceding problem. But G is clearly infinite. Therefore G cannot be finitely generated.

c. The type of a finitely generated abelian group

In Section 6.3b we proved that a finitely generated abelian group is a direct sum of cyclic groups. However, such a decomposition is not unique: first, the direct summands are not unique (see Problem 6.46 below); moreover, the number of direct summands can vary (see Problem 6.19, page 187).

We say that two decompositions *are of the same kind* if they have the same number of summands of each order. For example, two decompositions of a group into the direct sum of three cyclic groups of order 4 and two cyclic groups of infinite order are said to be of the same kind. A concrete example of two decompositions of the same kind is given in Problem 6.46.

As we remarked in Corollary 6.19, every finitely generated group can be decomposed into the direct sum of a finite number of cyclic groups of prime power or else infinite order.

Our aim is to prove

Theorem 6.21: Any two decompositions of a group G into the direct sum of a finite number of cyclic groups which are either of prime power order ($\neq 1$) or of infinite order, are of the same kind.

Proof: We shall separate the proof into four cases: (1) both decompositions involve only infinite cyclic groups, (2) both decompositions involve only cyclic groups of order a power of fixed prime p, (3) both decompositions involve no infinite cyclic groups, and (4) the general case.

Case 1.
$$G = I_1 \oplus \cdots \oplus I_k = \hat{I}_1 \oplus \cdots \oplus \hat{I}_l$$
where $I_j, \hat{I}_i$ for $j = 1, \ldots, k$ and $i = 1, \ldots, l$ respectively, are infinite cyclic groups.

From Corollary 6.12, page 193, we conclude that $k = l$. (Alternatively we may proceed as in Problem 6.52.)

Case 2.

Both decompositions involve only cyclic groups of order a power of a fixed prime p. We shall write for any integer n, $nG = \{ng \mid g \in G\}$. If G is a group, nG is a subgroup (Problem 6.53). To prove case 2 we will need the following lemma.

Lemma 6.22: Let $G = A \oplus B$. If n is any integer, then $nG = nA \oplus nB$.

Proof: As $nA \cap nB \subseteq A \cap B = \{0\}$, $gp(nA, nB) = nA \oplus nB$. If $g \in nG$, there exists $h \in G$ such that $nh = g$. Let $h = a + b$, $a \in A$ and $b \in B$. Then $g = nh = na + nb$. Accordingly $nG \subseteq nA \oplus nB \subseteq nG$ and so $nG = nA \oplus nB$.

Corollary 6.23: Let $G = A_1 \oplus \cdots \oplus A_k$. Let n be an integer. Then
$$nG = nA_1 \oplus \cdots \oplus nA_k$$

Proof: We apply Lemma 6.22 to one direct summand at a time. Then the result follows.

Corollary 6.24: Let G be expressed as the direct sum of k_i cyclic groups of order p^i, $1 \leqq i \leqq r$. Then pG is expressible as the direct sum of k_i cyclic groups of order p^{i-1} where $2 \leqq i \leqq r$.

Proof: This is an immediate consequence of Corollary 6.23 and the fact that if A is cyclic of order p^i, pA is cyclic of order p^{i-1}. Hence the corollary follows.

We are now in a position to prove case 2. We proceed by induction on the order of G. If $|G| = 1$ or p, then the result is immediate. If the result is assumed true for all groups of order less than n that satisfy the conditions of case 2, then let $|G| = n$. Suppose G is expressed as the direct sum of k_i cyclic groups of order p^i for $1 \leqq i \leqq r$, and also as l_j cyclic groups of order p^j for $1 \leqq j \leqq s$. Then pG is expressible (by Corollary 6.24) as the direct sum of k_i cyclic groups of order p^{i-1} for $2 \leqq i \leqq r$ on the one hand, and as the direct sum of l_i cyclic groups of order p^{i-1} for $2 \leqq i \leqq s$ on the other. As $|pG| < |G|$, it follows by the induction assumption that $r = s$ and $k_i = l_i$ for $2 \leqq i \leqq r$. Now we must still prove that $k_1 = l_1$. But $|G| = p^{k_1}(p^2)^{k_2} \cdots (p^r)^{k_r} = p^{l_1}(p^2)^{l_2} \cdots (p^r)^{l_r}$, and so $l_1 = k_1$. Thus we have proved both decompositions are of the same kind, as required.

Case 3.

G is expressed in two ways as the direct sum of a finite number of cyclic groups of prime power order.

We have dealt with the case where only one prime is involved. We proceed by induction on the number of primes involved. Let p be one of the primes involved. Let $A_1, \ldots, A_m$ be all the direct summands of order a power of p in the one decomposition, $B_1, \ldots, B_n$ the other direct summands involved, so that
$$G = A_1 \oplus \cdots \oplus A_m \oplus B_1 \oplus \cdots \oplus B_n$$
Putting $A = A_1 \oplus \cdots \oplus A_m$ and $B = B_1 \oplus \cdots \oplus B_n$, it follows that $G = A \oplus B$.

Let $X_1, \ldots, X_k$ be all the direct summands of order a power of p in the second decomposition, $Y_1, \ldots, Y_l$ the remaining direct summands, so that
$$G = X_1 \oplus \cdots \oplus X_k \oplus Y_1 \oplus \cdots \oplus Y_l$$
Put $X = X_1 \oplus \cdots \oplus X_k$, $Y = Y_1 \oplus \cdots \oplus Y_l$. Then $G = X \oplus Y$. We claim that $A = X$ and $B = Y$.

Let $g \in A$. Then $g = x + y$ where $x \in X$ and $y \in Y$. Now the order of any nonzero element of Y is coprime to p. As g is of order a power of p, $y = 0$ (Problem 6.54). Hence $g \in X$, and so $A \subseteq X$. Similarly $X \subseteq A$ and we conclude that $A = X$. By a similar argument $B = Y$.

Thus $A_1 \oplus \cdots \oplus A_m = X_1 \oplus \cdots \oplus X_k$ and $B_1 \oplus \cdots \oplus B_n = Y_1 \oplus \cdots \oplus Y_l$. By the induction hypothesis, $A_1 \oplus \cdots \oplus A_m$ and $X_1 \oplus \cdots \oplus X_k$ on the one hand, and $B_1 \oplus \cdots \oplus B_n$ and $Y_1 \oplus \cdots \oplus Y_l$ on the other, are of the same kind. Hence the two decompositions are of the same kind and the result follows.

Case 4.

Let G be expressed as the direct sum of cyclic groups of prime power order or of infinite order in two ways, say

$$G = I_1 \oplus \cdots \oplus I_m \oplus F_1 \oplus \cdots \oplus F_n = \hat{I}_1 \oplus \cdots \oplus \hat{I}_k \oplus \hat{F}_1 \oplus \cdots \oplus \hat{F}_l$$

where $I_j, \hat{I}_j$ are infinite cyclic groups and $F_j, \hat{F}_j$ are groups of prime power order.

Let $T(G)$ be the set of all elements of finite order (see Theorem 6.9, page 189). Then $T(G)$ is the direct sum of the direct summands of finite order in both cases (Problem 6.55). Thus

$$T(G) = F_1 \oplus \cdots \oplus F_n = \hat{F}_1 \oplus \cdots \oplus \hat{F}_l$$

Hence by case 3, $F_1 \oplus \cdots \oplus F_m$ and $\hat{F}_1 \oplus \cdots \oplus \hat{F}_l$ are of the same kind.

Also $G/T(G) \cong I_1 \oplus \cdots \oplus I_m \cong \hat{I}_1 \oplus \cdots \oplus \hat{I}_k$ (by Problem 6.11, page 181). By Problem 6.56, $I_1 \oplus \cdots \oplus I_m$ is the direct sum of k infinite cyclic groups. Then $k = m$ by case 1. Therefore we have proved that $I_1 \oplus I_2 \oplus \cdots \oplus I_m \oplus F_1 \oplus \cdots \oplus F_n$ and $\hat{I}_1 \oplus \hat{I}_2 \oplus \cdots \oplus \hat{I}_k \oplus \hat{F}_1 \oplus \cdots \oplus \hat{F}_l$ are of the same kind. This completes the proof of the theorem.

If a finitely generated group G is the direct sum of cyclic groups of orders $p_1^{r_1}, \ldots, p_k^{r_k}$ and s infinite cyclic groups, where $p_1, \ldots, p_k$ are primes, $p_1 \leq p_2 \leq \cdots \leq p_k$, $r_1, \ldots, r_k$ positive integers with $r_i \geq r_{i+1}$ if $p_i = p_{i+1}$, then the ordered $k+1$-tuple $(p_1^{r_1}, \ldots, p_k^{r_k}; s)$ is called the *type* of G. (The definition of type differs slightly from book to book. Usually it is applied only to p-groups.) By Theorem 6.21 the type of G is uniquely defined. We can now give a criterion for the isomorphism of two finitely generated abelian groups.

Theorem 6.25: If F and G are two finitely generated groups, then they are isomorphic if and only if they have the same type.

Proof: Let $F = A_1 \oplus \cdots \oplus A_k$. If $\phi: F \to G$ is an isomorphism, then $G = A_1\phi \oplus \cdots \oplus A_k\phi$ (Problem 6.56). As $A_i\phi \cong A_i$, it follows that F and G have the same type. Conversely, if F and G have the same type they are clearly isomorphic (Theorem 6.5, page 185).

Problems

6.46. Let $G = A \oplus B$ where A and B are cyclic of order 2. Find C and D such that $G = C \oplus D$ where C and D are cyclic of order 2 and $C \neq A$ and $C \neq B$.

Solution:

Let $A = \{0, a\}$, $B = \{0, b\}$. Put $C = \{0, a+b\}$. Then C is cyclic of order 2. Also put $D = B$. Then $C + D = \{0, a+b, b, a+b+b = a\}$, and so $C + D = G$. Also $C \cap D = \{0\}$. Thus $G = C \oplus D$.

6.47. If the type of F is $(f_1, \ldots, f_k; f)$ and that of G is $(g_1, \ldots, g_p; g)$ where $f_1 = p_1^{r_1}, f_2 = p_2^{r_2}, \ldots, f_k = p_k^{r_k}$, $g_1 = q_1^{s_1}, \ldots, g_l = q_l^{s_l}$ and $p_k < q_1$, where the p_i and q_i are primes, find the type of $F \oplus G$.

Solution:

We have $p_1 \leq p_2 \leq \cdots \leq p_k < q_1 \leq \cdots \leq q_l$. Hence the type of $F \oplus G$ is

$$(f_1, \ldots, f_k, g_1, \ldots, g_l; f+g)$$

6.48. If F, G and H are finitely generated abelian groups, show that $F \oplus G \cong F \oplus H$ implies that $G \cong H$.

Solution:

Express F, G and H as direct sums of cyclic groups of prime power and infinite orders. If the type of F is $(f_1, \ldots, f_k; f)$, and that of G is $(g_1, \ldots, g_l; g)$ while that of H is $(h_1, \ldots, h_m; h)$, then the type of $F \oplus G$ is $(a_1, \ldots, a_{k+l}; f+g)$ where $a_1, \ldots, a_{k+l}$ is $f_1, \ldots, f_k, g_1, \ldots, g_l$ in some order, while the type of $F \oplus H$ is $(b_1, \ldots, b_{k+m}; f+h)$ where $b_1, \ldots, b_{k+m}$ is $f_1, \ldots, f_k, h_1, \ldots, h_m$ in some order. For two abelian groups to be isomorphic we require that (by Theorem 6.25) their types are the same. Accordingly the types of G and H are the same and $G \cong H$.

6.49. Find up to isomorphism all abelian groups of order 1800.

Solution:

Observe that $1800 = 2^3 3^2 5^2$. So an abelian group of order 1800 is a direct sum of a group of order 2^3, a group of order 3^2 and a group of order 5^2. The possible types of a group of order 2^3 are $(2^3; 0)$, $(2^2, 2; 0)$, $(2, 2, 2; 0)$. Thus there are precisely 3 groups of order 8. The possible types of a group of order 3^2 are $(3^2; 0)$ and $(3, 3; 0)$, so there are 2 non-isomorphic groups of order 9. Similarly there are 2 non-isomorphic groups of order 25. Then the total number of non-isomorphic groups of order 1800 is $3 \times 2 \times 2 = 12$.

Compare the ease with which we solve this problem with the effort required to find *all* the groups (non-abelian as well as abelian) of order 8 which we have considered in Chapter 5.

6.50. Let p be any prime and let m be any integer. Prove that the number of groups of order p^m is equal to the number of ways of writing $m = r_1 + \cdots + r_k$ where $r_1, \ldots, r_k$ are positive integers and $r_1 \geqq r_2 \geqq \cdots \geqq r_k$.

Solution:

A group of order p^m has all its elements of order a power of p. Hence its type will be of the form $(p^{r_1}, p^{r_2}, \ldots, p^{r_k}; 0)$ with $r_1 \geqq r_2 \geqq \cdots \geqq r_k$. Since G is of order p^m, and

$$|G| = p^{r_1} p^{r_2} \cdots p^{r_k} = p^{r_1 + \cdots + r_k}$$

we conclude that $r_1 + \cdots + r_k = m$.

6.51. Prove that a cyclic group of order p^n, where p is a prime, is not expressible as the direct sum of nontrivial subgroups by the following two methods: (1) directly, (2) by using Theorem 6.21.

Solution:

(1) Suppose $G = A \oplus B$ where A, B are nontrivial subgroups of G. Clearly $|A| \leqq p^{n-1}$ and $|B| \leqq p^{n-1}$ as $|G| = p^n$. Let $G = gp(g)$. Then $g = a + b$, $a \in A$, and $b \in B$. As $p^{n-1}g = p^{n-1}a + p^{n-1}b = 0$, g is of order less than p^{n-1}. But $gp(g) = G$ and is of order p^n. Thus we have a contradiction.

(2) Since G is cyclic of order p^n, the type of G is $(p^n; 0)$. Hence by Theorem 6.21, only one of the direct summands is nonzero, i.e. G cannot be expressed as a direct sum of more than one nontrivial group.

6.52. Prove, by considering the direct sum of cyclic groups of order 2, that if G is the direct sum of k infinite cyclic groups and also the direct sum of l infinite cyclic groups, then $k = l$.

Solution:

Let $G = gp(x_1) \oplus \cdots \oplus gp(x_k)$. Let $H = gp(2x_1, \ldots, 2x_k)$. Then by Corollary 6.17, G/H is the direct sum of k cyclic groups of order 2. Thus $|G/H| = 2^k$. Clearly $H \subseteq 2G$. Also if $g \in G$, $g = r_1 x_1 + \cdots + r_k x_k$. Then $2g = r_1(2x_1) + \cdots + r_k(2x_k) \in H$, from which $2G \subseteq H$. Thus $H = 2G$.

Now by a similar argument we conclude that if G is the direct sum of l infinite cyclic groups, $|G/2G| = 2^l$. Thus $l = k$.

6.53. Prove that nG is a subgroup of G where n is a given integer.

Solution:

If $h, k \in nG$, $h = nf$, $k = ng$ where $f, g \in G$. Hence $h - k = n(f-g) \in nG$, and so nG is a subgroup.

6.54. Let G be an abelian group, $G = X \oplus Y$. Let $x \in X$, $y \in Y$. Prove that (1) if x and y are of finite order, then the order of $x + y$ is the least common multiple (lcm) of the orders of x and y; (2) if x is of infinite order, $x + y$ is of infinite order.

Solution:

(1) Let $l = \text{lcm}$ of the orders of x and y. Then $l(x+y) = lx + ly = 0$. Now if $m =$ order of $x + y$, then $m(x+y) = mx + my = 0$ implies $mx = 0$ and $my = 0$. This in turn implies that the order of x divides m and the order of y divides m. Thus we have the result.

(2) If x is of infinite order and $m(x+y) = 0$, then $mx + my = 0$. But by the uniqueness of such expressions in direct sums, $mx = my = 0$. Since x is of infinite order, $m = 0$.

6.55. Let $G = I_1 \oplus \cdots \oplus I_m \oplus F_1 \oplus \cdots \oplus F_n$ where each I_j is torsion-free and each F_i finite. Let $T(G)$ be the set of all elements of finite order. Prove that

$$T(G) = F_1 \oplus \cdots \oplus F_n$$

Solution:

Clearly $T(G) \supseteq F_1 \oplus \cdots \oplus F_n$. If $g \in T(G)$, $g = i_1 + \cdots + i_m + f_1 + \cdots + f_n$ where $i_1, \ldots, i_m$ are elements of $I_1, \ldots, I_m$ respectively, and $f_1, \ldots, f_n$ are elements of $F_1, \ldots, F_n$ respectively. As g is of finite order r, say,

$$rg = ri_1 + ri_2 + \cdots + ri_m + rf_1 + \cdots + rf_n = 0$$

By definition of the direct sum, it follows that

$$ri_1 = ri_2 = \cdots = ri_m = rf_1 = \cdots = rf_n = 0$$

Since $I_1, \ldots, I_m$ are torsion-free, we have $i_1 = i_2 = \cdots = i_m = 0$. Thus $g \in F_1 \oplus \cdots \oplus F_n$ and the result follows.

6.56. If $F = A_1 \oplus \cdots \oplus A_k$ and $\phi: F \to G$ is an isomorphism, then $G = A_1\phi \oplus \cdots \oplus A_k\phi$.

Solution:

We must show that every element of G is uniquely of the form $a_1\phi + \cdots + a_k\phi$ where $a_1, \ldots, a_k$ belong to $A_1, \ldots, A_k$ respectively. Now if $g \in G$, there exists $f \in F$ such that $f\phi = g$. But $f = a_1 + \cdots + a_k$ and so $g = a_1\phi + \cdots + a_k\phi$. If $a_1\phi + \cdots + a_k\phi = a_1'\phi + \cdots + a_k'\phi$, then

$$(a_1 - a_1')\phi + \cdots + (a_k - a_k')\phi = 0$$

Let $h = a_1 - a_1' + \cdots + a_k - a_k'$. h belongs to $\text{Ker}\,\phi$. Since ϕ is an isomorphism, $h = 0$.

By the uniqueness of expression of direct sums,

$$a_1 - a_1' = a_2 - a_2' = \cdots = a_k - a_k' = 0$$

so that $a_1 = a_1'$, $a_2 = a_2'$, ..., $a_k = a_k'$. Therefore each element of G is expressible in the form $a_1\phi + \cdots + a_k\phi$ in one and only one way.

d. Subgroups of finitely generated abelian groups

The purpose of this section is to decide which groups (up to isomorphism) can appear as subgroups of finitely generated abelian groups. We begin with

Theorem 6.26: Let G be free abelian of rank n. Then any subgroup H of G is free abelian of rank less than or equal to n.

Proof: By Lemma 6.15, page 196, there exist a basis $c_1, \ldots, c_n$ of G and integers $u_1, \ldots, u_n$ such that $H = gp(u_1c_1, \ldots, u_nc_n)$. If $u_1, \ldots, u_i$ are nonzero, and $u_{i+1} = u_{i+2} = \cdots = u_n = 0$, then

$$gp(u_1c_1, \ldots, u_nc_n) = gp(u_1c_1) \oplus \cdots \oplus gp(u_ic_i)$$

(See Problem 6.57.) Hence the result.

Corollary 6.27: Let A be a finitely generated abelian group. Then every subgroup of A is finitely generated.

Proof: As A is a finitely generated abelian group, it is isomorphic to some factor group of a finitely generated free abelian group G, say $A \cong G/N$. The subgroups of G/N are of the form H/N where H is a subgroup of G. By Theorem 6.26, H is finitely generated and therefore so is H/N. Consequently every subgroup of A is finitely generated.

From this corollary we see that only finitely generated abelian groups can occur as subgroups of finitely generated abelian groups.

Theorem 6.28: Let G be a finitely generated group and H a subgroup of G. Let G and H be expressed as direct sums of infinite cyclic groups and cyclic groups of prime power order. If the number of infinite cyclic groups in these decompositions for G and H are m and k respectively, then $k \leqq m$.

Proof: Let

$$G = I_1 \oplus \cdots \oplus I_m \oplus F_1 \oplus \cdots \oplus F_n, \quad H = \hat{I}_1 \oplus \cdots \oplus \hat{I}_k \oplus \hat{F}_1 \oplus \cdots \oplus \hat{F}_l$$

where the $I_i, \hat{I}_i$ are infinite cyclic groups and the $F_i, \hat{F}_i$ are cyclic groups of prime power order. Let $T(G)$ and $T(H)$ be the torsion subgroups of G and H respectively, i.e. the respective sets of elements of finite order (Theorem 6.9, page 189). Then $T(H) = H \cap T(G)$. Now $G/T(G) \cong I_1 \oplus \cdots \oplus I_m$ by Problem 6.11, page 181. Thus the rank of $G/T(G)$ is m. (See the remarks following Theorem 6.11, page 192.) Since $(H + T(G))/T(G) \subseteq G/T(G)$, $(H + T(G))/T(G)$ is free abelian of rank less than m, by Theorem 6.26. But

$$(H + T(G))/T(G) \cong H/H \cap T(G) \cong H/T(H)$$

by the subgroup isomorphism theorem (Theorem 4.23, page 125). It follows as before that $H/T(H) \cong I_1 \oplus \cdots \oplus I_k$. Thus $k \leqq m$.

Again let G be finitely generated and H a subgroup of G. (Recall that if F is any abelian group and p a prime, $F_p = \{f \mid f \in F$ and of order a power of $p\}$.) G_p, as a subgroup of a finitely generated abelian group, is finitely generated (Corollary 6.27) and so is H_p. Clearly $G_p \supseteq H_p$. Thus we are led to inquire what groups can occur as subgroups of finitely generated p-groups. Of course a finitely generated p-group is finite (Problem 6.44).

We first require a lemma.

Lemma 6.29: Let G have type $(p^{r_1}, \ldots, p^{r_n}; 0)$. Then the number of elements of order p in G is $p^n - 1$.

Proof: Let $G = C_1 \oplus \cdots \oplus C_n$ where each $C_i = gp(c_i)$ and the order of c_i is p^{r_i}. If $x \in G$ is of order p, and $x = t_1c_1 + \cdots + t_nc_n$ where $t_i \in Z$, then $p^{r_i - 1}$ divides t_i. Hence the elements of order p are a subset of H, where

$$H = p^{r_1-1}C_1 \oplus \cdots \oplus p^{r_n-1}C_n$$

On the other hand every element ($\neq 0$) of H is of order p, so $H - \{0\}$ is the set of all elements of order p. Accordingly, as $|H| = p^n$, the number of elements of order p in G is $p^n - 1$.

Theorem 6.30: Let G be a group with type $(p^{r_1}, p^{r_2}, \ldots, p^{r_m}; 0)$. Let H be any subgroup. If the type of H is $(p^{s_1}, \ldots, p^{s_n}; 0)$, then $n \leqq m$ and $0 < s_i \leqq r_i$, $i = 1, \ldots, n$.

Proof: We proceed by induction on $|G|$. If $|G| = 1$ or p, the result is trivial. Hence assume $|G| > p$, and the result holds for groups of order less than $|G|$. Now H has type $(p^{s_1}, p^{s_2}, \ldots, p^{s_n}; 0)$, and so the number of elements of order p in H is, by Lemma 6.29, $p^n - 1$. Similarly, the number of elements of order p in G is, by Lemma 6.29, $p^m - 1$. Clearly $p^n - 1 \leqq p^m - 1$ and consequently $n \leqq m$.

Now $|pG| < |G|$. We can therefore assume the result holds by induction for pG and its subgroup pH.

At this point we experience a minor notational inconvenience. If for example $r_m > 1$, pG is of type $(p^{r_1-1}, \ldots, p^{r_m-1}; 0)$. However, if $r_m = 1$ and $r_{m-1} > 1$, pG is of type $(p^{r_1-1}, \ldots, p^{r_{m-1}-1}; 0)$. Therefore we need additional notation. Define $m^* = m$ if $r_m > 1$; otherwise define m^* to be an integer such that $r_{m^*} > 1$, but $r_{m^*+1} = 1$. As $r_1 \geqq r_2 \geqq \cdots \geqq r_{m^*} > 1$ and, if $m^* \neq m$, $r_{m^*+1} = \cdots = r_m = 1$, then pG is of type $(p^{r_1-1}, \ldots, p^{r_{m^*}-1}; 0)$.

Similarly, let us define $n^* = n$ if $s_n > 1$; otherwise define n^* to be an integer such that $s_{n^*} > 1$ but $s_{n^*+1} = 1$. Arguing as in the paragraph above, pH is of type $(p^{s_1-1}, \ldots, p^{s_{n^*}-1}; 0)$. Then by the inductive hypothesis we have

$$n^* - 1 \leq m^* - 1 \quad \text{and} \quad s_i - 1 \leq r_i - 1 \quad \text{for } i = 1, \ldots, n^*$$

If $n^* = n$, the result follows immediately. If $n^* \neq n$, then $s_{n^*+1} = \cdots = s_n = 1$. Hence $r_{n^*+1} \geq s_{n^*+1}, \ldots, r_n \geq s_n$. Thus $s_i \leq r_i$ for $i = 1, \ldots, n$.

With Theorems 6.28 and 6.30 it is easy to determine, knowing the type of a given finitely generated abelian group G, the possible types of subgroups of G. (See Problem 6.60.)

It can be shown (Problems 6.62-65) that every factor group of a finite abelian group G is isomorphic to a subgroup of G. Therefore we know the types of homomorphic images of finite abelian groups.

Problems

6.57. Let $G = A \oplus B$ and let C, D be subgroups of A, B respectively. Show that $C + D = C \oplus D$. (This can obviously be generalized to the direct sum of any number of groups.)

Solution:

As $\{0\} = A \cap B \supseteq C \cap D$, we have $C \cap D = \{0\}$. Thus $C + D = C \oplus D$.

6.58. Let G have type $(p_1^{r_1}, \ldots, p_n^{r_n}; s)$. Suppose that for some $i \leq j$

$$p_i = p_{i+1} = \cdots = p_j, \qquad p_{i-1} \neq p_i \qquad \text{and} \qquad p_j \neq p_{j+1}$$

Put $p = p_i$. Show that the type of G_p is $(p^{r_i}, \ldots, p^{r_j}; 0)$.

Solution:

Decompose G into the direct sum of infinite cyclic groups and groups of prime power order. Clearly $G = A_i \oplus \cdots \oplus A_j \oplus R$ where A_k is of order p^{r_k} for $i \leq k \leq j$, and R is the direct sum of the cyclic groups which are not of order a power of this prime p in the given decomposition of G. Then $G_p \supseteq A_i \oplus \cdots \oplus A_j$. On the other hand, as any nonzero element of finite order from R is of order coprime to that of p, $G_p \subseteq A_i \oplus \cdots \oplus A_j$. Thus $G_p = A_i \oplus \cdots \oplus A_j$, and the result follows.

6.59. Let G be of type $(p^{r_1}, \ldots, p^{r_m}; u)$. Show that G has subgroups of type $(p^{s_1}, \ldots, p^{s_n}; v)$ where $n \leq m$ and $1 \leq s_i \leq r_i$ for $1 \leq i \leq n$ and $v \leq u$.

Solution:

Let $G = A_1 \oplus \cdots \oplus A_m \oplus I_1 \oplus \cdots \oplus I_u$ where A_i is cyclic of order p^{r_i}, and $I_1, \ldots, I_u$ are infinite cyclic groups. Now each A_i has a subgroup B_i of order p^{s_i}, $1 \leq i \leq n$. By repeated application of Problem 6.57,

$$B_1 + \cdots + B_n + I_1 + \cdots + I_v \;=\; B_1 \oplus \cdots \oplus B_n \oplus I_1 \oplus \cdots \oplus I_v$$

This is then a subgroup of G of type $(p^{s_1}, \ldots, p^{s_n}; v)$, as required.

6.60. Let G be of type $(3^3, 3^2, 5^2, 7^3; 1)$. Determine whether G has a subgroup H of type
(*a*) $(3, 3, 7^2, 7; 1)$, (*b*) $(3, 3, 5, 7; 2)$, (*c*) $(3^3, 3^3, 5^2, 7^3; 0)$, (*d*) $(3, 3, 7; 1)$.

Solution:

(*a*) No, as then G_7 is of type $(7^3; 0)$ whereas H_7 is of type $(7^2, 7; 0)$ by Problem 6.58. This is a contradiction to Theorem 6.30.

(*b*) No. A direct contradiction to Theorem 6.28.

(*c*) No. Compare G_3 and H_3 as in (*a*).

(*d*) Yes.

6.61. Give an infinite number of examples of an abelian p-group that contains exactly $p + 1$ subgroups of order p.

Solution:

If G is a group with $p + 1$ subgroups of order p, each of them contributes $p - 1$ distinct elements (the identity is common to all) of order p. Hence in all there are $(p-1)(p+1) = p^2 - 1$ elements of order p.

By Lemma 6.29 a p-group with p^2-1 distinct elements of order p contains p^2 summands which are cyclic groups of order a power of p. Let $1 \leq r_i$, where $i = 1, \ldots, p^2$, be integers. Then any group of type $(p^{r_1}, \ldots, p^{r_p}; u)$ where u is a nonnegative integer, has exactly p^2-1 elements of order p and thus exactly $p+1$ subgroups of order p.

6.62. Let G be a p-group. Suppose $G = gp(a) \oplus B$. Prove that $G = gp(a+b) \oplus B$ where $b \in B$ is of order less than or equal to the order of a.

Solution:

If $x \in gp(a+b) \cap B$, then $x = r(a+b) = b_1$ where $b_1 \in B$ and r is an integer. Thus $ra = b_1 - rb$. Since $gp(a) \cap B = \{0\}$, $ra = 0$. Then r is divisible by the power of p which is the order of a. Consequently $rb = 0$. Hence $b_1 = 0$ and $x = 0$. Clearly $gp(a+b) + B = G$ and the result follows.

6.63. Let G be a finite p-group. Prove that if $g \in G$ and g is of order p, g appears as an element of a cyclic direct summand of G.

Solution:

G is the direct sum of cyclic groups, say $G = gp(c_1) \oplus \cdots \oplus gp(c_n)$. If $g = 0$, the result follows immediately. Otherwise, without loss of generality, suppose that

$$g = r_1 p^{w_1} c_1 + \cdots + r_m p^{w_m} c_m$$

where $(r_i, p) = 1$, $r_i p^{w_i} c_i \neq 0$ and $w_1 \leq w_2 \leq \cdots \leq w_n$. Put $r_1 c_1 = c_1'$. Clearly $gp(c_1') = gp(c_1)$. Then $g = p^{w_1}(c_1' + d)$ where $d \in gp(c_2, \ldots, c_n)$. As g is of order p and $p^{w_1}c_1' \neq 0$, the order of d is less than or equal to the order of c_1'. By Problem 6.62, on putting $c = c_1' + d$, we obtain

$$G = gp(c) \oplus gp(c_2, \ldots, c_n)$$

Since $g \in gp(c)$, the result follows.

6.64. Let G be a finite p-group. Let $N = gp(g)$ be of order p. Prove that G/N is isomorphic to a subgroup of G.

Solution:

By Problem 6.63, $G = gp(c) \oplus B$ where $g \in gp(c)$. Then $G/N \cong (gp(c)/N) \oplus B$ (Lemma 6.16, page 197). But clearly $gp(pc) \cong gp(c)/N$. Thus $gp(pc, B) \cong G/N$.

6.65. Let G be a finite group. Prove by induction on $|G|$ that if N is a subgroup of G, G/N is isomorphic to a subgroup of G.

Solution:

Assume the result is true for all groups of order less than r. Let $|G| = r$ and let N be a subgroup of G. If $N = \{0\}$, $G/N \cong G$ and there is nothing to prove. If N is of order a prime p, $G = G_p \oplus E$ where G_p is the p-component of G, and $N \subseteq G_p$. Then $G/N \cong (G_p/N) \oplus E$ by Lemma 6.16. Now $G_p/N \cong H$, a subgroup of G_p, by the preceding problem. Hence $G/N \cong H \oplus E$ and $H \oplus E$ is a subgroup of G. If N is not of order a prime, there is an element n_0 of N of order a prime p by Proposition 5.9, page 137. Let $N_0 = gp(n_0)$. Then $(G/N_0)/(N/N_0) \cong G/N$. As $|G/N_0| < |G|$, G/N_0 has a subgroup $H/N_0 \cong G/N$. $H \neq G$, since otherwise $N = N_0$ which is not true. Thus $|H| < |G|$ and by induction it has a subgroup K such that $K \cong H/N_0 \cong G/N$. The result follows.

6.4 DIVISIBLE GROUPS

a. p-Prüfer groups. Divisible subgroups

A group G is said to be divisible if for each integer $n \neq 0$ and each element $g \in G$ there exists $h \in G$ such that $nh = g$. Both the additive group of rationals and p-Prüfer groups are divisible in this sense (Problem 6.66 below).

If the groups G_i, $i \in I$, are divisible, then $\sum_{i \in I} G_i$ is divisible. For if $n \neq 0$ is any integer and $g \in \sum_{i \in I} G_i$, then $g = g_1 + \cdots + g_k$, say. So there exist $h_1, \ldots, h_k$ such that $nh_1 = g_1, \ldots, nh_k = g_k$. Then

$$n(h_1 + \cdots + h_k) = g_1 + \cdots + g_k = g$$

It follows that direct sums of p-Prüfer groups and copies of the additive group of rationals are also divisible. We show that in fact this exhausts all divisible groups.

To prove this we need several facts.

First we remark that a homomorphic image of a divisible group is divisible. For suppose G is divisible and H is a subgroup of G. Let $g+H \in G/H$ ($g \in G$) and let n be a positive integer. Then there exists $g' \in G$ such that $ng' = g$. Accordingly $n(g'+H) = g+H$, and so G/H is divisible.

Secondly we remark that if $G = H \oplus K$ and G is divisible, so also are H and K, since $H \cong G/K$ (Problem 6.11, page 181) is a homomorphic image of a divisible group. Similarly K is divisible.

Next we need the following theorem which classifies p-Prüfer groups.

Theorem 6.31 (Main Theorem on *p*-Prüfer Groups): Let p be a prime. Let G be a group which is the union of an ascending sequence of subgroups $C_1 \subseteq C_2 \subseteq \cdots$ where C_r is cyclic of order p^r for $r = 1, 2, \ldots$. Then G is isomorphic to the p-Prüfer group.

Proof: We may suppose that $C_r = gp(c_r)$ and that $pc_{r+1} = c_r$ for $r = 1, 2, \ldots$ (see Problem 6.67 for the details). Define $\theta: G \to (Q/Z)_p$ by $(mc_r)\theta = m/p^r + Z$ for all integers m. We must prove that θ is an isomorphism. We are however not even certain that θ is a mapping. The snag is this:

If $g = mc_r$ and if also $g = nc_s$, is $g\theta = m/p^r + Z$ or is $g\theta = n/p^s + Z$?

We will show that $m/p^r + Z = n/p^s + Z$, thus proving that θ is uniquely defined.

Assume without loss of generality that $r \geqq s$. It follows that $p^{r-s}c_r = c_s$. Then

$$mc_r = np^{r-s}c_r \quad \text{from which} \quad (m - np^{r-s})c_r = 0$$

As c_r is of order p^r, $m - np^{r-s} = kp^r$ for some integer k. Thus $m = np^{r-s} + kp^r$ from which $m/p^r = np^{-s} + k$. We therefore conclude that $m/p^r + Z = n/p^s + Z$. Thus θ is a mapping.

Next we show that θ is a homomorphism. If $g, h \in G$, then $g, h \in C_r$ for some integer r, so that $g = sc_r$, $h = tc_r$, with $s, t \in Z$. Then

$$(g+h)\theta = ((s+t)c_r)\theta = (s+t)/p^r + Z = (s/p^r + Z) + (t/p^r + Z) = g\theta + h\theta$$

Finally θ is one-to-one as

$$\begin{aligned} \operatorname{Ker}\theta &= \{g \mid g\theta = Z\} \\ &= \{sc_r \mid s \text{ an integer}, r \text{ a positive integer and } (sc_r)\theta = s/p^r + Z = Z\} \\ &= \{sc_r \mid s/p^r \in Z\} = \{sc_r \mid s \text{ divisible by } p^r\} = \{0\} \end{aligned}$$

Thus the result follows.

In the future we will call any group isomorphic to $(Q/Z)_p$ a *p-Prüfer group*.

The following result is not only the main tool in Section 6.4b, but is also of interest in itself.

Theorem 6.32: Let G contain a divisible subgroup D. Then there exists a subgroup K of G such that $G = D \oplus K$, i.e. a divisible subgroup is a direct summand.

Proof: We accomplish this proof by Zorn's lemma. Let $\mathcal{P}$ be the collection of all subgroups L of G such that $L \cap D = \{0\}$. (Our idea is to pick a maximal subgroup K which meets D in $\{0\}$. Then $D + K = D \oplus K$ and we need only show that $D + K = G$ which will turn out to be true because of the maximality of K.) Can we apply Zorn's lemma to $\mathcal{P}$? Suppose $\{L_i \mid i \in I\}$ is a chain in $\mathcal{P}$. Is $\bigcup_{i \in I} L_i$ in $\mathcal{P}$? We require

(i) $D \cap \bigcup_{i \in I} L_i = \{0\}$.

(ii) $\bigcup_{i \in I} L_i$ is a subgroup of G.

Part (i) is true because $D \cap \bigcup_{i \in I} L_i \neq \{0\}$ implies $D \cap L_j \neq \{0\}$ for some L_j.

To prove (ii) we must show that if $g, h \in \bigcup_{i \in I} L_i$, then $g - h \in \bigcup_{i \in I} L_i$. Now $g, h \in \bigcup_{i \in I} L_i$ implies $g \in L_j$ and $h \in L_k$. Either $L_k \supseteq L_j$ or $L_j \supseteq L_k$, so without loss of generality assume that $L_k \supseteq L_j$. Then since L_k is a subgroup of G, $g - h \in L_k$. Therefore $g - h \in \bigcup_{i \in I} L_i$ and (ii) holds. So $\mathcal{P}$ has a maximal element, say K, which satisfies $D \cap K = \{0\}$. Thus $D + K = D \oplus K$.

Suppose now that $D + K \neq G$. Then $G/(D \oplus K)$ is nonzero. We prove first that $G/(D \oplus K)$ is a torsion group. Suppose the contrary. Then we can find $x \in G$ such that $gp(x + (D \oplus K))$ is of infinite order in $G/(D \oplus K)$. Now $x \notin K$. If we put $K_1 = gp(K, x)$, it consists of all elements of the form $nx + k$, where n is an integer and k is an element of K. If $nx + k \in D$, then $nx \in (K + D)$. So $n = 0$ is the only possibility. But $K \cap D = \{0\}$. So $k = 0$ follows also. Therefore $D \cap K_1 = \{0\}$. This contradicts the maximality of K. Thus we have proved that $G/(K \oplus D)$ is a torsion group.

Let $x \in G$, $x \notin K \oplus D$. Then $gp(x + (K \oplus D))$ is a subgroup of $G/(K \oplus D)$ of finite order. Suppose that $x + (K \oplus D)$ is of order w. It follows then that $wx \in K \oplus D$, but $rx \notin K \oplus D$ for $0 < r < w$. Suppose $wx = k + d$. Since D is divisible, we can find $d_1 \in D$ so that $wd_1 = d$. Put $x_1 = x - d_1$. Then $wx_1 = wx - wd_1 = k + d - d = k$. Notice that $x_1 \notin K$ but $wx_1 \in K$. Put $K_1 = gp(x_1, K)$. Then

$$K_1 = \{rx_1 + k \mid r = 0, 1, \ldots, w-1, \text{ and all } k \in K\}$$

We claim that $K_1 \cap D = \{0\}$. For if $rx_1 + k \in D$, $r \in \{0, 1, \ldots, w-1\}$, and $k \in K$, then

$$D \oplus K = rx_1 + (D \oplus K) = r(x_1 + (D \oplus K))$$

Since $x_1 + (D \oplus K) = x + (D \oplus K)$, we must have $r = 0$. So $k \in D$ and thus $k = 0$. Therefore $K_1 \cap D = \{0\}$. But K is maximal. This contradiction shows that our original assumption, i.e. $G \neq D \oplus K$, is false, thus proving the theorem.

Problems

6.66. Use a proof different from that of Problem 6.36, page 192, to prove that a p-Prüfer group is divisible.

Solution:

As Q is divisible, so is Q/Z. But as $Q/Z = \sum_{p \in \Pi} (Q/Z)_p$, each $(Q/Z)_p$ is itself divisible, since every direct summand of a divisible group is divisible.

6.67. Let G be as defined in Theorem 6.31. Prove that we can choose elements c_r such that $C_r = gp(c_r)$ and $pc_{r+1} = c_r$, $r = 1, 2, \ldots$.

Solution:

Assume by induction that $c_1, \ldots, c_n$ have been chosen with $pc_{r+1} = c_r$ for $r = 1, \ldots, n-1$ and $C_i = gp(c_i)$, $i = 1, \ldots, n$. Let $C_{n+1} = gp(c)$. Then $gp(pc)$ is a cyclic group of order p^n and, since cyclic groups have only one subgroup of any given order, $gp(pc) = C_n$. Hence $r(pc) = c_n$ for some integer r. As c_n is of order p^n, r and p are coprime. Thus $gp(rc) = C_{n+1}$. Now put $c_{n+1} = rc$. Then $pc_{n+1} = c_n$ and it is possible to choose the elements $c_1, c_2, \ldots$, as required.

6.68. A p-group G is a p-Prüfer group if and only if it has the following two properties:

(1) every proper subgroup of G is cyclic,

(2) there is a cyclic subgroup of G of order p^i for every $i = 1, 2, \ldots$. (Hard.)

Solution:

First we shall prove that if G satisfies (1) and (2), it is a p-Prüfer group.

Let $C_1 \subseteq C_2 \subseteq \cdots$ be a sequence of subgroups of G where each C_i is a cyclic group of order p^i. If the sequence is infinite, $D = \bigcup_{i=1}^{\infty} C_i$ is a p-Prüfer group (Theorem 6.31). Hence it is divisible and $G = D \oplus K$ by Theorem 6.32. However, if $K \neq \{0\}$, G has a subgroup which is the direct sum of two cyclic p-groups, and hence is not cyclic, contrary to the hypothesis. Accordingly, $G = D$ is a p-Prüfer group.

Suppose now that $C_1 \subseteq C_2 \subseteq \cdots \subseteq C_n$ is a sequence of subgroups, each C_i cyclic of order p^i, and there exists no subgroup $C_{n+1} \supseteq C_n$ where C_{n+1} is of order p^{n+1}. Let $C_n = gp(a)$. We know that there exists a subgroup $B = gp(b)$ of order p^{n+1}. Consider $gp(a, b)$. As a finitely generated abelian group, it is the direct sum of cyclic groups. If it is the direct sum of two or more cyclic groups, it is not cyclic (Theorem 6.21). Thus $gp(a, b)$ is cyclic, and as G is a p-group, $gp(a, b)$ is cyclic of order p^m where $m \geqq n+1$. But then $gp(a, b)$ contains a cyclic subgroup of order p^{n+1} containing C_n, contrary to the hypothesis. The result follows.

Next we note that the p-Prüfer group satisfies condition (2). As for condition (1), let H be a subgroup of $(Q/Z)_p$. If $H \neq \{Z\}$, then let $x \in H$, $x \neq 0 + Z$. Let $x = m/p^r + Z$ where m is an integer between 1 and $p^r - 1$. Clearly $gp(x) = gp(1/p^r + Z)$. If there is no integer n for which $1/p^r + Z \notin H$ for $r \geqq n$, then $H = (Q/Z)_p$. If there exists such an integer n, $H \subseteq gp(1/p^n + Z)$. Thus H is cyclic as required.

6.69. Show that if U is the subgroup of the multiplicative group of the complex numbers consisting of all the nth roots of unity, then $U \cong Q/Z$.

Solution:

$U = \sum_{p \in \Pi} U_p$ by Theorem 6.10, page 190. Now U_p is the union of cyclic groups of order p^i, namely U_p is the union of $C_i = gp\,(\{x \mid x \text{ is a } p^i\text{th root of unity}\})$. But C_i is cyclic of order p^i. Then by Theorem 6.31, $U_p \cong (Q/Z)_p$. Thus, as $Q/Z = \sum_{p \in \Pi} (Q/Z)_p$ (by Theorem 6.10), $Q/Z \cong U$ by Theorem 6.5, page 185.

6.70. Show that the additive group of rationals is the union of an ascending sequence of infinite cyclic groups.

Solution:

Let $Q_n = gp(1/n!)$. Then $Q_{n+1} \supseteq Q_n$ and $\bigcup_{n=1}^{\infty} Q_n = Q$ the additive group of rationals.

6.71. Show that if G is a p-Prüfer group and H, K are subgroups of G, then either $H \supseteq K$ or $K \supseteq H$.

Solution:

If one of H or K is G, the result is true. Assume both H and K are proper subgroups. Then by Problem 6.68, H and K are both finite cyclic groups. Suppose $p^r = |H| \geqq |K| = p^s$. Then H contains a subgroup of order $|K|$, say H_1. Now both H_1 and K are contained in some cyclic subgroup of G, as G is the union of cyclic subgroups. But then it follows that $H_1 = K$, as there is one and only one subgroup of order p^s in a cyclic p-group of order exceeding p^s. Therefore $H \supseteq K$.

6.72. Let G be an ascending union of infinite cyclic groups C_i such that $C_i = gp(c_i)$ and $(i+1)c_{i+1} = c_i$, for $i = 1, 2, \ldots$. Prove that G is isomorphic to the additive group of rationals. (Hard.)

Solution:

Let Q denote the additive group of rationals and let $Q_i = gp(1/i!)$, $i = 1, 2, \ldots$. Clearly, $Q_{i+1} \supset Q_i$ and $\bigcup_{i=1}^{\infty} Q_i = Q$. We shall prove that θ below defines a mapping of G to Q. Define $(zc_i)\theta = z/i!$ where $z \in Z$. We must prove that θ is uniquely defined.

Suppose $z_1c_i = z_2c_j$, z_1, z_2 and i, j integers. If $i \leqq j$, then $c_i = (j!/i!)c_j$. Hence $z_1c_i = z_1(j!/i!)c_j = z_2c_j$. Since C_j is infinite cyclic, $z_1(j!/i!)c_j = z_2c_j$ implies that $z_1(j!/i!) = z_2$.

To prove that θ is well defined, we must show that $z_1/i! = z_2/j!$, i.e. $z_1(j!/i!) = z_2$. But this is what we have just shown. Since $C_i\theta = Q_i = gp\,(1/i!)$, it follows that $G\theta \supseteq \bigcup_{i=1}^{\infty} Q_i = Q$. Hence θ is an onto mapping.

Is θ a homomorphism? Let $f, g \in G$. We may as well suppose that $f, g \in C_i$ for some integer i. Hence $f = z_1c_i$, $g = z_2c_i$, say. $f + g = (z_1 + z_2)c_i$. Then

$$(f+g)\theta \;=\; (z_1+z_2)\frac{1}{i!} \quad\text{and}\quad f\theta + g\theta \;=\; \frac{z_1}{i!} + \frac{z_2}{i!} \;=\; (z_1+z_2)\frac{1}{i!}$$

Thus θ is a homomorphism.

Finally, to show that θ is an isomorphism, it is sufficient to show that $\operatorname{Ker}\theta = \{0\}$. Suppose f is such that $f\theta = 0$. We have that $f = zc_i$ for some integers z and i. Then $f\theta = z/i! = 0$ only if $z = 0$. Hence $f = 0$ and the result follows.

b. Decomposition theorem for divisible groups

The results of Section 6.4a enable us to deduce the following decomposition theorem for divisible groups.

Theorem 6.33: A divisible group is the direct sum of p-Prüfer groups and copies of the additive group of rationals.

Proof: Let G be divisible and let T be the torsion subgroup of G. Now for any integer n and element $t \in T$, there exists an element $g \in G$ such that $ng = t$. Since t is of finite order, so is g, and hence $g \in T$. Thus T is itself divisible. A divisible subgroup is a direct summand (Theorem 6.32); so $G = T \oplus F$. Since $T \cap \{F\} = 0$, $F \cong (T \oplus F)/T$; hence F is torsion-free by Theorem 6.9, page 189. Moreover, as F is a direct summand, F is itself divisible. We now consider F and T separately.

(*a*) F.

We show that F is a direct sum of copies of the additive group of rationals. To this end let S be a maximal independent set (Theorem 6.13, page 194). For each $s \in S$ we shall define a subgroup C_s of F. Let $r_{1,s} = s$. For a given $s \in S$ and positive integer i, there exists by the divisibility of F an element $r_{i+1,s} \in F$ such that $(i+1)r_{i+1,s} = r_{i,s}$. We put $C_s = gp(r_{i,s} \mid i = 1, 2, \ldots)$. It follows then from Problem 6.72 that C_s is isomorphic to the additive group of rational numbers. Note that if $x \in C_s$, $x \neq 0$, then there is a nonzero multiple of x which is also a multiple of s, as $s \neq 0$. (This is true for any two nonzero rational numbers.)

We claim that F is actually the direct sum of these subgroups C_s as s ranges over S. To prove this, suppose that $s_1, s_2, \ldots, s_n$ are distinct elements of S and that $c_1 + c_2 + \cdots + c_n = 0$, where $c_j \in C_{s_j}$, $c_j \neq 0$ $(j = 1, \ldots, n)$.

As we remarked above, there exists a nonzero multiple k_j of each c_j which is then a multiple of s_j. Hence there exists a nonzero integer k, namely $k_1 \cdots k_n$, such that $kc_j = l_j s_j$, where l_j is a nonzero integer. As $kc_1 + \cdots + kc_n = 0$ is a consequence of $c_1 + c_2 + \cdots + c_n = 0$, we find therefore, on substituting for the elements kc_j the elements $l_j s_j$, that $l_1 s_1 + \cdots + l_n s_n = 0$. But the elements $s_1, \ldots, s_n$ are independent. From this contradiction it follows that the C_s generate the direct sum

$$C = gp(C_s \mid s \in S) = \sum_{s \in S} C_s$$

But C is divisible since each summand C_s is divisible. Hence C is a direct summand (Theorem 6.32), i.e. $F = C \oplus D$, say. If $D \neq \{0\}$, then let $d \in D$ $(d \neq 0)$. Clearly the set $S \cup \{d\}$ is definitely larger than S since $d \notin S$ (d does not even lie in C) and $S \cup \{d\}$ is independent. This is a contradiction as S is a maximal independent set. So $F = C$ is a direct sum of copies of the rationals.

(*b*) T.

First of all $T = \sum_{p \in \Pi} T_p$ by Theorem 6.10. Since T is divisible, so also are the summands T_p. It is sufficient then to assume that T is a divisible p-group and to prove that T is a direct sum of p-Prüfer groups.

Let P be the set of elements of T of order at most p. P is clearly a subgroup. Let S be a maximal independent subset of P (Theorem 6.13). For each $s \in S$ define $c_{1,s}, c_{2,s}, \ldots$ inductively as follows: (*a*) $pc_{1,s} = s$, (*b*) $pc_{i+1,s} = c_{i,s}$ for $i = 1, 2, \ldots$. This is possible since T is divisible. Clearly $gp(c_{1,s}) \subseteq gp(c_{2,s}) \subseteq \cdots$. Since $gp(c_{i,s})$ is cyclic of order p^{i+1}, $C_s = gp(c_{1,s}, c_{2,s}, \ldots)$ is an ascending union of cyclic groups of order p^i, one for each $i = 1, 2, \ldots$. Accordingly by Theorem 6.31, C_s is a p-Prüfer group. Then $C = gp(C_s \mid s \in S)$ is divisible. By Theorem 6.32, C is a direct summand of T, i.e. $T = C \oplus D$. If $D \neq \{0\}$, there is an element of $d \in D$ of order p. Since $S \subseteq C$, $S \cup \{d\}$ is an independent subset of P larger then S, and so we must have $D = \{0\}$ by the maximality of S. Thus $T = C$.

It remains only to prove that $C = \sum_{s \in S} C_s$. If $s_1, s_2, \ldots, s_n$ are distinct elements of S, and $c_1 + c_2 + \cdots + c_n = 0$ where $c_i \in C_{s_i}$ for $i = 1, \ldots, n$, then, similarly to (*a*) above, we arrive at a dependent relation between $s_1, \ldots, s_n$, unless $c_1 = c_2 = \cdots = c_n = 0$ (Problem 6.78).

The proof of the theorem is complete.

Problems

6.73. Prove that (*a*) every free abelian group is isomorphic to a subgroup of a divisible abelian group, and (*b*) every abelian group is isomorphic to a subgroup of a divisible group.

Solution:

(*a*) A free abelian group F is a direct sum of infinite cyclic groups C_i $(i \in I)$: $F = \sum_{i \in I} C_i$. Now we choose one copy of the rationals Q_i for each $i \in I$. Let $K = \sum_{i \in I} Q_i$ and let $d_i \neq 0$ be chosen in each Q_i. F is clearly isomorphic to $gp(\{d_i \mid i \in I\})$. But $\sum_{i \in I} Q_i$ is divisible since each Q_i is divisible. The result follows.

(*b*) If G is any group, $G \cong F/N$ for some free abelian group F and some subgroup N.

Now in (*a*) we have proved that there exists a divisible group D containing F. Hence D contains N, and so D/N contains as a subgroup F/N, i.e. G. Now D/N is divisible since a homomorphic image of a divisible group is divisible. Thus the result follows.

6.74. Suppose G has the property that if H is any group such that $H \supseteq G$, then G is a direct summand of H. Prove that G is divisible.

Solution:

G is a subgroup of some divisible group D by Problem 6.73. Thus $D = G \oplus T$. But every direct summand of a divisible group is divisible. Therefore G is divisible.

6.75. Let G be an infinite group whose proper subgroups are all finite. Prove that G is a p-Prüfer group by using the theorem which states: if G is a group such that for some integer $n \neq 0$, $nG = \{0\}$, then G is a direct sum of cyclic groups (of finite order). (This theorem is not proved in this text.)

Solution:

Consider the subgroups nG for all positive integers n. If $nG = G$ for all such n, then G is divisible, and so G is the direct sum of p-Prüfer groups and copies of the additive group of rationals. As all the proper subgroups of G are finite and the additive group of rationals has an infinite cyclic group as a proper subgroup, only p-Prüfer groups are involved. Since each p-Prüfer group is infinite, G must in fact be a p-Prüfer group.

If on the other hand nG is a proper subgroup of G for some n, then nG is a finite group of order m, say, and so $mnG = \{0\}$. Using the theorem quoted in the statement of the problem, G is the direct sum of finite cyclic groups. As G is infinite, it must be the direct sum of an infinite number of cyclic groups C_i. But then the subgroup generated by all but one of the C_i must be infinite. This contradiction proves the result.

6.76. (*a*) Let G be any group and let S be the subgroup generated by all the divisible subgroups of G. Prove that S is divisible.

(*b*) Prove that G is the direct sum of a divisible group and a group which has no divisible subgroups other than the identity subgroup $\{0\}$.

Solution:

(*a*) Let $s \in S$. Then $s = h_1 + h_2 + \cdots + h_n$ where each h_i belongs to a divisible subgroup H_i of G. Thus if $z \neq 0$ is any integer, there exist $k_i \in H_i$ such that $zk_i = h_i$. As $S \supseteq H_i$ for each $i \in I$,

$$k_1 + k_2 + \cdots + k_n \in S \quad \text{and} \quad z(k_1 + \cdots + k_n) = s$$

Hence S is divisible.

(b) Since S is divisible, we can apply our direct summand theorem for divisible subgroups (Theorem 6.32) to find that $G = S \oplus T$. As T contains no divisible subgroup other than $\{0\}$, the result follows.

6.77. (a) Prove that the additive group of rationals Q has a proper subgroup which is not free abelian.

(b) Let G be a torsion-free group, every proper subgroup of which is free abelian. Prove that G is free abelian.

Solution:

(a) Consider the subgroup H generated by $1/2, 1/4, 1/8, \ldots, 1/2^i, \ldots$ in the additive group of rationals. H is of rank 1 as Q is of rank 1 (Problem 6.38, page 194). So if H is free abelian, it must be infinite cyclic. But H is not infinite cyclic; for if it were, $H = gp(z/2^i)$ for some integers z and i. But then $1/2^{i+1} \in H$ and $1/2^{i+1} \notin gp(z/2^i)$. We have only to prove that H is not G. This is obvious since $1/3 \notin H$.

(b) Suppose that $nG = G$ for all positive integers n. Then G is divisible and is the direct sum of copies of the additive group of rationals. (As G is torsion-free, no p-Prüfer group is involved.) But by (a) above, G will have a non-free subgroup. Thus for some $n \neq 0$, $nG \neq G$, and nG is free abelian.

Now $\theta: g \to ng$ is a homomorphism of G onto nG. $\operatorname{Ker} \theta = \{g \mid ng = 0\} = \{0\}$ as G is torsion-free. Hence θ is an isomorphism, and so G is free abelian.

6.78. Prove the result stated at the end of the proof of Theorem 6.33, i.e. prove that if $c_1 + \cdots + c_n = 0$, then $c_1 = c_2 = \cdots = c_n = 0$.

Solution:

Assume the contrary. As the order of $c_1, \ldots, c_n$ is immaterial, we may assume that $c_1 \neq 0$ and c_1 is of highest order. Say c_1 is of order p^l. Then $p^{l-1}c_i = m_i s_i$ where m_i is an integer, $i = 1, 2, \ldots, n$. Thus we obtain

$$p^{l-1}(c_1 + \cdots + c_n) \;=\; m_1 s_1 + m_2 s_2 + \cdots + m_n s_n$$

Since $m_1 s_1 \neq 0$, we have a contradiction to the fact that S is an independent set.

A look back at Chapter 6

This chapter was mainly concerned with the structure of divisible and finitely generated abelian groups.

Direct sums of groups were discussed. Given a family A_i $(i \in I)$ of groups, there is always a group which is the direct sum of groups isomorphic to each of the groups A_i. Any homomorphism of the direct summands of a group extends to a homomorphism of the whole group. From this it follows that if two groups are direct sums of isomorphic subgroups, they are isomorphic. Direct sums of infinite cyclic groups are all the free abelian groups. An important fact is that every abelian group is a homomorphic image of a free abelian group.

The torsion group $T(G)$ of a group G was defined, and it was shown that $G/T(G)$ is torsion-free. It was proved that if G is a torsion group, it is the direct sum of its p-components. This led to the definition of the p-Prüfer group as the p-component of Q/Z.

An application of Zorn's lemma proves every group has a maximal independent subset. The rank of a group was defined and proved an invariant of the group by the Steinitz exchange theorem.

In the fundamental theorem of abelian groups, finitely generated abelian groups were shown to be expressible as the direct sum of cyclic groups. Two finitely generated abelian groups were shown to be isomorphic if and only if they have the same type. Finally, the type of a subgroup of a group was shown to be, roughly speaking, "less than" the type of the group.

Divisible groups were discussed. Any group which is the union of cyclic groups of order a power of p turns out to be a p-Prüfer group. Any divisible subgroup of a group is also a direct summand. This led to the proof that every divisible group is the direct sum of isomorphic copies of the additive group of rationals and p-Prüfer groups.

Supplementary Problems

DIRECT SUMS AND FREE ABELIAN GROUPS

6.79. If the mapping $a \to a^{-1}$, $a \in G$, is an automorphism of the group G, prove G is abelian.

6.80. Suppose that G is a finite group, $\alpha \in \text{aut}(G)$, α is of order 2, and $g\alpha \neq g$ for all $g\ (\neq 1) \in G$. Show that G is an abelian group. (*Hint*: First prove $G = \{g^{-1}(g\alpha) \mid g \in G\}$ and then use Problem 6.79.)

6.81. Denote the set of all homomorphism of an abelian group G into an abelian group H by $\text{Hom}(G, H)$. If $\phi, \Psi \in \text{Hom}(G, H)$, we define $\phi + \Psi$ by

$$g(\phi + \Psi) = g\phi + g\Psi \tag{1}$$

for all $g \in G$. Show that $\text{Hom}(G, H)$ with the operation defined by *(1)* is an abelian group.

6.82. If A is an abelian group and Z is the group of integers under addition, prove that $\text{Hom}(Z, A) \cong A$.

6.83. Prove that the group of rationals under addition is not the direct sum of cyclic groups.

6.84. If G is the direct sum of cyclic subgroups, show that a factor group of G is not necessarily a direct sum of cyclic subgroups. (*Hint*: Use the preceding problem and free abelian groups.)

6.85. Let N be a normal subgroup of G. Prove that if G/N is free abelian, N is a direct summand.

TORSION GROUP AND RANK

6.86. If G is a finite group, show that $\text{aut}(G) \cong \prod_{p \in \pi} \text{aut}(G_p)$, where π is the set of all primes.

6.87. Prove the first Sylow theorem, page 130, for abelian groups using the p-components.

6.88. Let G denote the group of rotations of the plane (see Section 3.4c, page 68). As G is an abelian group we may use additive notation. Thus the rotation ρ_{θ_1} followed by rotation ρ_{θ_2} is $\rho_{\theta_1} + \rho_{\theta_2} = \rho_{\theta_1 + \theta_2}$. Let $\mathcal{R} = \{\rho_\theta \mid \theta = 2\pi m/n$ radians where m, n are integers, $n > 0\}$. Prove that (*a*) $\mathcal{R}$ is a subgroup of G; (*b*) every element of $\mathcal{R}$ has finite order; (*c*) if $\mathcal{R}_p$ denotes the p component of $\mathcal{R}$ for any prime p, then $\mathcal{R}_p = \{\rho_\theta \mid \theta = 2\pi m/p^r$ radians where m and r are integers$\}$ and $\mathcal{R}_p \cong$ the p-Prüfer group.

6.89. Let a, b be elements of an abelian group. Let the order of a plus the order of b be n. Prove by induction on n that $a + b$ is of order the least common multiple of the orders of a and b.

6.90. Let G be an abelian group. Suppose every element of G is of order less than some fixed integer n and there are elements of order n in G. Prove that the elements of order n generate G.

FUNDAMENTAL THEOREM FOR FINITELY GENERATED ABELIAN GROUPS

6.91. Prove that if G is a finitely generated abelian group, then $G = I_1 \oplus \cdots \oplus I_m \oplus C_1 \oplus \cdots \oplus C_n$ where I_j is an infinite cyclic group $(j = 1, \ldots, m)$, C_i is a finite cyclic group of order v_i $(i = 1, \ldots, n)$, and v_i divides v_{i+1} for $i = 1, \ldots, n-1$. (*Hint*: Use the fundamental theorem and then first look at the highest power of each different prime in the decomposition.)

6.92. Prove that the automorphism group of a finitely generated abelian group is finite if and only if there is at most one infinite cyclic summand in a cyclic decomposition of G.

6.93. Find the type of the additive group of integers modulo m for any integer $m > 0$.

6.94. Let G be a non-cyclic finite abelian group. Show that G has a subgroup of type $(p, p; 0)$ for some prime p.

6.95. Prove that the automorphism group of a finite non-cyclic abelian group G is non-abelian. (*Hint*: Use Problem 6.91 to find suitable elements $a, b \in G$ such that the order of a divides the order of b. Then look at the mappings $\alpha_1: a^s b^t \to a^{s+t} b^t$; $\alpha_2: a^s b^t \to a^w b^{-t}$ and $\alpha_3: a^s b^t \to a^t b^s$ where s and t are integers.)

6.96. Let G be a finite abelian group. Suppose that for each divisor d of G there are at most d elements in G of order d. Prove that G is cyclic.

6.97. Let G be a finitely generated abelian group. Prove by induction on the number of generators of G that every subgroup of G is finitely generated.

DIVISIBLE GROUPS

6.98. Show that a divisible abelian group has no subgroup of finite index.

6.99. If G is a nondivisible abelian group, then G has a subgroup of prime index. (*Hint*: Use the following theorem (not proved in this book): An abelian group G for which $nG = \{0\}$, $n \neq 0$, is the direct sum of cyclic groups.)

6.100. Prove that the additive group of the real numbers is the direct sum of isomorphic copies of the additive group of rationals.

6.101. (*a*) Let $G = \text{Hom}(A, B)$ (see Problem 6.81) where B is a torsion-free divisible group. Prove that G is the direct sum of copies of the additive group of rationals.

(*b*) Let G be as in (*a*) but with B a divisible p-group. Prove that if A is finite, G is the direct sum of p-Prüfer groups.

6.102. A subgroup H of an abelian group A is a *pure* subgroup of A if whenever $na = h \in H$ for some $a \in A$, then there is an $h' \in H$ such that $nh' = h$. Prove that (*a*) a direct summand of an abelian group is a pure subgroup, and (*b*) the torsion subgroup of an abelian group is a pure subgroup.

6.103. Prove that all the subgroups of a group in which every element has square-free order is pure.

6.104. Let H be a pure subgroup of an abelian group G. Prove that if $g + A \in G/A$, there is an element $\bar{g} \in G$ such that $\bar{g} + A = g + A$ and the order of $\bar{g}$ is equal to the order of $g + A$.

Chapter 7

Permutational Representations

Preview of Chapter 7

There are three main divisions of this chapter. In the first we generalize Cayley's theorem, that every group is isomorphic to a permutation group. As consequences of this generalization we prove the following theorems for G, a group generated by a finite number of elements: (1) A subgroup of finite index in G is itself finitely generated. (2) The number of subgroups of fixed finite index in G is finite. (3) If the subgroups of finite index of G intersect in the identity, then every homomorphism of G onto G is an automorphism.

The second main division of this chapter appears in Section 7.7. We call a group G an extension of a group H by a group K if there is a normal subgroup $\bar{H}$ of G such that $G/\bar{H} \cong K$ and $\bar{H} \cong H$. We examine G to see how it is built up from H and K. The most general case is complicated and we restrict ourselves to a special extension called "the splitting extension."

Reversing our analysis, we are able to build a group G that is the splitting extension of a given group H by a given group K. A particular example of a splitting extension is the direct product, used in Chapter 5.

Our third division, which begins in Section 7.8, defines a homomorphism of a group into one of its abelian subgroups. This homomorphism is called the transfer. We use it to show that a group G with center of finite index has finite derived group.

7.1 CAYLEY'S THEOREM

a. Another proof of Cayley's theorem

We saw in Chapter 2 that every groupoid is isomorphic to a groupoid of mappings. In particular, every group is isomorphic to a group of permutations. The consequences of this theorem are important. We repeat the proof here for the case of groups alone.

Theorem 7.1 (Cayley): Every group is isomorphic to a group of permutations.

Proof: Let G be a group. Let ρ be the mapping which assigns to each element g in G the following mapping of G into G:

$$x \to xg \quad \text{for each} \quad x \in G$$

Thus the image of g in G under ρ is, by definition, $g\rho$ where

$$g\rho:\ x \to xg \ \ (x \in G)$$

The definition of ρ is unambiguous. To prove that ρ defines an isomorphism of G onto a subgroup of S_G, we have to check that:

(i) $g\rho$ is a permutation of G for every $g \in G$;

(ii) ρ is a homomorphism, i.e. if $g, h \in G$, then $(gh)\rho = g\rho \cdot h\rho$;

(iii) ρ is also an isomorphism, i.e. ρ is one-to-one.

We deal first with (i). Thus we must prove $g\rho$ is a one-to-one mapping of G onto G. If $x(g\rho) = y(g\rho)$, then $xg = yg$. So, multiplying by g^{-1} on the right, we find $x = y$. Next we prove $g\rho$ is a mapping of G onto G. Suppose $x \in G$; then $(xg^{-1})(g\rho) = (xg^{-1})g = x$ and so $g\rho$ is onto.

Secondly we prove (ii). For $x \in G$,

$$x((gh)\rho) \;=\; x(gh) \;=\; (xg)h \;=\; (x(g\rho))(h\rho) \;=\; x(g\rho h\rho)$$

Since $(gh)\rho$ and $g\rho h\rho$ have precisely the same effect on every element of G, $(gh)\rho = g\rho h\rho$ (by the definition of equality of mappings).

It remains to prove (iii), i.e. ρ is one-to-one. Suppose $g\rho = h\rho$; then if 1 is the identity element of G, $g = 1(g\rho) = 1(h\rho) = h$, i.e. $g = h$. Therefore ρ is one-to-one.

(*Note*: In the proof of Cayley's theorem ρ is a mapping of G into S_G, so that $g\rho$ is itself a mapping of G to G. Caution and patience are required to avoid confusion in some of our subsequent equations.)

b. Cayley's theorem and examples of groups

Cayley's theorem tells us that there is an isomorphic image of every group among the permutation groups of suitably chosen sets. If one demands that a permutation group satisfy further conditions, one frequently comes across interesting groups (see Chapter 3). Historically many important groups arose in precisely this way.

Problem

7.1. Describe in detail the isomorphisms given by Cayley's theorem for (i) a cyclic group of order 2, (ii) a cyclic group of order n $(n \geqq 3)$, (iii) the symmetric group on three letters.

Solution:

(i) Let G be cyclic of order 2. Then G consists of two elements, 1 and a, where $a^2 = 1$ and $1 \cdot a = a = a \cdot 1$. Let ρ be as in Theorem 7.1.

1ρ is the mapping $\qquad 1\rho: \; 1 \to 1, \; a \to a$

and $a\rho$ is the mapping given by $\qquad a\rho: \; 1 \to a, \; a \to 1$

Clearly ρ is one-to-one, as $a\rho \neq 1\rho$.

(ii) Let G be cyclic of order n. Then G consists of n elements $1, a, a^2, \ldots, a^{n-1}$, say (see Lemma 4.5, page 102). Then

$$\begin{aligned}
1\rho &: \; 1 \to 1, \quad a \to a, \quad \ldots, \; a^{n-1} \to a^{n-1} \\
a\rho &: \; 1 \to a, \quad a \to a^2, \quad \ldots, \; a^{n-1} \to 1 \\
a^2\rho &: \; 1 \to a^2, \quad a \to a^3, \quad \ldots, \; a^{n-2} \to 1, \; a^{n-1} \to a \\
&\;\;\vdots \\
a^{n-1}\rho &: \; 1 \to a^{n-1}, \quad a \to 1, \quad \ldots, \; a^{n-1} \to a^{n-2}
\end{aligned}$$

(iii) The symmetric group on the set $\{1, 2, 3\}$ consists of the permutations

$$\begin{aligned}
p_1 &: \; 1 \to 1, \; 2 \to 2, \; 3 \to 3 \qquad & p_4 &: \; 1 \to 2, \; 2 \to 1, \; 3 \to 3 \\
p_2 &: \; 1 \to 1, \; 2 \to 3, \; 3 \to 2 \qquad & p_5 &: \; 1 \to 3, \; 2 \to 2, \; 3 \to 1 \\
p_3 &: \; 1 \to 2, \; 2 \to 3, \; 3 \to 1 \qquad & p_6 &: \; 1 \to 3, \; 2 \to 1, \; 3 \to 2
\end{aligned}$$

Then

$$\begin{aligned}
p_1\rho &: \; p_1 \to p_1, \; p_2 \to p_2, \; p_3 \to p_3, \; p_4 \to p_4, \; p_5 \to p_5, \; p_6 \to p_6 \\
p_2\rho &: \; p_1 \to p_2, \; p_2 \to p_1, \; p_3 \to p_5, \; p_4 \to p_6, \; p_5 \to p_3, \; p_6 \to p_4 \\
p_3\rho &: \; p_1 \to p_3, \; p_2 \to p_4, \; p_3 \to p_6, \; p_4 \to p_5, \; p_5 \to p_2, \; p_6 \to p_1 \\
p_4\rho &: \; p_1 \to p_4, \; p_2 \to p_3, \; p_3 \to p_2, \; p_4 \to p_1, \; p_5 \to p_6, \; p_6 \to p_5 \\
p_5\rho &: \; p_1 \to p_5, \; p_2 \to p_6, \; p_3 \to p_4, \; p_4 \to p_3, \; p_5 \to p_1, \; p_6 \to p_2 \\
p_6\rho &: \; p_1 \to p_6, \; p_2 \to p_5, \; p_3 \to p_1, \; p_4 \to p_2, \; p_5 \to p_4, \; p_6 \to p_3
\end{aligned}$$

It is worth checking $(p_i\rho)(p_j\rho) = (p_ip_j)\rho$ for some i and j between 1 and 6.

7.2 PERMUTATIONAL REPRESENTATIONS

Definition of a permutational representation

A homomorphism of a group G into the symmetric group on the set X is called a *permutational representation of G on X.*

So if ρ is the isomorphism provided by Cayley's theorem for the group G, then ρ is a permutational representation of G on G.

Repeating the definition of a permutational representation of a group G in detail, we say that a mapping μ of G into the symmetric group on some set X is a permutational representation of G if

$$(gh)\mu = g\mu h\mu$$

for all g and h in G. The permutational representation provided by Cayley's theorem is called the *right-regular representation* (the adjective right is used because the representation is obtained by multiplication on the right).

Example 1: (i) Let G be the symmetric group on $\{1,2,3\}$, and let ρ be as in the solution to Problem 7.1(iii). Then ρ itself is a representation of G as a permutation group on six elements.

(ii) There is another representation of the symmetric group G on $\{1,2,3\}$, the most natural one. This is the identity isomorphism, for G is itself a permutation group on $\{1,2,3\}$.

(iii) Let G be the cyclic group of order 2 and let X be the set of all integers $\ldots,-1,0,1,\ldots$. Let α be the permutation of X defined by

$$\alpha: \quad 2i \to 2i+1, \quad 2i+1 \to 2i \quad \text{for } i = 0, \pm 1, \ldots$$

Thus α sends an even integer to the succeeding odd integer and an odd integer to the preceding even integer. Let ι denote the identity permutation. Then ι and α together constitute a subgroup Γ of the symmetric group on X since

$$\alpha^2 = \iota, \quad \iota\alpha = \alpha\iota = \alpha$$

Now suppose G consists of the elements 1 and a. Then the mapping $\mu: a \to \alpha$, $1 \to \iota$ is actually an isomorphism since both G and Γ are cyclic of order two. So μ is a permutational representation.

(iv) Suppose n is a positive integer and that G is the cyclic group of order n,

$$G = \{1, a, a^2, \ldots, a^{n-1}\}$$

Let α be the permutation of $X = Z$ the integers defined by

$$\alpha: \quad jn \to jn+1, \quad jn+1 \to jn+2, \quad \ldots, \quad jn+(n-1) \to jn \quad (j = 0, \pm 1, \ldots)$$

In particular we have

$$0\alpha = 1, \quad 1\alpha = 2, \quad \ldots, \quad (n-1)\alpha = 0$$

Note that α cyclically permutes the integers taken in blocks of n.

It is not difficult to see that α is of order n. First of all, $0\alpha^2 = 2$, $0\alpha^3 = 3$, $\ldots$, $0\alpha^{n-1} = n-1$. This means that the permutations $\iota, \alpha, \alpha^2, \ldots, \alpha^{n-1}$ are all distinct since they act differently on 0.

If j is any integer, $j\alpha^n = j$. So $\alpha^n = \iota$ and $\{\iota, \alpha, \alpha^2, \ldots, \alpha^{n-1}\}$ is a cyclic group Γ of order n. Hence the mapping μ of G into Γ defined by

$$1 \to \iota, \quad a \to \alpha, \quad \ldots, \quad a^{n-1} \to \alpha^{n-1}$$

is an isomorphism and therefore a permutational representation of G.

(v) Suppose that G is the dihedral group of degree 4. G is the group of symmetries of the square

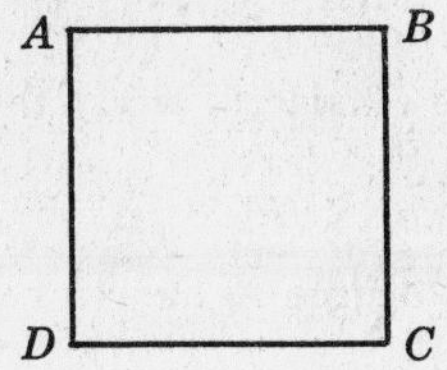

Let $g \in G$. By Lemma 3.12, page 75, g takes each vertex of $ABCD$ to a vertex. Since g is one-to-one, Ag, Bg, Cg and Dg are distinct vertices. If we define $X = \{A, B, C, D\}$ and γ_g by $x\gamma_g = xg$ for all $x \in X$, then $\gamma_g \in S_X$. Let $\theta: G \to S_X$ be defined by $g\theta = \gamma_g$. Then if $x \in X$, $g, h \in G$,

$$x(gh)\theta = x\gamma_{gh} = x(gh) = (xg)h = (x\gamma_g)\gamma_h = x(\gamma_g\gamma_h) = x(g\theta)(h\theta)$$

Thus $(gh)\theta = g\theta h\theta$, and so θ is a permutational representation of G. If $g\theta$ is the identity permutation of X, then g leaves every vertex of $ABCD$ unchanged. But by Lemma 3.7, page 71, $g = \iota$. Hence $\operatorname{Ker}\theta = \{\iota\}$ and θ is actually an isomorphism.

7.3 DEGREE OF A REPRESENTATION AND FAITHFUL REPRESENTATIONS

a. Degree of a representation

Definition: The *degree* of a permutational representation on X (more briefly referred to as a representation) is the number of elements in X.

Example 2: We inspect Example 1(i)-(v).

(i) This representation is of degree 6.

(ii) This representation is of degree 3.

Notice that in (i) and (ii) we have two representations of the same group, namely the symmetric group on $\{1, 2, 3\}$, of different degrees.

(iii) This representation is of infinite degree. Notice that here G is cyclic of order 2. Hence there are representations of *finite* groups which are of *infinite* degree.

(iv) This representation is of infinite degree.

(v) This representation is of degree 4.

Problems

7.2. Find a representation of degree 3 for a cyclic group of order 2.

Solution:

Let G be the cyclic group of order 2, say $G = \{1, a\}$. Let $X = \{1, 2, 3\}$. Then there are several possible representations of G on X. First of all there is the rather trivial representation

$$\tau: 1 \to \iota, \; a \to \iota$$

where ι is as usual the identity permutation. τ is clearly a representation. For no matter which elements g, h in G we choose,

$$(gh)\tau = \iota \quad \text{and} \quad (g\tau)(h\tau) = \iota\iota = \iota$$

from which $(gh)\tau = (g\tau)(h\tau)$. Another representation of G involves choosing a different homomorphism. Now observe that if μ is a homomorphism of G into the symmetric group S_X, then either $\mu = \tau$ or $G\mu$ is of order 2. Note also that $G\mu$ must be a subgroup. So in deciding on a choice of μ, we need subgroups of S_X of order 2. There are actually 3 of them. To see this let

$$\alpha_1: 1 \to 2, \; 2 \to 1, \; 3 \to 3 \qquad \alpha_2: 1 \to 3, \; 2 \to 2, \; 3 \to 1 \qquad \alpha_3: 1 \to 1, \; 2 \to 3, \; 3 \to 2$$

Then $\{\iota, \alpha_1\}, \{\iota, \alpha_2\}, \{\iota, \alpha_3\}$ are subgroups of S_X of order two, since $\alpha_i^2 = \iota$ for $i = 1, 2, 3$. Thus the mappings

$$\mu_1: 1 \to \iota, \; a \to \alpha_1 \qquad \mu_2: 1 \to \iota, \; a \to \alpha_2 \qquad \mu_3: 1 \to \iota, \; a \to \alpha_3$$

are representations of G on X.

The proof that there are precisely three subgroups of order 2 follows from an inspection of all the subgroups of S_X. Since we have no real need for such proof here, we leave the details to the reader.

The upshot of these considerations is that we have produced 4 representations of G of degree 3.

7.3. Find representations of degree 10 and 15 respectively of the cyclic group of order 5.

Solution:

Let G be cyclic of order 5. Then we can find $a \in G$ such that $G = \{1, a, a^2, a^3, a^4\}$. Let

$$X = \{1, 2, \ldots, 10\} \qquad Y = \{1, 2, \ldots, 15\}$$

and let $\alpha_1: \; 1 \to 2,\; 2 \to 3,\; 3 \to 4,\; 4 \to 5,\; 5 \to 1,\; 6 \to 7,\; 7 \to 8,\; 8 \to 9,\; 9 \to 10,\; 10 \to 6$

$$\alpha_2: \; 1 \to 2,\; 2 \to 3,\; 3 \to 4,\; 4 \to 5,\; 5 \to 1,\; j \to j, \;\text{ for } j > 5$$

It follows from a direct calculation that both α_1 and α_2 are of order 5. Then

$$\Gamma_1 = \{\iota, \alpha_1, \alpha_1^2, \alpha_1^3, \alpha_1^4\} \qquad \Gamma_2 = \{\iota, \alpha_2, \alpha_2^2, \alpha_2^3, \alpha_2^4\}$$

are subgroups of order 5 of S_X and S_Y respectively. Thus

$$\mu_1: \; 1 \to \iota, \; a \to \alpha_1, \; \ldots, \; a^4 \to \alpha_1^4$$

and

$$\mu_2: \; 1 \to \iota, \; a \to \alpha_2, \; \ldots, \; a^4 \to \alpha_2^4$$

are both representations of G. The first is of degree 10 and the second is of degree 15.

b. Faithful representations

Definition: A representation is termed faithful if it is one-to-one. Both faithful and non-faithful representations are useful, as we shall see later.

Problems

7.4. Are the representations in Example 1, page 216, faithful?

Solution:

(i) ρ is faithful.

(ii) The identity isomorphism is one-to-one, so this representation is also faithful. Notice that (i) and (ii) provide examples of faithful representations of the same group which are of different degrees.

(iii) μ is faithful.

(iv) μ is faithful.

(v) The representation is faithful.

7.5. Inspect the representations of (*a*) Problem 7.2 and (*b*) Problem 7.3 for faithfulness.

Solution:

(*a*) τ is not faithful since $a \neq 1$ and $a\tau = \iota$, i.e. τ is not one-to-one. However μ_1, μ_2, μ_3 are faithful.

(*b*) μ_1 and μ_2 are both faithful.

7.4 PERMUTATIONAL REPRESENTATIONS ON COSETS

Definition: Suppose that G is a group and that H is a subgroup of G. Then the right cosets of H in G are

$$Hx_1, Hx_2, \ldots \qquad (7.1)$$

If H consists of the identity element alone, then (*7.1*) is simply an enumeration of the elements of G. Here we will show how to obtain a permutational representation of G using the cosets (*7.1*) which coincides with the regular representation when $H = \{1\}$.

To describe our representation, let us choose a complete system X of representatives of the right cosets Hg of H in G, with 1 the representative of H. In other words, we select in each coset Hg an element which we term the *representative* of the coset, with 1 the representative of H. X is then simply the set of chosen representatives. We call such a set X a *right transversal* of H in G.

Given a right transversal X of H in G, we denote the representative of the coset Hg by $\bar{g}$. Thus $\bar{g}$ is an element of X. Since two right cosets are either identical or disjoint, it follows that $Hg = H\bar{g}$ because $\bar{g} \in Hg$. Notice that if $h \in H$, then $\overline{hg} = \bar{g}$ since $Hg = H\bar{g} = Hhg = H\overline{hg}$.

For example, if G is cyclic of order 4, say $G = \{1, a, a^2, a^3\}$, and if $H = \{1, a^2\}$ is a subgroup of order 2 of G, then $\{1, a\}$ is a right transversal of H in G. We take $X = \{1, a\}$ and note that $\bar{1} = 1$, $\bar{a} = a$, $\bar{a}^2 = 1$, $\bar{a}^3 = a$.

With each element g in G we associate a mapping γ_g of X into X, where γ_g is defined by

$$\gamma_g : x \to \overline{xg} \quad (x \in X) \tag{7.2}$$

In fact γ_g is a permutation of X. To prove this we first show that if $g_1, g_2 \in G$, then $\overline{\bar{g}_1 g_2} = \overline{g_1 g_2}$. For $\overline{g_1 g_2}$ is the representative of the coset Hg_1g_2, while $\overline{\bar{g}_1 g_2}$ is the representative of the coset $H\bar{g}_1 g_2$. But $H\bar{g}_1 = Hg_1$, and so

$$H\overline{\bar{g}_1 g_2} = Hg_1g_2 = H\overline{g_1 g_2}$$

Thus

$$\overline{\bar{g}_1 g_2} = \overline{g_1 g_2} \quad \text{for all } g_1, g_2 \in G \tag{7.3}$$

We use (*7.3*) to prove that the mapping γ_g is a permutation. First we prove γ_g is one-to-one. Assume $x\gamma_g = y\gamma_g$ $(x, y \in X)$. Then $\overline{xg} = \overline{yg}$, and so $\overline{\overline{xg}g^{-1}} = \overline{\overline{yg}g^{-1}}$. Using (*7.3*), we find that $\overline{\overline{xg}g^{-1}} = \overline{xgg^{-1}} = \bar{x} = x$ and similarly $\overline{\overline{yg}g^{-1}} = y$, from which $x = y$. Finally we prove γ_g is onto. Suppose $x \in X$; then $\overline{xg^{-1}} \in X$ and

$$\overline{xg^{-1}}\gamma_g = \overline{\overline{xg^{-1}}g} = \overline{xg^{-1}g} = x$$

Hence every element of X is an image. Thus γ_g is a permutation.

We now define a mapping π of G into S_X. π assigns to g in G the mapping γ_g, so that $g\pi$ is the permutation of X defined by (*7.2*). The aim of the discussion in this section is to prove that this mapping π is a permutational representation of G on X. We have only to verify that it is a homomorphism, i.e. if $g_1, g_2 \in G$, then $(g_1g_2)\pi = (g_1\pi)(g_2\pi)$. Note that $(g_1g_2)\pi$ is a permutation of X. To prove $(g_1g_2)\pi = (g_1\pi)(g_2\pi)$ we must show that the effect of the mapping $(g_1g_2)\pi$ is the same as the effect of the mapping $(g_1\pi)(g_2\pi)$. (Note that $(g_1\pi)(g_2\pi)$ is the product of two mappings, i.e. the result of first performing the mapping $(g_1\pi)$ and then $(g_2\pi)$.) If $x \in X$, using (*7.3*) we find

$$x((g_1g_2)\pi) = \overline{x(g_1g_2)} = \overline{\overline{xg_1}g_2} = (x(g_1\pi))(g_2\pi) = x((g_1\pi)(g_2\pi))$$

Hence $(g_1g_2)\pi = (g_1\pi)(g_2\pi)$ as claimed.

We shall refer to π as a *coset representation* of G (with respect to H). Of course π depends on H, G and X. In a sense π is independent of the choice of the transversal X (see Problem 7.10).

Problems

7.6. Choose right transversals for (*a*) the center Z in the dihedral group D of degree 4, and (*b*) the center Z in the quaternion group $\mathcal{H}$ of order 8 (see Table 5.1, page 151).

Solution:

(*a*) The dihedral group $D = \{1, a_1, a_2, a_3, a_4, a_5, a_6, a_7\}$ is most easily described by its multiplication table:

	1	a_1	a_2	a_3	a_4	a_5	a_6	a_7
1	1	a_1	a_2	a_3	a_4	a_5	a_6	a_7
a_1	a_1	a_2	a_3	1	a_5	a_6	a_7	a_4
a_2	a_2	a_3	1	a_1	a_6	a_7	a_4	a_5
a_3	a_3	1	a_1	a_2	a_7	a_4	a_5	a_6
a_4	a_4	a_7	a_6	a_5	1	a_3	a_2	a_1
a_5	a_5	a_4	a_7	a_6	a_1	1	a_3	a_2
a_6	a_6	a_5	a_4	a_7	a_2	a_1	1	a_3
a_7	a_7	a_6	a_5	a_4	a_3	a_2	a_1	1

(Here a_1 corresponds to a rotation of 90°, while a_4 corresponds to a reflection.) The center Z of G is given by $Z = \{1, a_2\}$. This can be checked by verifying that a_2 commutes with every element of D (and no element other than 1 and a_2 has this property).

Finally Z, Za_1, Za_4, Za_5 are the cosets of Z in G. Thus

$$X = \{1, a_1, a_4, a_5\}$$

is a right transversal of Z in G.

(*b*) The quaternion group $\mathcal{H}$ of order 8, with elements $1, a_1, a_2, a_3, a_4, a_5, a_6, a_7$, is given by the following multiplication table:

	1	a_1	a_2	a_3	a_4	a_5	a_6	a_7
1	1	a_1	a_2	a_3	a_4	a_5	a_6	a_7
a_1	a_1	a_2	a_3	1	a_5	a_6	a_7	a_4
a_2	a_2	a_3	1	a_1	a_6	a_7	a_4	a_5
a_3	a_3	1	a_1	a_2	a_7	a_4	a_5	a_6
a_4	a_4	a_7	a_6	a_5	a_2	a_1	1	a_3
a_5	a_5	a_4	a_7	a_6	a_3	a_2	a_1	1
a_6	a_6	a_5	a_4	a_7	1	a_3	a_2	a_1
a_7	a_7	a_6	a_5	a_4	a_1	1	a_3	a_2

(Comparing with Table 5.1, we have the following correspondence: $1 \to 1$, $a \to a_1$, $a^2 \to a_2$, $a^3 \to a_3$, $b \to a_4$, $ab \to a_5$, $a^2b \to a_6$, $a^3b \to a_7$.) The center Z of $\mathcal{H}$ is given by $Z = \{1, a_2\}$. One has only to check this from the multiplication table.

The cosets of Z in $\mathcal{H}$ are Z, Za_1, Za_4, Za_5. This again can be checked directly from the multiplication table. Thus

$$X = \{1, a_1, a_4, a_5\}$$

is a right transversal of Z in $\mathcal{H}$.

7.7. Find a coset representation of (*a*) D of Problem 7.6(a) with respect to Z, and (*b*) $\mathcal{H}$ of Problem 7.6(b) with respect to Z.

Solution:

(*a*) We work out a coset representation π using the right transversal X given in Problem 7.6. Thus we must find the permutations of X that π assigns to each element of D. We will calculate in detail the permutation $a_1\pi$. Now $X = \{1, a_1, a_4, a_5\}$. Then $\overline{1 \cdot a_1} = a_1$, and so $1(a_1\pi) = a_1$. $\overline{a_1 a_1} = 1$ since $Za_1^2 = Z = Z1$, and so $a_1(a_1\pi) = 1$. $\overline{a_4 a_1} = a_5$ since $Za_4a_1 = Za_7 = Za_5$, and so $a_4(a_1\pi) = a_5$. Finally $\overline{a_5 a_1} = a_4$ since $Z(a_5a_1) = Za_4$, and so $a_5(a_1\pi) = a_4$. Therefore

$$a_1\pi: \quad 1 \to a_1, \ a_1 \to 1, \ a_4 \to a_5, \ a_5 \to a_4$$

Similarly the other permutations are

$$1\pi: \quad 1 \to 1, \ a_1 \to a_1, \ a_4 \to a_4, \ a_5 \to a_5$$
$$a_2\pi: \quad 1 \to 1, \ a_1 \to a_1, \ a_4 \to a_4, \ a_5 \to a_5$$
$$a_3\pi: \quad 1 \to a_1, \ a_1 \to 1, \ a_4 \to a_5, \ a_5 \to a_4$$
$$a_4\pi: \quad 1 \to a_4, \ a_1 \to a_5, \ a_4 \to 1, \ a_5 \to a_1$$
$$a_5\pi: \quad 1 \to a_5, \ a_1 \to a_4, \ a_4 \to a_1, \ a_5 \to 1$$
$$a_6\pi: \quad 1 \to a_4, \ a_1 \to a_5, \ a_4 \to 1, \ a_5 \to a_1$$
$$a_7\pi: \quad 1 \to a_5, \ a_1 \to a_4, \ a_4 \to a_1, \ a_5 \to 1$$

It is instructive to check directly that this mapping π is a homomorphism of D into the permutation group on $\{1, a_1, a_4, a_5\}$.

(*b*) Again we must find the permutations that π assigns to each element of $\mathcal{H}$. The argument follows closely that of (*a*) above, using the set $X = \{1, a_1, a_4, a_5\}$. Here we give only the result which the reader is urged to check.

$$1\pi: \quad 1 \to 1, \ a_1 \to a_1, \ a_4 \to a_4, \ a_5 \to a_5$$
$$a_1\pi: \quad 1 \to a_1, \ a_1 \to 1, \ a_4 \to a_5, \ a_5 \to a_4$$
$$a_2\pi: \quad 1 \to 1, \ a_1 \to a_1, \ a_4 \to a_4, \ a_5 \to a_5$$
$$a_3\pi: \quad 1 \to a_1, \ a_1 \to 1, \ a_4 \to a_5, \ a_5 \to a_4$$
$$a_4\pi: \quad 1 \to a_4, \ a_1 \to a_5, \ a_4 \to 1, \ a_5 \to a_1$$
$$a_5\pi: \quad 1 \to a_5, \ a_1 \to a_4, \ a_4 \to a_1, \ a_5 \to 1$$
$$a_6\pi: \quad 1 \to a_4, \ a_1 \to a_5, \ a_4 \to 1, \ a_5 \to a_1$$
$$a_7\pi: \quad 1 \to a_5, \ a_1 \to a_4, \ a_4 \to a_1, \ a_5 \to 1$$

7.8. Consider the permutation group $S_{\{1,2\}}$. Its elements are $\phi_1 = \begin{pmatrix} 1 & 2 \\ 1 & 2 \end{pmatrix}$ and $\phi_2 = \begin{pmatrix} 1 & 2 \\ 2 & 1 \end{pmatrix}$. Consider now the permutation group $S_{\{a,b\}}$. Its elements are $\Psi_1 = \begin{pmatrix} a & b \\ a & b \end{pmatrix}$, $\Psi_2 = \begin{pmatrix} a & b \\ b & a \end{pmatrix}$. Ψ_1 and ϕ_1, Ψ_2 and ϕ_2 are essentially the same except for the elements they act on. Give a definition which will make this idea of "essentially the same" precise.

Solution:

Let F be a permutation group on a set X and let G be a permutation group on a set Y. We say that F and G are isomorphic as permutation groups if there exists a one-to-one onto correspondence $\alpha: X \to Y$ and an onto mapping $\theta: F \to G$ such that for all x in X and $f \in F$, $(xf)\alpha = (x\alpha)(f\theta)$. (In this problem $\alpha: 1 \to a, 2 \to b$ and $\theta: \phi_1 \to \Psi_1, \phi_2 \to \Psi_2$.)

7.9. Prove that if F and G are isomorphic as permutation groups, then they are isomorphic as groups. (Hard.)

Solution:

The θ of the solution of Problem 7.8 provides the isomorphism. First we show that it is a homomorphism. Let $f_1, f_2 \in F$. For any $x \in X$,

$$(x\alpha)((f_1f_2)\theta) = (x(f_1f_2))\alpha = ((xf_1)f_2)\alpha \quad \text{(by the definition of composition of mappings)}$$
$$= ((xf_1)\alpha)(f_2\theta) = ((x\alpha)(f_1\theta))(f_2\theta) = (x\alpha)((f_1\theta)(f_2\theta))$$

Since α is a mapping onto Y, it follows that as x ranges over X, $x\alpha$ ranges over Y. Hence as $(f_1f_2)\theta$ and $(f_1\theta)(f_2\theta)$ are permutations, $f_1\theta f_2\theta = (f_1\theta)(f_2\theta)$. Thus θ is a homomorphism.

Next we must show that θ is one-to-one. Suppose that $f_1\theta = f_2\theta$; then if $x \in X$, $(x\alpha)(f_1\theta) = (x\alpha)(f_2\theta)$. Thus

$$(xf_1)\alpha = (xf_2)\alpha$$

Since α is one-to-one, $xf_1 = xf_2$ from which $f_1 = f_2$. Hence θ is an isomorphism.

7.10. Let G be any group and H a subgroup of G. Let X_i be a transversal for H in G and π_i the corresponding coset representation, $i = 1, 2$. Prove that $G\pi_1$ and $G\pi_2$ are isomorphic as permutation groups. (Hard.)

Solution:

Since X_1 and X_2 contain one and only one element in each coset of H in G, we define $\alpha: X_1 \to X_2$ by sending $x_1 \in X_1$ to $x_2 \in X_2$ if $Hx_1 = Hx_2$. α is then a one-to-one correspondence. We define $\mu: G\pi_1 \to G\pi_2$ by $(g\pi_1)\mu = g\pi_2$. It is easy to check that μ is a mapping. Let $g\pi_1 = \beta$, $g\pi_2 = \gamma$. We need only verify that $(x\beta)\alpha = (x\alpha)\gamma$ for each $x \in X$. Now by definition of β and α,

$$Hxg = H(x\beta) = H((x\beta)\alpha)$$

Also,
$$Hxg = H(x\alpha)g = H((x\alpha)\gamma)$$

Hence $(x\alpha)\gamma = (x\beta)\alpha$ as they are elements of X_2 that belong to the same coset, i.e. Hxg. The result follows.

7.5 FROBENIUS' VARIATION OF CAYLEY'S THEOREM

a. The kernel of a coset representation

Is a coset representation π of a group G with respect to a subgroup H ever faithful? We know that the answer is yes if $H = \{1\}$. The object now is to find the kernel of π.

Theorem 7.2: Let G be a group and H a subgroup of G. Let π be a coset representation of G with respect to H. Then the kernel of π is the largest normal subgroup of G contained in H, i.e. if $N \triangleleft G$ and $H \supseteq N$, then $N \subseteq \operatorname{Ker} \pi$.

Before proving Theorem 7.2, it should be noted that from this theorem it follows that π is faithful if the only normal subgroup of G which is contained in H is the identity subgroup. This implies, for example, that if G is a simple group and $H \neq G$, π is automatically faithful. For then, by definition, the only normal subgroups of G are G and the identity subgroup. This observation has been useful in the theory of finite groups.

Proof: Let X be the right transversal of H in G from which π was defined. First we prove that if K is the kernel of π, i.e. the set of all elements g of G such that $g\pi = \iota$, the identity mapping of X onto itself, then K is contained in H.

If $a \in K$ and $x \in X$, then $a\pi = \iota$, i.e. $x = x(a\pi) = \overline{xa}$. In particular on putting $x = 1$, $1 = \bar{a}$. Hence $a \in H$. This means that K is contained in H. Of course K, as the kernel of the homomorphism π, is a normal subgroup of G. To complete the proof of the theorem, we must show that K is the *largest* normal subgroup of G contained in H. To do this it is sufficient to prove that if N is any normal subgroup of G contained in H, then $N\pi = \{\iota\}$.

Suppose that $a \in N$. If $x \in X$, then

$$Hxa = Hxax^{-1}x$$

Since N is normal, $xax^{-1} \in N$. But N is contained in H, and so $xax^{-1} \in H$. Accordingly $Hxa = Hx$. Thus $\overline{xa} = x$ which means

$$x(a\pi) = x \quad \text{for all } x \in X$$

Hence $a\pi = \iota$ and $N \subseteq K$ as required.

Problems

7.11. Let G be the symmetric group on $\{1,2,3\}$. Let $\sigma: 1\to 2, 2\to 3, 3\to 1$, and let $\tau: 1\to 2, 2\to 1, 3\to 3$. Then the elements of G are $\iota, \sigma, \sigma^2, \tau, \sigma\tau, \sigma^2\tau$. Let (a) $H = gp(\sigma) = \{\iota, \sigma, \sigma^2\}$ and (b) $H = gp(\tau) = \{\iota, \tau\}$. In each case find a coset representation of G with respect to H. Find also the kernels of both representations.

Solution:

(a) A right transversal of H in G is $\{\iota, \tau\}$. Notice that $\tau^{-1}\sigma\tau = \sigma^{-1} = \sigma^2$. So H is normal in G. Hence by Theorem 7.2, every coset representation has kernel precisely H. The coset representation π associated with the right transversal $\{\iota, \tau\}$ is given by

$$\begin{array}{ll} \iota\pi: \iota\to\iota,\ \tau\to\tau & \tau\pi: \iota\to\tau,\ \tau\to\iota \\ \sigma\pi: \iota\to\iota,\ \tau\to\tau & (\sigma\tau)\pi: \iota\to\tau,\ \tau\to\iota \\ \sigma^2\pi: \iota\to\iota,\ \tau\to\tau & (\sigma^2\tau)\pi: \iota\to\tau,\ \tau\to\iota \end{array}$$

Clearly $\operatorname{Ker}\pi = \{\iota, \sigma, \sigma^2\}$.

(b) Let $X = \{\iota, \sigma, \sigma^2\}$. Since $G = H \cup H\sigma \cup H\sigma^2$, X is a right transversal of H in G. The associated permutational representation π is

$$\begin{array}{ll} \iota\pi: \iota\to\iota,\ \sigma\to\sigma,\ \sigma^2\to\sigma^2 & \tau\pi: \iota\to\iota,\ \sigma\to\sigma^2,\ \sigma^2\to\sigma \\ \sigma\pi: \iota\to\sigma,\ \sigma\to\sigma^2,\ \sigma^2\to\iota & (\sigma\tau)\pi: \iota\to\sigma^2,\ \sigma\to\sigma,\ \sigma^2\to\iota \\ \sigma^2\pi: \iota\to\sigma^2,\ \sigma\to\iota,\ \sigma^2\to\sigma & (\sigma^2\tau)\pi: \iota\to\sigma,\ \sigma\to\iota,\ \sigma^2\to\sigma^2 \end{array}$$

It follows immediately that π is faithful, since the only element mapped to the identity permutation of X is ι. Hence the kernel of $\pi = \{\iota\}$.

7.12. Let G be the alternating group of degree 4 and let H be the subgroup consisting of the permutations

$$\iota,\ \begin{pmatrix}1&2&3&4\\2&1&4&3\end{pmatrix},\ \begin{pmatrix}1&2&3&4\\3&4&1&2\end{pmatrix},\ \begin{pmatrix}1&2&3&4\\4&3&2&1\end{pmatrix}$$

Find an associated coset representation of G with respect to H. Is this representation faithful?

Solution:

Let $\tau_1 = \begin{pmatrix}1&2&3&4\\2&1&4&3\end{pmatrix}$, $\tau_2 = \begin{pmatrix}1&2&3&4\\3&4&1&2\end{pmatrix}$, $\tau_3 = \begin{pmatrix}1&2&3&4\\4&3&2&1\end{pmatrix}$. Then if $\sigma = \begin{pmatrix}1&2&3&4\\2&3&1&4\end{pmatrix}$, G consists of the elements

$$\iota, \sigma, \sigma^2, \tau_1, \tau_1\sigma, \tau_1\sigma^2, \tau_2, \tau_2\sigma, \tau_2\sigma^2, \tau_3, \tau_3\sigma, \tau_3\sigma^2$$

It follows immediately that a right transversal of H in G is $X = \{\iota, \sigma, \sigma^2\}$. Let π be the associated coset representation. It is easy to check directly that H is normal in G. Then $\operatorname{Ker}\pi = H$ by Theorem 7.2, and so π is not faithful. Finally we list the permutations $g\pi$ with g in G:

$$\begin{array}{ll} \iota\pi: \iota\to\iota,\ \sigma\to\sigma,\ \sigma^2\to\sigma^2 & \tau_2\pi: \iota\to\iota,\ \sigma\to\sigma,\ \sigma^2\to\sigma^2 \\ \sigma\pi: \iota\to\sigma,\ \sigma\to\sigma^2,\ \sigma^2\to\iota & (\tau_2\sigma)\pi: \iota\to\sigma,\ \sigma\to\sigma^2,\ \sigma^2\to\iota \\ \sigma^2\pi: \iota\to\sigma^2,\ \sigma\to\iota,\ \sigma^2\to\sigma & (\tau_2\sigma^2)\pi: \iota\to\sigma^2,\ \sigma\to\iota,\ \sigma^2\to\sigma \\ \tau_1\pi: \iota\to\iota,\ \sigma\to\sigma,\ \sigma^2\to\sigma^2 & \tau_3\pi: \iota\to\iota,\ \sigma\to\sigma,\ \sigma^2\to\sigma^2 \\ (\tau_1\sigma)\pi: \iota\to\sigma,\ \sigma\to\sigma^2,\ \sigma^2\to\iota & (\tau_3\sigma)\pi: \iota\to\sigma,\ \sigma\to\sigma^2,\ \sigma^2\to\iota \\ (\tau_1\sigma^2)\pi: \iota\to\sigma^2,\ \sigma\to\iota,\ \sigma^2\to\sigma & (\tau_3\sigma^2)\pi: \iota\to\sigma^2,\ \sigma\to\iota,\ \sigma^2\to\sigma \end{array}$$

b. Frobenius' theorem

Let H be a subgroup of G, X a transversal and π the associated coset representation. Let ρ denote the right regular representation of G. Our idea is to express ρ in terms of π.

If $x \in X$ and $g \in G$, we have

$$x(g\rho) = xg$$

Of course xg need not lie in X.

We know, however, that xg belongs to the same coset as $\overline{xg}$. Hence $xg = a\overline{xg}$ where $a \in H$. Now a is clearly dependent on x and g, and we denote it by $a_{x,g}$. Substituting $a_{x,g}$ for a, we have

$$xg = a_{x,g}\overline{xg} \tag{7.4}$$

or

$$x(g\rho) = a_{x,g}\, x(g\pi) \tag{7.5}$$

Now any element of G can be expressed in the form hx for $h \in H$ and $x \in X$. Therefore $(hx)(g\rho) = hxg = h(x(g\rho)) = ha_{x,g}\, x(g\pi)$, i.e.

$$(hx)(g\rho) = (ha_{x,g})(x(g\pi)) \tag{7.6}$$

This equation suggests that the effect of $g\rho$ on an element hx of G can be explained by what happens to h (it goes to the element $ha_{x,g}$ of H) and what happens to x (it goes to $x(g\pi)$).

We express this formally in the following theorem which is due (essentially) to Frobenius.

Theorem 7.3: Let G be a group and H a subgroup of G. Let X be a right transversal of H in G. Then there is a faithful representation θ of G as a group of permutations of $H \times X$ (the cartesian product of H and X) defined by

$$(h, x)(g\theta) = (ha_{x,g}, x(g\pi))$$

Proof: The proof is an adaptation of the discussion of the last few paragraphs. First let π be the coset representation with respect to H with right-transversal X. For each $g \in G$ the permutation $g\pi$ gives rise to a permutation $g\lambda$ of $H \times X$ which is defined as follows:

$$(h, x)(g\lambda) = (h, x(g\pi)), \quad (h \in H,\ x \in X) \tag{7.7}$$

Note that $g\lambda$ is a permutation of $H \times X$. For if

$$(h, x)(g\lambda) = (h', x')(g\lambda)$$

then

$$(h, x(g\pi)) = (h', x'(g\pi))$$

Since $g\pi$ is a permutation of X, it follows from $x(g\pi) = x'(g\pi)$ that $x = x'$. Hence we have proved that $g\lambda$ is one-to-one. But $g\lambda$ is also onto since $g\pi$ is onto X. Clearly $g\lambda$ is then a permutation of $H \times X$.

Now as we saw in (7.5), if $g \in G$, then

$$xg = a_{x,g}(x(g\pi)), \quad (a_{x,g} \in H)$$

For each $g \in G$ define

$$(h, x)g\sigma = (ha_{x,g}, x)$$

We verify that $g\sigma$ is a permutation of $H \times X$. Suppose $(h, x) \in H \times X$. Then

$$(ha_{x,g}^{-1}, x)g\sigma = (ha_{x,g}^{-1}a_{x,g}, x) = (h, x)$$

Thus $g\sigma$ is a mapping of $H \times X$ onto $H \times X$. It remains to verify that $g\sigma$ is one-to-one. Suppose that $(h, x)g\sigma = (h', x')g\sigma$. This means that

$$(ha_{x,g}, x) = (h'a_{x',g}, x')$$

and therefore we find $x = x'$ and $ha_{x,g} = h'a_{x,g}$ from which $h = h'$ and $(h, x) = (h', x')$. Hence $g\sigma$ is a permutation of $H \times X$.

Finally we compute $g\sigma g\lambda$.

$$(h, x)(g\sigma g\lambda) = ((h, x)g\sigma)g\lambda = (ha_{x,g}, x)g\lambda = (ha_{x,g}, x(g\pi)) \tag{7.8}$$

Thus

$$(g\sigma)(g\lambda) = g\theta \tag{7.9}$$

As both $g\sigma$ and $g\lambda$ are permutations of $H \times X$, $g\theta$ is a permutation of $H \times X$.

Let $g_1, g_2 \in G$. We must show that

$$(g_1\theta)(g_2\theta) = (g_1g_2)\theta \tag{7.10}$$

(Note that the left-hand side of (*7.10*) is the product of two permutations.) To facilitate the proof of (*7.10*) we introduce the following notation:

$$\alpha_1 = g_1\pi, \quad \alpha_2 = g_2\pi, \quad \alpha_3 = (g_1g_2)\pi, \quad \alpha = g\pi \tag{7.11}$$

where g is an element of G.

Equations (*7.8*), (*7.9*) and (*7.11*) yield

$$(h, x)g\theta = (ha_{x,g}, x\alpha) \tag{7.12}$$

Note that as π is a homomorphism, $(g_1\pi)(g_2\pi) = (g_1g_2)\pi$ so that

$$\alpha_1\alpha_2 = \alpha_3 \tag{7.13}$$

Applying (*7.12*) twice and (*7.13*) once, we have

$$\begin{aligned}(h, x)(g_1\theta g_2\theta) &= ((h, x)g_1\theta)g_2\theta = (ha_{x,g_1}, x\alpha_1)g_2\theta = (ha_{x,g_1}a_{x\alpha_1,g_2}, (x\alpha_1)\alpha_2)\\ &= (ha_{x,g_1}a_{x\alpha_1,g_2}, x(\alpha_1\alpha_2)) = (ha_{x,g_1}a_{x\alpha_1,g_2}, x\alpha_3)\end{aligned} \tag{7.14}$$

On the other hand, again on using (*7.12*), we have

$$(h, x)(g_1g_2)\theta = (ha_{x,g_1g_2}, x\alpha_3) \tag{7.15}$$

To prove that θ is a homomorphism, we must show that the right-hand side of (*7.14*) is equal to the right-hand side of (*7.15*), i.e. we must show that

$$a_{x,g_1}a_{x\alpha_1,g_2} = a_{x,g_1g_2} \tag{7.16}$$

To accomplish this we use equations (*7.5*) and (*7.6*) and obtain

$$x(g_1g_2) = a_{x,g_1g_2}(x\alpha_3) \tag{7.17}$$

Also $\quad (xg_1)g_2 = (a_{x,g_1}(x\alpha_1))g_2 = a_{x,g_1}a_{x\alpha_1,g_2}((x\alpha_1)\alpha_2) = a_{x,g_1}a_{x\alpha_1,g_2}(x\alpha_3)$

from (*7.13*). This means

$$(xg_1)g_2 = a_{x,g_1}a_{x\alpha_1,g_2}(x\alpha_3) \tag{7.18}$$

Since $x(g_1g_2) = (xg_1)g_2$, the right-hand sides of (*7.17*) and (*7.18*) are equal. Thus (*7.16*) follows and therefore θ is a homomorphism.

It only remains to show that θ is one-to-one. Assume $g_1\theta = g_2\theta$. Then $(h, x)g_1\theta = (h, x)g_2\theta$ for all pairs $(h, x) \in H \times X$; and in particular, if $(h, x) = (1, 1)$,

$$(a_{1,g_1}, 1(g_1\pi)) = (a_{1,g_2}, 1(g_2\pi))$$

from which

$$a_{1,g_1} = a_{1,g_2} \quad \text{and} \quad 1(g_1\pi) = 1(g_2\pi) \tag{7.19}$$

Using equation (*7.5*) with $x = 1$, we see that

$$g_1 = 1(g_1\rho) = a_{1,g_1}(1(g_1\pi)), \quad g_2 = 1(g_2\rho) = a_{1,g_2}(1(g_2\pi))$$

Using (*7.19*) we conclude that $g_1 = g_2$. Thus θ is one-to-one.

This completes the proof of Theorem 7.3.

We call the homomorphism θ of Theorem 7.3 a *Frobenius representation* (*with respect to H*). Of course θ depends on X as well, but it can be shown that in a sense this dependence does not matter.

Problems

7.13. Describe in detail a Frobenius representation for the symmetric group on $\{1,2,3\}$ relative to the subgroups given in Problem 7.11(a) and (b).

Solution:

(*a*) The set $X = \{\iota, \tau\}$ is a right transversal of $H = \{\iota, \sigma, \sigma^2\}$ in G. Let θ be the faithful representation of G on $H \times X$ described above. The elements of G are

$$\iota,\ \sigma,\ \sigma^2,\ \tau,\ \sigma\tau,\ \sigma^2\tau$$

We use the formula $xg = a_{x,g}\overline{xg}$ to calculate $a_{x,g}$. Note that $\bar{\iota} = \iota$, $\bar{\sigma} = \iota$, $\overline{\sigma^2} = \iota$, $\bar{\tau} = \tau$, $\overline{\sigma\tau} = \tau$, $\overline{\sigma^2\tau} = \tau$.

As an illustration of the procedure we calculate $a_{\tau,\sigma}$. Now $\tau\sigma = a_{\tau,\sigma}\overline{\tau\sigma}$. Since $\tau\sigma = \sigma^2\tau$, we have $\overline{\tau\sigma} = \tau$ and $a_{\tau,\sigma} = \tau\sigma\tau = \sigma^2$. Similar calculations lead to

$$\begin{array}{llll} a_{\iota,\iota} = \iota & a_{\iota,\tau} = \iota & a_{\tau,\iota} = \iota & a_{\tau,\tau} = \iota \\ a_{\iota,\sigma} = \sigma & a_{\iota,\sigma\tau} = \sigma & a_{\tau,\sigma} = \sigma^2 & a_{\tau,\sigma\tau} = \sigma^2 \\ a_{\iota,\sigma^2} = \sigma^2 & a_{\iota,\sigma^2\tau} = \sigma^2 & a_{\tau,\sigma^2} = \sigma & a_{\tau,\sigma^2\tau} = \sigma \end{array}$$

We use the definition of θ given in the statement of Theorem 7.3. The effect of π on g is given in the solution of Problem 7.11(a). The results, repeated here for convenience, are

$$\iota\pi = \sigma\pi = (\sigma^2)\pi : \ \iota \to \iota,\ \tau \to \tau$$

$$\tau\pi = (\sigma\tau)\pi = (\sigma^2\tau)\pi : \ \iota \to \tau,\ \tau \to \iota$$

We can now calculate the effect of $g\theta$ for each g in G. In particular,

$$(h,\iota)\sigma\theta = (ha_{\iota,\sigma},\ \iota(\sigma\pi)) = (h\sigma, \iota)$$

and
$$(h,\tau)\sigma\theta = (ha_{\tau,\sigma},\ \tau(\sigma\pi)) = (h\sigma^2, \tau) \quad \text{for all } h \in H$$

We list the effect of the permutations $g\theta$ for the elements of G.

$$\begin{array}{rlll} \iota\theta: & (h,x) \to (h,x) & (h \in H,\ x \in X) & \\ \sigma\theta: & (h,\iota) \to (h\sigma,\iota), & (h,\tau) \to (h\sigma^2,\tau) & (h \in H) \\ \sigma^2\theta: & (h,\iota) \to (h\sigma^2,\iota), & (h,\tau) \to (h\sigma,\tau) & (h \in H) \\ \tau\theta: & (h,\iota) \to (h,\tau), & (h,\tau) \to (h,\iota) & (h \in H) \\ (\sigma\tau)\theta: & (h,\iota) \to (h\sigma,\tau), & (h,\tau) \to (h\sigma^2,\iota) & (h \in H) \\ (\sigma^2\tau)\theta: & (h,\iota) \to (h\sigma^2,\tau), & (h,\tau) \to (h\sigma,\iota) & (h \in H) \end{array}$$

One could check that θ is a homomorphism by inspecting $(g_1\theta)(g_2\theta)$ and $(g_1g_2)\theta$, where $(g_1g_2 \in G)$. The above description of θ immediately shows that θ is one-to-one.

(*b*) Here $H = \{1, \tau\}$; X, a right transversal of H in G, is given by $X = \{\iota, \sigma, \sigma^2\}$. The Frobenius representation θ is then given by

$$\begin{array}{rllll} \iota\theta: & (h,x) \to (h,x) & (h \in H,\ x \in X) & & \\ \sigma\theta: & (h,\iota) \to (h,\sigma), & (h,\sigma) \to (h,\sigma^2), & (h,\sigma^2) \to (h,\iota) & (h \in H) \\ \sigma^2\theta: & (h,\iota) \to (h,\sigma^2), & (h,\sigma) \to (h,\iota), & (h,\sigma^2) \to (h,\sigma) & (h \in H) \\ \tau\theta: & (h,\iota) \to (h\tau,\iota), & (h,\sigma) \to (h\tau,\sigma^2), & (h,\sigma^2) \to (h\tau,\sigma) & (h \in H) \\ (\sigma\tau)\theta: & (h,\iota) \to (h\tau,\sigma^2), & (h,\sigma) \to (h\tau,\sigma), & (h,\sigma^2) \to (h\tau,\iota) & (h \in H) \\ (\sigma^2\tau)\theta: & (h,\iota) \to (h\tau,\sigma), & (h,\sigma) \to (h\tau,\iota), & (h,\sigma^2) \to (h\tau,\sigma^2) & (h \in H) \end{array}$$

7.14. Describe in detail the Frobenius representations for the alternating group of degree 4 relative to the subgroup H given in Problem 7.12.

Solution:

Here G consists of

$$\iota,\ \sigma,\ \sigma^2,\ \tau_1,\ \tau_1\sigma,\ \tau_1\sigma^2,\ \tau_2,\ \tau_2\sigma,\ \tau_2\sigma^2,\ \tau_3,\ \tau_3\sigma,\ \tau_3\sigma^2$$

and $H = \{\iota, \tau_1, \tau_2, \tau_3\}$. A right transversal of H in G is $X = \{\iota, \sigma, \sigma^2\}$. The Frobenius representation θ is then described as follows:

$$\iota\theta: \quad (h,x)\to(h,x), \qquad (h\in H,\ x\in X)$$

$$\sigma\theta: \quad (h,\iota)\to(h,\sigma), \quad (h,\sigma)\to(h,\sigma^2), \quad (h,\sigma^2)\to(h,\iota) \qquad (h\in H)$$

$$\sigma^2\theta: \quad (h,\iota)\to(h,\sigma^2), \quad (h,\sigma)\to(h,\iota), \quad (h,\sigma^2)\to(h,\sigma) \qquad (h\in H)$$

$$\tau_1\theta: \quad (h,\iota)\to(h\tau_1,\iota), \quad (h,\sigma)\to(h\tau_2,\sigma), \quad (h,\sigma^2)\to(h\tau_3,\sigma^2) \qquad (h\in H)$$

$$(\tau_1\sigma)\theta: \quad (h,\iota)\to(h\tau_1,\sigma), \quad (h,\sigma)\to(h\tau_2,\sigma^2), \quad (h,\sigma^2)\to(h\tau_3,\iota) \qquad (h\in H)$$

$$(\tau_1\sigma^2)\theta: \quad (h,\iota)\to(h\tau_1,\sigma^2), \quad (h,\sigma)\to(h\tau_2,\iota), \quad (h,\sigma^2)\to(h\tau_3,\sigma) \qquad (h\in H)$$

$$\tau_2\theta: \quad (h,\iota)\to(h\tau_2,\iota), \quad (h,\sigma)\to(h\tau_3,\sigma), \quad (h,\sigma^2)\to(h\tau_1,\sigma^2) \qquad (h\in H)$$

$$(\tau_2\sigma)\theta: \quad (h,\iota)\to(h\tau_2,\sigma), \quad (h,\sigma)\to(h\tau_3,\sigma^2), \quad (h,\sigma^2)\to(h\tau_1,\iota) \qquad (h\in H)$$

$$(\tau_2\sigma^2)\theta: \quad (h,\iota)\to(h\tau_2,\sigma^2), \quad (h,\sigma)\to(h\tau_3,\iota), \quad (h,\sigma^2)\to(h\tau_1,\sigma) \qquad (h\in H)$$

$$\tau_3\theta: \quad (h,\iota)\to(h\tau_3,\iota), \quad (h,\sigma)\to(h\tau_1,\sigma), \quad (h,\sigma^2)\to(h\tau_2,\sigma^2) \qquad (h\in H)$$

$$(\tau_3\sigma)\theta: \quad (h,\iota)\to(h\tau_3,\sigma), \quad (h,\sigma)\to(h\tau_1,\sigma^2), \quad (h,\sigma^2)\to(h\tau_2,\iota) \qquad (h\in H)$$

$$(\tau_3\sigma^2)\theta: \quad (h,\iota)\to(h\tau_3,\sigma^2), \quad (h,\sigma)\to(h\tau_1,\iota), \quad (h,\sigma^2)\to(h\tau_2,\sigma) \qquad (h\in H)$$

7.6 APPLICATIONS TO FINITELY GENERATED GROUPS

a. Subgroups of finite index

Frobenius' representation, although only a variation of Cayley's, is very useful. Here we shall give one application of this representation. First we recall a definition given in Chapter 4.

Definition: A group G is *finitely generated* if it can be generated by a finite set, i.e. if there is a finite subset S $(\neq \emptyset)$ of G such that for each $g \in G$ there are elements $s_1, s_2, \ldots, s_n \in S$ and integers $\epsilon_1, \ldots, \epsilon_n$ $(\epsilon_i = \pm 1)$ such that

$$g = s_1^{\epsilon_1} \cdots s_n^{\epsilon_n}$$

Theorem 7.4 (O. Schreier): A subgroup of finite index in a finitely generated group is finitely generated.

Proof: Let G be a finitely generated group. Let S be a finite set of generators of G, with $|S| = m$. Suppose H is a subgroup of finite index in G. Choose a right transversal $X = \{x_1, \ldots, x_j\}$ of H in G, with x_1 the identity. Notice that $j < \infty$ by assumption. Let θ be a Frobenius representation of G with respect to H given in Theorem 7.3. If $h \in H$, then $\bar{h} = 1$ and $a_{1,h} = h$ so that we have $(1,1)(h\theta) = (h,1)$. But h can be written as

$$h = s_1^{\epsilon_1} \cdots s_n^{\epsilon_n} \quad (\epsilon_i = \pm 1)$$

with $s_1, \ldots, s_n \in S$. Put $t_i = s_i^{\epsilon_i}$, $i = 1, \ldots, n$; then $h = t_1 \cdots t_n$. Since θ is a homomorphism,

$$(1,1)(h\theta) = (1,1)(t_1\theta)(t_2\theta)\cdots(t_n\theta)$$

Let $t_i\pi = \alpha_i$, $i = 1, \ldots, n$. By repeated applications of the definition of the action of $g\theta$ in $H \times X$ (see Theorem 7.3) we have

$$\begin{aligned}(1,1)(h\theta) &= (a_{1,t_1}, 1\alpha_1)(t_2\theta)\cdots(t_n\theta) = (a_{1,t_1}a_{1\alpha_1,t_2}, 1(\alpha_1\alpha_2))(t_3\theta)\cdots(t_n\theta) = \cdots \\ &= (a_{1,t_1}a_{1\alpha_1,t_2}a_{1\alpha_1\alpha_2,t_3}\cdots a_{1\alpha_1\cdots\alpha_{n-1},t_n}, 1(\alpha_1\cdots\alpha_n)) = (h,1)\end{aligned}$$

Hence $h = a_{1,t_1}a_{1\alpha_1,t_2}\cdots a_{1\alpha_1\cdots\alpha_{n-1},t_n}$. Since $\alpha_1, \ldots, \alpha_n \in S_X$, $1(\alpha_1\cdots\alpha_i) \in X$ for each i, $1 \leq i \leq n$. In other words we have expressed h as the product of elements of the form

$$a_{x,t} \text{ where } x \in X, \text{ and } t = s^{\pm 1} \text{ where } s \in S$$

As $|X| = j$ and $|S| = m$, the number of such elements is at most $2mj$. This means that H is finitely generated.

b. Remarks about the proof of Theorem 7.4

We have actually proved that if G can be generated by m elements and H is of index j in G, then H can be generated by $2jm$ elements. However, it is not difficult to reduce this number to jm. To do so observe that we have proved that the elements $a_{x,t}$ generate H, where x is an element of a right transversal X of H in G. We recall that if $\bar{g}$ denotes the representative of the coset Hg, then $a_{x,g}$ is defined by

$$xg = a_{x,g}\overline{xg}$$

(equation (*7.4*), page 224). This means that

$$a_{x,g} = xg(\overline{xg})^{-1}$$

Therefore the elements $xt(\overline{xt})^{-1}$ generate H, with $t=s$ or $t=s^{-1}$, where $s \in S$. Note that

$$\overline{\overline{xs^{-1}}s} = \overline{xs^{-1}s} = \bar{x} = x$$

by equation (*7.3*), page 219. Let $y = \overline{xs^{-1}}$; then $\overline{ys} = x$ and

$$a_{x,s^{-1}}^{-1} = \overline{xs^{-1}}sx^{-1} = ys(\overline{ys})^{-1} = a_{y,s}$$

i.e. $a_{x,s^{-1}}$ is the inverse of $a_{y,s}$. Thus H is actually generated by the elements $a_{x,s}$, $x \in X$ and $s \in S$. The number of these elements is jm. (In fact one can lower this bound and prove that H can be generated by $1+(m-1)j$ elements. For a proof of this result see Theorem 8.13, page 264.)

Note that in Section 7.6a we have actually proved that H is generated by $a_{x,s}$, $x \in X$, $s \in S$, *without* the assumption that $|X|$ and $|S|$ are finite.

Problems

7.15. Let G be the symmetric group of degree 3 and let H be a subgroup of index 2 in G. Find the set of generators of H described above.

Solution:

We use the description of G in Problem 7.11(a). Now a subgroup of index 2 in any group is normal. Thus if H is of index 2 in G, H is normal in G. Moreover, $|H| = 3$. Hence $H = \{\iota, \sigma, \sigma^2\}$, since this is the only subgroup of order 3 in G. A right transversal of H in G is $X = \{\iota, \tau\}$. Clearly G can be generated by σ and τ, and so H is generated by

$$a_{\iota,\sigma} = \sigma\bar{\sigma}^{-1} = \sigma \qquad a_{\tau,\sigma} = \tau\sigma(\overline{\tau\sigma})^{-1} = \tau\sigma\tau^{-1} = \sigma^2$$

$$a_{\iota,\tau} = \tau\bar{\tau}^{-1} = \iota \qquad a_{\tau,\tau} = \tau\tau(\overline{\tau\tau})^{-1} = \tau^2 = \iota$$

Thus we find that ι, σ, σ^2 generate H. Of course H is actually generated by σ alone.

7.16. Let G be any group and let H be a subgroup of G of index 2. Prove that if G can be generated by two elements, then H can be generated by three elements.

Solution:

Suppose that G is a group generated by c and d and that H is a subgroup of index 2 in G. If both c and d are in H, then $H \supseteq G$ and H is not of index 2 as initially assumed. Without loss of generality we may suppose that $c \notin H$. Then the cosets of H in G are H and Hc. Thus every element of G can be written in the form hc or h $(h \in H)$. This means that $\{1, c\}$ is a right transversal of H in G. Therefore G is generated by the elements

$$a_{1,c} = 1c(\overline{1c})^{-1} = cc^{-1} = 1,\ a_{1,d},\ a_{c,c},\ a_{c,d}$$

Hence H is generated by the three elements $a_{1,d}$, $a_{c,c}$ and $a_{c,d}$.

7.17. Let G be generated by a and b and suppose that N is a normal subgroup of G such that G/N is generated by Na and G/N is infinite cyclic. Find a set of generators for N in terms of a and b. (Hard.)

Solution:

It is clear that $X = \{1, a^{\pm 1}, a^{\pm 2}, \ldots\}$ is a right transversal of N in G. If $b \notin N$ then since $Nb = (Na)^w$ for some integer w, $ba^{-w} \in N$. Put $c = ba^{-w}$. On the other hand if $b \in N$, put $c = b$. Clearly $gp(a, c) = G$. We therefore take $S = \{a, c\}$. The generators of N are the elements

$$a_{x,s} = xs(\overline{xs})^{-1} \quad (x \in X,\ s \in S)$$

If $x = a^j$ and $s = a$, $$a_{x,s} = a^j a(\overline{a^j a})^{-1} = a^j a(a^{j+1})^{-1} = 1$$

If $x = a^j$ and $s = c$, $$a_{x,s} = a^j c(\overline{a^j c})^{-1} = a^j c a^{-j}$$

Hence the elements $\{\ldots, a^{-1}ca, c, aca^{-1}, \ldots\}$ are a set of generators for N. Since either $c = ba^{-w}$ or $c = b$, we can restate this set of generators in terms of a and b thereby obtaining a set of generators of G of the desired kind.

7.18. Let G be generated by a and b. Find generators for all possible subgroups of index 2. Hence show G has at most three subgroups of index 2.

Solution:

Let H be a subgroup of index 2 in G. Then we may have

(1) $a \notin H$, $b \in H$, (2) $a \in H$, $b \notin H$, (3) $a \notin H$, $b \notin H$.

In case (1) take $X = \{1, a\}$ and $S = \{a, b\}$. Then generators of H are the $a_{x,s}$ ($x \in X$ and $s \in S$). Thus H is generated by b, a^2 and aba^{-1}.

In case (2), proceeding as in (1) with $X = \{1, b\}$, H is generated by a, b^2 and bab^{-1}.

In case (3), $ab = c \in H$. Take $X = \{1, a\}$, and $S = \{a, c\}$. Then H is generated by c, aca^{-1} and a^2.

So the possible subgroups of index 2 are:

$$\text{(1)}\ gp(b, aba^{-1}, a^2), \quad \text{(2)}\ gp(a, bab^{-1}, b^2), \quad \text{(3)}\ gp(ab, a^2ba^{-1}, a^2)$$

7.19. Let $G = gp(a, b, c)$ and let N be a normal subgroup of G of index 3 with $G/N = gp(Na)$. Suppose N contains b and c. Find a set of generators for N in terms of a, b and c.

Solution:

Choose $S = \{a, b, c\}$ and $X = \{1, a, a^2\}$. Then the elements $a_{x,s}$, with $x \in X$ and $s \in S$, generate N. Thus N is generated by a^3, b, c, aba^{-1}, aca^{-1}, a^2ba^{-2}, a^2ca^{-2}.

7.20. Prove that if a group G contains a subgroup H of index 2 which is cyclic, then every subgroup of G of index 2 can be generated by two elements. (Hard.)

Solution:

Let H be generated by b and suppose that $a \notin H$. It follows that $G = gp(a, b)$. From Problem 7.18 the possible subgroups of index 2 are

$$\text{(1)}\ gp(b, aba^{-1}, a^2) = H_1, \quad \text{(2)}\ gp(a, bab^{-1}, b^2) = H_2, \quad \text{(3)}\ gp(ab, a^2ba^{-1}, a^2) = H_3$$

Clearly $b \notin H_2$ or H_3, as then each of them would actually be equal to G. Thus $H = H_1$. Since H is the cyclic group generated by b, $aba^{-1} = b^r$ and $a^2 = b^s$ for some integers r and s. We have

$$ba = a^{-1}b^r \tag{7.20}$$

and

$$a^2 = b^s \tag{7.21}$$

The generators of H_2 are a, bab^{-1} and b^2. $bab^{-1} = (ba)b^{-1} = a^{-1}b^{r-1}$ from (7.20), and so H_2 is generated by a, b^{r-1}, b^2. But $gp(b^{r-1}, b^2)$ is cyclic generated by c, say, as it is a subgroup of a cyclic group. Thus H_2 can be generated by two elements.

The generators of H_3 are ab, a^2ba^{-1} and a^2. Hence ab, $a^2ba^{-1}(ab)$ and a^2 are generators for H_3. Using (7.21) we conclude that ab, b^{s+2} and b^s are generators for H_3. But $gp(b^{s+2}, b^s)$ is cyclic generated by c, say; so H_3 is generated by two elements and the proof is complete.

c. Marshall Hall's theorem

The second application of permutational representations is due to Marshall Hall.

Theorem 7.5: The number of subgroups of finite index j in a finitely generated group is finite.

This theorem may be restated as follows: Let G be a group which is generated by $a_1, \ldots, a_n$ where $n < \infty$. Suppose j is a fixed positive integer. Then the number of subgroups of G of index j is finite.

Proof: Let S_j be the symmetric group on $1, \ldots, j$. For each subgroup H of index j in the finitely generated group G choose a right transversal X_H (we emphasize that X_H contains the identity of G). To avoid confusion between the number 1 and the identity of G, we shall write the identity (for this proof alone) as e. Thus we have $e \in X_H$. Let π_H be the coset representation of G with respect to the transversal X_H (see Section 7.4). Then π_H is a homomorphism of G into S_{X_H}. Since $|X_H| = j$, it is easy to prove that there exists an isomorphism $\phi_H : S_{X_H} \to S_j$ such that $\theta \in S_{X_H}$ moves e or leaves it fixed according as to whether $\theta\phi_H$ moves 1 or leaves it fixed (see Problems 7.21 and 7.22 below).

Note that $\Psi_H = \pi_H\phi_H : G \to S_j$ is a homomorphism of G into S_j since it is the composition of two homomorphisms. Note also that if H and K are two subgroups of index j and $H \neq K$, then $\Psi_H \neq \Psi_K$, for there exists an element $g \in H$ but $g \notin K$ (or vice versa). Then $e(g\pi_H) = e$ for $e(g\pi_H) = \overline{eg} = e$ as $g \in H$ (see Section 7.4). Accordingly $1\Psi_H = 1$. On the other hand, as $g \notin K$, $e(g\pi_K) \neq e$ and hence $1\Psi_H \neq 1$. Thus $\Psi_H \neq \Psi_K$ if $H \neq K$.

We have therefore found that the number of subgroups of index j in G is certainly not greater than the number of homomorphisms of G into S_j. This is where the fact that G is finitely generated comes in. For suppose G is generated by $a_1, \ldots, a_n$. If ϕ, θ are homomorphisms of G into S_j such that $a_i\phi = a_i\theta$ for $i = 1, \ldots, n$, then $\phi = \theta$. To prove this, observe that if $g \in G$, then

$$g = a_{i_1}^{\epsilon_1} \cdots a_{i_k}^{\epsilon_1} \qquad \epsilon_i = \pm 1, \; i_j \in \{1, \ldots, n\}$$

and

$$g\phi = (a_{i_1}\phi)^{\epsilon_1} \cdots (a_{i_k}\phi)^{\epsilon_k} = (a_{i_1}\theta)^{\epsilon_1} \cdots (a_{i_k}\theta)^{\epsilon_k} = g\theta$$

Since ϕ and θ agree on every element of G, $\phi = \theta$. This means that the number of homomorphisms of G into S is finite since the number of possible images of the generators of G is finite (at most $(j!)^n$).

This completes the proof of the theorem.

d. One consequence of Theorem 7.5

Let G be a finite group and θ a homomorphism of G onto G. It follows that θ is an isomorphism, for $|G| = |G\theta| = |G/\text{Ker}\,\theta|$ and hence $\text{Ker}\,\theta = \{1\}$. If G is not finite, is it possible to have a homomorphism θ of G onto G with θ not an isomorphism? For example, if P is a p-Prüfer group (see Section 6.2c, page 191), let $\theta : P \to P$ be defined by $x\theta = px$, $x \in P$. Then $P\theta = P$; but as P has an element of order p, θ is not an isomorphism.

In the following theorem we prove a result which tells us that for a special class of groups every onto homomorphism is an isomorphism.

Theorem 7.6 (A. I. Mal'cev): Let G be any finitely generated group whose subgroups of finite index have intersection 1. Then every homomorphism ϕ of G onto G is an automorphism.

Proof: Let K be the kernel of ϕ and let L be any subgroup of finite index in G. If L is of index j, then the number of subgroups of G of index j is finite. Let these subgroups be

$$L = L_1, L_2, \ldots, L_k$$

Now, by Theorem 4.18, page 117,

$$G = G\phi \cong G/K$$

and so the number of subgroups of index j in G/K is precisely k, the number of subgroups of index j in G. Let $M_1/K, \ldots, M_k/K$ be these subgroups of index j in G/K. Then $M_1, \ldots, M_k$ are k distinct subgroups of index j in G, by Corollary 4.20, page 121. Thus the M_i's are simply a rearrangement of the L_i. Therefore every L_i is an M_i and so contains K. In particular

$$L \supseteq K$$

This means that *every* subgroup of finite index contains K. Hence K is contained in the intersection of the subgroups of finite index. By hypothesis, this intersection is 1. So $K = 1$. Accordingly ϕ is one-to-one, and ϕ is an automorphism.

This theorem is important in current research in group theory.

Problems

7.21. Prove that there exists an isomorphism μ between $S_{\{x_1, x_2\}}$ and S_2 such that if $\theta \in S_{\{x_1, x_2\}}$ and $x_i\theta = x_j$, then $i(\theta\mu) = j$. (*Hint*: see Problems 7.8 and 7.9, page 221.)

Solution:

Let $\alpha: \{x_1, x_2\} \to \{1, 2\}$ be defined by $x_j\alpha = j$, $j = 1, 2$. Let μ be defined by

$$\begin{pmatrix} x_1 & x_2 \\ x_1 & x_2 \end{pmatrix}\mu = \begin{pmatrix} 1 & 2 \\ 1 & 2 \end{pmatrix}, \quad \begin{pmatrix} x_1 & x_2 \\ x_2 & x_1 \end{pmatrix}\mu = \begin{pmatrix} 1 & 2 \\ 2 & 1 \end{pmatrix}$$

Then α, μ define a permutation isomorphism (see Problems 7.8 and 7.9) and therefore μ is an isomorphism with the required effect.

7.22. Prove that in general there exists an isomorphism μ between $S_{\{x_1, \ldots, x_n\}}$ and S_n such that if $\theta \in S_{\{x_1, \ldots, x_n\}}$ and $x_i\theta = x_j$, then $i(\theta\mu) = j$. (Use Problems 7.8 and 7.9.)

Solution:

Let $\alpha: \{x_1, \ldots, x_n\} \to \{1, 2, \ldots, n\}$ be defined by $x_i\alpha = i$. If $\theta \in S_{\{x_1, \ldots, x_n\}}$, define $\theta\mu \in S_n$ to be the mapping that sends $i \to j$ if $x_i\theta = x_j$ (as θ is a permutation of $\{x_1, \ldots, x_n\}$, $\theta\mu$ is a permutation of $\{1, 2, \ldots, n\}$). It is clear then that μ is onto S_n, and hence α, μ provide an isomorphism of permutation groups. Thus μ is the required isomorphism.

7.23. Prove that if H and K are subgroups of G, then each coset of H intersects a coset of K either in the empty set or in a coset of $H \cap K$. Hence prove that if H and K are of finite index in G, so is $H \cap K$.

Solution:

If a coset of H and a coset of K have an element g in common, then the two cosets are Hg and Kg. $x \in Hg \cap Kg$ if and only if $x = hg = kg$, for some $h \in H$ and $k \in K$. But $hg = kg$ if and only if $h = k$, i.e. $h \in H \cap K$. Thus $x \in Hg \cap Kg$ if and only if $x \in (H \cap K)g$. Then $Hg \cap Kg = (H \cap K)g$ and the two cosets meet in a coset of $H \cap K$.

If H is of index n and K is of index m, at most nm cosets can be found as intersections of a coset of H with a coset of K. Furthermore these are all the cosets of $H \cap K$, for any coset $(H \cap K)g = Hg \cap Kg$ and so is the intersection of a coset of H by a coset of K. Therefore $H \cap K$ is of finite index in G if both H and K are.

7.24. Let G be finitely generated with a subgroup of index j. Prove that the intersection of all subgroups of index j in G is a normal subgroup of finite index. (Hard.)

Solution:

By Theorem 7.5, G has only a finite number of subgroups, $M_1, \ldots, M_n$ say, of index j. Now if M is any subgroup of index j, it is easy to prove that $x^{-1}Mx = \bar{M}$ is also of index j. (If the cosets of M are $Mg_1, \ldots, Mg_j$, then the cosets of $\bar{M}$ are $\bar{M}x^{-1}g_1x, \bar{M}x^{-1}g_2x, \ldots, \bar{M}x^{-1}g_jx$. For if $g \in G$, $x^{-1}gx \in Mg_i$, for some i, implies that $g \in x^{-1}Mxx^{-1}g_ix = \bar{M}x^{-1}g_ix$.) Hence $M_1 \cap M_2 \cap \cdots \cap M_n = K$ is a normal subgroup of G, for if $g \in K$ and $x \in G$, $x^{-1}gx \in x^{-1}M_1x, x^{-1}M_2x, \ldots, x^{-1}M_nx$. But $x^{-1}M_1x, \ldots, x^{-1}M_nx$ are n distinct subgroups of index j. Hence they must be all the subgroups of index j (perhaps in a different order). Thus $x^{-1}gx \in K$ and so $K \lhd G$. K is of finite index by repeated application of Problem 7.23.

7.25. Let G and H be two groups and suppose that G satisfies the conditions of Theorem 7.6. Let $\theta: G \to H$ and $\phi: H \to G$ be epimorphisms. Prove that $G \cong H$.

Solution:

$\theta\phi$ is an epimorphism of G to itself and so by Theorem 7.6, $\theta\phi$ is an isomorphism. Then if $g \neq 1$, $g \in G$, $g(\theta\phi) \neq 1$ and thus $g\theta \neq 1$. Therefore θ is one-to-one and θ is an isomorphism of G to H.

7.26. Let N and M be unequal normal subgroups of a finitely generated group G, $M \supseteq N$. Suppose the intersection of the subgroups of finite index in G/N is the identity. Prove that G/M is not isomorphic with G/N.

Solution:

Let $\theta: G/M \to G/N$ be such an isomorphism. Let $\mu: G/N \to G/M$ be defined by $Ng \to Mg$. It is easy to verify that μ is an epimorphism. Then $\mu\theta$ is a homomorphism of G/N onto itself. By Theorem 7.6, $\mu\theta$ is an isomorphism. Now if $g \in M - N$, $(Ng)(\mu\theta) = M\theta$ is the identity of G/N. Thus $\mu\theta$ is not an isomorphism. This contradiction yields the required result.

7.7 EXTENSIONS

a. General extension

Suppose G is a group with a normal subgroup H and that $G/H \cong K$. Then, using the terminology introduced in Chapter 5, G is an extension of H by K. It is convenient to generalize this concept and to say that G is an extension of H by K if G has a subgroup $\bar{H}$ with $\bar{H} \cong H$ and $G/\bar{H} \cong K$. It is our aim to investigate how a group is built as an extension of one group by another.

In this section let G be a fixed group and H a normal subgroup of G. Let ϕ be an isomorphism of G/H onto K. Let X be a *left* transversal of H in G, i.e. a set of elements of G containing one and only one element from each left coset of H in G with $1 \in X$.

If $g \in G$, $g = xh$ for some $x \in X$ and some $h \in H$. It is easy to see that this expression for g is unique. Let $g \in G$, $x \in X$; then gx belongs to some coset of H in G, say the coset yH where $y \in X$. Therefore

$$gx = yh$$

for some $h \in H$. Now h is uniquely determined by g and x; we denote h by $m_{g,x}$. Thus

$$gx = ym_{g,x} \tag{7.22}$$

The elements $m_{g,x}$ correspond to the elements $a_{x,g}$ introduced in Section 7.5b. (We use $m_{g,x}$ instead of $a_{x,g}$ because here we are dealing with left instead of right cosets. We will explain in Section 7.7c the minor reason why we use left cosets here.)

Note that ϕ, the isomorphism of G/H onto K, is a one-to-one mapping of the set of left cosets of H onto K. Therefore we can unambiguously denote the representative of the coset gH by x_k if $(gH)\phi = k$. In particular then, $x_1 = 1$. With this notation,

$$X = \{x_k \mid k \in K\}$$

Notice that as $(x_k x_{k'} H)\phi = \{(x_k H)(x_{k'} H)\}\phi = (x_k H)\phi(x_{k'} H)\phi = kk'$ where $k, k' \in K$, the representative of the coset $x_k x_{k'} H$ is $x_{kk'}$. Then, from (7.22) we have

$$x_k x_{k'} = x_{kk'} m_{x_k, x_{k'}} \quad \text{where } m_{x_k, x_{k'}} \in H$$

We suppress the x's and write $m_{k,k'}$ for $m_{x_k, x_{k'}}$. Thus,

$$x_k x_{k'} = x_{kk'} m_{k,k'} \quad \text{where } k, k' \in K, \; m_{k,k'} \in H \tag{7.23}$$

Every element g in G can be written uniquely in the form $x_k h$ where $x_k \in X$, $h \in H$.

To express the product of two elements $x_k h$ and $x_{k'} h'$ as the product of an element of X by an element of H, we proceed as follows:

$$x_k h \cdot x_{k'} h' = x_k x_{k'} \cdot x_{k'}^{-1} h x_{k'} h' = x_{kk'} m_{k,k'} x_{k'}^{-1} h x_{k'} h' \tag{7.24}$$

Observe that $x_{k'}^{-1} h x_{k'} \in H$ since H is a normal subgroup of G. So $x_{kk'} \in X$ and $m_{k,k'} x_{k'}^{-1} h x_{k'} h' \in H$. The right-hand-side of (7.24) looks less complicated if we introduce the notation $h^{k'}$ for $x_{k'}^{-1} h x_{k'}$; then

$$x_k h \cdot x_{k'} h' = x_{kk'} m_{k,k'} h^{k'} h' \tag{7.25}$$

It appears from equation (7.25) that the extension G of H by K that we have been inspecting is determined by the $m_{k,k'}$ and by the images $h^{k'}$ of the elements h obtained by conjugation by the $x_{k'}$, i.e. by forming $x_{k'}^{-1} h x_{k'}$. One may conveniently think of the elements $m_{k,k'}$ in H as the images of a function m of two variables (coming from K) with values in H. In other words, we may think of m as a mapping from the cartesian product $K \times K$ into H, where we use $m_{k,k'}$ to denote the image of $(k, k') \in K \times K$ under this mapping m. Continuing with this analysis, let us turn to the elements h^k. For each $k \in K$ we have a mapping, $k\alpha$ say, of H into H, namely the mapping which sends an element h in H to the element h^k. In a way then the group G is made up of two mappings:

(1) a mapping m from $K \times K$ into H,

(2) a mapping α of K into a set of mappings of H into H. (The effect of $k\alpha$ is to map h to h^k.)

Indeed m and α determine G up to isomorphism (see Problem 7.27). If we add enough conditions to these mappings, one can reverse the procedure we have been outlining and construct from H, K and the mappings m and α an extension G of H by K. (See A. G. Kurosh, *The Theory of Groups*, Vol. II, Chelsea, 1960, translated by K. A. Hirsch, for details.) We will not tackle the general problem but we will consider only a particular case (in Section 7.7c).

Problems

7.26. Let G be the dihedral group of degree 3. G is an extension of a cyclic group of order 3 by a cyclic group of order 2. After choosing a suitable left transversal, find the mappings m and α introduced above.

Solution:

Using the notation of Section 3.4f, page 75, let $H = \{\sigma_1, \sigma_2, \sigma_3\}$. Then, as can be easily checked, $H \lhd G$. Since $|G| = 6$, G/H is of order 2 and thus a cyclic group of order 2. Let $\{\iota, \tau\}$ be a left transversal for H in G. Then

$$m_{\iota,\iota} = \iota, \quad m_{\iota,\tau} = \tau, \quad m_{\tau,\iota} = \tau, \quad m_{\tau,\tau} = \iota$$

$\iota\alpha$ is the identity mapping of H onto H, while $\tau\alpha$ sends σ_1 to σ_2, σ_2 to σ_1 and σ_3 to σ_3.

7.27. Let $G, \bar{G}$ be two groups, both extensions of H by K. Assume that H is actually a subgroup of both G and $\bar{G}$. Let $X, \bar{X}$ be transversals of H in $G, \bar{G}$ respectively. Let $m, \bar{m}$ and $\alpha, \bar{\alpha}$ be the mappings obtained above. Prove that if $m = \bar{m}$ and $\alpha = \bar{\alpha}$, then $G \cong \bar{G}$.

Solution:

The elements of G are uniquely of the form $x_k h$, where $x_k \in X$, $h \in H$, while the elements of $\bar{G}$ are uniquely of the form $\bar{x}_k h$, where $\bar{x}_k \in \bar{X}$. Let $\theta: G \to \bar{G}$ be defined by $(x_k h)\theta = \bar{x}_k h$. Then θ is a one-to-one onto mapping. To prove θ is a homomorphism, we consider the product of two elements of G.

$$(x_k h x_{k'} h')\theta = (x_{kk'} m_{k,k'} h^{k'} h')\theta = (x_{kk'} m_{k,k'} h(k'\alpha) h')\theta = \bar{x}_{kk'} m_{k,k'} h(k'\alpha) h'$$

$$= \bar{x}_{kk'} \bar{m}_{k,k'} h(k'\bar{\alpha}) h' = \bar{x}_k h \bar{x}_{k'} h' = (x_k h)\theta (x_{k'} h')\theta$$

Thus $G \cong \bar{G}$.

b. The splitting extension

Suppose G is as in Section 7.7a. Consider the particular case where $m_{x,x'} = 1$ for all $x, x' \in X$. By examining equation (7.22) $xx' = x''$, i.e. the product of two elements in X is again in X. Furthermore, we have $1 \in X$. Let $x \in X$, and let $y \in X$ be such that $x^{-1}H = yH$; then $xy \in H$. Accordingly

$$xy = 1m_{x,y} \quad \text{where } m_{x,y} \in H$$

However $m_{x,y} = 1$, and so $xy = 1$. Thus x has an inverse in X, and so X is a subgroup of G. Since $H \lhd G$, XH is a subgroup (Theorem 4.23, page 125). But every element of G is of the form xh, $x \in X$, $h \in H$. Hence $XH = G$. Since distinct elements of X lie in distinct cosets of H in G and $1 \in H$, we have

$$X \cap H = \{1\}$$

Since X and H are subgroups of G with $XH = G$, $H \cap X = \{1\}$ and $H \lhd G$, G is said to *split* over H. X is called a *complement* of H.

Note that if G splits over H, we can choose any complement X of H as a transversal for H in G, since two distinct elements of X belong to distinct cosets of H. Now X is a subgroup of G, and it follows that if we define $m_{g,x}$ as in Section 7.7a with X as transversal, then $m_{x,x'} = 1$ for all x, x' in X.

If G splits over H and X is a complement of H, then

$$G/H = HX/H \cong X/H \cap X \cong X$$

In other words, G is an extension of H by X; that is, if G splits over H, G/H is isomorphic with any complement of H.

It is convenient to introduce the following definition. We say that a group G is a *splitting extension of H_1 by X_1* if there exists a normal subgroup H of G isomorphic to H_1 such that G splits over H and $G/H \cong X_1$.

Problems

7.28. Prove that the dihedral group D of order 8 is a splitting extension of a cyclic group of order 4 by a cyclic group of order 2.

Solution:

We use the multiplication table for D given in Problem 7.6(a), page 220. Let $H = \{1, a_1, a_2, a_3\}$; then H is cyclic of order 4 since

$$a_1^2 = a_2, \quad a_1^3 = a_3, \quad a_1^4 = 1$$

Moreover H is a normal subgroup of G. This can be verified either by direct calculation, checking that if $d \in D$ and $h \in H$ then $d^{-1}hd \in H$, or by noting that H is of index 2 in D. Now letting $X = \{1, a_4\}$, $a_4^2 = 1$ and so X is a subgroup of D. Since $a_4 \notin H$, it follows that the cosets H, a_4H are disjoint. Therefore, as there are 8 elements in $H \cup a_4H$,

$$D = H \cup a_4H \quad \text{or} \quad D = XH$$

Finally

$$H \cap X = \{1\}$$

So D is a splitting extension of H by X as required.

7.29. Prove that neither the dihedral group of order 8 nor the quaternion group of order 8 is a splitting extension of a group of order 2 by a group of order 4.

Solution:

Let E stand for either the dihedral group of order 8 or the quaternion group of order 8. Suppose E is a splitting extension of a subgroup H of order 2 with complement X of order 4. Then H is a normal subgroup of G. Now suppose $H = \{1, h\}$. If $e \in E$, then $e^{-1}he \in H$. Since $h \neq 1$, $e^{-1}he \neq 1$. Thus $e^{-1}he = h$ for all $e \in E$. In particular if $x \in X$, then $x^{-1}hx = h$. Now $E = XH$. Since X is of order 4, X is abelian (see Problem 5.19, page 140).

Now suppose $e, f \in E$; then

$$e = x'h', \quad f = x''h'' \qquad (x', x'' \in X,\ h', h'' \in H)$$

Recall that every element of H commutes with every element of X and that X is abelian; then

$$ef = x'h' \cdot x''h'' = x'x''h'h'' = x''x'h''h' = x''h''x'h' = fe$$

Thus E is abelian. But neither the dihedral group of order 8 nor the quaternion group of order 8 is abelian. Hence we have a contradiction to the assumption that either of these groups is a splitting extension of the type described.

Alternate proof: If E is a splitting extension of H by X of order 2 and 4 respectively, since $H \cap X = \{1\}$, $H \lhd E$ and $X \lhd E$ (as X is of index 2), it follows that $E = X \times H$, the direct product of X and H by Corollary 5.17, page 145. From this it again follows that E is abelian, thus producing a contradiction.

7.30. Prove that the quaternion group of order 8 is an extension of a group of order 4 by a group of order 2, but is not a splitting extension.

Solution:

We use the multiplication table for $\mathcal{H}$, the quaternion group of order 8 given in Problem 7.6(b), page 220.

First let $K = gp(a_3)$. Then K is of order 4 and therefore of index 2. Thus K is a normal subgroup of $\mathcal{H}$, and it follows that $\mathcal{H}$ is an extension of K by $\mathcal{H}/K$. Clearly $\mathcal{H}/K$ is cyclic of order 2, and so $\mathcal{H}$ is an extension of a group of order 4 by a group of order 2.

Now suppose $\mathcal{H}$ splits over any subgroup K of order 4. Then $\mathcal{H}/K$ is of order 2 and hence abelian. Therefore K contains the commutator subgroup of $\mathcal{H}$ by Problem 4.68, page 116. In particular, K contains a_2 since

$$a_2 = a_4^{-1}a_7^{-1}a_4a_7$$

Now we must check that if x is any element except 1 and a_2 of $\mathcal{H}$, then x is of order 4. This can be done directly, using the multiplication table for $\mathcal{H}$. Suppose now, if possible, that $\mathcal{H}$ is a splitting extension of K by X. Then the subgroup X is of order 2, say $X = \{1, x\}$. But as we saw above, X is of order 4 since $x \neq 1$, $x \neq a_2$. So the subgroup X is not of order 2. This is a contradiction and so the desired result follows.

Alternate proof: If $\mathcal{H}$ is a splitting extension of a subgroup K of order 4 by a subgroup X of order 2, then $K \cap X = \{1\}$ and $K \lhd \mathcal{H}$. But every subgroup of the quaternion group is normal (Problem 5.43, page 158). Therefore $X \lhd \mathcal{H}$. It follows, by Theorem 5.16′, page 146, that $\mathcal{H} = K \times X$. This implies $\mathcal{H}$ is abelian, which is a contradiction.

7.31. Is the alternating group of degree 4 a splitting extension of a group of order 6 by a group of order 2?

Solution:

By Problem 5.1, page 131, the alternating group of degree 4 does not contain a subgroup of order 6. Thus the result follows.

c. An analysis of splitting extensions

Suppose now that G is a splitting extension of H by K with complement X. Then by Section 7.7b, we may use X as a transversal for H in G, and $m_{x,x'} = 1$ for all $x, x' \in X$. Consequently each element $g \in G$ is uniquely expressible in the form

$$g = x_k h \quad \text{where } k \in K,\ h \in H \tag{7.26}$$

and equation (7.25) becomes

$$x_k h x_{k'} h' = x_{kk'} h^{k'} h' \tag{7.27}$$

For each $k \in K$ the mapping

$$h \to h^k \quad (h \in H) \tag{7.28}$$

is an automorphism of H, for $H \lhd G$ and $h^k = x_k^{-1} h x_k$ imply, by Problem 3.57, page 85, that the mapping $h \to h^k$ $(h \in H)$ is an automorphism of H.

Finally we remark that if we let α be the mapping which assigns to each element $k \in K$ the automorphism

$$k\alpha: h \to h^k \quad (h \in H)$$

of H, then α is itself a homomorphism of K into the group of automorphisms of H. (This explains why we used a left transversal in Section 7.7(a), namely so that the mapping α be a homomorphism.) We have only to prove that

$$(kk')\alpha = k\alpha k'\alpha \tag{7.29}$$

To verify (7.29), let us take an arbitrary element $h \in H$ and apply the automorphism $(kk')\alpha$ to h:

$$\begin{aligned} h[(kk')\alpha] &= x_{kk'}^{-1} h x_{kk'} = (x_k x_{k'})^{-1} h (x_k x_{k'}) \qquad \text{since } x_{kk'} = x_k x_{k'} \text{ by (7.27)} \\ &= (x_{k'}^{-1} x_k^{-1}) h (x_k x_{k'}) = x_{k'}^{-1} (x_k^{-1} h x_k) x_{k'} \\ &= x_{k'}^{-1} [h(k\alpha)] x_{k'} = [h(k\alpha)](k'\alpha) \\ &= h[(k\alpha)(k'\alpha)] \qquad \text{by the definition of the product of two automorphisms} \end{aligned}$$

Thus we have

$$(kk')\alpha = (k\alpha)(k'\alpha)$$

We now replace $h^{k'}$ by $h(k'\alpha)$ in (7.27). Then (7.27) becomes

$$x_k h \cdot x_{k'} h' = x_{kk'} \cdot h(k'\alpha) h' \tag{7.30}$$

What is the situation we have arrived at? We started from a group G which splits over H. Then we chose a subgroup X of G such that $G = XH$ and $X \cap H = \{1\}$. We were given an implicit isomorphism ϕ of G/H with K and we denoted the element in X which corresponds to k in K by x_k. Then we observed that the elements of G were uniquely expressible in the form $x_k h$ with $k \in K$, $h \in H$. The way in which elements of G are multiplied was then computed by making use of the existence of a homomorphism α of K into the automorphism group of H and by applying (7.30). Therefore we may suspect that if we are given

(*a*) a group H,

(*b*) a group K,

(*c*) a homomorphism α of K into the automorphism group of H,

then we can create a splitting extension of H by K.

Indeed this is the case. All we have to do is to reverse the process we have described. To be precise, starting with the data (*a*), (*b*) and (*c*), we let G be the cartesian product of K and H, i.e.

$$G = K \times H = \{(k, h) \mid k \in K,\ h \in H\}$$

We define a binary operation in G according to the formula

$$(k, h)(k', h') = (kk', h(k'\alpha)h') \tag{7.31}$$

The reader will note the strong similarity between (7.30) and (7.31).

Before verifying that G is a group and a splitting extension of H by K, we will make the similarity of equations (7.30) and (7.31) even more evident. Let $\mathbf{x}_k = (k, 1)$ and $\mathbf{h} = (1, h)$, and define $k\boldsymbol{\alpha}$ by $\mathbf{h}(k\boldsymbol{\alpha}) = (1, h(k\alpha))$. Then from (7.31) obtain

$$\mathbf{x}_k\mathbf{h}\mathbf{x}_{k'}\mathbf{h}' = \mathbf{x}_{kk'}\mathbf{h}(k'\boldsymbol{\alpha})\mathbf{h}'$$

This equation resembles equation (7.30) even more closely than (7.31) does.

We will now prove that G is a group and that it is a splitting extension of H by K.

(i) We note first that (7.31) defines a binary operation in G.

(ii) The binary operation defined by (7.31) is associative:

$$\begin{aligned} ((k, h)(k', h'))(k'', h'') &= (kk', h(k'\alpha)h')(k'', h'') \\ &= ((kk')k'', \{[h(k'\alpha)h'](k''\alpha)\}h'') \\ &= ((kk')k'', \{[(h(k'\alpha))(k''\alpha)][h'(k''\alpha)]\}h'') \end{aligned} \tag{7.32}$$

Now we work out

$$\begin{aligned} (k, h)((k', h')(k'', h'')) &= (k, h)(k'k'', h'(k''\alpha)h'') \\ &= (k(k'k''), [h(k'k'')\alpha][h'(k''\alpha)h'']) \end{aligned}$$

By (*c*) above, α is a homomorphism, and so $(k'k'')\alpha = k'\alpha \cdot k''\alpha$. Thus

$$\begin{aligned} (k, h)((k', h')(k'', h'')) &= (k(k'k''), [h((k'\alpha)(k''\alpha))][h'(k''\alpha)h'']) \\ &= ((kk')k'', [(h(k'\alpha))(k''\alpha)][h'(k''\alpha)h'']) \end{aligned} \tag{7.33}$$

The associative law immediately yields the equality of (7.32) and (7.33). Therefore (7.31) is an associative operation.

(iii) There exists an identity in G: $(1, 1)$. Notice, of course, that the left-hand 1 in $(1, 1)$ is the identity of K while the right-hand 1 of $(1, 1)$ is the identity of H. To check that $(1, 1)$ is an identity, let $(k, h) \in G$. Then

$$(1, 1)(k, h) = (k, 1(k\alpha)h) = (k, h)$$

since $k\alpha$ is an automorphism of H, and so maps the 1 of H to itself. Similarly

$$(k, h)(1, 1) = (k, h(1\alpha)1) = (k, h)$$

since 1α is the identity automorphism of H, and so leaves H identically fixed.

(iv) Finally we must check that every element of G has an inverse. Let $(k, h) \in G$. We claim that

$$(k^{-1}, h^{-1}(k^{-1}\alpha))$$

is the inverse of (k, h). To prove this, we simply observe that

$$(k, h)(k^{-1}, h^{-1}(k^{-1}\alpha)) = (1, [h(k^{-1}\alpha)][h^{-1}(k^{-1}\alpha)]) = (1, (hh^{-1})(k^{-1}\alpha))$$

since $k^{-1}\alpha$ is an automorphism of H. Thus

$$(k, h)(k^{-1}, h^{-1}(k^{-1}\alpha)) = (1, 1)$$

Similarly $$(k^{-1}, h^{-1}(k^{-1}\alpha))(k, h) = (1, 1)$$

We have thus verified that $(k^{-1}, h^{-1}(k^{-1}\alpha))$ is the inverse of (k, h).

(i), (ii), (iii) and (iv) above establish in every detail the proof that G is a group.

Next we verify that G is a splitting extension of H by K. To accomplish this, put

$$\bar{H} = \{(1,h) \mid h \in H\} \quad \text{and} \quad \bar{K} = \{(k,1) \mid k \in K\}$$

Then it is easy to prove that $\bar{H}$ and $\bar{K}$ are subgroups of G and that

$$h \to (1,h) \quad \text{and} \quad k \to (k,1)$$

are isomorphisms of H onto $\bar{H}$ and K onto $\bar{K}$ respectively.

Now to prove that $\bar{H}$ is a normal subgroup of G, observe again that if $g \in G$, then

$$g = (k,h) = (k,1)(1,h)$$

If $(1,h') \in \bar{H}$, then

$$\begin{aligned} g^{-1}(1,h')g &= (k,h)^{-1}(1,h')(k,h) \\ &= ((k,1)(1,h))^{-1}(1,h')((k,1)(1,h)) \\ &= [(1,h)^{-1}(k,1)^{-1}](1,h')((k,1)(1,h)) \end{aligned}$$

Since $(k,1)^{-1} = (k^{-1},1)$, we have $(k,1)^{-1}(1,h')(k,1) = (1,h'(k\alpha))$. Thus

$$g^{-1}(1,h')g = (1,h^{-1})(1,h'(k\alpha))(1,h) \in \bar{H}$$

since the product of elements of $\bar{H}$ belongs to $\bar{H}$. Hence $\bar{H}$ is normal in G as claimed. Clearly $\bar{H} \cap \bar{K} = \{(1,1)\}$ and

$$G = \bar{K}\bar{H}$$

as we saw earlier, since $(k,h) = (k,1)(1,h)$. Consequently we have constructed from the data (a), (b) and (c) a splitting extension G of H with complement $\bar{K}$. Therefore G is a splitting extension of H by K.

The group G that we have constructed is called the *splitting extension of H by K via α.*

We emphasize the importance of the above discussion and the related problems which follow.

Problems

7.32. Construct a non-abelian group of order 6 as a splitting extension of a group of order 3 by a group of order 2.

Solution:

Recall that if we are given (a) a group H, (b) a group K, (c) a homomorphism α of K into the automorphism group of H, then we can construct a group from this data as follows. Consider the set G of all the pairs (k,h) $(k \in K,\ h \in H)$ and define a binary operation in G by

$$(k,h)(k',h') = (kk',\ h(k'\alpha)h')$$

Then G becomes a group which is a splitting extension of H by K. So in the case at hand, we have the group H (the cyclic group of order 3) and the group K (the cyclic group of order 2). We need the homomorphism α. This means in the first place that we need to know more about the automorphism group of H, the cyclic group of order 3. Now if H is generated by h, then

$$H = \{1, h, h^2\}$$

The mapping which sends each element of an abelian group into its inverse is an automorphism (check this). So if $\eta: H \to H$ is this automorphism,

$$\eta: \quad 1 \to 1, \quad h \to h^{-1} = h^2, \quad h^2 \to h^{-2} = h$$

Now $\eta^2 = \iota$, and so $gp(\eta)$ is cyclic of order 2. Thus the groups K and $gp(\eta)$ are isomorphic, and accordingly we can take α to be the isomorphism of K onto $gp(\eta)$. Let G be the splitting extension of H by K via α. To see how some of the elements of G combine, let $K = \{1, k\}$. Then

$$(k, h)(k, 1) \;=\; (k^2, h(k\alpha)1) \;=\; (1, h^2)$$

Now $(k, 1)(k, h) = (k^2, 1(k\alpha)h) = (1, h)$, and so $(k, h)(k, 1) \neq (k, 1)(k, h)$. Thus G is non-abelian and is an extension of H by K.

7.33. Are there non-abelian groups of order 2×777?

Solution:

There are non-abelian groups of order 2×777. To produce one such group, let H be cyclic of order 777. Then H has an automorphism τ of order 2, namely the mapping τ which sends every element of H to its inverse. So there is a homomorphism α of K, the cyclic group on k, of order 2, into the automorphism group of H, namely the one which sends k to τ. Let G be the splitting extension of H by K via α. Then G is the required group.

7.34. Construct two non-isomorphic non-abelian groups of order 168.

Solution:

Let $H_1 = gp(h_1)$ be cyclic of order 84 and $K_1 = gp(k_1)$ cyclic of order 2. Let α be the homomorphism of K_1 into the automorphism group of H_1 defined by $k_1\alpha: h_1^i \to h_1^{-i}$. Then G_1, the splitting extension of H_1 by K_1, is of order 168. The center Z_1 of G_1 is of order 2, consisting of $(1,1)$ and $(1, h_1^{42})$ which can be easily checked by direct calculation.

Now we construct a second group of order 168. Here we take $H_2 = gp(h_2)$ to be of order 42 and $K_2 = gp(k_2)$ to be cyclic of order 4.

As usual, H_2 has an automorphism of order 2, i.e. the mapping defined by

$$\tau: h_2^i \to h_2^{-i}, \quad i = 0, 1, \ldots, 42$$

Let β be the homomorphism of K_2 onto $gp(\tau)$ defined by

$$k_2^j\beta \;=\; \tau^j, \quad j = 0, 1, 2, 3$$

Thus $k_2^2\beta = \iota$. Let G_2 be the splitting extension of H_2 by K_2 via β. Then, as the reader may check by direct calculation, the center Z_2 of G_2 consists of $(1,1)$, $(1, h_2^{21})$, $(k_2^2, 1)$, (k_2^2, h_2^{21}) and is therefore of order 4.

Now both G_1 and G_2 are non-abelian since $|Z_1| = 2$, $|Z_2| = 4$. Moreover if G_1 and G_2 are isomorphic groups, they have isomorphic centers. Therefore G_1 and G_2 are not isomorphic.

7.35. Construct all possible groups of order 30 which are extensions of a cyclic group of order 10 by a cyclic group of order 3.

Solution:

The solution of this problem requires knowledge of the automorphism group of the cyclic group $H = gp(h)$ of order 10. If τ is an automorphism of H, then $h\tau$ is of order 10. The possibilities for $h\tau$ are therefore

$$h\tau = h^3, \quad h\tau = h^7, \quad h\tau = h^9$$

Let τ_i be the automorphisms defined by the possibilities listed above $(i = 1, 2, 3)$. Now $\tau_1^2 = 1$, $\tau_2^4 = 1$, $\tau_3^4 = 1$; thus none of the automorphisms of H is of order 3. So the only possible homomorphism α of a cyclic group K of order 3 into the automorphism group of H is the one which sends every element of K onto the identity automorphism. The resultant splitting extension of H by K is then abelian (indeed it is isomorphic to the cyclic group of order 30).

7.36. Construct a non-abelian group of order 222 by using splitting extensions.

Solution:

Form a splitting extension of a cyclic group H of order 111 by a group $K = gp(k)$ of order 2 via the homomorphism taking k to the automorphism which sends every element of H into its inverse.

7.37. Construct a non-abelian group of order p^3 for each prime p by using splitting extensions. (Hard.)

Solution:

Let H be the direct product of a cyclic group $gp(a)$ of order p with a cyclic group $gp(b)$ of order p. Each element of H is uniquely of the form

$$a^r b^s \quad \text{with} \quad 0 \leqq r < p, \; 0 \leqq s < p$$

Let $\tau: a^r b^s \to a^r b^{s+r}$. We will show that τ is an automorphism of H of order p. First let m and n be any integers. We claim that

$$(a^m b^n)\tau = a^m b^{m+n}$$

Let $m = m' + qp$, $n = n' + sp$ where $0 \leqq m' < p$, $0 \leqq n' < p$. Then

$$(a^m b^n)\tau = a^{m'} b^{m'+n'} = a^m b^{m'} b^{n'} = a^m b^m b^n = a^m b^{m+n}$$

To verify that τ is a homomorphism, observe that

$$(a^{m_1} b^{n_1} \cdot a^{m_2} b^{n_2})\tau = a^{m_1+m_2} b^{m_1+m_2+n_1+n_2}$$
$$= a^{m_1} b^{m_1+n_1} \cdot a^{m_2} b^{m_2+n_2} = (a^{m_1} b^{n_1})\tau (a^{m_2} b^{n_2})\tau$$

Clearly $\operatorname{Ker} \tau = \{1\}$, and so τ is one-to-one. It is easy to check that τ is also onto. Thus τ is an automorphism. Note that $a\tau^p = ab^p = a$, so that τ^p acts as the identity on a and b, which form a set of generators of H. Hence τ is of order p.

Now let $K = gp(k)$ be of order p. The mapping $\alpha: k^i \to \tau^i$ is an isomorphism. Then we form the splitting extension of H by K via α. This gives a group G of order p^3. G is non-abelian since $(1, a)(k, 1) = (k, ab)$, but $(k, 1)(1, a) = (k, a)$.

d. Direct product

Consider the special case of a splitting extension G of a group H by a group K via a homomorphism α in which α takes K onto the identity group of automorphisms of H. In this case G consists of the pairs (h, k) $(h \in H, k \in K)$ with binary operation given by $(h, k)(h', k') = (hh', kk')$. So G is, in the terminology of Section 5.3a, page 143, simply the external direct product of H and K. The obvious usefulness of this construction is that we do not require any knowledge of the automorphism group of H to construct the direct product. Notice that if $\bar{H} = \{(1, h) \mid h \in H\}$ and $\bar{K} = \{(k, 1) \mid k \in K\}$, then G is the internal direct product of its subgroups $\bar{H}$ and $\bar{K}$, again in the sense of Chapter 5. We will not pursue this concept of direct product here any further.

7.8 THE TRANSFER

a. Definition

Suppose G is a group with an *abelian* subgroup A of finite index. The transfer is a special homomorphism of G into A. The use of such transfer homomorphisms has been important in the theory of finite groups. Here we will examine one application of the transfer and briefly mention another (at the end of Section 7.8d).

To define the transfer τ of G into A, choose a right transversal X of G in A. We repeat that A is an abelian subgroup of G of finite index n, say. Therefore $|X| = n$. Recall that if $g \in G$, then $\bar{g}$ is the element of X in the coset Ag. Let $x_1, x_2, \ldots, x_n$ be the elements of X; then if $g \in G$, we define a mapping τ of G into A by

$$g\tau = x_1 g(\overline{x_1 g})^{-1} \cdot x_2 g(\overline{x_2 g})^{-1} \cdot \cdots \cdot x_n g(\overline{x_n g})^{-1}$$

It is clear that $g\tau \in A$ since, as $x_i g$ and $\overline{x_i g}$ belong to the same coset of A,

$$x_i g(\overline{x_i g})^{-1} \in A \quad (i = 1, \ldots, n)$$

This mapping τ is the homomorphism of G into A mentioned above; it is called the transfer of G into A. There are two items to be verified: (1) τ is a homomorphism and (2) τ is independent of the choice of the transversal X.

b. Proof that τ is a homomorphism

We compute $(gh)\tau$, where $g, h \in G$:

$$(gh)\tau = (x_1gh(\overline{x_1gh})^{-1}) \cdot (x_2gh(\overline{x_2gh})^{-1}) \cdot \cdots \cdot (x_ngh(\overline{x_ngh})^{-1})$$

$$= (x_1gh(\overline{\overline{x_1g}h})^{-1}) \cdot (x_2gh(\overline{\overline{x_2g}h})^{-1}) \cdot \cdots \cdot (x_ngh(\overline{\overline{x_ng}h})^{-1})$$

$$\text{(since } \overline{x_igh} = \overline{\overline{x_ig}h} \text{ by } (7.3)\text{, page 219)}$$

$$= (x_1g(\overline{x_1g})^{-1} \cdot \overline{x_1g}h(\overline{\overline{x_1g}h})^{-1}) \cdot (x_2g(\overline{x_2g})^{-1} \cdot \overline{x_2g}h(\overline{\overline{x_2g}h})^{-1}) \cdot \cdots$$
$$\cdot (x_ng(\overline{x_ng})^{-1} \cdot \overline{x_ng}h(\overline{\overline{x_ng}h})^{-1})$$

Now every one of the elements $x_ig(\overline{x_ig})^{-1}$, $\overline{x_jg}h(\overline{\overline{x_jg}h})^{-1}$ lies in A; since A is abelian, they must commute. So we can rewrite $(gh)\tau$ in the form

$$(gh)\tau = [x_1g(\overline{x_1g})^{-1} \cdot x_2g(\overline{x_2g})^{-1} \cdot \cdots \cdot x_ng(\overline{x_ng})^{-1}]$$
$$\cdot [\overline{x_1g}h(\overline{\overline{x_1g}h})^{-1} \cdot \overline{x_2g}h(\overline{\overline{x_2g}h})^{-1} \cdot \cdots \cdot \overline{x_ng}h(\overline{\overline{x_ng}h})^{-1}] \qquad (7.34)$$

But observe that $\qquad x_i \to \overline{x_ig} \quad (i = 1, \ldots, n)$

is a permutation of X (Section 7.4a). This means that

$$\overline{x_1g}h(\overline{\overline{x_1g}h})^{-1} \cdot \overline{x_2g}h(\overline{\overline{x_2g}h})^{-1} \cdot \cdots \cdot \overline{x_ng}h(\overline{\overline{x_ng}h})^{-1}$$

simply consists of the n elements $x_ih(\overline{x_ih})^{-1}$ $(i = 1, \ldots, n)$ multiplied together in some order. Since A is abelian, the order of such a product is immaterial. Thus

$$\overline{x_1g}h(\overline{\overline{x_1g}h})^{-1} \cdot \overline{x_2g}h(\overline{\overline{x_2g}h})^{-1} \cdot \cdots \cdot \overline{x_ng}h(\overline{\overline{x_ng}h})^{-1} = x_1h(\overline{x_1h})^{-1} \cdot x_2h(\overline{x_2h})^{-1} \cdot \cdots \cdot x_nh(\overline{x_nh})^{-1}$$
$$= h\tau, \text{ by definition}$$

Then it follows from equation (7.34) that

$$(gh)\tau = (g\tau)(h\tau)$$

for all g, h in G. Thus τ is a homomorphism.

c. Proof that τ is independent of the choice of transversal

We now prove that τ is independent of the choice of the transversal X.

The proof depends on an analysis of the product

$$g\tau = x_1g(\overline{x_1g})^{-1} \cdot x_2g(\overline{x_2g})^{-1} \cdot \cdots \cdot x_ng(\overline{x_ng})^{-1}$$

where again $\{x_1, x_2, \ldots, x_n\} = X$ is a right transversal of A in G. We recall from Section 7.4a that the mapping $x_i \to \overline{x_ig}$ is a permutation of X. Now every permutation of a finite set can be written as a product of disjoint cycles (Theorem 5.26, page 167). So, after relabeling the elements of X if necessary, we can assume that

$$\overline{x_1g} = x_2, \quad \overline{x_2g} = x_3, \quad \ldots, \quad \overline{x_{k-1}g} = x_k, \quad \overline{x_kg} = x_1$$

$$\overline{x_{k+1}g} = x_{k+2}, \quad \overline{x_{k+2}g} = x_{k+3}, \quad \ldots, \quad \overline{x_{k+l-1}} = x_{k+l}, \quad \overline{x_{k+l}g} = x_{k+1}$$

$$\cdots\cdots\cdots\cdots\cdots\cdots\cdots\cdots\cdots\cdots\cdots\cdots\cdots\cdots\cdots\cdots\cdots\cdots\cdots\cdots$$

$$\overline{x_{n-m+1}g} = x_{n-m+2}, \quad \overline{x_{n-m+2}g} = x_{n-m+3}, \quad \ldots, \quad \overline{x_{n-1}g} = x_n, \quad \overline{x_ng} = x_{n-m+1}$$

(Note that $k + l + \cdots + m = n$, where n is the index of A in G.) Then

$$g\tau = (x_1g(\overline{x_1g})^{-1} \cdot x_2g(\overline{x_2g})^{-1} \cdot \cdots \cdot x_kg(\overline{x_kg})^{-1})$$
$$\cdot (x_{k+1}g(\overline{x_{k+1}g})^{-1} \cdot x_{k+2}g(\overline{x_{k+2}g})^{-1} \cdot \cdots \cdot x_{k+l}g(\overline{x_{k+l}g})^{-1}) \cdot \cdots$$
$$\cdot (x_{n-m+1}g(\overline{x_{n-m+1}g})^{-1} \cdot x_{n-m+2}g(\overline{x_{n-m+2}g})^{-1} \cdot \cdots \cdot x_n(\overline{gx_ng})^{-1})$$

$$= (x_1gx_2^{-1} \cdot x_2gx_3^{-1} \cdot \cdots \cdot x_kgx_1^{-1})$$
$$\cdot (x_{k+1}gx_{k+2}^{-1} \cdot x_{k+2}gx_{k+3}^{-1} \cdot \cdots \cdot x_{k+l}gx_k^{-1}) \cdot \cdots$$
$$\cdot (x_{n-m+1}gx_{n-m+2}^{-1} \cdot x_{n-m+2}gx_{n-m+3}^{-1} \cdot \cdots \cdot x_ngx_{n-m+1}^{-1})$$

and thus $$g\tau = (x_1 g^k x_1^{-1})\cdot(x_{k+1}g^l x_{k+1}^{-1})\cdot\cdots\cdot(x_{n-m+1}g^m x_{n-m+1}^{-1}) \tag{7.35}$$

Note that $x_1 g^k x_1^{-1} = (x_1 g(\overline{x_1 g})^{-1})\cdot(x_2 g(\overline{x_2 g})^{-1})\cdot\cdots\cdot(x_k g(\overline{x_k g})^{-1}) \in A$, since it is the product of factors $x_i g(\overline{x_i g})^{-1}$ which belong to A. Similarly, $x_{k+1}g^l x_{k+1}^{-1}, \ldots, x_{n-m+1}x^m x_{n-m+1}^{-1} \in A$.

We are now able to prove

Lemma 7.7: Let A be an abelian subgroup of finite index in a group G and let $X = \{x_1, x_2, \ldots, x_n\}$, $Y = \{y_1, y_2, \ldots, y_n\}$ be two left transversals of A in G. Furthermore let τ and $\tilde{\tau}$ be mappings of G into A defined respectively by

$$g\tau = x_1 g(\overline{x_1 g})^{-1}\cdot x_2 g(\overline{x_2 g})^{-1}\cdot\cdots\cdot x_n g(\overline{x_n g})^{-1} \text{ (where } g\in G)$$

and $$g\tilde{\tau} = y_1 g(\widetilde{y_1 g})^{-1}\cdot y_2 g(\widetilde{y_2 g})^{-1}\cdot\cdots\cdot y_n g(\widetilde{y_n g})^{-1} \text{ (where } g\in G)$$

where, if $h \in G$, $\bar{h}$ denotes the element of X in the coset Ah and $\tilde{h}$ denotes the element of Y in the coset Ah. Then

$$\tau = \tilde{\tau}$$

Proof: We may assume on suitably reordering Y that

$$y_i = a_i x_i \quad (a_i \in A)$$

Now if $\overline{x_i g} = x_j$, then $Ay_j = Aa_j x_j = Ax_j = Ax_i g = Aa_i x_i g = Ay_i g$. This means that

$$\widetilde{y_i g} = y_j$$

It follows therefore as in equation (7.35) that if $g \in G$, then

$$g\tilde{\tau} = (y_1 g^k y_1^{-1})\cdot(y_{k+1}g^l y_{k+1}^{-1})\cdot\cdots\cdot(y_{n-m+1}g^m y_{n-m+1}^{-1}) \tag{7.36}$$

But $y_i = a_i x_i$, and we know that the elements $x_1 g^k x_1^{-1}, x_{k+1}g^l x_{k+1}^{-1}, \ldots, x_{n-m+1}g^m x_{n-m+1}^{-1}$ lie in the abelian subgroup A. Therefore

$$y_1 g^k y_1^{-1} = a_1(x_1 g^k x_1^{-1})a_1^{-1} = x_1 g^k x_1^{-1}$$

$$y_{k+1}g^l y_{k+1}^{-1} = a_{k+1}(x_{k+1}g^l x_{k+1}^{-1})a_{k+1}^{-1} = x_{k+1}g^l x_{k+1}^{-1}$$

$$\cdots\cdots\cdots\cdots\cdots\cdots\cdots\cdots\cdots$$

$$y_{n-m+1}x^m y_{n-m+1}^{-1} = a_{n-m+1}(x_{n-m+1}g^m x_{n-m+1}^{-1})a_{n-m+1}^{-1} = x_{n-m+1}g^m x_{n-m+1}^{-1}$$

Thus it follows from (7.35), (7.36) and the above remarks that $g\tau = g\tilde{\tau}$.

This lemma establishes that the transfer homomorphism τ is independent of the choice of transversal. Accordingly we may speak of *the* transfer of G into A.

d. A theorem of Schur

Using the transfer, we now prove the following important theorem of I. Schur.

Theorem 7.8: Let G be a group whose center A is of finite index. Then the derived group G' of G is finite.

Proof: Suppose $|G/A| = n$. Let $X = \{x_1, \ldots, x_n\}$ be a transversal of A in G and let τ be the transfer of G into A. Now if $g \in G$, then by equation (7.35)

$$g\tau = x_1 g^k x_1^{-1}\cdot x_{k+1}g^l x_{k+1}^{-1}\cdot\cdots\cdot x_{n-m+1}g^m x_{n-m+1}^{-1}$$

But $x_1 g^k x_1^{-1} \in A$ and A is the center of G. So

$$x_1 g^k x_1^{-1} = x_1^{-1}(x_1 g^k x_1^{-1})x_1 = g^k$$

Similarly, $$x_{k+1}g^l x_{k+1}^{-1} = g^l, \ \ldots, \ x_{n-m+1}g^m x_{n-m+1}^{-1} = g^m$$

Thus, for each $g \in G$, $g\tau = g^n$.

Notice that the image of G under τ is a subgroup of A and is therefore abelian. Then by Theorem 4.18, page 117, G/K is abelian, where K is the kernel of τ. But by Problem 4.68, page 116, this implies $G' \subseteq K$.

Now as G/A is finite, so is $G'A/A$. This means that $G'/G' \cap A$ is also finite since by Theorem 4.23, page 125,

$$G'A/A \cong G'/G' \cap A \tag{7.37}$$

Since $G'/G' \cap A$ is finite, G' itself is finite if $G' \cap A$ is finite. Since $g\tau = g^n$ for every $g \in G$, every element of the kernel has finite order. It follows that the elements of $G' \cap A$ have finite order. We will show that $G' \cap A$ is finitely generated. Assuming this true for the moment, it follows that $G' \cap A$ is finite (Problem 6.44, page 198). Thus G' is also finite. This completes the proof of Schur's theorem but for the verification that $G' \cap A$ is finitely generated.

We accomplish this by showing first that G' is finitely generated. If $g, h \in G$, then $g = ax_i$ and $h = bx_j$, where $a, b \in A$, $x_i, x_j \in X$. Then since A is the center of G, we have

$$g^{-1}h^{-1}gh = x_i^{-1}a^{-1}x_j^{-1}b^{-1}ax_ibx_j = x_i^{-1}x_j^{-1}x_ix_ja^{-1}ab^{-1}b = x_i^{-1}x_j^{-1}x_ix_j$$

This means that there are at most n^2 distinct commutators in G. Therefore G' is finitely generated since it is generated by commutators.

Finally $G' \cap A$ is finitely generated since it is of finite index in a finitely generated group (Theorem 7.4, page 227). This completes the proof of Schur's theorem.

It is worth noting one fact that emerged in this proof: the transfer into the center is simply the mapping that takes each element g to g^n where n is the index of the center in G.

We end our discussion of the transfer by mentioning that if all the Sylow subgroups of a finite group G are cyclic, then G is *metacyclic* (i.e. an extension of a cyclic group by a cyclic group). This theorem can be proved by using the transfer. The proof is not too complicated; however, it is lengthy and will not be given here. Reference to a proof may be found in Section 5.2a, page 139.

A look back at Chapter 7

We re-proved Cayley's theorem, namely that every group is isomorphic to a group of permutations. The ideas that arose from Cayley's theorem were generalized. In particular we called a homomorphism of a group G into the symmetric group on a set X a *permutational representation* of G (on X). Permutational representations were roughly classified. We explained an important permutational representation of a group G called a coset representation. This representation allowed us to provide a variation of Cayley's theorem due to Frobenius. Then we used Frobenius' theorem and the coset representation to prove three theorems: (1) a subgroup of finite index in a finitely generated group is finitely generated; (2) the number of subgroups of fixed finite index in a finitely generated group is finite; (3) if G is a finitely generated group whose subgroups of finite index have only the identity subgroup in common, then every homomorphism of G onto itself is also one-to-one, i.e. an automorphism.

We called a group G an extension of a group H by K if there is a normal subgroup $\bar{H}$ of G such that $G/\bar{H} \cong K$ and $\bar{H} \cong H$. An analysis of this situation was made simpler because of our discussion of both coset representations and Frobenius' theorem. This analysis of extensions was specialized to splitting extensions, where we provided a method of constructing a splitting extension of two groups H and K.

Finally we defined a special kind of homomorphism of a group into an abelian subgroup, called the transfer. We then used the transfer to prove that the derived group of a group whose center is of finite index is finite.

Supplementary Problems

PERMUTATION REPRESENTATIONS, COSET REPRESENTATIONS, FROBENIUS THEOREM

7.38. Let D_n be the dihedral group of order $2n$ and let C_n be its cyclic normal subgroup of order n. Find explicitly
(1) the representation of D_n onto itself given by Cayley's theorem,
(2) a coset representation of D_n using C_n as the subgroup,
(3) the representation provided by the Frobenius theorem.

7.39. Give a permutation representation of A_n of degree $2n$.

7.40. Find a faithful representation of $G \times H$ on $n+m$ letters if $G \subseteq S_n$, $H \subseteq S_m$.

EXTENSIONS

7.41. Construct a non-abelian group which is an extension of an infinite cyclic group by a group of order 2.

7.42. Prove that if G is a non-abelian group with an infinite cyclic normal subgroup of index 2, then G is a splitting extension of an infinite cyclic group by a group of order 2.

7.43. Prove that there are precisely three non-isomorphic extensions of an infinite cyclic group by a group of order 2.

7.44. Prove that an extension of a cyclic group of even order by a group of order 3 splits.

7.45. Construct a non-abelian group of order 36.

7.46. Construct five non-isomorphic groups of order $5^5 \times 3$.

7.47. Let D be the set of infinite sequences of integers, i.e. D consists of the sequences $a = \ldots, a_{-1}, a_0, a_1, \ldots$ $(a_i \in Z)$. If $b = \ldots, b_{-1}, b_0, b_1, \ldots,$ define $a+b = \ldots, a_{-1}+b_{-1}, a_0+b_0, a_1+b_1, \ldots.$ D is an abelian group under the operation of addition of sequences. For each integer n define a mapping α_n of D by putting $a\alpha_n = \ldots, b_{-1}, b_0, b_1, \ldots,$ where $b_i = a_{i-n}$. Prove that
(1) α_n is an automorphism of D, (2) $\alpha_m\alpha_n = \alpha_{m+n}$,
(3) $A = \{\alpha_i \mid i \in Z\}$ is an infinite cyclic group generated by α_1.

7.48. Let W be the splitting extension of D of Problem 7.47 by the infinite cyclic group $C = gp(c)$ via the mapping that sends c to α_1. Prove that W is a non-abelian group. Find a proper subgroup of W which is isomorphic to W. Prove that W/W' is infinite cyclic. (Hard.)

TRANSFER. MAL'CEV'S THEOREM, MARSHALL HALL'S THEOREM.

7.49. Prove that if A is an abelian subgroup of finite index in a simple group G, then the transfer of G into A sends G into 1.

7.50. Prove that if G is a finite group whose center Z has order co-prime to its index, then the transfer of G into Z is onto. (Hard.)

7.51. Prove that an infinite group with a subgroup of finite index is not simple.

7.52. A group is said to be residually finite if the intersection of all its normal subgroups of finite index is the identity. Prove that an extension of a residually finite group by a finite group is residually finite. (*Hint*: The preceding problem gives a clue.)

7.53. Let G be a cyclic extension of a cyclic group N of order n by a cyclic group of order m. Let $N = gp(a)$ and $G/N = gp(bN)$. Prove that $b^{-1}ab = a^j$, where j is co-prime to n and $b^m = a^k$ (where j and k are integers). Prove that $jk \equiv k$ modulo n.

7.54. Prove that a cyclic extension of a finitely generated residually finite group (defined in Problem 7.52) is residually finite. (*Hint*: Use Marshall Hall's theorem. Then use the fact that the automorphism group of a finite group is finite.) (Hard.)

7.55. Prove that if G is a group and $G_1, G_2, \ldots$ are subgroups of G such that $G_1 \neq G_2$, $G_1 \subseteq G_2$, $G_2 \neq G_3$, $G_2 \subseteq G_3, \ldots,$ then $\cup G_i$ is a subgroup of G which is not finitely generated.

7.56. Let G be a residually finite group (defined in Problem 7.52) and suppose every subgroup of G is finitely generated. Prove that if H is a subgroup of G such that $H/N \cong G$ for some normal subgroup N of G, then $N = \{1\}$. (*Hint*: Use Problem 7.55.)

7.57. G is a finitely generated group, every element of which has only a finite number of conjugates. Prove that $|G'| < \infty$. $\left(\textit{Hint: } \bigcap_{i=1}^{n} C(g_i) = Z(G) \text{ if } g_1, \ldots, g_n \text{ are the generators of } G.\right)$

7.58. G is a group in which every element has only a finite number of conjugates. Prove that every element of G' is of finite order. (*Hint*: Use Problem 7.57.)

Chapter 8

Free Groups and Presentations

Preview of Chapter 8

We begin with a property of the infinite cyclic group and generalize this property to define free groups. We ask questions similar to those we asked in Chapter 4 concerning cyclic groups:

(1) Do free groups exist?

(2) When are two free groups isomorphic?

(3) What are the homomorphisms of free groups?

(4) What are the subgroups of free groups?

In answering (3) we will learn that every group is a homomorphic image of a free group. This provides a new way of describing a group, i.e. as a factor group of a free group. Such a description of a group is called a presentation.

8.1 ELEMENTARY NOTIONS

a. Definition of a free group

Recall that if G is a group and X $(\neq \emptyset)$ a subset of G,

$$gp(X) \;=\; \{x_1^{\epsilon_1} \cdots x_n^{\epsilon_n} \mid x_i \in X,\ \epsilon_i = \pm 1\}$$

If $x_1^{\epsilon_1} \cdots x_n^{\epsilon_n}$ and $y_1^{\eta_1} \cdots y_m^{\eta_m}$ are two products with $x_i, y_i \in X$ and $\epsilon_i = \pm 1$, $\eta_i = \pm 1$, then they are said to be *identical* if $n = m$, $x_i = y_i$ and $\epsilon_i = \eta_i$ for $i = 1, \ldots, m$. Two products are said to be *different* if they are not identical.

It is easy to see that two different products of the form $x_1^{\epsilon_1} \cdots x_n^{\epsilon_n}$ can give rise to the same element of G. For example, if $X = \{x, y\}$, then xy and $xx^{-1}xy$ are different products of the form $x_1^{\epsilon_1} \cdots x_n^{\epsilon_n}$ but they give rise to the same element of G, i.e. xy. To avoid redundancy, we introduce the concept of a *reduced product*:

A product $x_1^{\epsilon_1} \cdots x_n^{\epsilon_n}$, where $\epsilon_i = \pm 1$ and $x_i \in X$, is said to be a *reduced X-product* if $x_i = x_{i+1}$ implies $\epsilon_i \neq -\epsilon_{i+1}$.

Synonyms for reduced X-product are *reduced product* (X being understood) and reduced product in X.

(Examples of reduced products are easily given. Let $X = \{x, y\}$; then xy, $x^{-1}yxyx^{-1}$ and $x^{-1}yyxy^{-1}$ are reduced products. However, $yxyxx^{-1}$ and $x^{-1}yxyy^{-1}$ are not reduced products.)

Lemma 8.1: $gp(X) = \{w \mid w = 1 \text{ or } w = \text{a reduced product in } X\}$.

Proof: Let $\{w \mid w = 1 \text{ or } w = \text{a reduced product in } X\} = R$. Clearly $R \subseteq gp(X)$. If $u \in gp(X)$, then $u = x_1^{\epsilon_1} \cdots x_k^{\epsilon_k}$ where $x_i \in X$ and $\epsilon_i = \pm 1$.

We proceed to show that $u \in R$. If $k = 1$, u is a reduced product in X, and so $u \in R$. Assume then that any product $x_1^{\epsilon_1} \cdots x_k^{\epsilon_k} \in R$ for $k \leqq n-1$. Suppose $k = n$. If $u = x_1^{\epsilon_1} \cdots x_k^{\epsilon_k}$ is a reduced product in X, then $u \in R$. If u is not a reduced product, there exists an integer i such that $x_i = x_{i+1}$ and $\epsilon_i = -\epsilon_{i+1}$. Suppose $n > 2$; then we can delete $x_i^{\epsilon_i} x_{i+1}^{\epsilon_{i+1}}$ to obtain u as a product involving $n-2$ elements of X. By the inductive hypothesis, u then belongs to R. If $n = 2$, then $u = x_1^{\epsilon_1} x_2^{\epsilon_2}$, where $x_1 = x_2$ and $\epsilon_1 = -\epsilon_2$, from which $u = 1 \in R$. Thus $gp(X) \subseteq R$ and accordingly $gp(X) = R$.

Consider now the infinite cyclic group generated by the element x. The reduced products in $\{x\}$ are of two kinds:

$$x \cdots x = x^r \quad \text{or} \quad x^{-1}x^{-1} \cdots x^{-1} = x^{-r}$$

where r is a positive integer. From what has been said about the infinite cyclic group, we know that if m and n are integers, $x^m = x^n$ implies that $m = n$. Thus different reduced $\{x\}$-products give rise to different elements. (This is by no means the usual situation. For example in a cyclic group of order 2 generated by y, we have $yyyy = yy$.) G is said to be freely generated by $\{x\}$. More generally we have the following definition:

A group G is said to be *freely generated* by the set $X \subseteq G$ if $X \neq \emptyset$, the empty set, and

(i) $gp(X) = G$;

(ii) two different reduced X-products define two different nonunit elements of G.

Notice that it follows from (ii) that if $x \in X$, $x^{-1} \notin X$. For if not, there exist x and y, both elements of X, with $y^{-1} = x$. But then x and y^{-1} are two different reduced X-products which are equal. It also follows from (ii) that $1 \notin X$.

A set X of generators of G satisfying (ii) is often called a *free set of generators* of G. A group G is *free* if it is the identity group or if it possesses a free set of generators. G is also said to be *free on X*. If G is a free group freely generated by X, then the study of G is facilitated by the fact that we know exactly whether two X-products are equal. All we need to do is to express each of the products in reduced product form. If the reduced products are distinct the elements are not equal. This process of expressing an element as a reduced X-product can be carried out in a finite number steps. To illustrate, let F be freely generated by $\{x, y\}$. Are the products $f = xyx^{-1}y^{-1}xxx^{-1}y^{-1}y^3$ and $g = xyy^{-2}y^3$ equal? We convert f to a reduced product by deleting *inverse pairs* (i.e. two adjacent inverse factors). Thus

$$f = xyx^{-1}y^{-1}x(xx^{-1})(y^{-1}y)y^2 = xyx^{-1}y^{-1}xy^2 = xyx^{-1}y^{-1}xyy$$

Similarly $g = xyy$. Hence as f and g are equal to different reduced products, they are not equal. The fact that we can determine in a finite number of steps whether or not two elements are equal is often expressed by saying that the word problem is decidable for free groups. (The interested reader may find more details in, e.g., J. J. Rotman, *The Theory of Groups*, Allyn and Bacon, 1965.)

Problems

8.1. Let G be freely generated by the set $X = \{x, y, z\}$. (*a*) Write down three distinct elements of G. (*b*) Is $xyz(yz)^{-1}$ a reduced product? (*c*) Is $xyy^{-1}z^{-1}yx$ equal to $xz^{-1}yzz^{-1}x$? (*d*) Express $x^2y^3(y^3x^2)^{-1}y^{-3}$ and $(xzy)^{-1}xzy^2$ as reduced products.

Solution:

(*a*) (1) x, y, z, (2) x, x^2, x^3, (3) x, xy, xz

In fact any three different reduced products in $\{x, y, z\}$ are distinct elements of G.

(*b*) No, because it is not written in the required $x_1^{\epsilon_1} \cdots x_n^{\epsilon_n}$ form, since it involves $(yz)^{-1}$. However, even if we replace $(yz)^{-1}$ by its equivalent $z^{-1}y^{-1}$ to get $xyzz^{-1}y^{-1}$, this, though in the form $x_1^{\epsilon_1} \cdots x_n^{\epsilon_n}$, is not a reduced product because of the inverse pair zz^{-1}.

(c) Yes, for on expressing each as a reduced product we get $xz^{-1}yx$.

(d)
$$x^2y^3(y^3x^2)^{-1}y^{-3} = x^2y^3x^{-2}y^{-3}y^{-3} = xxyyyx^{-1}x^{-1}y^{-1}y^{-1}y^{-1}y^{-1}y^{-1}y^{-1}$$
$$(xzy)^{-1}xzy^2 = y^{-1}z^{-1}x^{-1}xzy^2 = y^{-1}z^{-1}zy^2 = y^{-1}y^2 = y$$

8.2. Prove that if G is a free group, $G \neq \{1\}$, then the infinite cyclic group is a subgroup of G.

Solution:

Suppose G is freely generated by X. If $x \in X$, consider $gp(\{x\})$. This is a cyclic group. Now $x \cdot x \cdot \cdots \cdot x = x^r$, with r a positive integer, is a reduced product in X. Hence $x^r = x^s$, where r and s are positive integers implies $r = s$. Thus $gp(\{x\})$ is infinite and is infinite cyclic and the result follows.

8.3. Prove that if G is freely generated by X, where X contains at least two elements, then G is not abelian.

Solution:

There exist two distinct elements x, y of X. Now xy and yx are two different reduced products, so $xy \neq yx$. But this implies that G is not abelian.

8.4. Prove that a finite group G is not free if $G \neq \{1\}$.

Solution:

In Problem 8.2 we proved that except for the identity, every free group has as a subgroup the infinite cyclic group. Consequently if G is free it must have the infinite cyclic group as a subgroup. This is absurd.

8.5. The direct product of two infinite cyclic groups is not a free group.

Solution:

The direct product of two infinite cyclic groups is abelian. But we have proved in Problem 8.3 that a free group is not abelian if it is freely generated by a set X which contains at least two elements. Hence if the direct product of two infinite cyclic groups is free, it is freely generated by some set $X = \{x\}$, i.e. it is infinite cyclic. But a free abelian group of rank two is not cyclic. This follows immediately either from the uniqueness of the type of an abelian group (Theorem 6.21, page 197) or directly.

8.6. Prove that a free group freely generated by X with $|X| \geqq 2$ has no center (i.e. its center consists of the identity element). (Hard.)

Solution:

Suppose G has an element $z \neq 1$ in its center. Let z be expressed as a reduced product $z = x_1^{\epsilon_1} \cdots x_n^{\epsilon_n}$. Let $y \in X$, $y \neq x_1$. Consider
$$yz = yx_1^{\epsilon_1} \cdots x_n^{\epsilon_n}$$
This is a reduced X-product. On the other hand
$$zy = x_1^{\epsilon_1} \cdots x_n^{\epsilon_n}y$$
If $x_n^{\epsilon_n} \neq y^{-1}$, then $x_1^{\epsilon_1} \cdots x_n^{\epsilon_n}y$ is a reduced product and so clearly $zy \neq yz$ as yz begins as a reduced product with y but zy begins with $x_1^{\epsilon_1} \neq y$. If $x_n^{\epsilon_n} = y^{-1}$ then for $n > 1$ (for otherwise $x_1 = y$ contrary to the choice of y),
$$zy = x_1^{\epsilon_1} \cdots x_{n-1}^{\epsilon_{n-1}}$$
and again it is clear that zy and yz are two different reduced products, so $zy \neq yz$. But as z is in the center, we must have $zy = yz$. Therefore this contradicts the assumption that G has a center.

8.7. Generalize Problem 8.2 by showing that if F is freely generated by $\{x_1, \ldots, x_n\}$ with $n > 1$, then $\{x_1, \ldots, x_i\}$ freely generates $F_i = gp(x_1, \ldots, x_i)$ for $i = 1, 2, \ldots, n$.

Solution:

Put $X_i = \{x_1, \ldots, x_i\}$. By definition, X_i generates F_i. We have only to prove that two different reduced X_i-products define different elements of F_i. But a reduced X_i-product is obviously also a reduced X-product and two different reduced X-products define different elements of F. Hence the result follows immediately.

b. Length of an element. Alternative description of a free group

Suppose F is freely generated by a set X. If $f \in F$, $f \neq 1$, it can be expressed in one and only one way as a reduced product $x_1^{\epsilon_1} \cdots x_n^{\epsilon_n}$. We define the *length of f* with respect to this free set of generators X to be n. The length of the identity is defined conventionally to be 0. For example, if F is freely generated by $X = \{x, y\}$, then the length of $x^2y^2x^{-1}$ is 5, because

$$x^2y^2x^{-1} = xxyyx^{-1}$$

A very useful technique in arguments involving free groups is to prove results by induction on the length of elements (see, for example, Lemma 8.9, page 261).

Lemma 8.2: A group F is freely generated by a set $X \neq \emptyset$, the empty set, if and only if (*a*) $gp(X) = F$, and (*b*) no reduced X-product is equal to the identity element.

Proof: Assume first (*a*) and (*b*) above. To show F is freely generated by X we must show that two different reduced products in X are not equal and are not the identity. Let $x_1^{\epsilon_1} \cdots x_n^{\epsilon_n}$ and $y_1^{\eta_1} \cdots y_m^{\eta_m}$ (where $\epsilon_i = \pm 1$, $\eta_i = \pm 1$, and $x_i, y_i \in X$) be two different reduced products. Suppose they are equal. We can assume without loss of generality that $x_n^{\epsilon_n} \neq y_m^{\eta_m}$; for if $x_n^{\epsilon_n} = y_m^{\eta_m}$, $x_1^{\epsilon_1} \cdots x_{n-1}^{\epsilon_{n-1}}$ and $y_1^{\eta_1} \cdots y_{m-1}^{\eta_{m-1}}$ are two different reduced products. Since $x_1^{\epsilon_1} \cdots x_n^{\epsilon_n} = y_1^{\eta_1} \cdots y_m^{\eta_m}$ and $x_n^{\epsilon_n} = y_m^{\eta_m}$, it follows that

$$x_1^{\epsilon_1} \cdots x_{n-1}^{\epsilon_{n-1}} = y_1^{\eta_1} \cdots y_{m-1}^{\eta_{m-1}}$$

If again $x_{n-1}^{\epsilon_{n-1}} = y_{m-1}^{\eta_{m-1}}$, we can delete them. But we cannot continue indefinitely this way as $x_1^{\epsilon_1} \cdots x_n^{\epsilon_n}$ and $y_1^{\eta_1} \cdots y_m^{\eta_m}$ are assumed to be different reduced products. So we may assume $x_n^{\epsilon_n} \neq y_m^{\eta_m}$. It follows then that $x_1^{\epsilon_1} \cdots x_n^{\epsilon_n} y_m^{-\eta_m} \cdots y_1^{-\eta_1}$ is a reduced product equal to 1. But by (*b*) it is not 1. Hence we have a contradiction, and so it follows that F is freely generated by X.

If on the other hand F is given freely generated by X, do (*a*) and (*b*) hold? Of course, since this is entailed by the definition of a group freely generated by a set X.

Problem

8.8. Let F be freely generated by $X = \{x, y, z\}$. Determine the length of (i) 1, (ii) $xzyz^{-1}$, and (iii) $f = x^{-1}yxx^{-2}y^2x^3z^{-1}$.

Solution:

(i) The length of 1 is taken to be 0. (This is just the convention mentioned earlier.)

(ii) The length of $xzyz^{-1}$ is 4.

(iii) The length of f is calculated by expressing it first in reduced form, that is,

$$f = x^{-1}yx^{-1}yyxxxz^{-1}$$

Hence the length of f is 9.

c. Existence of free groups

An obvious question is: Do free groups exist? Up to now we have tacitly assumed that they do.

Theorem 8.3: Let n be any positive integer. There exists a free group freely generated by a set of n elements.

Proof: Since every group is isomorphic to a subgroup of the symmetric group of some set, it is natural to look for a suitable subgroup of some symmetric group in order to find a free group of rank n.

Our plan is as follows: we shall introduce a set T consisting of certain ordered 1-tuples of integers, ordered 2-tuples of integers, and so on. We then choose a set $X = \{\theta_1, \ldots, \theta_n\}$ of permutations of T such that X freely generates $gp(X)$.

The elements of T are the ordered m-tuples $(r_1, r_2, \ldots, r_m)$ with $r_1 = 0$ and $r_2, \ldots, r_m$ nonzero integers with $r_i + r_{i+1} \neq 0$ for $i = 1, \ldots, m-1$. Thus $(0, 1, 2, -3) \in T$ but $(1, 2) \notin T$ and $(0, 1, 2, -2) \notin T$.

We define now for each integer $i = \pm 1, \ldots, \pm n$ a permutation θ_i of T as follows. If $(r_1, \ldots, r_m) \in T$ we define

(i) $(r_1, \ldots, r_m)\theta_i = (r_1, \ldots, r_m, i)$ if $r_m \neq -i$.

(ii) $(r_1, \ldots, r_m)\theta_i = (r_1, \ldots, r_{m-1})$ if $r_m = -i$.

It is clear that θ_i is a well-defined mapping of T into T. Moreover $\theta_i\theta_{-i} = \iota = \theta_{-i}\theta_i$ where ι is the identity permutation of T (Problem 8.9). Thus each θ_i is a permutation of T (Theorem 2.4, page 36), i.e. $\theta_i \in S_T$.

Let $G = gp(X)$, where $X = \{\theta_1, \ldots, \theta_n\}$. (As S_T is a group and $X \subseteq S_T$, it makes sense to talk of $gp(X)$.) We prove that X freely generates G, thereby completing the proof of the theorem.

We have only to verify that every reduced X-product is not ι. Let then

$$f = \theta_{1'}^{\epsilon_1} \cdots \theta_{m'}^{\epsilon_m} \quad (\text{where } 1', \ldots, m' \in \{1, \ldots, n\} \text{ and } \epsilon_j = \pm 1)$$

be any reduced X-product (so if $r' = (r+1)'$, $\epsilon_r + \epsilon_{r+1} \neq 0$). We compute the effect of f on (0). Now $\theta_i^{-1} = \theta_{-i}$, and so $\theta_i^{\pm 1} = \theta_{\pm i}$. Then

$$f = \theta_{\epsilon_1 1'}\theta_{\epsilon_2 2'} \cdots \theta_{\epsilon_m m'}$$

By the definition of the product of permutations, we have

$$(0)f = ((0)\theta_{\epsilon_1 1'})(\theta_{\epsilon_2 2'} \cdots \theta_{\epsilon_m m'}) = ((0, \epsilon_1 1')\theta_{\epsilon_2 2'})(\theta_{\epsilon_3 3'} \cdots \theta_{\epsilon_m m'})$$

Now if $1' = 2'$, then $\epsilon_1 + \epsilon_2 \neq 0$ or $\epsilon_1 1' \neq -\epsilon_2 2'$. Therefore

$$(0, \epsilon_1 1')\theta_{\epsilon_2 2'} = (0, \epsilon_1 1', \epsilon_2 2')$$

It follows in this way that

$$(0)f = (0, \epsilon_1 1', \epsilon_2 2', \ldots, \epsilon_m m') \neq (0)$$

Thus $f \neq \iota$ since it does not leave (0) fixed. This completes the proof of the theorem.

Note: It is possible to prove similarly that there exist free groups freely generated by sets of arbitrary cardinality. Since we have not introduced cardinal numbers, we cannot prove this more general theorem here. The reader who has a knowledge of cardinal numbers may read the account in J. J. Rotman's *The Theory of Groups*, Allyn and Bacon, 1965.

Problems

8.9. Prove that $\theta_i\theta_{-i} = \iota$ and $\theta_{-i}\theta_i = \iota$, where θ_i and ι are as defined above.

Solution:

Let $(r_1, \ldots, r_m) \in T$. Suppose first that $r_m \neq -i$; then

$$(r_1, \ldots, r_m)\theta_i\theta_{-i} = (r_1, \ldots, r_m, i)\theta_{-i} = (r_1, \ldots, r_m)$$

If $r_m = -i$, $\quad (r_1, \ldots, r_m)\theta_i\theta_{-i} = (r_1, \ldots, r_{m-1})\theta_{-i}$

Now if $r_{m-1} = -(-i) = i$, then $(r_1, \ldots, r_m)$ is of the form $(r_1, \ldots, i, -i)$. Since such an element does not belong to T, $r_{m-1} \neq -(-i)$. Therefore

$$(r_1, \ldots, r_{m-1})\theta_{-i} = (r_1, \ldots, r_{m-1}, -i) = (r_1, \ldots, r_m)$$

and so
$$(r_1, \ldots, r_m)\theta_i\theta_{-i} = (r_1, \ldots, r_m)$$

Thus $\theta_i\theta_{-i}$ leaves all elements of T unchanged and hence $\theta_i\theta_{-i} = \iota$. Similarly $\theta_{-i}\theta_i = \iota$.

8.10. Prove that there exists a free group freely generated by a set Y such that there is a one-to-one mapping of Y onto the positive integers.

Solution:

We proceed exactly as in the proof of Theorem 8.3 except that we define θ_i for all nonzero integers i and put $Y = \{\theta_1, \theta_2, \ldots\}$. The mapping $\rho: Y \to Z^+$, the positive integers, defined by $(\theta_i)\rho = i$ is a one-to-one onto mapping, for as none of the θ_i have the same effect on (0), they must be distinct. If $H = gp(Y)$, then H is freely generated by Y.

d. Homomorphisms of free groups

Now it is a fact (see Problem 8.11 below) that if a group G is generated by a set X, then a homomorphism θ is uniquely determined by its effect on X, because each element of G is a product of elements and inverses of elements from X.

Consider conversely what would happen if we had a map θ of X into a group H. Could we find a homomorphism of the whole group G into H whose effect on X was the same as that of θ? In general the answer is no (see Problem 8.12). However, we have the following result for free groups.

Theorem 8.4: Let F be freely generated by a set X, let H be any group, and let θ be a mapping of X into H. Then there exists a homomorphism $\hat{\theta}$ of F into H such that $\hat{\theta}$ agrees with θ on X. $\hat{\theta}$ is called an *extension* of θ.

Proof: Any nonunit element of F is *uniquely* expressible as a reduced X-product

$$f = x_1^{\epsilon_1} \cdots x_n^{\epsilon_n} \quad \text{where } x_i \in X,\ \epsilon_i = \pm 1$$

and $x_i = x_{i+1}$ implies $\epsilon_i \neq -\epsilon_{i+1}$.

Define $f\hat{\theta} = (x_1\theta)^{\epsilon_1}(x_2\theta)^{\epsilon_2} \cdots (x_n\theta)^{\epsilon_n}$ and $1\hat{\theta} = 1$ (the latter 1 being the identity of H). Clearly $\hat{\theta}$ is a mapping of F into H agreeing with θ on X. To conclude the proof we must prove that $\hat{\theta}$ is a homomorphism.

To do this we shall show that if $f = x_1^{\epsilon_1} \cdots x_n^{\epsilon_n}$ where $x_i \in X$ and $\epsilon_i = \pm 1$, then

$$f\hat{\theta} = (x_1\theta)^{\epsilon_1} \cdots (x_n\theta)^{\epsilon_n}$$

whether or not $x_1^{\epsilon_1} \cdots x_n^{\epsilon_n}$ is a reduced product. If $n = 1$, this is true by the definition of $\hat{\theta}$. Assume it is true for all positive integers $n < k$ and consider $f = x_1^{\epsilon_1} \cdots x_n^{\epsilon_n}$ when $n = k$. If this is a reduced product then $f\hat{\theta} = (x_1\theta)^{\epsilon_1} \cdots (x_n\theta)^{\epsilon_n}$ by definition. If it is not a reduced product, then there exists an integer i such that $x_i = x_{i+1}$ and $\epsilon_i = -\epsilon_{i+1}$. Consequently,

$$\begin{aligned}(x_1\theta)^{\epsilon_1} \cdots (x_n\theta)^{\epsilon_n} &= (x_1\theta)^{\epsilon_1} \cdots (x_i\theta)^{\epsilon_i}(x_{i+1}\theta)^{\epsilon_{i+1}} \cdots (x_n\theta)^{\epsilon_n} \\ &= (x_1\theta)^{\epsilon_1} \cdots (x_{i-1}\theta)^{\epsilon_{i-1}}(x_{i+2}\theta)^{\epsilon_{i+2}} \cdots (x_n\theta)^{\epsilon_n} \\ &= (x_1^{\epsilon_1} \cdots x_{i-1}^{\epsilon_{i-1}} x_{i+2}^{\epsilon_{i+2}} \cdots x_n^{\epsilon_n})\hat{\theta}\end{aligned}$$

by our inductive hypothesis. Therefore

$$(x_1\theta)^{\epsilon_1} \cdots (x_n\theta)^{\epsilon_n} = (x_1^{\epsilon_1} \cdots x_{i-1}^{\epsilon_{i-1}}(x_i^{\epsilon_i}x_{i+1}^{\epsilon_{i+1}})x_{i+2}^{\epsilon_{i+2}} \cdots x_n^{\epsilon_n})\hat{\theta} = f\hat{\theta}$$

So if $f = x_1^{\epsilon_1} \cdots x_n^{\epsilon_n}$ and $g = y_1^{\eta_1} \cdots y_m^{\eta_m}$, $\epsilon_i = \pm 1$, $\eta_i = \pm 1$, $x_i, y_i \in X$, then

$$\begin{aligned}(fg)\hat{\theta} &= (x_1^{\epsilon_1} \cdots x_n^{\epsilon_n} y_1^{\eta_1} \cdots y_m^{\eta_m})\hat{\theta} = (x_1\theta)^{\epsilon_1} \cdots (x_n\theta)^{\epsilon_n}(y_1\theta)^{\eta_1} \cdots (y_m\theta)^{\eta_m} \\ &= [(x_1\theta)^{\epsilon_1} \cdots (x_n\theta)^{\epsilon_n}][(y_1\theta)^{\eta_1} \cdots (y_m\theta)^{\eta_m}] = f\hat{\theta}\, g\hat{\theta}\end{aligned}$$

Hence $\hat{\theta}$ is a homomorphism and the theorem follows.

Corollary 8.5: Let G be any finitely generated group (i.e. there exists a finite set Y such that $gp(Y) = G$). Then G is a homomorphic image of some free group.

Proof: Let G be finitely generated by a set $\{y_1, \ldots, y_n\}$. There exists a free group F freely generated by $\{x_1, \ldots, x_n\}$, say. Define a map θ from $\{x_1, \ldots, x_n\}$ to G by $x_i\theta = y_i$ for $i = 1, \ldots, n$. By Theorem 8.4 there exists a homomorphism θ of F into G which agrees with θ on each x_i. We know that $F\theta$ is a subgroup of G. But as $F\theta$ contains $y_1, \ldots, y_n$, it contains $gp(y_1, \ldots, y_n) = G$. Therefore $F\theta = G$, and so G is a homomorphic image of a free group.

Note: The same result applies whether G is finitely generated or not. However, to prove the more general result we must use some of the ideas involving cardinal numbers. For this reason we have chosen only to consider the finitely generated case.

Our last theorem reveals the importance of free groups. As every group is a homomorphic image of a free group, from the knowledge of the properties of a free group we may achieve some understanding of other groups.

Problems

8.11. Let θ_1 and θ_2 be homomorphisms of $G \to H$. Let $G = gp(X)$ and suppose $\theta_1|_X = \theta_2|_X$. Prove that $\theta_1 = \theta_2$.

Solution:

Let $g \in G$. Then $g = x_1^{\epsilon_1} \cdots x_n^{\epsilon_n}$ where $x_i \in X$ and $\epsilon_i = \pm 1$, and

$$g\theta_1 = (x_1\theta_1)^{\epsilon_1} \cdots (x_n\theta_1)^{\epsilon_n} = (x_1\theta_2)^{\epsilon_1} \cdots (x_n\theta_2)^{\epsilon_n} = g\theta_2$$

Thus θ_1 and θ_2 have the same effect on the elements of G, and so $\theta_1 = \theta_2$.

8.12. Find an example of a group G generated by a set X, a group H and a map $\theta: X \to H$, such that there exists no homomorphism $\hat{\theta}: G \to H$ with $\hat{\theta}|_X = \theta$.

Solution:

Let G be cyclic of order 2, $G = gp(\{x\})$. Put $X = \{x\}$ and let $H = gp(\{y\})$ be infinite cyclic. Define $\theta: X \to H$ by $x\theta = y$. Suppose there exists a homomorphism $\hat{\theta}: G \to H$, such that $x\hat{\theta} = y$. Then $G\hat{\theta}$ is a subgroup of H which contains $gp(y)$ and hence $G\hat{\theta} = H$. But $G\hat{\theta}$, being the image of a finite group, must be finite. Since H is infinite, this is clearly impossible.

8.13. Let H be the cyclic group of order 2, say $H = \{1, a\}$. Let F be freely generated by $\{x\}$. The map $\theta: x \to a$ gives rise to a homomorphism of $F \to H$. Describe the effect of this homomorphism on all elements of F. Check directly that it is a homomorphism, and find its kernel.

Solution:

The free group on a single generator is infinite cyclic. Its elements are uniquely of the form x^n, n an integer. Now $\hat{\theta}$ maps $x^n \to a^n$. If n is even, say $n = 2r$, $a^n = 1$. Hence $\hat{\theta}$ maps $x^n \to 1$. If n is odd, say $n = 2r+1$, $\hat{\theta}$ maps x^n to a.

To check that $\hat{\theta}$ is a homomorphism, consider whether $(x^n x^m)\hat{\theta} = x^{n+m}\hat{\theta}$. Now $x^{n+m}\hat{\theta}$ is a if $n+m$ is odd, and is 1 if $n+m$ is even. If $n+m$ is odd, then one of the integers n and m is odd and the other even. Hence $x^n\hat{\theta}x^m\hat{\theta} = a$, as one of $x^n\hat{\theta}, x^m\hat{\theta}$ is a while the other is 1. If $n+m$ is even, then either n and m are both even, or they are both odd; if n and m are both even, $x^n\hat{\theta}x^m\hat{\theta} = 1 \cdot 1 = 1$; if both are odd, then $x^n\hat{\theta}x^m\hat{\theta} = a \cdot a = 1$. Hence $\hat{\theta}$ is a homomorphism.

The kernel of $\hat{\theta} = \{x^{2r} \mid \text{all integers } r\}$.

8.14. Find a free group that has as a homomorphic image S_n, n any given positive integer.

Solution:

Let F be free on X where $|X| = n!$. Let $\theta: X \to S_n$ map each element of X onto a distinct element of S_n. Then by Theorem 8.4 there is a homomorphism $\hat{\theta}$ of F onto S_n that agrees with θ.

8.15. Prove that a group F freely generated by $n+1$ elements has as a homomorphic image a group G freely generated by n elements, where n is any given positive integer and G is a subgroup of F.

Solution:

Let F be freely generated by X where $|X| = n+1$. Let $X = X_1 \cup \{a\}$ where $|X_1| = n$. Then, as we have shown before, $gp(X_1) = G$ is freely generated by X_1. (Problem 8.7, page 247.) Let $\theta : X \to G$ be defined as follows: If $x \in X$, $x \neq a$, put $x\theta = x$. If $x = a$, put $x\theta = 1$, the identity. Then there exists a homomorphism $\hat{\theta}$ of F onto G by Theorem 8.4.

8.16. Let F be a free group with free generating set $\{a, b\}$. Let G be the direct product of two infinite cyclic groups generated by c and d respectively. The map which takes a to c and b to d extends to a homomorphism of F onto G (since $G = gp(\{c, d\})$). Prove that the kernel of this homomorphism is F', the derived group of F.

Solution:

First we show that if θ is the homomorphism $F \to G$ for which $a\theta = c$, $b\theta = d$, then $F' \subseteq \text{Ker}\,\theta$.

F' is generated by the set of all commutators, so it is sufficient to show that each commutator belongs to Ker θ. Now a commutator is of the form $[f_1, f_2] = f_1^{-1} f_2^{-1} f_1 f_2$ where $f_1, f_2 \in F$. Then $[f_1, f_2]\theta = [f_1\theta, f_2\theta] = 1$, as G is abelian. Hence $F' \subseteq \text{Ker}\,\theta$.

Now every element of F that does not belong to F' is of the form $a^r b^s e$ where not both r and s are zero and $e \in F'$, as F/F' is abelian and aF', bF' generate it. Under θ such an element goes onto $c^r d^s$ and $c^r d^s = 1$ only if both r and s are 0. Thus only the elements of F' belong to Ker θ. Therefore $\text{Ker}\,\theta = F'$.

8.17. Prove that a free group freely generated by n elements, n any positive integer, has a subgroup of index m for each positive integer m.

Solution:

We will exploit the homomorphism property. Let C_m be the cyclic group of order m generated by an element a, say. Let F be freely generated by $X = \{x_1, \ldots, x_n\}$. Then there exists a homomorphism θ of F onto C_m for which $x_i\theta = a$, $i = 1, \ldots, n$. Hence by the homomorphism theorem (Theorem 4.18, page 117),

$$F/\text{Ker}\,\theta \cong C_m$$

Since $|C_m| = m$, the number of cosets of Ker θ in F is m. Hence F has a subgroup of index m.

8.18. Let F be freely generated by X. Let Y be a subset of F such that $gp(Y) = F$, and $|X| = |Y| < \infty$. Use Theorem 8.4 to find an epimorphism of F onto itself.

Solution:

Let $x_1, \ldots, x_n$ be the elements of X, and $y_1, \ldots, y_n$ the elements of Y. Define a homomorphism $\theta : F \to F$ by $x_i\theta = y_i$ where $i = 1, \ldots, n$, using Theorem 8.4. Since $F\theta$ contains Y, $F\theta = F$. Hence θ is an epimorphism of F onto itself.

8.19. Let θ be an isomorphism of F with G. Let F be freely generated by X. Prove that G is freely generated by $X\theta$.

Solution:

Let $(x_1\theta)^{\epsilon_1} \cdots (x_n\theta)^{\epsilon_n}$ be a reduced product in $X\theta$. Then we must show that this reduced product is not the identity of G. Clearly

$$(x_1^{\epsilon_1} \cdots x_n^{\epsilon_n})\theta = (x_1\theta)^{\epsilon_1} \cdots (x_n\theta)^{\epsilon_n}$$

and as F is freely generated by X and θ is an isomorphism, the result follows.

8.20. Let F be freely generated by X and G freely generated by Y. If $\theta : X \to Y$ is a one-to-one correspondence, prove that $F \cong G$.

Solution:

Let $\hat{\theta}$ be the homomorphism of F into G which is an extension of θ. (Theorem 8.4.) Clearly $\hat{\theta}$ is onto G. Let $\phi : Y \to X$ be the mapping such that $\theta\phi$ is the identity mapping on X and $\phi\theta$ is the identity mapping on Y. Let $\hat{\phi}$ be the extension of ϕ to $G \to H$. Then $\hat{\theta}\hat{\phi}$ is the identity on F and $\hat{\phi}\hat{\theta}$ is the identity on G. It follows readily that $\hat{\theta}$ is an isomorphism of F with G.

8.21. Let F be freely generated by a, b, c. Let N be the normal subgroup generated by c, i.e. the intersection of all normal subgroups of F containing c. Prove that F/N is freely generated by aN and bN. (Hard.)

Solution:

Let G be free on x, y. Let θ be the homomorphism of F onto G defined by $a\theta = x$, $b\theta = y$ and $c\theta = 1$. (Since F is free on a, b, c, such a homomorphism θ exists by Theorem 8.4.) Let K be the kernel of θ. First we shall prove that $K = N$, the normal subgroup generated by c. Then we shall prove that aN and bN freely generate F/N.

To see that $K = N$ first observe that as $c\theta = 1$, N is contained in K (K is a normal subgroup containing c). On the other hand, suppose $f \in F$, $f \notin N$. Then f can be expressed in the form

$$f = f_1 n_1$$

where f_1 is a reduced $\{a, b\}$-product and $n_1 \in N$. But then

$$f\theta = f_1\theta$$

and $f_1\theta$ is a reduced $\{x, y\}$-product. So $f_1\theta \neq 1$, which means $f\theta \neq 1$ and so $f \notin K$. Thus if $f \notin N$, $f \notin K$ which implies K is contained in N. Hence as we observed earlier that N is contained in K, we find $K = N$ as desired.

Now a, b, c generate F. So aN, bN and cN generate F/N. Since $cN = N$, aN and bN generate F/N. We want to prove that aN and bN freely generate F/N. Suppose $(y_1N)^{\epsilon_1} \cdots (y_kN)^{\epsilon_k}$ is a reduced $\{aN, bN\}$-product.

Now θ gives rise to an isomorphism ν of F/K with G, i.e. the mapping defined by $(fK)\nu = f\theta$ (the homomorphism theorem, Theorem 4.18, page 117).

Observe that $(aK)\nu = x$ and $(bK)\nu = y$ so that

$$((y_1N)^{\epsilon_1} \cdots (y_kN)^{\epsilon_k})\nu = x_1^{\epsilon_1} \cdots x_k^{\epsilon_k}$$

where $(y_iN)\nu = x_i \in \{x, y\}$. The product $x_1^{\epsilon_1} \cdots x_k^{\epsilon_k}$ is a reduced $\{x, y\}$-product. But $\{x, y\}$ freely generates G. Thus $x_1^{\epsilon_1} \cdots x_k^{\epsilon_k} \neq 1$. Consequently

$$(y_1N)^{\epsilon_1} \cdots (y_kN)^{\epsilon_k} \neq N$$

and the proof is complete.

Instead of completing the proof this way we could refer to Problem 8.19 using the isomorphism ν^{-1}.

8.2 PRESENTATIONS OF GROUPS

a. Definitions

We have shown (Theorem 8.4) that if F is freely generated by X, then for every group G and every mapping θ of X into G there is a homomorphism of F into G which extends θ, i.e. which agrees with θ on X. This fact will enable us to "present" a given group in terms of a free group. This idea of a presentation is especially important in topology and analysis where groups arise in just this way, as the "groups of certain presentations".

First we need a definition. If S is any subset of a group, then the *normal closure of* S is defined to be the intersection of all normal subgroups of G containing S. Clearly the normal closure of S is a normal subgroup of G containing S. Thus the normal closure of S is often called the *normal subgroup generated by* S. It is easy to prove that the normal closure of S is $gp(g^{-1}sg \mid g \in G \text{ and } s \in S)$ (see Problem 8.22 below).

A *presentation* is defined to be a pair $(X; R)$ where X is a free set of generators of a free group F and R is a subset of F. The *group* of the *presentation* $(X; R)$ is F/N where N is the normal subgroup of F generated by R; we usually denote the group of a presentation $(X; R)$ by $|X; R|$. Finally a *presentation of a group* G consists, by definition, of a presentation $(X; R)$ and an isomorphism θ between $|X; R|$ and G. A presentation $(X; R)$ is *finite* if *both* X and R are finite, and a group G is termed *finitely presented* if it has a finite presentation. Not all groups are finitely generated (a necessary prerequisite for being finitely presented) and not all finitely generated groups are finitely presented. For a more detailed discussion of these notions the reader may consult R. H. Crowell and R. H. Fox, *An Introduction to Knot Theory,* Blaisdell, 1963.

b. Illustrations of presentations

We shall work some examples to illustrate the definitions of Section 8.2a.

First of all, suppose F is a free group freely generated by $X = \{a, b\}$. Then the following

(i) $(\{a, b\}; \{a, b\})$

(ii) $(\{a, b\}; \{a^2b^2, a^3b^3, a^4b^4, \ldots\})$

(iii) $(\{a, b\}; \{[a, b]\})$

are patently presentations (by the very definition). The presentations (i) and (iii) are finite, but the presentation (ii) is not.

The obvious question is: what are the groups of these presentations?

Clearly the group of (i) is a group of order 1.

What about (ii)? Well, we are interested in F/N where N is the normal subgroup generated by $a^2b^2, a^3b^3, a^4b^4, \ldots$. We have, since $a^2b^2 \in N$, $a^2N = b^{-2}N$. Furthermore $a^3N = b^{-3}N$ since $a^3b^3 \in N$. Thus it follows that $a^3N = a^2aN = b^{-2}aN = b^{-3}N$. Cancelling $b^{-2}N$ from both sides yields $aN = b^{-1}N$. Hence $ab \in N$. Indeed we would like to prove that N is the normal subgroup generated by ab. Let K be the normal subgroup generated by ab. Since $ab \in N$, we have $K \subseteq N$. Now observe that $aK = b^{-1}K$. Then

$$a^ib^iK \;=\; (a^iK)(b^iK) \;=\; (b^{-i}K)(b^iK) \;=\; K$$

This implies that $a^ib^i \in K$ $(i = 2, 3, \ldots)$. Hence N is contained in K and so we have proved that $K = N$. Therefore F/N is cyclic (because aN and bN clearly generate F/N; but $aN = b^{-1}N$). In fact F/N is infinite. To see this suppose G is an infinite cyclic group generated by g. Let θ be the mapping of $\{a, b\}$ into G defined by

$$a\theta \;=\; g, \quad b\theta \;=\; g^{-1}$$

Let $\hat{\theta}$ be the homomorphism of F into G defined by θ (Theorem 8.4). Clearly $\hat{\theta}$ is onto and $(ab)\hat{\theta} = 1$. So if L is the kernel of $\hat{\theta}$, $L \supseteq N$. But as $F/L \cong G$, F/L is infinite cyclic. Therefore F/N is also infinite cyclic since, as we have already noted, F/N is cyclic. (Actually, $N = L$; however, we don't need this fact here.)

Finally we come to (iii). In fact $|\{a, b\}; \{[a, b]\}|$ is free abelian of rank 2. The reader may attempt to prove this before we do so in a more general case. At this point we simplify our notation. Instead of using our set-theoretic notation which encloses the elements of a given set in braces { }, we shall omit the braces in writing presentations. Thus we write $(a, b; [a, b])$ for $(\{a, b\}; \{[a, b]\})$, $|a, b; [a, b]|$ for $|\{a, b\}; \{[a, b]\}|$, etc.

Let F be freely generated by $a_1, \ldots, a_n$. We shall prove that

$$|a_1, a_2, \ldots, a_n;\ [a_i, a_j] \text{ where } 1 \leqq i \leqq j \leqq n|$$

is free abelian of rank n (from which it follows that $|a, b; [a, b]|$ is free abelian of rank 2).

To this end let H be the free abelian group of rank n. Then H is the direct product of infinite cyclic groups generated by h_i, $i = 1, \ldots, n$, say. We define a homomorphism $\theta: F \to H$ by $a_i\theta = h_i$. Now $[a_i, a_j]\theta = [a_i\theta, a_j\theta] = [h_i, h_j] = 1$. Hence $[a_i, a_j] \in \operatorname{Ker}\theta$. Let N be the normal subgroup generated by the $[a_i, a_j]$, $1 \leqq i \leqq j \leqq n$. Clearly $N \subseteq \operatorname{Ker}\theta$. Note that $[a_iN, a_jN] = [a_i, a_j]N = N$. As the generators a_iN of F/N commute, the elements f of F are of the form

$$f \;=\; a_1^{r_1} \cdots a_n^{r_n}c \quad \text{where } c \in N$$

and if $f \notin N$, at least one of the $r_i \neq 0$. Hence as $c\theta = 1$, $f\theta = h_1^{r_1} \cdots h_n^{r_n}$. Now as one of the $r_i \neq 0$, $f\theta \neq 1$. It follows that $f \notin \text{Ker}\,\theta$, and so $N = \text{Ker}\,\theta$.

Therefore F/N is the free abelian group of rank n, i.e.

$$|a_1, \ldots, a_n;\ [a_i, a_j] \text{ with } 1 \leq i \leq j \leq n|$$

is the free abelian group of rank n.

Note that as N is generated as a normal subgroup by commutators, $N \subseteq F'$. But as F/N is abelian, $N \supseteq F'$. Thus $N = F'$ and we therefore conclude that F/F' is free abelian of rank n.

We state this fact as

Theorem 8.6: Let F be a free group freely generated by a set of n elements $(n < \infty)$. Then F/F' is a free abelian group of rank n.

Corollary 8.7: Let F be a free group. Suppose X and Y both freely generate F. If $|X|$ is finite, then so is $|Y|$ and $|X| = |Y|$.

Proof: By Theorem 8.6 we know that F/F' is a free abelian group of rank $n = |X|$. If $|Y| < \infty$, then F/F' is free abelian of rank $|Y|$, again by Theorem 8.6. Consequently $|X| = |Y|$ since the rank of an abelian group is unique (see Section 6.2d, page 193). It remains only to prove that $|Y|$ cannot be infinite. To do this, suppose $y_1, y_2, \ldots, y_{n+1} \in Y$. Now in a free abelian group of rank n every $n+1$ elements are dependent. Hence there exist integers $m_1, \ldots, m_{n+1}$, not all zero, such that

$$(y_1F')^{m_1} \cdots (y_{n+1}F')^{m_{n+1}} = F'$$

i.e. $y_1F', \ldots, y_{n+1}F'$ are dependent. Let A be the free abelian group on $a_1, \ldots, a_{n+1}$ and let θ be the homomorphism of F to A defined by

$$y_i\theta = a_i \ (i = 1, \ldots, n+1), \quad y\theta = 1 \text{ if } y \notin \{y_1, \ldots, y_{n+1}\} \text{ and } y \in Y$$

The kernel K of θ contains F' since F/K is abelian. Then

$$1 = (y_1\theta)^{m_1} \cdots (y_{n+1}\theta)^{m_{n+1}} = a_1^{m_1} \cdots a_{n+1}^{m_{n+1}}$$

But A is free abelian on $a_1, \ldots, a_{n+1}$. Hence $m_1 = m_2 = \cdots = m_{n+1} = 0$, a contradiction. Thus $|Y| < \infty$ and the corollary has been proved.

It follows from this corollary that if F is a finitely generated free group, then every pair of sets which freely generate F have the same number of elements. We define the *rank* of F to be this common number, i.e. the number of elements in any set which freely generates F. Note that free groups of the same rank are isomorphic (Problem 8.20).

We can easily give a presentation of A, a free abelian group on $a_1, \ldots, a_n$, with the results we now have. Let F be free on $x_1, \ldots, x_n$ and let θ be the isomorphism defined by $(x_jF')\theta = a_j$ $(j = 1, \ldots, n)$. Then

$$(x_1, \ldots, x_n;\ [x_i, x_j] \text{ with } 1 \leq i \leq j \leq n)$$

together with θ is a presentation of A.

In some of the following problems we will often be dealing with factor groups. A simple convention makes the arguments simpler to follow.

Let G be a group and N a normal subgroup of G. If we use some phrase such as "let us calculate modulo N," then by G we mean the factor group G/N. We shall mean by $g = h$ that $Ng = Nh$. If we say let M be a subgroup of G, what we really mean is "let M/N be a subgroup of G/N." In other words, we must remember that we are talking of a factor group and instead of writing the cosets, we will simply write the coset representative. (See Problem 8.24 for an example.)

Problems

8.22. If G is a group, show that the normal closure of a subset R ($\neq \emptyset$) of G is $gp(\{g^{-1}rg \mid g \in G$ and $r \in R\})$.

Solution:

Clearly $N = gp(\{g^{-1}rg \mid g \in G \text{ and } r \in R\})$ is a subgroup of G containing R. Also N is a normal subgroup of G. Finally any normal subgroup containing R must contain N. Thus the result follows.

8.23. Let G be a group with subgroups H and K and let $[G:H] = n < \infty$ and $[G:K] \leqq n$. Prove that if $H \supseteq K$, then $H = K$.

Solution:

Let $x_1 = 1$ and let the distinct cosets of H in G be $H = Hx_1, Hx_2, \ldots, Hx_n$. As $H \supseteq K$, it follows that $Kx_1, Kx_2, \ldots, Kx_n$ are distinct. As $[G:K] \leqq n$, $G = Kx_1 \cup Kx_2 \cup \cdots \cup Kx_n$. If $h \in H$, then $h \in Kx_i$ for some integer i. If $i \neq 1$, $H \cap Kx_i = \emptyset$ since $H \cap Hx_i = \emptyset$ for $i \neq 1$. Hence $i = 1$ and $h \in Kx_1 = K$. Accordingly $H \subseteq K$ and thus $H = K$.

8.24. Rewrite the first part of the argument of presentation (ii), page 254, using the modulo N convention introduced above. Also calculate modulo K and stop after proving $K = N$.

Solution:

We are interested in F/N where N is the normal subgroup generated by $a^2b^2, a^3b^3, a^4b^4, \ldots$. Let us calculate modulo N. Since $a^2b^2 = 1$, $a^2 = b^{-2}$. Furthermore $a^3 = b^{-3}$ since $a^3b^3 = 1$. Thus it follows that $a^3 = a^2a = b^{-2}a = b^{-3}$. Cancelling b^{-2} from both sides yields $a = b^{-1}$, and so $ab = 1$.

As we are calculating modulo N, this means that $ab \in N$. Indeed we would like to prove that N is the normal subgroup generated by ab. Let K be the normal subgroup generated by ab. Since $ab \in N$, we have $K \subseteq N$.

Now we calculate modulo K. $a = b^{-1}$, and so $a^ib^i = b^{-i}b^i = 1$ $(i = 2, 3, \ldots)$. This means as we are calculating modulo K that $a^ib^i \in K$ $(i = 2, 3, \ldots)$. Therefore we have proved that $K = N$.

8.25. Let $G = |a; a^2|$. Prove that G is cyclic of order 2.

Solution:

The free group F generated by a is infinite cyclic. Now $gp(a^2)$ is normal in F since F is abelian. Hence $gp(a^2)$ is the normal subgroup generated by a^2. It can be easily seen that $gp(a)/gp(a^2)$ is the cyclic group of order 2. Thus G is cyclic of order 2.

8.26. Prove that $G = |a; a^n|$, where n is any positive integer, is the cyclic group of order n (again with F freely generated by a).

Solution:

The free group F on a is infinite cyclic and $N = gp(a^n)$ is the normal subgroup generated by a^n. Therefore $F/N = G$ is cyclic of order n (Theorem 4.9, page 105).

8.27. Prove that $|a, b; a^2, b^2, [a, b]|$ is the direct product of two cyclic groups of order 2.

Solution:

Although it has not been stated, a and b are (as usual) free generators of a free group F. Let N be the normal subgroup generated by a^2, b^2 and $[a, b]$. In F/N, Na and Nb commute as $[Na, Nb] = N[a, b] = N$, since $[a, b] \in N$. Since F/N is generated by Na and Nb, it is therefore abelian. Also $(Na)^2 = N$ and $(Nb)^2 = N$. Let $A = \{N, Na\}$ and $B = \{N, Nb\}$. As A is a normal subgroup of F/N, AB is a subgroup of F/N which contains both Na and Nb. It follows that $AB = F/N$. Thus $|F/N| \leqq |A||B| \leqq 2 \cdot 2 = 4$. On the other hand there is a homomorphism θ of F onto the direct product of two cyclic groups of order 2. Clearly $\operatorname{Ker}\theta$ contains a^2, b^2 and $[a, b]$. Hence $\operatorname{Ker}\theta \supseteq N$. It follows that $N = \operatorname{Ker}\theta$ (Problem 8.23). Thus F/N is the direct product of two cyclic groups of order 2.

8.28. Find a presentation for S_3. (Hard.)

Solution:

Let $\rho = \begin{pmatrix} 1 & 2 & 3 \\ 2 & 1 & 3 \end{pmatrix}$ and $\sigma = \begin{pmatrix} 1 & 2 & 3 \\ 2 & 3 & 1 \end{pmatrix}$. Then $\sigma^2 = \begin{pmatrix} 1 & 2 & 3 \\ 3 & 1 & 2 \end{pmatrix}$. So $\rho, \rho\sigma, \rho\sigma^2$ and $1, \sigma, \sigma^2$ are six distinct elements and hence the whole of S_3. Now

$$\rho^{-1}\sigma\rho = \begin{pmatrix}1 & 2 & 3\\2 & 1 & 3\end{pmatrix}\begin{pmatrix}1 & 2 & 3\\2 & 3 & 1\end{pmatrix}\begin{pmatrix}1 & 2 & 3\\2 & 1 & 3\end{pmatrix} = \begin{pmatrix}1 & 2 & 3\\3 & 1 & 2\end{pmatrix} = \sigma^2$$

Thus the equation $\rho^{-1}\sigma\rho\sigma^{-2} = 1$ holds. Also $\rho^2 = 1$ and $\sigma^3 = 1$. We use only these equations and hope that they will give rise to a presentation for S_3.

Let F be the free group on a_1, a_2 and let $G = |a_1, a_2;\, a_1^2, a_2^3, a_1^{-1}a_2a_1a_2^{-2}|$. Furthermore let N be the normal subgroup of F generated by $a_1^2, a_2^3, a_1^{-1}a_2a_1a_2^{-2}$, and let $\theta: F \to S_3$ be defined by $a_1\theta = \rho$ and $a_2\theta = \sigma$. Then $N \subseteq \operatorname{Ker}\theta$.

Now we calculate modulo N. We see that $a_1^2 = 1$ and so $a_1^{-1} = a_1$. Since $a_2^3 = 1$, $a_2^{-1} = a_2^2$, $a_1^{-1}a_2a_1a_2^{-2} = 1$ from which $a_2a_1 = a_1a_2^2$. Now F is generated by a_1, a_2. Let $M = gp(a_2)$; then $M \lhd F$. It follows from Problem 4.62, page 114, that $M\{1, a_1\}$ is a subgroup of F and as it contains a_1 and a_2, $M\{1, a_1\} = F$. Thus the elements of F are $\{1, a_2, a_2^2\}\{1, a_1\}$, i.e. $1, a_2, a_2^2, a_1, a_2a_1, a_2^2a_1$ (although we do not know if they are distinct).

It follows that $|F/N| \leqq 6$. But $|F/\operatorname{Ker}\theta| = 6$ and $N \subseteq \operatorname{Ker}\theta$, and so $\operatorname{Ker}\theta = N$ (Problem 8.23). Then

$$|a_1, a_2;\, a_1^2, a_2^3, a_1^{-1}a_2a_1a_2^{-2}| \cong S_3$$

under the isomorphism $\phi: a_1N \to \rho,\ a_2N \to \sigma$ (by the homomorphism theorem, Theorem 4.18, page 117).

8.29. Prove that $|a, b;\, a^2, b^n, a^{-1}bab|$ is isomorphic to the dihedral group of order $2n$.

Solution:

Let G be the dihedral group of order $2n$. Then G is the symmetry group of the regular n-gon S (see Section 3.4f, page 75). Recall that σ_2 rotates S in a clockwise direction through an angle of $2\pi/n$. It follows that σ_2 is of order n. Put $\sigma = \sigma_2$. If τ is the reflection about A_1O, where A_1 is any vertex of S, and O the center of S, then $\tau^2 = \iota$. Moreover every element of G can be written in the form $\tau^\epsilon\sigma^\delta$, where $\epsilon = 0, 1$ and $\delta = 0, 1, \ldots, n-1$ (see Section 3.4f and note that $\sigma^i = \sigma_i$ for $1 \leqq i \leqq n$). Let θ be the homomorphism of the free group F, freely generated by a and b, onto G defined by $a\theta = \tau$ and $b\theta = \sigma$. Then $a^2\theta = 1$, $b^n\theta = 1$ and

$$(a^{-1}bab)\theta = \tau^{-1}\sigma\tau\sigma = \sigma^{-1}\sigma = \iota$$

Thus a^2, b^n and $a^{-1}bab$ lie in the kernel K of θ. Then N, the normal subgroup of F generated by a^2, b^n and $a^{-1}bab$, is contained in K. Moreover since θ is onto, $F/K \cong G$. If we can show that $N = K$, then the proof follows.

We calculate (compare Problem 8.28) modulo N. We shall show that $|F/N| \leqq 2n$. Since $|F/K| = 2n$ and $K \supseteq N$, this will establish that $K = N$ by Problem 8.23. Modulo N, $a^{-1}ba = b^{-1}$. Let $M = gp(b)$. Then $M \lhd F$. Thus $M\{1, a\}$ is a subgroup of F by Problem 4.62, page 114. As it contains both a and b, $M\{1, a\} = F$. The elements of M are $1, b, b^2, \ldots, b^{n-1}$. Therefore the elements of F are

$$a^\epsilon b^\delta \quad (\epsilon = 0, 1,\ \delta = 0, 1, \ldots, n-1)$$

Since we are calculating modulo N, F really stands for F/N. This implies $|F/N| \leqq 2n$. Thus the proof is complete.

8.30. Prove that $G = |a, b;\, a^2, a^{-1}bab|$ is infinite. (This group is called the infinite dihedral group. *Hint*: Show that each dihedral group D_n is a homomorphic image of G.)

Solution:

Let F be the free group on a and b and let θ_n be a homomorphism of F onto D_n (as D_n is a two generator group (see, for example, Problem 8.29), we know such a homomorphism exists). Also $\operatorname{Ker}\theta_n \supseteq \{a^2, a^{-1}bab\}$. Thus $\operatorname{Ker}\theta_n \supseteq N$, the normal subgroup generated by $a^2, a^{-1}bab$. Therefore $G = F/N$ has each D_n as a homomorphic image $(D_n \cong (F/N)/(\operatorname{Ker}\theta_n/N))$. If G were finite of order k, say, we would have a contradiction. For then $G\theta_k$ as a homomorphic image of a group of order k is of order $\leqq k$. But $G\theta_k = D_k$ is of order $2k$. This contradiction proves that G is of infinite order.

8.31. Prove that every finitely generated group has a presentation.

Solution:

In Section 8.1d we proved that every finitely generated group is a factor group of a free group. We will use this fact to prove that every finitely generated group has a presentation. Let G be an arbitrary group. Let X be chosen so that X freely generates a free group F and, furthermore, so that there is a mapping θ of X onto one of the finite sets of generators for G. Let ϕ be the homomorphism of F onto G such that ϕ agrees with θ on X and let $N = \text{Ker}\,\phi$. Then $(X; N)$ together with the isomorphism $\mu: fN \to f\phi$ is a presentation of G.

8.32. Let G be a finite group with elements $1 = x_1, x_2, \ldots, x_n$. Suppose $x_i x_j = x_{(i,j)}$, where $x_{(i,j)} \in \{x_1, \ldots, x_n\}$ (in other words (i,j) is an integer between 1 and n). Let F be the free group freely generated by $a_1, \ldots, a_n$, N be the normal subgroup of F generated by $a_i a_j a_{(i,j)}^{-1}$ $(1 \leq i, j \leq n)$, and θ be the homomorphism from

$$|a_1, \ldots, a_n;\ a_i a_j a_{(i,j)}^{-1} \text{ where } 1 \leq i, j \leq n|$$

to G defined by $(a_k N)\theta = x_k$. Show that

$$(a_1, \ldots, a_n;\ a_i a_j a_{(i,j)}^{-1} \text{ where } 1 \leq i, j \leq n)$$

together with θ is a presentation of G.

Solution:

Since $a_i a_j N = a_{(i,j)} N$, every product of a's in which there are no negative exponents is equal, modulo N, to some a_k. Now if $x_i^{-1} = x_j$, then $x_i x_j = x_{(i,j)} = x_1$, and so $a_i a_j = a_1$ modulo N. But $x_1 x_1 = x_{(1,1)} = x_1$. Thus

$$a_1 a_1 = a_1 \text{ modulo } N$$

This means that $a_1 \in N$ and so $a_i^{-1} = a_j$ modulo N. Therefore every product of the a_k involving both positive and negative exponents can be replaced, modulo N, by a product involving only positive exponents. Accordingly every product of the a_k and their inverses is equal, modulo N, to an a_l. It follows that

$$|F/N| \leq n = |G|$$

Now let ϕ be the homomorphism from F to G defined by $a_i\phi = x_i$ (for $1 \leq i \leq n$). Then

$$(a_i a_j a_{(i,j)}^{-1})\phi = x_i x_j x_{(i,j)}^{-1} = 1$$

and so $\text{Ker}\,\phi$ includes $a_i a_j a_{(i,j)}^{-1}$. Therefore $\text{Ker}\,\phi \supseteq N$. Since

$$|F/\text{Ker}\,\phi| = n \ (< \infty)$$

and $|F/N| \leq n$, it follows by Problem 8.23 that $N = \text{Ker}\,\phi$. Thus the mapping

$$\theta:\ a_k N \to x_k$$

is indeed an isomorphism (Theorem 4.18, page 117).

8.33. Prove that $|x, y;\ x^2y^2, y^2| = |x, y;\ x^2, y^2|$.

Solution:

We have a free group F freely generated by x and y and two normal subgroups N and M generated, as normal subgroups, respectively by x^2y^2 and y^2, and x^2 and y^2. We must prove $N = M$.

Since N contains x^2y^2 and y^2, it contains $x^2y^2(y^2)^{-1} = x^2$. Thus $N \supseteq M$. But $M \supseteq \{x^2y^2\}$ since $M \supseteq \{x^2, y^2\}$. Hence $M \supseteq N$ and $M = N$.

8.34. Show that a free group of rank $n < \infty$ cannot be generated by $n - 1$ elements.

Solution:

Let G be free of rank n. Then by Theorem 8.6, G/G' is free abelian of rank n. If G can be generated by $n - 1$ elements, G/G' can be generated by $n - 1$ elements. But this contradicts Problem 6.41(b), page 195.

8.3 THE SUBGROUP THEOREM FOR FREE GROUPS: AN EXAMPLE

The object of the next sections of this chapter is to prove that every subgroup of a free group is a free group. This theorem is one of the more difficult in this book. So in order to give the reader a chance to become accustomed to the ideas involved, we first work an example.

Example 1: Let F be the free group freely generated by two elements a and b. Consider the set $Y = \{a^{-1}ba, a^{-2}ba^2, \ldots\}$. Show that $H = gp(Y)$ is freely generated by Y. (Thus we see that a free group freely generated by two elements has a subgroup which is freely generated by infinitely many elements.)

Proof: Let $a^{-i}ba^i = y_i$ $(i = 0, 1, 2, \ldots)$. Consider any Y-reduced product, where $Y = \{y_i \mid i = 0, 1, \ldots\}$. Say $f = y_{1'}^{\epsilon_1} \cdots y_{n'}^{\epsilon_n}$, where $1', \ldots, n'$ are positive integers, and $y_{j'} = y_{(j+1)'}$ implies $\epsilon_j \neq -\epsilon_{j+1}$. (Given, for example, $y_3y_4^{-1}y_1y_2$, then $1' = 3$, $2' = 4$, $3' = 1$ and $4' = 2$; $\epsilon_1 = 1$, $\epsilon_2 = -1$, $\epsilon_3 = 1$ and $\epsilon_4 = 1$.) We will prove that $f \neq 1$ by induction on n, showing that H is freely generated by Y.

Our inductive hypothesis (on n, the number of y_i's that go into a given reduced product f) is that f when expressed as a reduced product in $\{a, b\}$ ends in $b^{\epsilon_n}a^{n'}$. (For example,

$$\begin{aligned} y_3y_4^{-1}y_1y_2 &= a^{-3}ba^3 \cdot a^{-4}b^{-1}a^4 \cdot a^{-1}ba \cdot a^{-2}ba^2 \\ &= a^{-3}ba^{-1}b^{-1}a^3ba^{-1}ba^2 \end{aligned}$$

as a reduced product in $\{a, b\}$, and, as asserted, it ends in ba^2.)

If $n = 1$ this is certainly true.

If it is true for $n = k$, let $n = k+1$. Then $y_{1'}^{\epsilon_1} \cdots y_{k'}^{\epsilon_k}$ ends in $b^{\epsilon_k}a^{k'}$ when expressed as a reduced product in $\{a, b\}$, i.e. $y_{1'}^{\epsilon_1} \cdots y_{k'}^{\epsilon_k} = zb^{\epsilon_k}a^{k'}$, where z is a reduced product in a and b such that $zb^{\epsilon_k}a^{k'}$ is a reduced product (i.e. z does not end in b^{-1} if $\epsilon_k = 1$ or, if $\epsilon_k = -1$, z does not end in b). If now $k' = (k+1)'$, then, as f is a reduced product, $\epsilon_k \neq -\epsilon_{k+1}$. Since ϵ_k and ϵ_{k+1} are the numbers 1 or -1, $\epsilon_k = \epsilon_{k+1}$. Thus

$$\begin{aligned} y_{1'}^{\epsilon_1} \cdots y_{(k+1)'}^{\epsilon_{k+1}} &= zb^{\epsilon_k}a^{k'}a^{-(k+1)'}b^{\epsilon_{k+1}}a^{(k+1)'} \\ &= zb^{\epsilon_k}a^{k'}a^{-k'}b^{\epsilon_{k+1}}a^{(k+1)'} = zb^{\epsilon_k}b^{\epsilon_{k+1}}a^{(k+1)'} \end{aligned}$$

Since $\epsilon_{k+1} = \epsilon_k$, this last expression is a reduced product in $\{a, b\}$ and f ends in $b^{\epsilon_{k+1}}a^{(k+1)'}$. If however $k' \neq (k+1)'$, then

$$y_{1'}^{\epsilon_1} \cdots y_{(k+1)'}^{\epsilon_{k+1}} = zb^{\epsilon_k}a^{k'}a^{-(k+1)'}b^{\epsilon_{k+1}}a^{(k+1)'} = zb^{\epsilon_k}a^{(k'-(k+1)')}b^{\epsilon_{k+1}}a^{(k+1)'}$$

Since $k' - (k+1)' \neq 0$, the last expression is a reduced product and so f expressed as a reduced product, ends in $b^{\epsilon_{k+1}}a^{(k+1)'}$. It follows therefore that in both situations $f \neq 1$. Thus Y freely generates H. Similar arguments will help to prove in general that a subgroup of a free group is free.

Problems

8.35. Verify both the inductive assumption of the preceding example and that $f \neq 1$ where

$$f = y_2^{-1}y_3^{-1}y_1y_1y_3$$

Solution:

$$f = a^{-2}b^{-1}a^2 \cdot a^{-3}b^{-1}a^3 \cdot a^{-1}ba \cdot a^{-1}ba \cdot a^{-3}ba^3 = a^{-2}b^{-1}a^{-1}b^{-1}a^2bba^{-2}ba^3$$

Clearly $f \neq 1$. The inductive assumption of the preceding example is that f when expressed as a reduced product should end in ba^3, which it does.

8.36. Given the existence of a free group freely generated by two elements, prove the existence of free groups freely generated by any finite number of elements.

Solution:

In Example 1 we have proved that a free group freely generated by two elements has a subset Y such that Y is infinite and $H = gp(Y)$ is freely generated by Y. If n is any positive integer let $y_1, \ldots, y_n$ be n distinct elements of Y. Then $gp(\{y_1, \ldots, y_n\})$ is easily shown to be freely generated by $y_1, \ldots, y_n$. Thus there exist free groups of rank n for each positive integer n.

8.37. Let F be freely generated by a and b. Prove that the subgroup of F generated by aba^3 and a^2b is freely generated by aba^3 and a^2b. (Hard.)

Solution:

Let $Y = \{aba^3, a^2b\}$ and let $y_1 = aba^3$, $y_2 = a^2b$. Consider a reduced Y-product

$$f = y_{1'}^{\epsilon_1} \cdots y_{n'}^{\epsilon_n} \quad (i' \in \{1,2\})$$

Then $j' = (j+1)'$ implies $\epsilon_j \neq -\epsilon_{j+1}$.

We will show by induction on n that if $n' = 1$, then f ends in a^3 if $\epsilon_n = 1$ and ends in $b^{-1}a^{-1}$ if $\epsilon_n = -1$; while if $n' = 2$, then f ends in b if $\epsilon_n = 1$ and ends in $b^{-1}a^{-2}$ if $\epsilon_n = -1$.

For $n = 1$ this is certainly true.

If it is true for $n = k$, we must proceed case by case in proving it true for $k+1$.

(i) $k' = 1$.

(*a*) $\epsilon_k = 1$. Then by the inductive hypothesis $y_{1'}^{\epsilon_1} \cdots y_{k'}^{\epsilon_k}$ ends in a^3. If now $(k+1)' = 1$, then as f is a reduced product we must have $\epsilon_{k+1} = 1$ and f ends in $a^3 \cdot aba^3 = aaaabaaa$. Thus f ends in a^3 as required. If $(k+1)' = 2$ and $\epsilon_{k+1} = 1$, then f ends in $a^3 \cdot a^2b$, and so f ends in b as required. If $\epsilon_{k+1} = -1$, then f ends in $a^3 \cdot b^{-1}a^{-2}$ and so f ends in $b^{-1}a^{-2}$ as required.

(*b*) $\epsilon_k = -1$. Then by the inductive hypothesis $y_{1'}^{\epsilon_1} \cdots y_{k'}^{\epsilon_k}$ ends in $b^{-1}a^{-1}$. If now $(k+1)' = 1$, then as f is a reduced product $\epsilon_{k+1} = -1$ and f ends in

$$b^{-1}a^{-1} \cdot a^{-3}b^{-1}a^{-1} = b^{-1}a^{-4}b^{-1}a^{-1}$$

Hence f ends in $b^{-1}a^{-1}$ as required. If $(k+1)' = 2$ and $\epsilon_{k+1} = 1$, then f ends in $b^{-1}a^{-1}a^2b = b^{-1}ab$, and so f ends in b as required. If $(k+1)' = 2$ and $\epsilon_{k+1} = -1$, then f ends in $b^{-1}a^{-1} \cdot b^{-1}a^{-2}$ and so f ends in $b^{-1}a^{-2}$ as required.

(ii) $k' = 2$.

(*a*) $\epsilon_k = 1$. Then by our inductive hypothesis $y_{1'}^{\epsilon_1} \cdots y_{k'}^{\epsilon_k}$ ends in b. If $(k+1)' = 1$ and $\epsilon_{k+1} = 1$, f ends in $baba^3$ and so f ends in a^3 as required. If $(k+1)' = 1$ and $\epsilon_{k+1} = -1$, then f ends in $ba^{-3}b^{-1}a^{-1}$ and f ends in $b^{-1}a^{-1}$ as required. If $(k+1)' = 2$, then as f is a reduced product $\epsilon_{k+1} = 1$ and f ends in ba^2b, and so f ends in b as required.

(*b*) $\epsilon_k = -1$. Then by our inductive hypothesis $y_{1'}^{\epsilon_1} \cdots y_{k'}^{\epsilon_k}$ ends in $b^{-1}a^{-2}$. If $(k+1)' = 1$ and $\epsilon_{k+1} = 1$, then f ends in $b^{-1}a^{-2}aba^3 = b^{-1}a^{-1}ba^3$, and so f ends in a^3 as required. If $(k+1)' = 1$ and $\epsilon_{k+1} = -1$, then f ends in $b^{-1}a^{-2} \cdot a^{-3}b^{-1}a^{-1} = b^{-1}a^{-5}b^{-1}a^{-1}$, and so f ends in $b^{-1}a^{-1}$ as required. If $(k+1)' = 2$, then as f is a reduced product $\epsilon_{k+1} = -1$. Thus f ends in $b^{-1}a^{-2} \cdot b^{-1}a^{-2}$ and so f ends in $b^{-1}a^{-2}$ as required.

Thus we have proved by induction that any reduced product always ends in one of $a^3, b^{-1}a^{-1}, b$ and $b^{-1}a^{-2}$. Hence $f \neq 1$ and Y freely generates $gp(Y)$.

8.4 PROOF OF THE SUBGROUP THEOREM FOR FREE GROUPS

a. Plan of the proof

The subgroup theorem for free groups (due to J. Nielsen and O. Schreier) may be stated as follows.

Theorem 8.8: Every subgroup H of a free group F is free.

Suppose that F is freely generated by S. We know from Section 7.6b, page 228, that if X is any right transversal of H in F, then the nonunit elements

$$a_{x,s} = xs(\overline{xs})^{-1} \quad (x \in X,\ s \in S)$$

(where, for $f \in F$, $\bar{f}$ is the unique element of X in the coset Hf) generate H. Let

$$Y = \{a_{x,s} \mid x \in X,\ s \in S,\ a_{x,s} \neq 1\}$$

Then we shall prove that H is actually freely generated by Y provided X *is chosen appropriately.* Thus there are two main steps in the proof:

(i) Choose X appropriately.

(ii) Prove that Y freely generates H.

Step (ii) of the proof will be broken into two parts: the first part requires a careful look at the elements $a_{x,s}$ and the second involves looking at the way in which products of these $a_{x,s}$ and their inverses interact.

b. Schreier transversals

Suppose that X is a transversal for H in F, where F is freely generated by a set S. Every element x $(x \neq 1)$ in X may be expressed uniquely as a reduced S-product

$$x = a_1a_2 \cdots a_n \quad (n \geq 1)$$

where a_i is an element of S or the inverse of an element of S. Recall that n is termed the *length* of x and that the length of 1 is 0. We shall call the elements

$$1, a_1, a_1a_2, \ldots, a_1a_2 \cdots a_n$$

initial segments of x.

Definition: A right transversal X is called a right Schreier transversal if every initial segment of an element in X is also in X.

Notice that it follows that if X is a right Schreier transversal, then $1 \in X$.

The main result of this section is the following.

Lemma 8.9: Suppose that F is a free group freely generated by S and that H is a subgroup of F. Then we can always find a right Schreier transversal X for H in F.

Proof: We say that a right coset Hf is of length n if there is an element in Hf of length n but no element of length less than n. We shall choose X inductively using the lengths of the cosets of H in F.

First if Hf is of length 0, then $1 \in Hf$ and so $Hf = H$. We choose 1 to be the representative of H.

Suppose that $n > 0$ and that for each coset of length less than n, representatives have already been chosen so that every initial segment of a representative is again a representative. We choose now representatives for the cosets of length n. Let Hf be a coset of length n and let $a_1a_2 \cdots a_n$ be an element in Hf of length n. The element $a_1a_2 \cdots a_{n-1}$ is of length $n-1$. Thus the coset $Ha_1a_2 \cdots a_{n-1}$ is of length at most $n-1$ (since $a_1a_2 \cdots a_{n-1} \in Ha_1a_2 \cdots a_{n-1}$). This means that the representative of $Ha_1a_2 \cdots a_{n-1}$ has already been chosen, by our induction assumption. Suppose this representative is $b_1b_2 \cdots b_m$. Now

$$H(b_1b_2 \cdots b_ma_n) = (Hb_1b_2 \cdots b_m)a_n = (Ha_1a_2 \cdots a_{n-1})a_n = Ha_1a_2 \cdots a_n$$

We select $b_1b_2 \cdots b_ma_n$ to be the representative of the coset $Ha_1a_2 \cdots a_n$. The initial segments of $b_1b_2 \cdots b_ma_n$, excluding $b_1b_2 \cdots b_ma_n$, are

$$1, b_1, \ldots, b_1b_2 \cdots b_m$$

which we know have already been chosen as representatives. In the same way we select suitable representatives for all the cosets of length n.

We have therefore verified the induction hypothesis and so we are able, in this way, to complete the choice of a right Schreier transversal for H in F.

c. A look at the elements $a_{x,s}$

Suppose that we have chosen a right Schreier transversal X for H in F. Consider a nonunit element $a_{x,s}$ where $x \in X$ and $s \in S$:

$$a_{x,s} = xs(\overline{xs})^{-1}$$

Now let the reduced S-product for x be $x = a_1 \cdots a_k$ where a_i or its inverse belongs to S. We will allow $k = 0$; this will be interpreted as $x = 1$. Thus we may write

$$a_{x,s} = a_1 a_2 \cdots a_k s(\overline{a_1 a_2 \cdots a_k s})^{-1}$$

Let $\overline{a_1 \cdots a_k s} = b_1 \cdots b_l$ where the right-hand side is a reduced S-product with b_i or $b_i^{-1} \in S$, $i = 1, \ldots, l$. We assert that $a_1 \cdots a_k s$ and $b_1 \cdots b_l s^{-1}$ are not elements of X. For suppose $a_1 \cdots a_k s \in X$, then $\overline{a_1 \cdots a_k s} = a_1 \cdots a_k s$ and so $a_{x,s} = 1$; whereas if $b_1 \cdots b_l s^{-1} \in X$, we utilize equation (*7.3*), page 219. and conclude that

$$\begin{aligned}(a_{x,s})^{-1} &= \overline{xs}s^{-1}x^{-1} = \overline{xs}s^{-1}(\overline{\overline{xs}s^{-1}})^{-1} = \overline{xs}s^{-1}(\overline{\overline{xs}s^{-1}})^{-1} \\ &= b_1 \cdots b_l s^{-1}(\overline{b_1 \cdots b_l s^{-1}})^{-1} = b_1 \cdots b_l s^{-1}(b_1 \cdots b_l s^{-1})^{-1} = 1\end{aligned}$$

From this it follows that the reduced S-product for $a_{x,s}$ is $a_1 \cdots a_k s b_l^{-1} \cdots b_1^{-1}$. For if not, s cancels either with a_k or with b_l^{-1}. But as every initial segment of an element of X belongs to X, either $a_1 \cdots a_k s \in X$ or $b_1 \cdots b_l s^{-1} \in X$, which is a contradiction. Note we have proved $a_{x,s} \neq 1$ implies x does not end in s^{-1} and $\overline{xs}$ does not end in s.

Let $W = \{w \mid w \in S \text{ or } w^{-1} \in S\}$. Then we have the following

Lemma 8.10: Let $\epsilon = 1$ or -1. Then if $a_{x,s} \neq 1$,

$$a_{x,s}^{\epsilon} = c_1 \cdots c_m w d_n^{-1} \cdots d_1^{-1}$$

where the right-hand side is a reduced product and both $c_1 \cdots c_m$ and $d_1 \cdots d_n$ are elements of X and both $c_1 \cdots c_m w$ and $d_1 \cdots d_n w^{-1}$ lie outside X.

Proof: The lemma is an immediate consequence of the preceding remarks. We need only note that

$$a_{x,s}^{-1} = b_1 \cdots b_l s^{-1} a_k^{-1} \cdots a_1^{-1}$$

for the case $\epsilon = -1$.

Corollary 8.11: Suppose $\epsilon = \pm 1$ and

$$a_{x,s}^{\epsilon} = c_1 \cdots c_m w d_n^{-1} \cdots d_1^{-1}$$

where the right-hand side is a reduced product, $c_1 \cdots c_m \in X$ and $c_1 \cdots c_m w \notin X$. Then

(i) if $w \in S$, then $\epsilon = 1$, $x = c_1 \cdots c_m$ and $s = w$,

(ii) if $w \notin S$, then $\epsilon = -1$, $x = d_1 \cdots d_n$ and $s = w^{-1}$.

Proof: It follows immediately from the preceding argument.

Corollary 8.12: Let $\epsilon = \pm 1$, $\eta = \pm 1$, $x, y \in X$, and $s, t \in S$. If

$$a_{x,s}^{\epsilon} = c_1 \cdots c_m w d_n^{-1} \cdots d_1^{-1} \quad \text{and} \quad a_{y,t}^{\eta} = c_1 \cdots c_m w e_p^{-1} \cdots e_1^{-1}$$

with the right-hand sides reduced products, and $c_1 \cdots c_m \in X$ but $c_1 \cdots c_m w \notin X$, then $\epsilon = \eta$, $x = y$ and $t = s$.

Proof: It follows from Corollary 8.11.

d. The proof of the subgroup theorem

Suppose again that F is a free group freely generated by S and that H is a subgroup of F. Choose a right Schreier transversal X for H in F. Then

$$Y = \{a_{x,s} \mid x \in X,\ s \in S,\ a_{x,s} \neq 1\}$$

generates H. It remains only to prove that Y freely generates H. For this it suffices to prove that any reduced Y-product is not the identity (Lemma 8.2, page 248).

Let $g = a^{\epsilon_1}_{x_1,s_1} \cdots a^{\epsilon_r}_{x_r,s_r}$ $(x_1, \ldots, x_r \in X$ and $s_1, \ldots, s_r \in S)$ be a reduced Y-product. Let $a^{\epsilon_r}_{x_r,s_r}$ be expressed in the form of Lemma 8.10 as $e_1 \cdots e_m v f_n^{-1} \cdots f_1^{-1}$, where $v \in W$ and $e_1 \cdots e_m \in X$ but $e_1 \cdots e_m v \notin X$. We will show that g ends in $vf_n^{-1} \cdots f_1^{-1}$ by induction on r, the case $r = 1$ being of course proved in Lemma 8.10.

If the result is assumed true for $r - 1$ and $a^{\epsilon_{r-1}}_{x_{r-1},s_{r-1}}$ is expressed as $c_1 \cdots c_k w d_l^{-1} \cdots d_1^{-1}$ in the form of Lemma 8.10 with $c_1 \cdots c_k \in X$ but $c_1 \cdots c_k w \notin X$ $(w \in W)$, then $a^{\epsilon_1}_{x_1,s_1} \cdots a^{\epsilon_{r-1}}_{x_{r-1},s_{r-1}}$ ends in $wd_l^{-1} \cdots d_1^{-1}$.

Consider the product

$$wd_l^{-1} \cdots d_1^{-1} e_1 \cdots e_m v f_n^{-1} \cdots f_1^{-1}$$

We can convert it into a reduced product by successively deleting inverse pairs. If $m = l$ and $wd_l^{-1} \cdots d_1^{-1} e_1 \cdots e_m v = 1$, then by Corollary 8.12, $a^{\epsilon_{r-1}}_{x_{r-1},s_{r-1}} = (a^{\epsilon_r}_{x_r,s_r})^{-1}$, contrary to the assumption that g is a reduced Y-product. If $e_1 \cdots e_m v$ is removed as a result of deleting inverse pairs in the product $d_l^{-1} \cdots d_1^{-1} e_1 \cdots e_m v$, then $e_1 \cdots e_m v$ is an initial segment of $d_1 \cdots d_l$ and hence an element of X. But this is contrary to Lemma 8.10. Similarly $wd_l^{-1} \cdots d_1^{-1}$ is not removed as a result of deleting inverse pairs in the product $wd_l^{-1} \cdots d_1^{-1} e_1 \cdots e_m$. Thus

$$wd_l^{-1} \cdots d_1^{-1} e_1 \cdots e_m v f_n^{-1} \cdots f_1^{-1} = w \cdots v f_n^{-1} \cdots f_1^{-1}$$

when expressed as a reduced product by deleting inverse pairs (the $\cdots$ between w and v represent the factors $d_l^{-1} \cdots d_1^{-1} e_1 \cdots e_m$ left after deleting inverse pairs). Consequently g ends in $w \cdots v f_n^{-1} \cdots f_1^{-1}$ and the inductive assertion follows. Therefore $g \neq 1$, and the result follows.

e. Subgroups of finite index

In Section 7.6b, page 228, we proved that a subgroup of index n in a group generated by r elements is generated by nr elements. We shall now find the rank of a subgroup of index n in a free group. To find the rank of the subgroup we use the result of Section 8.4d, i.e. the nonunit $a_{x,s}$ freely generate the subgroup.

Let F be freely generated by $s_1, \ldots, s_r$ and let $X = \{x_1, \ldots, x_n\}$ with $x_1 = 1$ be a Schreier transversal for a subgroup H of index n. Consider the elements

$$a_{x_i,s_j} = x_i s_j (\overline{x_i s_j})^{-1}$$

$i = 1, \ldots, n$ and $j = 1, \ldots, r$. The number of such elements is nr. To find the rank of H, we wish to determine how many of the a_{x_i,s_j} are unit elements. By line 13, page 262:

(1) If x_i ends in s_j^{-1}, then $a_{x_i,s_j} = 1$.

(2) If $\overline{x_i s_j}$ ends in s_j, then $a_{x_i,s_j} = 1$.

We show that (1) and (2) are mutually exclusive. Suppose that x_i ends in s_j^{-1}. Then $x_i = w_1 \cdots w_m s_j^{-1}$ (where $w_1, \ldots, w_m \in W$) is a reduced product. Consequently $w_m \neq s_j$ (for otherwise $w_1 \cdots w_m s_j^{-1}$ is not a reduced product). As x_i is in the Schreier transversal X, $w_1 \cdots w_m$ also belong to the Schreier transversal. Thus $\overline{x_i s_j} = \overline{w_1 \cdots w_m} = w_1 \cdots w_m$ and $\overline{x_i s_j}$ does not end in s_j. Therefore (1) and (2) are mutually exclusive.

Note that if neither x_i ends in s_j^{-1} nor $\overline{x_i s_j}$ ends in s_j, then $a_{x_i,s_j} = x_i s_j (\overline{x_i s_j})^{-1} \neq 1$ as s_j remains when a_{x_i,s_j} is expressed as a reduced product.

For fixed j, let α_j = number of $x_i \in X$ for which $a_{x_i,s_j} = 1$. Clearly α_j = number of $x_i \in X$ for which x_i ends in s_j^{-1} plus the number of x_i for which $\overline{x_i s_j}$ ends in s_j. As x_i runs through X, $\overline{x_i s_j}$ runs through X $(x \to \overline{x s_j},\ x \in X$, is a permutation, page 219). Thus the number of x_i such that $\overline{x_i s_j}$ ends in s_j is the number of elements in X which end in s_j. We conclude α_j = number of x_i that end in s_j or s_j^{-1}. So the *total* number (i.e. with $j = 1, \ldots, r$) of $a_{x_i,s_j} = 1$ is $\alpha_1 + \cdots + \alpha_r$ = number of elements of X that end in s_j or s_j^{-1}, $j = 1, \ldots, r$. But except for $x_1 = 1$, every element of X ends in some s_j or s_j^{-1}. Hence $\alpha_1 + \cdots + \alpha_r = n - 1$.

Thus there are exactly $n-1$ unit elements among the a_{x_i, s_j}, and we have proved

Theorem 8.13: Let F be a free group of rank r and let H be a subgroup of index n. Then H is a free group of rank $n(r-1)+1$.

Problems

8.38. Let F be freely generated by $s_1, \ldots, s_r$. Let H be a subgroup of index 2 such that $s_1 \notin H$, but $s_i \in H$ for $i = 2, \ldots, r$. Find a set of free generators for H by using the method of Section 8.4d. Verify that the number of free generators agrees with the number given by Theorem 8.13.

Solution:

We choose $\{1, s_1\}$ for the Schreier transversal. Then the nonunit elements among a_{1,s_j} and a_{s_1,s_j} $(j = 1, \ldots, r)$ freely generate H. Now $a_{1,s_j} = 1s_j(\overline{1s_j})^{-1}$. If $j = 1$, $a_{1,s_j} = 1$. If $j \neq 1$, $a_{1,s_j} = 1s_j(\overline{1s_j})^{-1} = s_j$ as $s_j \in H$ implies $\overline{s_j} = 1$. Now

$$\begin{aligned} a_{s_1,s_j} &= s_1 s_j(\overline{s_1 s_j})^{-1} = s_1 s_j s_1^{-1} \text{ if } j \neq 1 \text{ (for then } s_1 s_j \in Hs_1) \\ &= s_1^2 \text{ if } j = 1 \text{ (as } \overline{s_1^2} = 1) \end{aligned}$$

Thus the subgroup H is freely generated by $s_1^2, s_2, \ldots, s_r$ and $s_1 s_2 s_1^{-1}, \ldots, s_1 s_r s_1^{-1}$. Thus H is of rank $2(r-1)+1$, which agrees with Theorem 8.13.

8.39. Let F be freely generated by x and y. Find a set of free generators for F', the derived group of F.

Solution:

Let $X = \{x^r y^s \mid r \text{ and } s \text{ integers}\}$. Then to show that X is a Schreier transversal we need only show that (1) $x^r y^s, x^{r_1} y^{s_1}$ belong to the same coset of F' only if $r = r_1$, $s = s_1$, and (2) X is a set of coset representatives. Both (1) and (2) follow easily on using the fact that F/F' is free abelian with basis xF', yF' (by Theorem 8.6, page 255). The free generators of F' are the elements $a_{x^r y^s, x}$ and $a_{x^r y^s, y}$ which are nonunit. Now

$$a_{x^r y^s, x} = x^r y^s x(\overline{x^r y^s x})^{-1} = x^r y^s x(x^{r+1} y^s)^{-1} = x^r y^s x y^{-s} x^{-r-1} \neq 1 \quad \text{for } s \neq 0$$

On the other hand,

$$a_{x^r y^s, y} = x^r y^s y(\overline{x^r y^{s+1}})^{-1} = 1$$

Thus a set of free generators for F' are the elements $x^r y^s x y^{-s} x^{-r-1}$ for all integers r and all integers $s \neq 0$.

8.40. Let F be a free group of rank r and H and K subgroups of index n. Prove that $H \cong K$.

Solution:

By Theorem 8.13, H and K are free of the same rank. Thus they are isomorphic.

8.41. Let F be a free group on generators x and y. Suppose that $R \lhd F$, $y \in R$ and $F/R = gp(xR)$ is infinite cyclic. Prove that the group R/R' is freely generated as an abelian group by the elements $x^n y x^{-n} R'$, where $n \in Z$. Then prove that for no integer n, is $x^n y x^{-n} R'$ in the center of F/R'.

Solution:

The method of Section 8.4d with $X = \{x^n \mid n \in Z\}$ gives for the free generators of R the elements $a_{x^n, y} = x^n y x^{-n}$, $n \in Z$. By an argument similar to Theorem 8.6 (which deals only with free groups of finite rank), R/R' is free abelian with basis $x^n y x^{-n} R'$ for all integers n.

Now as

$$(xR')(x^n y x^{-n} R')(xR')^{-1} = x^{n+1} y x^{-(n+1)} R' \neq x^n y x^{-n} R'$$

$x^n y x^{-n} R'$ does not belong to the center of F/R'.

f. Intersection of finitely generated subgroups

We prove here that if H and K are subgroups of finite rank of a free group F, then $H \cap K$ is also a subgroup of finite rank. (This result is due to Howson.)

To simplify the problem we note that we may assume that F is finitely generated. For if not, we could consider $gp(H, K)$, which is certainly finitely generated (as H and K are), instead of F.

However, as we have seen in Example 1, page 259, a free group of rank 2 has a subgroup of infinite rank. Therefore we may as well assume that F is free of rank 2. Let F be freely generated by a and b.

We say that a *coset C is single-ended* if every element of C when expressed as a reduced product has the same last factor.

For example, if all the elements of C end in a, C is said to be single-ended. (We might have for instance $baaa \in C$.) But if C contains for example the elements $ababab$ and ab^{-1}, then C is not single-ended.

A coset which is not single-ended is called *double-ended.*

The following lemma is crucial:

Lemma 8.14: A subgroup H of the group freely generated by a and b is of finite rank if and only if it has a finite number of double-ended cosets.

Proof: Choose a Schreier transversal X for H in F. Put $S = \{a, b\}$.

(1) Let H be of finite rank. Then only a finite number of the elements $a_{x,s} \neq 1$ (where $x \in X$ and $s \in S$). But we proved in Section 8.4d that every element of H ended in the form $wd_l^{-1} \cdots d_1^{-1}$ where $wd_l^{-1} \cdots d_1^{-1}$ is either $s(\overline{xs})^{-1}$ or $s^{-1}x^{-1}$, where $a_{x,s} \neq 1$. Thus we concluded that all the elements of H end in the form $wd_l^{-1} \cdots d_1^{-1}$ where there are only a finite number of $wd_l^{-1} \cdots d_1^{-1}$ and $d_1 \cdots d_l w^{-1} \notin X$ although $d_1 \cdots d_l \in X$.

Let $x \in X$. Then if Hx is double-ended, there exists an element $h \in H$ such that in hx, x cancels completely. As h ends in some $wd_l^{-1} \cdots d_1^{-1}$ and $d_1 \cdots d_l w^{-1} \notin X$, it follows that if x cancels completely, x is an initial segment of $d_1 \cdots d_l$ (otherwise $wd_1 \cdots d_l$ is an initial segment of x and hence in X). It follows that as the number of $d_1 \cdots d_l$ that appear is finite, the number of initial segments is finite, and thus the number of double-ended cosets is finite.

(2) Let H be of infinite rank. If $a_{x,s} = xs(\overline{xs})^{-1} \neq 1$, it follows that x does not end in s^{-1} (line 13, page 262) and, of course, $xs(\overline{xs})^{-1} \in H$. The coset Hx contains x and $(xs(\overline{xs})^{-1})^{-1}x = \overline{xs}s^{-1}$. Now s^{-1} appears in the reduced product for $\overline{xs}s^{-1}$ (line 13, page 262). As x does not end in s^{-1}, Hx is double-ended.

Now as there are an infinite number of x such that $a_{x,s} \neq 1$, there are an infinite number of double-ended cosets.

Theorem 8.15: The intersection of two subgroups H and K of finite rank is again of finite rank.

Proof: By Lemma 8.14, H has only a finite number of double-ended cosets, say $H_1, \ldots, H_n$, and K has only a finite number of double-ended cosets, say $K_1, \ldots, K_m$. Now the cosets of $H \cap K$ are intersections of cosets of H and K (Problem 7.23, page 231). Also the intersection of a single-ended coset of H with any coset of K is single-ended (and vice versa). Thus the double-ended cosets of $H \cap K$ are among the cosets $H_i \cap K_j$, $i = 1, \ldots, n$ and $j = 1, \ldots, m$. Therefore $H \cap K$ has at most mn double-ended cosets and accordingly $H \cap K$ is of finite rank by Lemma 8.14.

A look back at Chapter 8

We defined free groups and gave an alternative definition. The existence of free groups was established and homomorphisms of free groups investigated. The main result was that if F is a free group freely generated by X, then for every group H and every mapping θ of X into H there is a homomorphism ϕ of F into H which coincides on X with θ. As a consequence every group is a homomorphic image of a free group.

We next discussed presentations of groups. The important notion of rank of a free group then arose naturally.

We discussed and proved the subgroup theorem for free groups. Using this, the rank of a subgroup of finite index was calculated. Finally we proved that the intersection of two subgroups of finite rank in a free group is of finite rank.

Supplementary Problems

ELEMENTARY NOTIONS

8.42. Suppose F is freely generated by a and b. Find elements $u, v \in F$ so that $u \neq a$ and

$$a^{-1}b^{-1}ab = u^{-1}v^{-1}uv$$

8.43. Prove that a free group of finite rank has only a finite number of elements of length $\leqq n$ for any fixed integer n.

8.44. Prove that if $N \lhd G$ and G/N is free, then G splits over N.

8.45. Let F be a free group freely generated by a and b. Let $N = gp(\ldots, a^2ba^{-2}, aba^{-1}, b, a^{-1}ba, a^{-2}ba^2, \ldots)$. Prove $N \lhd F$ and verify that F splits over N.

8.46. Using the notation of the preceding problem show that if N' is the derived group of N, then $N' \lhd F$. Verify that F/N' is a splitting extension of N/N' by an infinite cyclic group. Construct an isomorphic copy of F/N' directly as an extension of a free abelian group by an infinite cyclic group.

PRESENTATIONS OF GROUPS

8.47. Prove that the group $G = |a, b; a^{-1}ba = b^2|$ is not free. (*Hint*: Prove that G/G' is cyclic. So if G is free it must be free cyclic. Then show $G' \neq \{1\}$ by mapping G into a suitable factor group.)

8.48. Let $G = |x, y; x^2, y^2|$. Show that G is infinite. (*Hint*: Let F be the free group freely generated by x and y. Let $\theta: x \to a,\ y \to ab$ be a homomorphism onto the group of Problem 8.30, page 257.)

8.49. Let $G = |a, b, c, d; [a, c], [a, d], [b, c], [b, d]|$. Prove G is the direct product of two free groups each of rank 2.

8.50. Prove that if $G = |x_1, x_2, \ldots; 2x_2 = x_1, \ldots, (i+1)x_{i+1} = x_i, \ldots|$, then G is isomorphic to the additive group of rationals. (*Hint*: Note that G has the additive group of rationals as homomorphic image. Also it is abelian. Use Problem 6.72, page 208.)

8.51. Prove that if p is a prime and $G = |x_1, x_2, \ldots; x_1, x_2^p x_1^{-1}, \ldots, x_{i+1}^p x_i^{-1}, \ldots|$, then G is isomorphic to the p-Prüfer group.

8.52. What is the group $|x_1, x_2, \ldots, x_n; x_1x_1^{-1}|$?

THE SUBGROUP THEOREM FOR FREE GROUPS

8.53. Prove that if F is free on x, y, z then the elements $xyz, x^2y^2z^2, x^3y^3z^3, x^4y^4z^4, \ldots$ freely generate a subgroup of F.

PROOF OF THE SUBGROUP THEOREM

8.54. Show that if $a^m = b^n$ where a and b are elements of a free group $(mn \neq 0)$, then a and b generate a cyclic group.

8.55. Let N and M be nonidentity normal subgroups of G, a free group of rank greater than 1. Prove that $N \cap M \neq \{1\}$.

8.56. Let F be a free group. Prove that there is no sequence of subgroups $F_1 \subseteq F_2 \subseteq \cdots$ of F with $F_i \neq F_{i+1}$ and rank of $F_i = 2$. (*Hint*: Let $G = \cup F_i$. Then G is a free group of infinite rank. Consider G/G' which is free abelian of infinite rank. Obtain a contradiction by showing that G/G' is of finite rank.)

8.57. Find generators for F^2 where F is a free group of rank two and $F^2 = gp(x^2 \mid x \in F)$. (Note that F/F^2 is the Klein 4-group.)

8.58. Let F be a group which can be generated by two elements with a free subgroup of rank 3 which is of index two. Prove F is free by comparing it with a free group of rank 2. (Hard.)

8.59. Let F, R be as in Problem 8.41, page 264. Prove that F/R' has no element other than 1 in its center.

Appendix A

Number Theory

In this book we assume the reader knows the following:

1. The meaning of a divides b, for which we use the notation $a \mid b$. The notation $a \nmid b$ is read as "a does not divide b".

2. The definition of a prime, i.e. an integer not equal to 1 which is divisible only by 1 and itself.

3. If a, b are integers, then there exist an integer q and an integer r such that

$$a = bq + r$$

where $0 \leq r < b$.

4. The definitions of greatest common divisor of two numbers a and b (the largest integer which divides both a and b) and lowest common multiple of a and b (the smallest integer divisible by both a and b). We write the greatest common divisor of a and b as (a, b), the lowest common multiple of a and b as l.c.m.(a, b). If $(a, b) = 1$, then a and b are said to be co-prime.

5. Given integers a and b, there exist integers p and q such that

$$(a, b) = pa + qb$$

6. The fundamental theorem of arithmetic which says every integer is expressible in one and only one way (ignoring order) as the product of primes.

7. $a \equiv b$ modulo n means $a - b$ is divisible by n. Also some of the simpler properties such as $a \equiv b$ and $x \equiv y$ implies $a + x \equiv b + y$ and $ax \equiv by$.

The reader who does not know this material can consult

(*a*) Niven, I., and H. S. Zuckerman, *An Introduction to the Theory of Numbers*, Wiley, 1966.

(*b*) Upsensky, J. V., and M. A. Heaslet, *Elementary Number Theory*, McGraw-Hill, 1939.

(*c*) Birkhoff, G., and S. MacLane, *A Survey of Modern Algebra*, Macmillan, 1953.

Appendix B

A Guide to the Literature

Note: Numbers in brackets refer to the bibliography on pages 272 and 273.

General

Books which contain much of the material of this text (and frequently more) are, in the order of complexity, [46], [15], [16]. Useful books in algebra are [47] and [23].

Chapter 1

The theory of sets was established by G. Cantor in the last quarter of the 19th century [1]. A central notion in his work is that of cardinality – two sets have the same cardinality if there is a one-to-one mapping from the one onto the other. This leads to the idea of a transfinite number, and much of Cantor's work was devoted to developing these transfinite numbers (see [2] and [3]).

In the study of sets a number of paradoxes arose ([4] and [5]). This made it desirable to put set theory on a firm axiomatic footing. Such axiomatic approaches have led to a number of interesting developments ([4], [5] and [6]).

Recently a different approach to the foundations of mathematics has been originated by Lawvere, which is based on the notion of category [44] (see [7] for the axioms of a category).

Chapter 2

The study of groupoids arose in part from a desire to understand more clearly the axioms for group theory. The theory of groupoids can be subdivided into systems which satisfy various "natural" conditions. Thus, for example, among the various classes of groupoids one has semigroups, loops, groups and quasi-groups. Two major references are [8] and [9].

Chapter 3

It is clear from this chapter that groups arise in a great many mathematical disciplines. Groups arose initially from 19th century algebra, analysis and geometry. It was hoped that much of geometry could be handled by associating a group with each geometrical object. To some extent this aspect of group theory has been discussed in the section on isometry groups (see also [10], [11], and [14]). For the applications to topology see [37] and [39], and for knot theory see [40].

Groups arise also in quantum physics, crystallography [12] and chemistry [13]. This omnipresence of groups is part of the reason for its importance.

In addition, the study of groups has been carried forward for its own sake (see e.g. [15], [16], [17], [18], [19], [20], [21], [22], [43]). Finally we refer the reader to the study of groups with a topology: [41], [42].

Chapter 4

Chapter 4 is concerned mainly with cyclic groups and homomorphisms.

As we have dealt exhaustively with cyclic groups, there is not really any further information available except perhaps for finding the automorphism group of a cyclic group. If G is cyclic of order n, then the automorphism group of G is the group of integers co-prime to n, with multiplication modulo n; see e.g. [43].

The concept of homomorphism is basic. One can define homomorphisms for other algebraic concepts, such as rings. Factor rings can be defined and very similar theorems obtained [23], [24], [25]. (Indeed, once the group theoretic ones are known, it is routine to obtain the others.)

Some work, prompted by algebraic topology, has been done on more complex interaction of group homomorphisms [26].

Chapter 5

Much of this chapter is "arithmetical" in content, i.e. it deals with the order of a finite group. Thus for example we have the theorem of Lagrange: The order of a subgroup of a finite group divides the order of the group. There are a great number of important theorems of this kind, embodying various generalizations of the Sylow theorems (see [27], [17]). A very important result of a slightly different kind is the remarkable theorem of W. Feit and J. Thompson, viz. a finite group of odd order is solvable [45].

The classification of groups of small order has not been too successful. This is due to the extraordinary complexity of these groups. The reader might consult [21], [17] and [28] for a discussion of this classification problem.

Today the study of finite groups has proceeded at an extraordinary pace. One of the main aims of this study is the classification of finite simple groups. This classification is by no means complete, but much progress has been made (see [29]).

Chapter 6

Our proof of the fundamental theorem of abelian groups is nonconstructive, i.e. we do not have a definite procedure for finding a basis. Such procedures exist, e.g. [30].

There are also other criteria for deciding if an abelian group is a direct sum of cyclic groups, e.g. if every element is of bounded order [31]. A more comprehensive result is that of Kulikov [16].

In this chapter we found invariants for finitely generated abelian groups. The invariants for countable torsion groups are subtler and appear in Ulm's theorem ([31], [32]).

Some of the theorems of abelian groups extend to wider classes of algebraic structures, e.g. modules over rings ([31]). There has also been an attempt to prove theorems about classes of groups that are quite close to abelian groups, e.g. solvable and nilpotent groups ([17], [19]).

Excellent sources for abelian groups appear in [16] and [31]; a more encyclopedic account appears in [32].

Chapter 7

The aim of this chapter is to show how representing groups as permutation groups leads to information about the groups themselves. There are other types of representation. The most important of these is the representation of finite groups as $n \times n$ matrices. (These $n \times n$ matrices constitute a natural generalization of the 2×2 matrices we have discussed in Chapter 3.) The reader might consult [33], [34].

In infinite group theory the permutational representation of Frobenius has led to significant progress (§2 of Chapter 2, etc., of [35]).

The transfer has been used a great deal in finite group theory. A good source of information is [27]. See also [17].

Chapter 8

Our proof of the existence of free groups is really motivated by Cayley's theorem, as a glance at the usual proof will reveal [16].

The property of Theorem 8.4, i.e. the property of being able to extend a mapping of the free generators to a homomorphism of the free group, is of great importance. Indeed, it is often used as the definition of a free group. With this approach, the definition of free group is analogous to the definition of free abelian group (Chapter 6). There are even further generalizations and we can speak of the free group in a variety, where a variety is a collection of groups satisfying certain conditions. A detailed account appears in [35].

One can also regard a free group as being the free product of infinite cyclic groups. The free product of two groups is a way of putting the groups together in, roughly speaking, the freest way ([16], [17]).

Groups occur frequently as presentations. This is unfortunate as it is always difficult to say what the group of a presentation is or what its properties are. Indeed, there is not even a general and effective procedure for deciding whether the group of any given presentation is of order 1. All this is connected with the word problem, which is, roughly speaking, to determine in a finite number of steps if two products in the generators of a given presentation are equal. The word problem is unsolvable in general (see [15]).

An important concept which enables us to change from one presentation to another is that of Tietze transformations [36].

There are many proofs of the subgroup theorem, including topological ones (e.g. [37]). An important one is by Nielsen transformations ([16], [17], [36]). One of the advantages of the method we used in Chapter 8 is that it extends readily to provide generators and defining relations for subgroups of a group of a presentation [36].

More properties of free groups, free products and presentations of groups appear in [36]. See also [38] for presentations.

BIBLIOGRAPHY

[1] Cantor, G., *Contributions to the Founding of the Theory of Transfinite Numbers* (translated by P. E. B. Jourdain), Open Court, 1915.

[2] Kamke, E., *Theory of Sets* (translated from the 2nd German edition by F. Bagemihl), Dover, 1950.

[3] Halmos, P. R., *Naive Set Theory*, Van Nostrand, 1960.

[4] Kleene, S. C., *Introduction to Metamathematics*, Van Nostrand, 1952.

[5] Cohen, P. J., and R. Hersh, "Non-Cantorian Set Theory", *Scientific American*, December 1967.

[6] Suppes, P. C., *Axiomatic Set Theory*, Van Nostrand, 1960.

[7] Freyd, P., *Abelian Categories, An Introduction to the Theory of Functors*, Harper and Row, 1964.

[8] Bruck, R. H., *A Survey of Binary Systems*, Springer, 1958.

[9] Clifford, A. H., and G. B. Preston, "The Algebraic Theory of Semigroups", Mathematical Surveys 7, 1961, American Mathematical Society.

[10] Weyl, H., *Symmetry*, Princeton University Press, 1952.

[11] Hilbert, D., and S. Cohn-Vossen, *Geometry and the Imagination*, translated by P. Nemenyi, Chelsea, 1956.

[12] Hammermesh, M., *Group Theory and its Application to Physical Problems*, Addison-Wesley, 1962.

[13] Cotton, F. A., *Chemical Applications of Group Theory*, Interscience, 1963.

[14] Yale, P. B., *Geometry and Symmetry*, Holden Day, 1968.

[15] Rotman, J. J., *The Theory of Groups: An Introduction*, Allyn and Bacon, 1965.

[16] Kurosh, A., *The Theory of Groups*, (translated by K. A. Hirsch), Chelsea, 1960.

[17] Hall, Jr., M., *The Theory of Groups*, Macmillan, 1959.

[18] Wielandt, H., *Finite Permutation Groups*, Academic Press, 1964.

[19] Schenkman, E., *Group Theory*, Van Nostrand, 1965.

[20] Scott, W., *Group Theory*, Prentice-Hall, 1964.

[21] Burnside, W., *Theory of Groups of Finite Order*, Dover, 1955.

[22] Huppert, B., *Endliche Gruppen*, Springer-Verlag, 1967.

[23] Herstein, I., *Topics in Algebra*, Blaisdell, 1964.

[24] Van der Waerden, B. L., *Modern Algebra*, (translated by F. Blum), Ungar, 1953.

[25] Cohen, P. M., *Universal Algebra*, Harper and Row, 1965.

[26] Northcott, D. G., *An Introduction to Homological Algebra*, Cambridge U. P., 1960.

[27] Wielandt, H., and B. Huppert, "Arithmetical and Normal Structure of Finite Groups," *1960 Institute on Finite Groups*, Editor, M. Hall, Jr., American Mathematical Society, 1962.

[28] Hall, Jr., M., and J. K. Seniour, *The Groups of Order* 2^n ($n \leq 6$), Macmillan, 1964.

[29] Carter, R. W., "Simple Groups and Simple Lie Algebras", *Journal of the London Mathematical Society*, April 1965, pp. 193-240.

[30] Schreier, O., and E. Sperner, *Introduction to Modern Algebra and Matrix Theory*, (translated by M. Davis and M. Haussner), Chelsea, 1959.

[31] Kaplansky, I., *Infinite Groups*, U. of Michigan Press, 1954.

[32] Fuchs, L., *Abelian Groups*, Akademiai Kiado, 1958.

[33] Burrow, M., *Representation Theory of Finite Groups*, Academic Press, 1965.

[34] Curtiss, C. W., and I. Reiner, *Representation Theory of Finite Groups and Associative Algebras*, Interscience, 1962.

[35] Neumann, H., *Varieties of Groups*, Springer, 1967.

[36] Magnus, W., A. Karass and D. Solitar, *Combinatorial Group Theory: Presentations of Groups in Terms of Generators and Relations*, Interscience, 1966.

[37] Massey, W. S., *Algebraic Topology; An Introduction*, Harcourt, Brace, 1967.

[38] Coxeter, H. S. M., and W. O. J. Moser, *Generators and Relators for Discrete Groups*, Springer-Verlag, 1965.

[39] Wallace, A. H., *An Introduction to Algebraic Topology*, Pergamon, 1957.

[40] Crowell, R. H., and R. H. Fox, *Introduction to Knot Theory*, Ginn, 1963.

[41] Pontriagin, L. S., *Topological Groups*, Gordon and Breach, 1966.

[42] Cohn, P., *Lie Groups*, Cambridge U. P., 1957.

[43] Zassenhaus, H., *The Theory of Groups*, Chelsea, 1949.

[44] Lawvere, F. W., "The Category of Categories as a Foundation for Mathematics", Proceedings of the Conference on Categorical Algebra, La Jolla, 1965, Springer-Verlag, 1966.

[45] Feit, W., and J. G. Thompson, "Solvability of groups of odd order", *Pacific Journal of Mathematics*, vol. 3, no. 3, pp. 773-1029, 1963.

[46] Lederman, W., *Introduction to the Theory of Finite Groups*, Oliver and Boyd, 1957.

[47] Birkhoff, G., and S. MacLane, *A Survey of Modern Algebra*, Macmillan, 1953.

INDEX

Symbols and Notations

Symbols

aut (G)	Automorphism group of G, 84
$a_{x,g}$	$xg(\overline{xg})^{-1}$, 224
A_n	Alternating group of degree n, 61
C	Complex numbers, 1
C^*	Nonzero complex numbers, 51
$C(A)$	Centralizer of A, 112
C_n	Cyclic group of order n, 148
D_n	Dihedral group of degree n, 76
$gp(X)$	Subgroup generated by X, 98
h^k	Conjugation by x_k, 233
I	Isometries of plane, 67
$I(R)$	Group of isometries of R, 64
I_S	Symmetry group of S, 73
$(I : Z)$	Group of isometries of R that move integers to integers, 66
K_4	Klein four group, 148
Ker θ	Kernel of θ, 117
l.c.m. (a, b)	Lowest common multiple of a and b, 269
$L_n(V, F)$	Full linear group of dimension n, 89
$m_{g,x}$	$(\overline{gx})^{-1}gx$, 232
$m_{k,k'}$	$m_{x_k, x_{k'}}$, 233
M	Group of Möbius transformations, 78
$\mathcal{M}$	Group of 2×2 matrices over complex numbers, 81
M_X	Semigroup of mappings of X to X, 36
N	Nonnegative integers, 1
$N(A)$	Normalizer of A, 112
$N_H(A)$	Normalizer of A in H, 133
P	Positive integers, 1
Q	Rationals, 1
Q^*	Nonzero rationals, 20
R	Real numbers, 1
R^+	Nonnegative real numbers, 48
R^2	Euclidean plane, 2
$\mathcal{R}$	Set of representatives of the equivalence classes, 134
$\mathcal{R}^*$	Representatives whose intersection with the center is the empty set, 135
R-class	R-equivalence class, 9
s_p	Number of distinct Sylow p-subgroups, 131
S_n	Symmetric group of degree n, 56
S_X	Symmetry group on X, 56
$T(G)$	Torsion subgroup of G, 189
W	Elements of S and their inverses, 262
x_k	Coset representative, 232
Z	Integers, 1
$Z(G)$	Center of G, 112

Greek Symbols

Notations